Environmental Science and Engineering

Environmental Engineering

Series editors

Ulrich Förstner, Hamburg, Germany
Robert J. Murphy, Tampa, FL, USA
W. H. Rulkens, Wageningen, The Netherlands

For further volumes:
http://www.springer.com/series/3172

Jaime Klapp · Abraham Medina
Editors

Experimental and Computational Fluid Mechanics

Springer

Editors
Jaime Klapp
Instituto Nacional de Investigaciones Nucleares and Cinvestav-Abacus
México
Mexico

Abraham Medina
ESIME Unidad Azcapotzalco Instituto Politécnico Nacional
México
Mexico

ISSN 1431-2492
ISBN 978-3-319-00115-9 ISBN 978-3-319-00116-6 (eBook)
DOI 10.1007/978-3-319-00116-6
Springer Cham Heidelberg New York Dordrecht London

Library of Congress Control Number: 2013955881

Printed on acid-free paper

Springer is part of Springer Science+Business Media (www.springer.com)

Preface

The understanding of fluid flow has become a very important field of research because of its relevance in many human activities. The study of fluid flow is essential in many fields such as physics, biology, medicine, engineering, environmental sciences, energy, atmospheric flow processes that influence the weather and the climate, and so on. Many industrial processes are governed by the equations of fluid dynamics and so its study is valuable. Numerical simulations of fluid flow is reducing the high cost of very expensive experimental tests. Dynamical fluid dynamics are also very important for understanding fundamental physical processes at all scales.

This book presents recent experimental and theoretical advances in fluid dynamics in physics and engineering. It begins with invited lectures given during the International Enzo Levi Spring School 2012 held at the Universidad Autónoma Metropolitana-Azcapotzalco, May 17–18, 2012, and invited seminars presented in the 18th National Congress of the Fluid Dynamics Division of the Mexican Physical Society, held at the Centro de Investigación Científica y Educación Superior de Ensenada (CICESE), Ensenada, Baja California, Mexico, November 21–23, 2012.

The Spring School is organized every year in honor of Prof. Enzo Levi, a well-known Mexican scientist that dedicated his research to the study of fluids. He was one of the founders of the Instituto de Ingeniería (Engineering Institute) of the Universidad Nacional Autónoma de México (UNAM), and of the Instituto Mexicano de Tecnología del Agua (Mexican Institute for Water Technology) of the National Water Commission. He was the mentor of several generations of Mexican Engineers.

During the 2 days school, lectures were given by well-known national and international scientists. In 2012, many people attended the meeting with 50 researchers and more than a hundred graduate and undergraduate students. A wide variety of topics were presented by young national researchers that included asymptotic methods in fluids, convection, computational methods for biological systems, interfacial fluid dynamics, and problems related to colloidal dispersion and fractured porous media. Among the lectures we want to mention a very interesting description of the Alya Cardiac Computational Model given by Mariano Vazquez of the Barcelona Supercomputing Center in Spain, two lectures by Rouslan Kretchetnikov of the University of California in Santa Barbara entitled Walking with coffee: why it splits and A few puzzles in interfacial fluid dynamics, two lectures by Cesar Treviño (UNAM) on asymptotic method in fluids and linear stability of mixed convection flows. Other interesting lectures were given on the Turbulence of marine currents by Angel Ruíz (UNAM), on the hydrodynamic interactions of heterogeneous colloidal dispersions by Catalina Haro (UAM-A), on the Boundary element method in Fluid Mechanics by Abel López Villa (IPN), and on An Experimental Analogy of Pedestrians under Panic Situations by Abraham Medina (IPN). Several of these lectures were included in Part I of the Book.

The Annual Fluid Dynamics Congress has a different format. In 2012, it lasted 3 days and was composed of five plenary lectures and many short oral presentations of students and researchers.

In part I, we also included the plenary lectures given during the congress by international well-known invited speakers and some of the most interesting short oral contributions. Among the plenary lectures, we can mention those given on the basic concepts of ignition and fuel through diffusion flames by Amable Liñan of the Universidad Politécnica de Madrid, on the Transition to turbulence in stratified wakes by Patrice Meunier of the Aix-Marseille University, France, on flow driven by harmonic forcing in planetary atmospheres and cores by Michael Le Bars, also of the Aix-Marseille University, France, on the role of symmetry in biomimetic wakes for propulsion by Verónica Raspa of the University of Paris Diderot, France, on toroidal vortex with chaotic stream lines by Oscar Velasco of CICESE, Baja California, México, on the symmetry instability in a mixed convection problem by Cesar Treviño of UNAM, and on The boundary element method for bubble formation by Abel López of the IPN, México.

The other short presentations are organized in topics: Multiphase Flow and Granular Media (Part II), Convection and Diffusion (Part III), Vortex, Oceanography and Meteorology (Part IV), and General Fluid Dynamics and Applications (Part V).

In Part II, Multiphase Flow and Granular Media, we have focused on petroleum-related applications, you can find interesting contributions on the nitrogen dispersion in the neighborhood of a well, on a phenomenological description of heavy oil, on an experimental study of the growth of bubbles in corrugated tubes, which has applications for foam injection in fractured porous media, on surface tension and interfacial tension measurements in water-surfactant-oil systems, and other related problems. Other contributions in this section are in granular media and mass flow in a silo.

Convection and Diffusion can be found in Part III, with interesting contributions on conjugate convection in an open cavity, and on transport of particles in a periodically forced flow. We also found two applications to porous media flow: experimental studies of a vaporization front, and the mass and mass transfer during steam injection. There is also an interesting numerical study with SPH of the Kelvin–Helmholtz instability.

In Part IV, Vortex, Oceanography, and Meteorology, one can find three contributions on atmospheric fluid dynamics, the first on forced dynamics by normal wind to the boundary, the second on numerical experiments of wind circulation off the Baja California coast, and the third one is a review on environmental fluid dynamics. Two further contributions are on the effect of the inlet flow angle on the vortex induced vibration of a collinear array of flexible cylinders and the wake patterns behind a flapping foil.

Finally, in Part V, General Fluid Dynamics and Applications, we find several contributions of fluid dynamics to various fields such as magnetohydrodynamics, evaporation of a liquid layer, a study of the drag coefficient in the Navier–Stokes fractional equation, interactions of fluids, numerical simulation of biological systems with the DPD method, and soil transport of contaminants.

The book is aimed to fourth-year undergraduate and graduate students, and to scientists in the field of physics, engineering, and chemistry that have interest in fluid dynamics from the experimental and theoretical point of view. The material includes recent advances in experimental and theoretical fluid dynamics and is adequate for both teaching and research. The invited lectures are introductory and avoid the use of complicated mathematics. The other selected contributions are also adequate to fourth year undergraduate and graduate students.

The editors are very grateful to the institutions who made possible the realization of the International Enzo Levi Spring School 2012 and the 18th National Congress of the Fluid Dynamics Division of the Mexican Physical Society, especially the Consejo Nacional de Ciencia y Tecnología (CONACYT), the Sociedad Mexicana de Física, the Universidad Autónoma Metropolitana-Azcapotzalco, the Centro de Investigación Científica y Educación Superior de Ensenada (CICESE), the Universidad Autónoma de México (UNAM), the Instituto Politécnico Nacional (IPN), Cinvestav-Abacus, and the Instituto Nacional de Investigaciones Nucleares (ININ).

We acknowledge the help of the Edition Committee: Carlos A. Vargas, Gerardo Ruíz Chavarría, Catalina Stern, Salvador Galindo Uribarri, Estela Mayoral Villa, Elizabeth Teresita Romero Gúzman, Eduardo de la Cruz Sánchez, Abel López Villa, and in particular Fernando Aragón and Ruslan Gabbasov for their important and valuable contribution to the final manuscript.

Mexico City, July 2013

Jaime Klapp
Abraham Medina

Acknowledgments

The production of this book has been sponsored by the Consejo Mexiquense de Ciencia y Tecnología (COMECYT), Consejo Nacional de Ciencia y Tecnología (CONACYT), Universidad Autónoma Metropolitana-Azcapotzalco, Instituto Politécnico Nacional (IPN), Instituto Nacional de Investigaciones Nucleares (ININ), CISESE, Centro de Investigación y de Estudios Avanzados del Instituto Politécnico Nacional (CINVESTAV-Abacus), and Sociedad Mexicana de Física (SMF).

Contents

Contributors

A. Aguilar Corona Facultad de Ingeniería Mecánica, Universidad Michoacana de san Nicolás de Hidalgo, Ciudad Universitaria, Av. Francisco J. Mujica s/n, C.P. 58030 Morelia, Michoacán, México

Gerardo Alcalá Instituto de Energías Renovables, Universidad Nacional Autónoma de México, A.P. 34, 62580 Temixco, Morelos, Mexico, e-mail: gealp@cie.unam.mx

Rafael Almanza Instituto de Ingeniería, Universidad Nacional Autónoma de México, Ciudad Universitaria, Coyoacán, 04510 Mexico, D.F., Mexico

F. Aragón SEPI ESIME Zacatenco, Unidad Profesional "Adolfo López Mateos", Instituto Politécnico Nacional s/n, Col. Lindavista, Del., Gustavo A. Madero, C.P. 07738 Mexico, D.F., Mexico

R. A. Ávalos Zúñiga CICATA-Querétaro, Instituto Politécnico Nacional, Cerro Blanco No. 141, Colinas del Cimatario, C.P. 76090 Santiago de Querétaro, Querétaro, Mexico, e-mail: ravalosz@ipn.mx

Sergio Baz Rodríguez Faculty of Chemical Engineering, Universidad Autónoma de Yucatán, Campus de Ciencias Exactas e Ingenierías, Periférico Norte. Km. 33.5, Tablaje Catastral, 13615 Col. Chuburná de Hidalgo Inn, Mérida, Mexico, e-mail: ser-gio.baz@uady.mx

D. Cabrera SEPI ESIME Azcapotzalco, Instituto Politécnico Nacional, Av. de las Granjas 682, Col. Santa Catarina, 02250 Azcapotzalco, D.F., Mexico

J. C. Cajas Facultad de Ciencias, Universidad Nacional Autónoma de México, 04510 Mexico, D.F, Mexico, e-mail: jc.cajas@gmail.com

J. Callejas Fernández Departamento de Física Aplicada, Universidad de Granada, Granada, Spain

G. Carbajal De la Torre Facultad de Ingeniería Mecánica, Universidad Michoacana de san Nicolás de Hidalgo, Av. Francisco J. Mujica s/n, Ciudad Universitaria, C.P. 58030 Morelia, Michoacán, Mexico

A. Columbié Navarro Centro de Estudio de Energía y Tecnología de Avanzada de Moa, Instituto Superior Minero Metalúrgico de Moa, Mexico, Mexico

A. H. Cortés Estrada Facultad de Ingeniería Mecánica, Universidad Michoacana De San Nicolás De Hidalgo, Av. Francisco J. Mujica s/n, Ciudad Universitaria, C.P. 58030 Morelia, Michoacán, Mexico

Anne Cros Departamento de Física, Universidad de Guadalajara, Av. Revolución #1500, Col. Olímpica, Guadalajara, Jalisco, Mexico

J. Cruz Sampedro UAM-A, Av. San Pablo 180, Col. Reynosa, 02200 Mexico, D.F., Mexico, e-mail: jacs@correo.azc.uam.mx

Sergio Cuevas Instituto de Energías Renovables, Universidad Nacional Autónoma de México, A.P. 34, Temixco, Morelos, Mexico

Eduardo De la Cruz Sánchez Instituto Nacional de Investigaciones Nucleares, Carretera México-Toluca s/n, La Marquesa, 52750 Ocoyoacac, Estado de México, Mexico, e-mail: eduardo.delacruz@inin.gob.mx

María Dolores Durán Facultad de Ingeniería, Universidad Autónoma del Estado de México, Cerro de Coatepec s/n, Ciudad Universitaria, 50100 Toluca, Mexico

María Dolores Duran García Facultad de Ingeniería, Universidad Autónoma del Estado de México, Ciudad Universitaria, 50100 Toluca, Estado de México, Mexico, e-mail: mddg_2210@hotmail.com

G. Domínguez Zacarías Coordinación Tecnológica de Ingeniería Yacimientos, Eje Central Lázaro Cárdenas Norte 152, Col. San Bartolo Atepehuacán Del., C.P. 07730 Mexico, D.F., Mexico; Programa de Recuperación de Hidrocarburos, Eje Central Lázaro Cárdenas Norte 152, Col. San Bartolo Atepehuacán Del., Gustavo A. Madero, C.P. 07730 Mexico, D.F., Mexico

R. Durazo Facultad de Ciencias Marinas, Universidad Autónoma de Baja California, Ensenada, Baja California, Mexico

Alejandro Camilo Espinosa Ramírez Departamento de Física, Universidad de Guadalajara, Av. Revolución #1500, Col. Olímpica, Guadalajara, Jalisco, Mexico

Ruslan F. Gabbasov Instituto de Ciencias Básicas e Ingeniería, Universidad Autónoma del Estado de Hidalgo, Ciudad Universitaria, Km 4.5 Carretera Pachuca—Tulancingo, 42184 Mineral de la Reforma, Hidalgo, Mexico, e-mail: ruslan.gabb@gmail.com

Salvador Galindo Uribarri Nacional de Investigaciones Nucleares, Carretera México-Toluca Km. 36.5, 52750 La Marquesa, Estado de México, Mexico, e-mail: salvador.galindo@inin.gob.mx

Armando Gama Goicochea Departamento de Ciencias Naturales, DCNI, Universidad Autónoma Metropolitana Unidad Cuajimalpa, Av. Pedro Antonio de los Santos 84, 11850 Mexico, D.F., Mexico, e-mail: agama@alumni.stanford.edu

Jaime García Toscano Centro de Datos Oceanográficos de México/IIO-UABC, Km. 107 Carretera Tijua-na-Ensenada, 22860 Ensenada, B.C., Mexico

Eduardo Gil Silva Centro de Datos Oceanográficos de México/IIO-UABC, Km. 107 Carretera Tijuana-Ensenada, 22860 Ensenada, B.C., Mexico

Ever Góngora Leyva Centro de Estudio de Energía y Tecnología de Avanzada de Moa, Instituto Superior Minero Metalúrgico de Moa, Mexico, Mexico, e-mail: egongora@ismm.edu.cu

Roberto González Galán Instituto Nacional de Investigaciones Nucleares, Carretera México-Toluca s/n, La Marquesa, 52750 Ocoyoacac, Estado de México, Mexico; Facultad de Ciencias, Universidad Autónoma del Estado de México, El Cerrillo, C.P. 50200 Piedras Blancas, Estado de México, Mexico, e-mail: rgonzalez470@yahoo.com.mx

N. I. González SEPI-ESIME U.P. Adolfo López Mateos, Instituto Politécnico Nacional, Av. Instituto Politécnico Nacional s/n, Edificio 5, 3er Piso, C.P. 07738 Col. Lindavista, D.F., Mexico, e-mail: igonzalezn0900@alumno.ipn.mx

P. Guido Aldana Mexican Institute of Water Technology, Paseo Cuauhnáhuac 8532, Progreso, C.P. 62550 Jiutepec, Morelos, Mexico

G. J. Gutiérrez SEPI ESIME Azcapotzalco, Instituto Politécnico Nacional, Av. de las Granjas 682, Col. Santa Catarina, 02250 Azcapotzalc, D.F, Mexico

G. Gutiérrez Urueta Facultad de Ingeniería, Universidad Autónoma de San Luis Potosí, Av. Manuel Nava 8, Zona Universitaria Poniente, 78290 San Luis Potosí, S.L.P., Mexico

C. Haro Pérez Departamento de Ciencias Básicas, Universidad Autónoma Metropolitana-Azcapotzalco, Av. San Pablo 180, 02200 Mexico, D.F., Mexico, e-mail: cehp@correo.azc.uam.mx

J. A. Hernández Zarate Instituto Tecnológico de Veracruz, Departamento de Metal Mecánica y Mecatrónica, Calzada Miguel A. de Quevedo 2779, Col. Formando Hogar, 91860 Veracruz, Veracruz, Mexico

Sergio Hernández Zapata Facultad de Ciencias, Universidad Nacional Autónoma de México, Av. Universidad 3000, Circuito Exterior s/n, Delegación Coyoacán, C.P. 04510 Mexico, D.F., Mexico

R. Hidalgo Álvarez Departamento de Física Aplicada, Universidad de Granada, E-18071 Granada, Spain

L. A. Ibarra Bracamontes Facultad de Ingeniería Mecánica, Universidad Michoacana De San Nicolás De Hidalgo, Av. Francisco J. Mujica s/n, Ciudad Universitaria, C.P. 58030 Morelia, Michoacán, Mexico, e-mail: laibarrab@gmail.com

Jaime Klapp Escribano Instituto Nacional de Investigaciones Nucleares, Carretera México-Toluca Km. 36.5, 52750 La Marquesa, Estado de México, Mexico; Departamento de Matemáticas, Cinvestav del I.P.N., 07360 Mexico, D.F., Mexico, e-mail: jaime.klapp@inin.gob.com

Sergio Larios Castillo Centro de Datos Oceanográficos de México/IIO-UABC, Km. 107 Carretera Tijuana-Ensenada, 22860 Ensenada, B.C., Mexico

Michael Le Bars Rotating and Geophysical Flows, Institut de Recherche sur les Phénoménes Hors Equilibre, UMR 7342, CNRS—Aix-Marseille Université, 49 rue F. Joliot Curie, 13384 Marseille, Cédex 13, France, e-mail: lebars@irphe.univ-mrs.fr

Patrice Le Gal Institut de Recherche sur les Phénoménes Hors Equilibre, UMR 7342, CNRS—Aix-Marseille Université, 49 rue F. Joliot Curie, 13384 Marseille, Cédex 13, France, e-mail: legal@irphe.univ-mrs.fr

Francisco López Serrano Facultad de Química, Departamento de Ingeniería Química, Universidad Nacional Autónoma de México, CP 04510 Mexico, Mexico, e-mail: lopezserrano@unam.mx

A. López Villa ESIME Azcapotzalco, Instituto Politécnico Nacional, Av. de las Granjas No. 682, Col. Santa Catarina, 02550 Mexico, D.F., Mexico, e-mail: abelvilla77@hotmail.com

Erick Javier Lopez-Sanchez Facultad de Ciencias, Universidad Nacional Autonoma de Mexico, Circuito exterior s/n, Ciudad Universitaria, Coyoacan, C.P. 04510 Mexico, Mexico, e-mail: lsej@ciencias.unam.mx

Mariana Macias Carballo Centro de Datos Oceanográficos de México/IIO-UABC, Km. 107 Carretera Tijuana-Ensenada, 22860 Ensenada, B.C., Mexico

Iván Martínez Facultad de Ingeniería, Universidad Autónoma del Estado de México, Cerro de Coatepec s/n, Ciudad Universitaria, 50100 Toluca, Mexico, e-mail: igmartinezc@uaemex.mx

L. Martínez Suástegui ESIME Azcapotzalco, Instituto Politécnico Nacional, Avenida de las Granjas No. 682, Colonia Santa Catarina, Delegación Azcapotzalco, 02250 Mexico, D.F., Mexico, e-mail: martinezlorenzo@gmail.com

Sergio Martínez Delgadillo Basic Sciences Department, Universidad Autónoma Metropolitana-Azcapotzalco, Avenida San Pablo 180, Col. Reynosa Tamaulipas, 02200 Mexico, D.F., Mexico, e-mail: samdmas@gmail.com

J. A. Martínez Facultad de Ciencias Marinas, Universidad Autónoma de Baja California, Ensenada, Baja California, Mexico

Estela Mayoral Villa Instituto Nacional de Investigaciones Nucleares, Carretera México-Toluca S/N, La Marquesa, 52750 Ocoyoacac, Estado de México, Mexico, e-mail: estela.mayoral@inin.gob.mx

Abraham Medina ESIME Azcapotzalco, Instituto Politécnico Nacional, Av. de las Granjas No. 682, Col. Sta. Catarina, 02550 Mexico, D.F., Mexico, e-mail: amedinao@ipn.mx

Adán Mejia Trejo Centro de Datos Oceanográficos de México/IIO-UABC, Km. 107 Carretera Tijuana-Ensenada, 22860 Ensenada, B.C., Mexico

H. Méndez Azúa Facultad de Ingeniería, Universidad Autónoma de San Luis Potosí, Av. Manuel Nava 8, Zona Universitaria Poniente, 78290 San Luis Potosí, S.L.P., Mexico

J. R. Mercado Escalante Mexican Institute of Water Technology, Paseo Cuauhnáhuac 8532, Progreso, C.P. 62550 Jiutepec, Morelos, Mexico, e-mail: rmercado@tlaloc.imta.mx

Patrice Meunier Institut de Recherche sur les Phénoménes Hors Equilibre, UMR 6594, CNRS Aix-Marseille Université, 49 rue F. Joliot Curie, 13384 Marseille, Cédex 13, France, e-mail: meunier@irphe.univ-mrs.fr

P. L. A. Moreno SEPI, ESIME U.P. Adolfo López Mateos, Instituto Politécnico Nacional, Av. Instituto Politécnico Nacional S/N, Edif. 5, 3er Piso, Col. Lindavista, C.P. 07738 Mexico, D.F., Mexico

M. J. Nieto Pérez CICATA-Querétaro, Instituto Politécnico Nacional, Cerro Blanco No. 141, Colinas del Cimatario, C.P. 76090 Santiago de Querétaro, Querétaro, Mexico

W. Ojeda Bustamante Mexican Institute of Water Technology, Paseo Cuauhnáhuac 8532, Progreso, C.P. 62550 Jiutepec, Morelos, Mexico

Arnulfo Ortíz Gómez División de Ciencias Básicas, Facultad de Ingeniería, UNAM, Ciudad Universitaria, Coyoacán México, CP 04510. Mexico City, D.F., Mexico, e-mail: arnulfo64@gmail.com

Alma Delia Ortíz Bañuelos Department of Physics, University of Guadalajara, Av. Juárez 976, Col. Centro, 44100 Guadalajara, Jalisco, Mexico, e-mail: fisicalmaortiz@gmail.com

G. E. Ovando Chacon Departamento de Metal Mecánica y Mecatrónica, Instituto Tecnológico de Veracruz, Calzada Miguel A. de Quevedo 2779, Col. Formando Hogar, 91860 Veracruz, Veracruz, Mexico, e-mail: geoc@itver.edu.mx

S. L. Ovando Chacon Departamento de Química y Bioquímica, Instituto Tecnológico de Tuxtla Gutierrez, Carretera Panamericana Km. 1080, 29000 Tuxtla Gutierrez, Chiapas, Mexico

F. Oviedo Tolentino Facultad de Ingeniería, Universidad Autónoma de San Luis Potosí, Av. Manuel Nava 8, Zona Universitaria Poniente, 78290 San Luis Potosí, S.L.P., Mexico, e-mail: francisco.oviedo@uaslp.mx

Peralta Salomón Instituto Mexicano del Petróleo, Eje Central Lázaro Cárdenas 152, Col. Atepehuacán, 07730 Mexico, D.F., Mexico, e-mail: peraltasalomon@hotmail.com

F. G. Pérez Gutiérrez Facultad de Ingeniería, Universidad Autónoma de San Luis Potosí, Av. Manuel Nava 8, Zona Universitaria Poniente, 78290 San Luis Potosí, S.L.P., Mexico

S. Sánchez Pérez Moreno Facultad de Ciencias, Universidad Nacional Autónoma de México, Ciudad Universitaria, C.P. 04510 Mexico, D.F., Mexico, e-mail: sebasanper@gmail.com

Salomon Pérez Programa de Recuperación de Hidrocarburos, Coordinación Técnologica de Ingeniería de Yacimientos, Eje Central Lázaro Cárdenas Norte 152 Col. San Bartolo Atepehuacán Del. Gustavo A. Madero, C.P. 07730 Mexico, D.F., Mexico

J. C. Prince Avelino Departamento de Metal Mecánica y Mecatrónica, Instituto Tecnológico de Veracruz, Calzada Miguel A. de Quevedo 2779, Col. Formando Hogar, 91860 Veracruz, Veracruz, Mexico

M. Quesada Pérez Escuela Politécnica Superior de Linares, Universidad de Jaén, 23700 Jaén, Spain

Jorge Ramírez Muñoz Energy Department, Universidad Autónoma Metropolitana Azcapotzalco, Avenida San Pablo 180, Col. Reynosa Tamaulipas, 02200 Mexico, D.F., Mexico, e-mail: jrm@correo.azc.uam.mx

G. A. Ramos López CICATA-Querétaro, Instituto Politécnico Nacional, Cerro Blanco No. 141, Colinas del Cimatario, C.P. 76090 Santiago de Querétaro, Querétaro, Mexico

Yoalbis Retirado Mediacejas Centro de Estudio de Energía y Tecnología de Avanzada de Moa, Instituto Superior Minero Metalúrgico de Moa, Mexico, Mexico

Lázaro Raymundo Reyes Gutiérrez Departamento de Matemáticas, Cinvestav del I.P.N., 07360 Mexico, D.F., Mexico; Universidad Autónoma Metropolitana, Av. Hidalgo Poniente 46. Col. La Estación, Lerma de Villada, Elizabeth Teresita Romero-Guzmán, 52006, Estado de México, Mexico, e-mail: raregu@gmail.com

L. F. Rojas Ochoa Departamento de Física, CINVESTAV-IPN, Av. Instituto Politécnico Nacional 2508, 07360 Mexico, D.F., Mexico

Elizabeth Teresita Romero Guzmán Instituto Nacional de Investigaciones Nucleares, Carretera México-Toluca S/N, La Marquesa, 52750 Ocoyoacac, Estado de México, Mexico, e-mail: elizabeth.romero@inin.gob.mx

R. Romero Méndez Facultad de Ingeniería, Universidad Autónoma de San Luis Potosí, Av. Manuel Nava 8, Zona Universitaria Poniente, 78290 San Luis Potosí, S.L.P., Mexico

U. Romero ESIME Azcapotzalco, Instituto Politécnico Nacional, Av. de las Granjas No. 682, Col. Santa Catarina, 02550 Mexico, D.F., Mexico, e-mail: bich14nos@hotmail.com

Julio César Rubén Romo Cruz Facultad de Ciencias, Universidad Nacional Autónoma de México, Av. Universidad 3000, Circuito Exterior S/N Delegación, Coyoacán, Ciudad Universitaria, C.P. 04510 Mexico, D.F., Mexico, e-mail: juliocesar@fisica.unam.mx

Gerardo Ruiz Chavarría Facultad de Ciencias, Departamento de Física, Universidad Nacional Autónoma de México, Mexico, e-mail: gruiz@ unam.mx

S. Ruiz Chavarría Facultad de Ciencias, Universidad Nacional Autónoma de México, Ciudad Universitaria, C.P. 04510 Mexico, D.F., Mexico

Elizabeth Salinas Rodríguez Process and Hydraulics Engineering Department, Universidad Autónoma Metropolitana Iztapalapa, San Rafael Atlixco #186, Col. Vicentina, 09340 Mexico, D.F., Mexico, e-mail: sabe@xanum.uam.mx

Miriam Sánchez Facultad de Ingeniería, Universidad Autónoma del Estado de México, Cerro de Coatepec s/n, Ciudad Universitaria, 50100 Toluca, Mexico

J. Sánchez Sesma Mexican Institute of Water Technology, Paseo Cuauhnáhuac 8532, Progreso, C.P. 62550 Jiutepec, Morelos, Mexico

D. A. Serrano SEPI ESIME Azcapotzalco, Instituto Politécnico Nacional, Av. de las Granjas 682, Col. Santa Catarina, 02250 Azcapotzalco, D.F., Mexico, e-mail: arman2390@hotmail.com

A. Servin Martínez Departamento de Metal Mecánica y Mecatrónica, Instituto Tecnológico de Veracruz, Calzada Miguel A. de Quevedo 2779, Col. Formando Hogar, 91860 Veracruz, Mexico

Leonardo G. Di Sigalotti Centro de Física, Instituto Venezolano de Investigaciones Científicas, IVIC, Apartado Postal 20632, Caracas 1020-A, Caracas, Venezuela, e-mail: leonardo.sigalotti@gmail.com

Alberto Soria Process and Hydraulics Engineering Department, Universidad Autónoma Metropolitana-Iztapalapa, San Rafael Atlixco #186, Col. Vicentina, 09340 Mexico, D.F., Mexico, e-mail: asor@xanum.uam.mx

M. Tetlalmatzi Montiel UAEH, Carretera Pachuca-Tulancingo km 4.5, Pachuca, 42090 Mexico, Hidalgo, Mexico, e-mail: tmontiel@uaeh.edu.mx

Ayax Torres Coordinación del Posgrado, Instituto Mexicano del Petróleo, Eje Central Lázaro Cárdenas No. 152, Col. Atepehuacan, 07730 Mexico, D.F., Mexico, e-mail: higherintellect@hotmail.com

Carlos Torres Navarrete Centro de Datos Oceanográficos de México/IIO-UABC, Km. 107 Carretera Tijua-na-Ensenada, 22860 Ensenada, B.C., Mexico, e-mail: ctorres@uabc.edu.mx

C. Treviño Facultad de Ciencias, Universidad Nacional Autónoma de México Sisal, Mexico, Yucatán, Mexico, e-mail: ctrev@correo.unam.mx

E. G. Tolentino Instituto Politécnico Nacional, SEPI, ESIME U. P. Adolfo López Mateos, Av. Instituto Politécnico Nacional s/n, Edificio 5, 3er Piso, Col. Lindavista, C.P. 07738 Mexico, D.F., Mexico

René O. Vargas SEPI ESIME Azcapotzalco, Instituto Politécnico Nacional, Av. de las Granjas 682, Col. Santa Catarina, 02250 Azcapotzalco, D.F., Mexico, e-mail: reneosvargas@yahoo.com.mx

Oscar Velasco Fuentes CICESE, Ensenada, Mexico, e-mail: ovelasco@cicese.mx

F. A. Velazquez Muñoz Departamento de Física, Universidad de Guadalajara, Guadalajara, Jalisco, Mexico, e-mail: federico.velazquez@red.cucei.udg.mx

Federico Ángel Velázquez Muñoz Department of Physics, University of Guadalajara, Av. Juárez 976, Col. Centro, 44100 Guadalajara, Jalisco, Mexico

G. Viramontes Gamboa Facultad de Ciencias Físico-Matemáticas, Universidad Michoacana De San Nicolás De Hidalgo, Av. Francisco J. Mujica s/n, Ciudad Universitaria, C.P. 58030 Morelia, Michoacán, Mexico

F. Wong ESIME-Azcapotzalco, Instituto Politécnico Nacional, Av. de las Granjas 682, Col. Santa Catarina, C.P. 02250 Azcapotzalco, D.F., Mexico, e-mail: francis-cowong@hotmail.com

Luis S. Zamudio Programa de Recuperación de Hidrocarburos Coordinación Tecnológica de Ingeniería de Yacimientos, Eje Central Lázaro Cárdenas Norte 152, Col. San Bartolo Atepehuacán Del. Gustavo A. Madero, C.P. 07730 Mexico, D.F., Mexico

Part I
Invited Lectures

Symmetry Breaking Instability in a Mixed Convection Problem

J. C. Cajas, L. Martínez-Suástegui and C. Treviño

Abstract In this work, the stability of a strongly non-parallel symmetrical counterflow mixed convection problem is studied, using numerically generated eigenfunctions. The base flow is numerically obtained for each value of the buoyancy parameter (Richardson number), and the stability of this flow is analyzed by increasing its value while all the others remained fixed. The perturbed linear functions are numerically generated by introducing a transient modulated asymmetrical buoyancy, relaxing at later times to 'numerical eigenfunctions'. The time evolution of the amplitude of these perturbations is used to obtain the stability characteristics. Symmetry breaking instability occurs, for fixed geometry, Reynolds and Prandtl numbers, for values of the buoyancy parameter larger than a critical one. However, there is also a window for the buoyancy parameter below this critical value, where the system shows instability, producing a slightly asymmetric thermal and flow response.

1 Introduction

Mixed convection is defined as heat transfer situations where both natural and forced convection mechanisms interact. In particular, the oscillatory behavior in mixed convection flows is of great interest because of its rich dynamical features and useful

J. C. Cajas (✉)
Facultad de Ciencias, Universidad Nacional Autónoma de México, 04510 México, D. F., México
e-mail: jc.cajas@gmail.com

L. Martínez-Suástegui
ESIME Azcapotzalco, Instituto Politécnico Nacional, Avenida de las Granjas No. 682, Colonia Santa Catarina, Delegación Azcapotzalco, 02250 México, D. F., México
e-mail: lamartinezs@ipn.mx

C. Treviño
Facultad de Ciencias, UMDI Universidad Nacional Autónoma de México Sisal, Yucatán, México
e-mail: ctrev@correo.unam.mx

J. Klapp and A. Medina (eds.), *Experimental and Computational Fluid Mechanics*, Environmental Science and Engineering, DOI: 10.1007/978-3-319-00116-6_1,

results to applied problems. Chang and Lin (1993) studied the steady laminar and transient oscillatory mixed convection in a symmetrically heated vertical plane channel, subjected to an opposing buoyancy assuming a fully developed velocity profile at the inlet and discrete heat sources that are maintained at uniform and equal heat fluxes. The authors pointed out that an oscillatory flow with a single fundamental frequency is found when the buoyancy parameter, or Richardson number, exceeds a critical value. Lin et al. (1993) investigated numerically the detailed flow and thermal characteristics in transient laminar opposing mixed convection in a vertical plane channel subjected to a symmetrical heat input. Their results show that at high opposing buoyancy, sudden flow asymmetry and oscillation occur simultaneously in an early steady flow after the initial transient. Evans and Greif (1997) showed the strong effects of buoyancy, even for small temperature differences, on the downward flow of nitrogen in a partially heated tall vertical channel and reported time-dependent oscillations, including periodic flow reversals along the channel walls. Martínez-Suástegui and Treviño (2007, 2008) investigated the transient laminar mixed convection in an asymmetrically and differentially heated vertical channel of finite length subjected to an opposing buoyancy. Their results show that a final steady or oscillatory flow response is obtained depending on the value of the Reynolds and Richardson numbers, and that the critical value of the buoyancy strength between the two regimes strongly depends on the value of the Reynolds number.

Stability analyses in natural and mixed convection flows have been developed through the years, since they provide further insight of the instability mechanisms present in such situations and give quantitative information about the defining and critical parameters involved. Ever since the pioneering work of Lorenz (1963), who studied the instability of finite systems of deterministic ordinary non-linear differential equations representing forced dissipative hydrodynamic flows, stability analyses of fluid flow phenomena experienced a rapid growth and became the starting point of the chaos theory. Carey and Gebhart (1983) studied theoretically and experimentally the stability and disturbance amplification produced by the combination of a natural convection flow induced by a vertical uniform heat flux surface and a forced convection flow given by a uniform free stream by assuming a weak forced convection flow with strong buoyancy effects. The authors presented stability planes and constant amplification contours, and found very good agreement between the two approaches. Daniels (1989) studied the stationary instability of the convective flow between differentially heated vertical planes and determined the subsequent structure of the neutral curve for stationary disturbances. Later, Chait and Korpela (1989) studied numerically the multicellular flow between two vertical parallel plates using a time-splitting pseudo spectral method for a steady flow of air and a time-periodic flow of oil. In the case of air, the authors analyzed parametrically the three-dimensional linear stability of the flow and found that the domain of stable two-dimensional cellular motion is constrained by the Eckhaus instability and by two types of monotone instabilities. Their results show that the two-dimensional multicellular flow is unstable above a Grashof number of about 8,550. Hence, the authors conclude that the flow of air in a sufficiently tall enclosure should be considered three-dimensional for most practical applications. Two decades ago, Rogers et al. (1993) studied the finite amplitude instability of mixed convection of air in a vertical concentric annulus with each cylinder maintained at a different temperature by use of weakly non-linear instability theory and direct numerical simulation. They found three different instabilities in the parameter space of Grashof and Reynolds numbers: one due to shear and the two others induced by thermal effects. Suslov and Paolucci (1995) studied the stability of mixed convection flow in a tall vertical channel under non-Boussinesq conditions and showed that the stability characteristics, such as the critical Grashof number and the disturbance wave speed, depend strongly on the temperature difference when fluid properties are allowed to vary. Chen and Chung (1996; 1998) studied the stability of a differentially heated vertical channel for various Prandtl numbers and showed that both the Prandtl and Reynolds numbers hold very important effects on the instability mechanism for high Prandtl number fluids. More recently, van Putten et al. (2001) presented experimental results of heat transfer processes in mixed convection from a ducted vertical hot-plate thermal flow sensor for aiding and opposing flows for three different values of the Grashof numbers, $Gr = 289, 411$ and 456 for a range or Reynolds number from 0 to 120. The authors found, in the transition from free to mixed and to essentially forced convection, distinct sequences of instabilities in the flow that lead to several local minima and maxima in the heat transfer from the plate. Benoit Cushman-Roisin (2005) proposed a non-usual approach to the Kelvin-Helmholtz instability by considering the latter as a boundary-value problem instead of using the traditional approach of an initial-value problem where wave perturbations of a two-layer shear flow grow over time into billows and eventually generate vertical mixing. His results show that although the wavelength and period of the critical wave differ from the classical problem, the outcome of the boundary-value problem is the same as for the initial-value configuration. In the last years, Guillet et al. (2007) considered the case of laminar mixed convection flow between vertical parallel plates heated uniformly, they used a method based on the center manifold theorem to reduce the Navier-Stokes equations to ordinary differential equations in the vicinity of a trivial stationary solution and pointed out that when the forcing parameter or Rayleigh number increases beyond a critical value, the stationary solution is a pitchfork bifurcation point of the system. In the same year, Bera and Khalili (2007) used the linear theory of stability analysis to study numerically the impact of permeability on the stability of a buoyancy-opposed mixed convection in a vertical channel. They found that two main instability modes (Rayleigh-Taylor and buoyant instability) appear, and that for Darcy numbers $\leq 10^{-9}$, the Rayleigh-Taylor instability dominates within the entire Reynolds number range considered. In addition, they also found that for the same Re, the fully developed base flow is highly unstable (stable) for porous media with high (low) permeability, whilst the introduction of a one order of reduction in the permeability in the main flow direction made the system approximately twenty times more stable.

Although the oscillatory behavior of Navier-Stokes type systems in mixed convection has received relatively little attention, it is known that these flows can exhibit interesting dynamical phenomena. In many cases, the transition to turbulence is a chain of oscillatory states separated by bifurcations of different types. Examples where the bifurcation structure of such systems has been studied follow.

Meron and Procaccia (1987) showed that in dynamical systems described by critical flows, the onset of chaos is via gluing bifurcations, and that these systems can be analyzed using discontinuous maps of the interval. The gluing bifurcation is a class of global bifurcations where, as an external control parameter is varied, two symmetrically related time-periodic states simultaneously become homoclinic to an unstable saddle state and result in a single symmetric time-periodic state (Epstein and Pojman 1998; Ambruster et al. 1996; Abshagen et al. 2001) . Arneodo et al. (1981) performed a study for one route to chaos via a cascade of bifurcations involving homoclinic orbits. Rucklidge (1993) described the transition to chaos through gluing processes in a three-dimensional magnetoconvection model. Marques et al. (2001) studied numerically a one-dimensional route in parameter space of a periodically forced flow with symmetry and provided a comprehensive analysis of the route to chaos, which involves a new and convoluted symmetry breaking that includes heteroclinic, homoclinic and gluing bifurcations. Lopez and Marques (2000) obtained three-tori solutions of the Navier-Stokes equations and their dynamics by use of a global Poincaré map. Their results show that these solutions undergo global bifurcations that include a new gluing bifurcation associated with homoclinic and heteroclinic connections to unstable solutions (two-tori), that act as organizing centers for the three-tori dynamics.

The foregoing survey of literature reveals that there are relatively few studies that address the thermal and flow stability in internal mixed convection for opposing flow in situations where flow reversal occurs. In the present study, a detailed numerical investigation is done of the symmetry breaking instability for laminar opposing mixed convection flow in a vertical channel of finite length subjected to isothermal and discrete heat inputs. For fixed Reynolds, Prandtl and geometry, the stability characteristics are obtained for increasing values of the buoyancy parameter.

2 Problem Description

The loss of stability is studied for a symmetrical counter-current mixed convection problem inside a vertical channel of finite length with a flat velocity distribution at the channel entrance, with the channel walls heated discretely and symmetrically. The schematic view of the geometry considered is shown in Fig. 1. The forced flow is driven by gravitational force acting vertically downward, entering the duct with a uniform velocity u_0 and ambient temperature T_0. Axial distances from the entrance section are measured by the x coordinate (positive downward), while transverse distances are measured by y ($y = 0$ at the left wall). Both walls, separated by a distance h have discrete heat sources of length l_2 located at $x = l_1$, with uniform wall temperature T_w, where $T_w > T_0$. All other surfaces of the channel walls are assumed adiabatic insulators. Flow rectifiers are placed at the channel entrance and exit, thus producing a parallel flow at $x = 0$ and $x = l_1 + l_2 + l_3$. The viscous dissipation in the energy equation is neglected and the thermophysical properties of the fluid are assumed to be constant except for the density in the buoyancy term, which

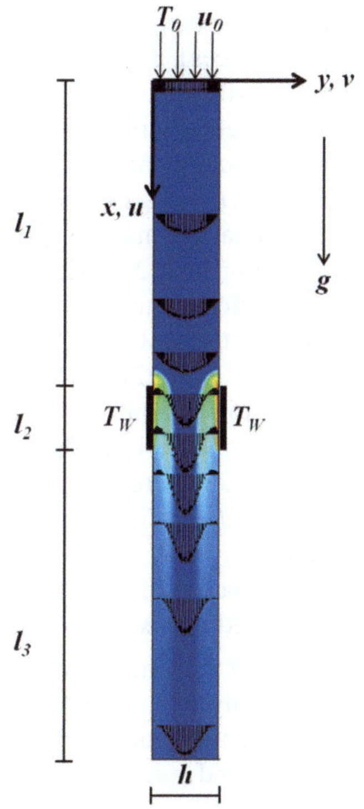

Fig. 1 Schematic representation of the channel

is treated according to the Boussinesq approximation. By the use of the vorticity ($\Omega = \partial V/\partial X - \partial U/\partial Y$) and stream function formulation ($U = \partial\psi/\partial Y,\ V = -\partial\psi/\partial X$), the flow is described by the nondimensional equations

$$\frac{\partial^2\psi}{\partial X^2} + \frac{\partial^2\psi}{\partial Y^2} = -\Omega, \tag{1}$$

$$\frac{\partial\Omega}{\partial\tau} + \frac{\partial\psi}{\partial Y}\frac{\partial\Omega}{\partial X} - \frac{\partial\psi}{\partial X}\frac{\partial\Omega}{\partial Y} = \frac{1}{Re}\left(\frac{\partial^2\Omega}{\partial X^2} + \frac{\partial^2\Omega}{\partial Y^2}\right) - Ri\frac{\partial\theta}{\partial Y}, \tag{2}$$

$$\frac{\partial\theta}{\partial\tau} + \frac{\partial\psi}{\partial Y}\frac{\partial\theta}{\partial X} - \frac{\partial\psi}{\partial X}\frac{\partial\theta}{\partial Y} = \frac{1}{RePr}\left(\frac{\partial^2\theta}{\partial X^2} + \frac{\partial^2\theta}{\partial Y^2}\right). \tag{3}$$

where $\mathbf{V} = (U, V)$ is the dimensionless velocity vector and θ is the dimensionless temperature. In the above equations, all velocity components (U in the X-direction and V in the Y-direction) are scaled with the inflow velocity, u_0, $U = u/u_0$ and $V = v/u_0$; the longitudinal coordinates are scaled with the channel width h, $X = x/h$ and $Y = y/h$; the time is scaled with the residence time h/u_0, $\tau = tu_0/h$; the temperature is normalized as $\theta = (T - T_0)/(T_w - T_0)$. The non-dimensional parameters appearing in the above equations are the Reynolds number, $Re = u_0 h/\nu$ (inertial to viscous forces), the Prandtl number, $\Pr = \nu/\alpha$ (viscous to thermal diffusivities) and the Richardson number, $Ri = g\beta\,(T_w - T_0)\,h/u_0{}^2$ (buoyancy induced potential to forced kinetic energy). Here, g is the gravity acceleration and β is the thermal expansion coefficient. Another buoyancy parameter employed frequently is the Grashof number, $Gr = Ri\,Re^2$, which relates buoyancy to viscous forces. Additional non-dimensional geometrical parameters arise through the boundary conditions, $L_i = l_i/h$, with $i = 1, 2, 3$. Equations (1–3) have to be solved with the following boundary conditions. Uniform flow at the channel entrance: $\psi(0, Y) - Y = \Omega(0, Y) = 0$, no slip at the walls, $\psi(X, 0) = \Omega(X, 0) - 2(\psi(X, 0) - \psi(X, \Delta Y))/\Delta Y^2 = 0$, $\psi(X, 1) - 1 = \Omega(X, 1) - 2(\psi(X, 1) - \psi(X, 1 - \Delta Y))/\Delta Y^2 = 0$, and relaxed parallel flow conditions at the channel exit: $\partial\psi/\partial X|_{X=L} = \partial^2\psi/\partial X\partial Y\big|_{X=L} = 0$, where L is the total length of the channel, $L = L_1+L_2+L_3$. The boundary conditions for temperature are the following. Fixed temperature at the heated sections $\theta = 1$ at $Y = Y_w$ at $L_1 \le X \le L_1 + L_2$. Adiabatic channel walls are considered, $\partial\theta/\partial Y = 0$ except at $L_1 < X < L_1 + L_2$ and $Y = Y_w$. The dynamical properties of the system are described using the average non-dimensional heat fluxes or Nusselt numbers at both heated plates, $\mathrm{Nu}_{L,R} = \left|\overline{q}_{L,R}\right| h/(k(T_w - T_0))$ and the non-dimensional first moment of the longitudinal velocity, $Y_p = (1/h^2 u_0) \int_0^h y u dy$. Here k is the thermal conductivity of the fluid. $\mathrm{Nu}_{L,R}$ depend only on time and Y_p is a function of the longitudinal position and time. Due to baroclinity -last term of Eq. (2)-, vorticity is produced and vortices (large recirculation bubbles) are generated. The position of the recirculation zones are represented by a stagnation point at $X = X_s(\tau)$, defined by the maximum value of X, where the longitudinal velocity component is non-negative in the vortex region. Eqs. (1–3) are numerically solved using a strongly non-uniform staggered grid system with a denser clustering near the heated plate. The technique employed has been described elsewhere Mart?nez-Su?stegui et al. (2011). In this work the Reynolds number is $Re = 100$, with $l = 12h$. The length of the heated slabs is $l_2 = h$, and they are located at $x = l_1 = 5.5h$, that is $l_1 = l_3 = 5.5h$. The non-dimensional time step $\Delta\tau = u_0\Delta t/h$ has been set to 5×10^{-4}, and computation is terminated when the time evolution of the system state reaches steady-state or a final self-sustained oscillatory state.

3 Numerical Solution

After switching buoyancy on and for relatively small values of the Richardson number, the flow reverses close to the heated slabs and a pair of symmetric vortices develop due to baroclinity. Both vortices reach a maximum position represented by the stagnation point, which decreases (higher positions) as the buoyancy parameter increases. There is an equilibrium between buoyancy in the hot recirculation bubble and the dynamical pressure and drag from the cold downward fluid. For a range of values of the Richardson number of $Ri < 5.2$, the transient response leads to a

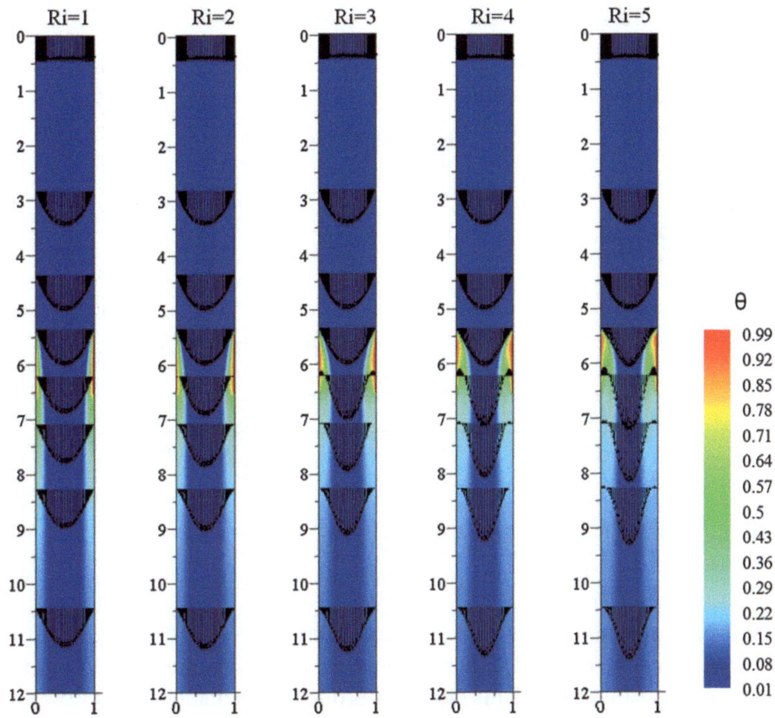

Fig. 2 Final states for a steady and symmetrical dynamical response of the system

symmetrical steady state solution. For these relatively low Richardson numbers there is only a weak interaction in the dynamics of both recirculation bubbles. Figure 2 shows the velocity and temperature profiles after the dynamical system has reached a symmetrical steady state. For this range of the buoyancy parameter, the final states of the system correspond to steady symmetric flow. As the Richardson number increases, small amplitude flow oscillations appear in the downstream region of the channel and their amplitude increase for increasing values of the buoyancy parameter. These flow oscillations are a manifestation of the Kelvin-Helmholtz instability due to the strong shear between the ascending hot fluid and the descending cold fluid located in the middle of the channel. This triggers a symmetry-breaking bifurcation and one of the two vortices (assumed to be close to the left heated surface, for simplicity) climbs while the other is pushed down by the downward flow with increased longitudinal momentum. If one vortex climbs, the cold fluid deflects to the other side of the channel increasing the longitudinal velocity and thus the momentum, pushing down the other vortex. As the fluid reaches the top of the lower vortex, the cold downward fluid switches sides again, thus supporting the upper recirculation bubble. Crude numerical simulations show that the symmetry breaking bifurcation occurs for a Richardson number of $Ri = 5.2$. As a result, a stable non-symmetric pattern develops after a relatively short transient. Figure 3 shows the resulting evolution of

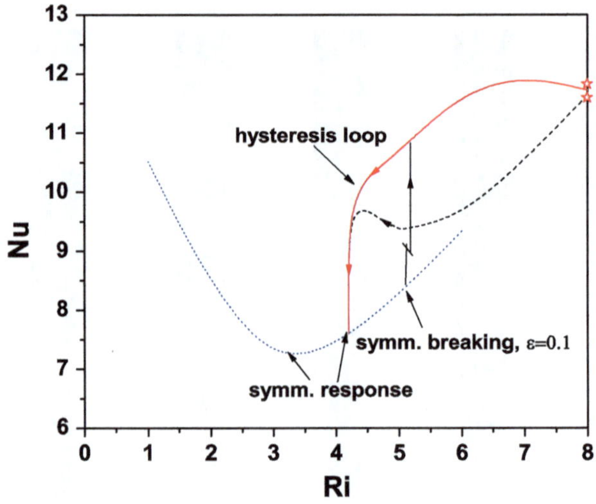

Fig. 3 Hysteresis behavior for $Re = 100$ and $Pr = 7$

the overall Nusselt number for the specific case of a Reynolds number of 100 and a Prandtl number of 7. As the Richardson number increases, the Nusselt number first decreases and later increases. This behavior is due to the change of the flow direction close to the wall. For a Richardson number close to 5.2 symmetry breaks and one of the recirculation bubble climbs while the other is pushed down, producing two solution branches for both hot surfaces. If now, the Richardson number decreases, the response continues to be asymmetric for values down to 4.1, where suddenly the symmetric behavior is recovered. This hysteresis loop is a clear indication of a subcritical bifurcation process to be analyzed below.

4 Linear Stability Analysis

For a given fixed Reynolds number and small values of the Richardson number, a symmetrical flow response is obtained. In order to study the stability of the symmetrical flow, a symmetrical strongly non-parallel base flow with variables denoted by ψ_0, Ω_0 and θ_0 as functions of the Richardson number, is numerically generated, given by Eqs. (1–3), but with the following symmetrical boundary conditions: Uniform flow at the channel entrance: $\psi_0(0, Y) - Y = \Omega_0(0, Y) = \theta_0(0, Y) = 0$,; no slip at the wall, $\psi_0(X, 0) = \Omega_0(X, 0) - 2(\psi_0(X, 0) - \psi_0(X, \Delta Y))/\Delta Y^2 = 0$; symmetry conditions at the symmetry plane ($Y = 0.5$), $\psi_0(X, 0.5) - 0.5 = \Omega_0(X, 0.5) = \partial\theta_0/\partial Y = 0$ and relaxed parallel flow conditions at the channel exit: $\partial\psi_0/\partial X|_{X=L} = \partial^2\psi_0/\partial X\partial Y|_{X=L} = 0$. A solution of the form $\Omega = \sum_{i=0}^{\infty} \delta^n \Omega_n$, $\psi = \sum_{i=0}^{\infty} \delta^n \psi_n$ and $\theta = \sum_{i=0}^{\infty} \delta^n \theta_n$, is assumed, with $\delta \to 0$. The linearized first order equations are

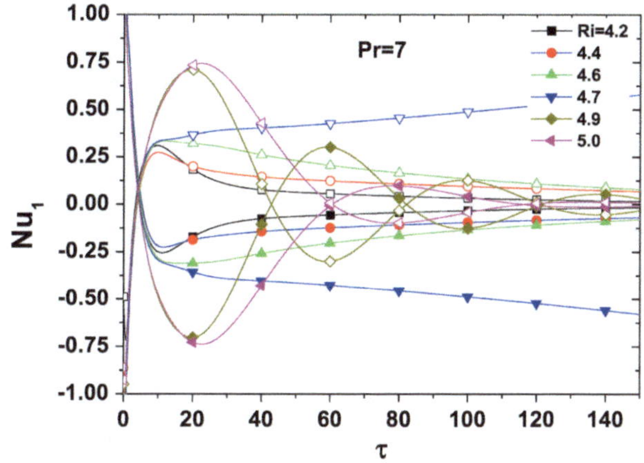

Fig. 4 Time evolution of the perturbed Nusselt number for $Pr = 7$ and different values of the Richardson numbers from 4.2 to 5.0

$$\frac{\partial \Omega_1}{\partial \tau} + \frac{\partial \psi_0}{\partial Y}\frac{\partial \Omega_1}{\partial X} - \frac{\partial \psi_0}{\partial X}\frac{\partial \Omega_1}{\partial Y} - \frac{1}{Re}\nabla^2 \Omega_1 = -\frac{\partial \psi_1}{\partial Y}\frac{\partial \Omega_0}{\partial X} + \frac{\partial \psi_1}{\partial X}\frac{\partial \Omega_0}{\partial Y} - Ri\,\frac{\partial \theta_1}{\partial Y}, \quad (4)$$

$$\nabla^2 \psi_1 = -\Omega_1, \quad (5)$$

$$\frac{\partial \theta_1}{\partial \tau} + \frac{\partial \psi_0}{\partial Y}\frac{\partial \theta_1}{\partial X} - \frac{\partial \psi_0}{\partial X}\frac{\partial \theta_1}{\partial Y} - \frac{1}{RePr}\nabla^2 \theta_1 = -\frac{\partial \psi_1}{\partial Y}\frac{\partial \theta_0}{\partial X} + \frac{\partial \psi_1}{\partial X}\frac{\partial \theta_0}{\partial Y}, \quad (6)$$

with the required homogeneous boundary conditions in the whole channel. A non trivial solution can be obtained from numerically generated eigenfunctions, using the perturbed Richardson number

$$Ri_\varepsilon = Ri\,(1 + \varepsilon e^{-\tau/\tau_0} f(y)),$$

with two different modes: $f(y) = sin(2\pi y)$ or $f(y) = cos(\pi y)$. In the above relation τ_o is set of order unity and ε is a parameter with a value very small compared with unity. The non-symmetric forcing relaxes the solution in a short time to a numerical eigenfunction, which evolves with time, giving the stability conditions by detecting the amplitude evolution of the overall Nusselt perturbation and its oscillation frequency, $Nu_1 = \int_{L_1}^{L_1+1} \partial\theta_1/\partial Y|_{Y=0}\, dX \sim \exp(\sigma + iS)\tau$, where $i = (-1)^{1/2}$. Positive values of σ means instability. *S* is then the non-dimensional oscillation frequency (Strouhal number). Figure 4 shows the time evolution of the perturbed Nusselt number Nu_1 for a Prandtl number of $Pr = 7$ at both heated surfaces, for different values of the Richardson numbers up to $Ri = 5.0$. For Richardson numbers up to 4.6 the perturbed solution decays without oscillating. On the other hand for values between 4.9 and 5.0 the response solution decays oscillating with different frequencies and

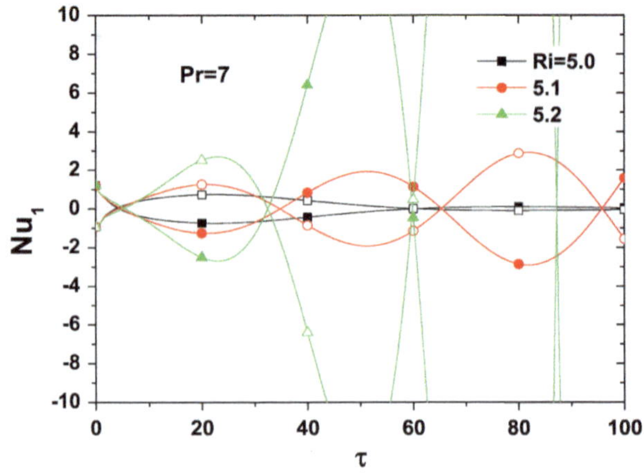

Fig. 5 Time evolution of the perturbed Nusselt number for $Pr = 7$ and different values of the Richardson numbers from 5.0 to 5.2

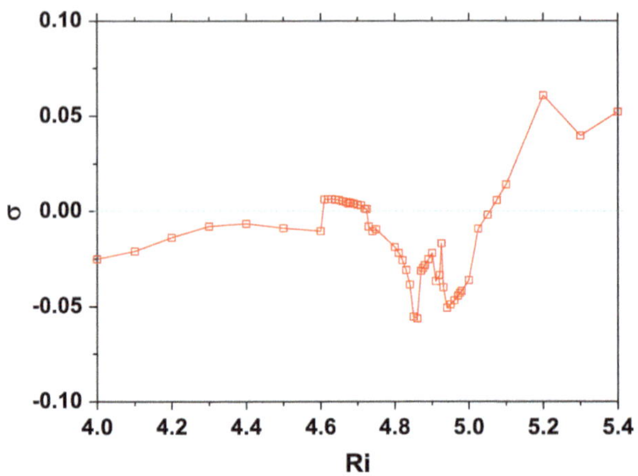

Fig. 6 Growth rate as a function of the Richardson number for $Pr = 7$. Positive values means instability

decay rates. However, surprisingly, for a Richardson number of 4.7, the amplitude perturbed solution increases monotonically with time, which is indication that for this particular case, the symmetrical flow is unstable. Using the non-linear equations it can be shown that a slightly non-symmetric steady state flow response is obtained for this Richardson number. For Richardson numbers above 5.0, Fig. 5 shows the time evolution of the perturbed Nusselt number. Here again, for $Ri = 5.0$ the stable decaying oscillating response is depicted. As the Richardson number increases, the oscillating frequency also increases. For Richardson numbers $Ri \geqq 5.1$ the system is now unstable. Figures 6 and 7 show the the real (σ) and the imaginary (S) parts, respectively, of the exponential growth, for a Prandtl number of $Pr = 7$. The sys-

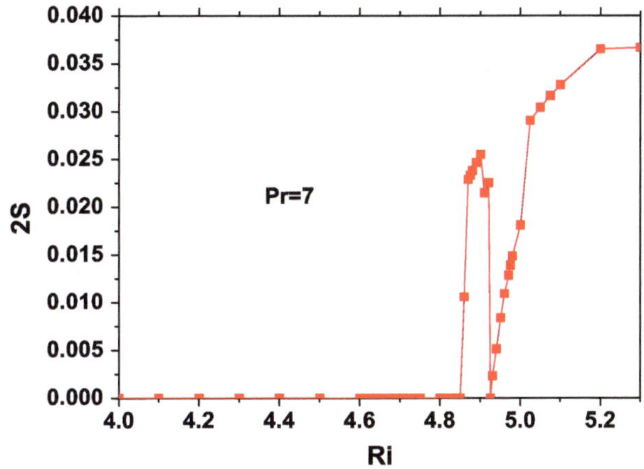

Fig. 7 Oscillation frequency of the linear perturbation as a function of the Richardson number for $Pr = 7$

tem is linearly unstable in the ranges $4.606 < Ri < 4.72$ and $Ri > 5.06$. Instability leads in both cases to steady nonsymmetric thermal and flow configurations with two vortex structures at different heights. Although the symmetrical flow configuration is unstable for values of $4.606 < Ri < 4.72$, soon non-linearities limit the growth of perturbations producing a slight modified asymmetrical steady thermal and flow pattern, where the recirculation bubbles only move apart slightly in the vertical position. However, for values of $Ri > 5.06$, the difference of the maximum height of both vortical structures at the final steady state is of the order of the channel width, that is of order unity in non-dimensional units. Figure 7 shows that the time evolution of the perturbation is monotonic for $Ri < 4.85$ and oscillatory for larger values of the Richardson number, except for a very thin region around $Ri = 4.92$, where the oscillation frequency reduces drastically. This atypical response arises because for each Richardson number, a different base flow is needed.

5 Conclusion

In this work, the linear stability of a two dimensional symmetrical counter-current mixed convection system is analyzed using numerically generated eigenfunctions. The problem depends on several non-dimensional parameters like the flow Reynolds number, the Prandtl number of the fluid, the Richardson number (buoyancy parameter) and geometrical parameters of the vertical channel. The symmetry breaking instability has been obtained for fixed geometry, Reynolds ($Re = 100$) and Prandtl ($Pr = 7$) numbers, by increasing the buoyancy parameter. For low Richardson numbers ($Ri < 4.606$), the resulting flow is steady and symmetric with two vortical flow structures anchored slightly above the heated sources. The system shows an instability for the range $4.606 < Ri < 4.72$, where the two vortical structures separate only slightly in the vertical position. The steady-state symmetrical response is again obtained for larger Richardson numbers up to $Ri = 5.06$, where again the system looses stability, producing a much larger separation of the vortices in the vertical coordinate.

Acknowledgments This work has been supported by the DGAPA, UNAM through IXTLI with grant: IXTLI100010 and by the Consejo Nacional de Ciencia y Tecnología (CONACYT), Grant number 167474.

References

Abshagen J, Pfister G, Mullin T (2001) Gluing bifurcations in a dynamically complicated extended flow. Phys Rev Lett 87:224501

Ambruster D, Nicolaenko B, Smaoui N, Chossat P (1996) Symmetries and dynamics for 2d navier-stokes flow. Physica D 95:81–93

Arneodo A, Coullet P, Tresser C (1981) A possible new mechanism for the onset of turbulence. Phys Lett 81A:197

Bera P, Khalili A (2007) Stability of buoyancy opposed mixed convection in a vertical channel and its dependence on permeabilty. Adv Water Res 30:2296–2308

Carey VP, Gebhart B (1983) The stability and disturbance-amplification characteristics of vertical mixed convection flow. J Fluid Mech 127:185–201

Chait A, Korpela SA (1989) The secondary flow and its stability for natural convection in a tall vertical enclosure. J Fluid Mech 200:189–216

Chang T-S, Lin T-F (1993) Steady and oscillatory opposing mixed convection in a symmetrically heated vertical channel with a low-prandtl number fluid. Int J Heat Mass Transfer 36:3783–3795

Chen Y-C, Chung J (1996) The linear stability of mixed convection in a vertical channel flow. J Fluid Mech 325:29–51

Chen Y-C, Chung J (1998) Stability of mixed convection in a differentially heated vertical channel. ASME J Heat Transfer 120:127–132

Cushman-Roisin B (2005) Kelvin-helmholtz instability as a boundary-value problem. Environ Fluid Mech 5:507–525

Daniels P (1989) Stationary instability of the convective flow beetween differentially heated vertical planes. J Fluid Mech 203:525–540

Epstein I, Pojman J (1998) An introduction to nonlinear chemical dynamics: oscillations, waves, patterns, and chaos, Oxford University Press, Oxford

Evans G, Greif R (1997) Buoyant instabilities in downward flow in a symmetrically heated vertical channel. Int J Heat Mass Transfer 40:2419–2425

Guillet C, Mare T, Nguyen CT (2007) Application of a non-linear local analysis method for the problem of mixed convection instability. Int J Non-Linear Mech 42:981–988

Lin T-F, Chang T-S, Chen Y-F (1993) Development of oscillatory asymmetric recirculating flow in transient laminar opposing mixed convection in a symmetrically heated vertical channel. ASME J Heat Transfer 115:342–352

Lopez J, Marques F (2000) Dynamics of three-tori in a periodically forced navier- stokes flow. Phys Rev Lett 85:972–975

Lorenz EN (1963) Deterministic non-periodic flow. J Atmos Sci 20:130–141

Marques F, Lopez J, Shen J (2001) A periodically forced flow displaying symmetry breaking via a three-tori gluing bifurcation and two-tori resonances. Physica D 156:81–97

Martínez-Suástegui L, Treviño C (2007) Particle image velocimetry measurements for opposing flow in a vertical channel with a differential and asymmetric heating condition. Exp Therm Fluid Sci 32:262–275

Martínez-Suástegui L, Treviño C (2008) Transient laminar opposing mixed convection in a differentially and asymmetrically heated vertical channel of finite length. Int J Heat Mass Transfer 51:5991–6005

Martínez-Suástegui L, Treviño C, Cajas J (2011) Thermal nonlinear oscillator in mixed convection. Phys Rev E 84:046310

Meron E, Procaccia I (1987) Gluing bifurcations in critical flows: the route to chaos in parametrically excited surface waves. Phys Rev A 35:4008–4011

Rogers BB, Moulic SG, Yao LS (1993) Finite-amplitude instability of mixed convection. J Fluid Mech 254:229–250

Rucklidge A (1993) Chaos in a low-order model of magnetoconvection. Physica D 62:323–337

Suslov S, Paolucci S (1995) Stability of mixed-convection flow in a tall vertical channel under non-boussinesq conditions. J Fluid Mech 302:91–115

van Putten MJAM, Kleijn CR, van den Akker HEA (2001) Heat transfer and temporal behavior of the laminar mixed-convection flow around a ducted flat-plate thermal flow sensor. Exp Heat Transf 14:229–250

The Boundary Element Method in Fluid Mechanics: Application to Bubble Growth

A. López-Villa, Luis S. Zamudio and A. Medina

Abstract The origin of the numerical implementation of boundary integral equations can be traced from fifty years earlier, when the electronic computers had become available. The full emergence of the numerical technique known as the boundary element method occurred in the late 1970s. In implementing the method, only the boundary of the solution domain has to be discretized into elements. In the case of a two-dimensional problem, this is really easy to do: put closely packed points on the boundary (a curve) and join up two consecutive neighboring points to form straight line elements. In this chapter we present one of the applications of this method, namely, the growth and detachment of bubbles generated by the continuous injection of gas into a quiescent liquid and the effect of partial confinement on the shape and volume of bubbles generated by injection of a constant flow rate of gas. In the problem of bubble generation, the contours are the surfaces of the bubbles and the solid surfaces of the reservoir, which are all surfaces of revolution. The unknowns involved in the formulation of the boundary element are fluid particle velocities that define surfaces of the bubbles and the stresses on the vessel wall. First, we neglect viscous effects and assume the flow to be irrotational so that a velocity potential exists. In second case we solve the Stokes equations for the liquid and the evolution equation for the surface of a bubble. Experiments with two different liquids show that cylindrical and conical walls and cylinder walls with periodic concentric corrugations with a gas injected through an orifice at the bottom of the liquid may strongly affect the shape and volume of the bubbles, and can be used to control the size of the generated bubbles without changing the flow rate of gas.

A. López-Villa (✉) · A. Medina
ESIME Azcapotzalco, Instituto Politécnico Nacional, Av. de las Granjas
No. 682, Col. Sta. Catarina, 02550 México, D.F., México
e-mail: abelvilla77@hotmail.com

Luis S. Zamudio
Programa de Ingeniería Molecular, Grupo de Química Aplicada a la Industria Petrolera, Instituto Mexicano del Petróleo, eje Central Lázaro Cárdenas
No. 152, col, San Bartolo Atepehuacan, 07730 México, D.F., México

J. Klapp and A. Medina (eds.), *Experimental and Computational Fluid Mechanics*, Environmental Science and Engineering, DOI: 10.1007/978-3-319-00116-6_2,

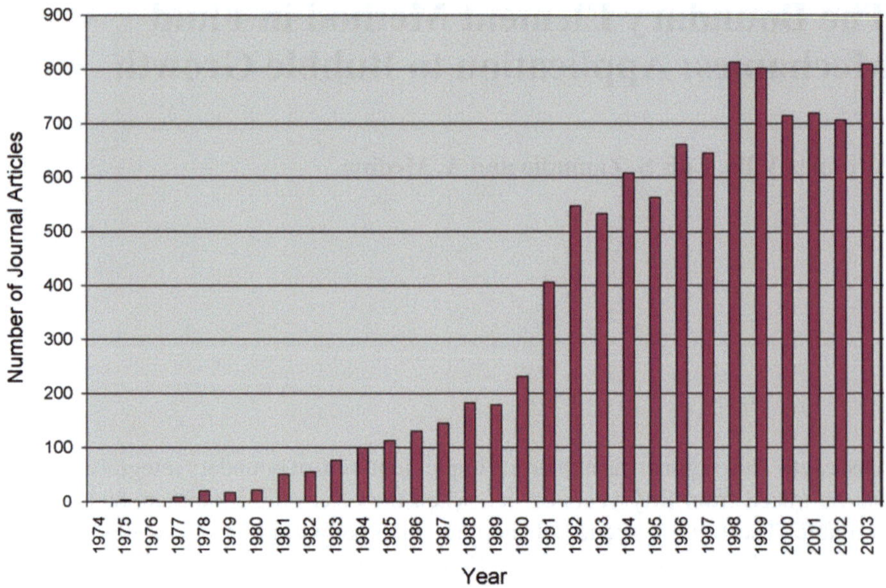

Fig. 1 Number of journal articles published by the year on the subject of BEM, based on the Web of Science search. Refer to Appendix for the search criteria (Search date: May 3, 2004)

1 Introduction

Mathematicians from the eighteenth to twentieth centuries, whose contributions were the key to the theoretical development, are honored with short biographies. The origin of the numerical implementation of boundary integral equations can be traced back to the 1960s, when the electronic computers had become available.

The full emergence of the numerical technique known as the boundary element method occurred in the late 1970s. This article reviews the early history of the boundary element method up to the late 1970s.

After three decades of development, the boundary element method (BEM) has found a firm footing in the area of numerical methods for partial differential equations. Comparing to the more popular numerical methods, such as the Finite Element Method (FEM) and the Finite Difference Method (FDM), which can be classified as the domain methods, the BEM distinguish itself as a boundary method, meaning that the numerical discretization is conducted at reduced spatial dimension. For example, for problems in three spatial dimensions, the discretization is performed on the bounding surface only; and in two spatial dimensions, the discretization is on the boundary contour only. This reduced dimension leads to smaller linear systems, less computer memory requirements, and more efficient computation. This effect is most pronounced when the domain is unbounded. Unbounded domain needs to be truncated and approximated in domain methods. The BEM, on the other hand, automatically models the behavior at infinity without the need of deploying a mesh to approximate it. Figure 1 presents the histogram of the number of journal papers published annually, containing BEM as a keyword (Alexander and Daisy 2005; Costabel Martin 1986).

Because a given set of boundary and initial conditions uniquely define the solution in the domain, the value of the function at any point in the interior can be expressed as a sole contribution of boundary values, what is achieved mathematically by the Green-Stokes-Gauss-divergence theorem, which is the foundation of the boundary elements method (BEM). With this method, first the full solution (function and derivatives) at the boundary points are computed by a kind of finite-element method where the base functions are the fundamental solutions of the PDE at the boundary nodes, then solving a set of algebraic equations at the nodes, and finally, if needed, the value at any internal point is directly computed by a quadrature (without interpolation).

The problem with the boundary element method is that the local integration in the boundary is more involved than in the standard FEM because there are singular points that require more elaborated computations. Other handicap is that the BEM only applies to regions of constant properties. The great advantage is that for bulky domains the number of nodes significantly decreases, particularly for infinite domains (what explains its massive use in external fluid mechanics and geomechanics). Incidentally, for infinite domains, besides the BEM, one may also resort to classical FEM with a truncated domain progressively enlarged, or matched to an asymptotic analytical expansion, or stretching the external elements with a log-transformation.

On the other hand, the growth and detachment of bubbles generated by the continuous injection of gas into a quiescent liquid has been very much studied in conditions where the viscosity of the liquid plays no important role (Davidson and Schuler 1960; Longuet-Higgins et al. 1991; Marmur and Rubin 1976; Clift et al. 1978; Corchero et al. 2006). Results of these studies are of interest in metallurgical and chemical industries among others, where liquids of low viscosity, such as liquid metals and aqueous solutions, need to be handled. Bubbles in these liquids can be used to modify the concentrations of different substances and promote chemical reactions between them, to clean liquids from impurities captured by adhesion or diffusion processes, and for many other purposes (López-Villa et al. 2011).

The generation and dynamics of bubbles in very viscous liquids is also of interest but has not been so much studied. Thus, while many aspects of the dynamics of bubbles in unbounded viscous liquids are well understood, the formation and detachment of bubbles in confined systems has received less attention. Bubbles in very viscous liquids are commonly found when dealing with polymers in their liquid phases, in the flows of lava, and in processes of oil extraction from production pipelines, among others. The latter example is a motivation of the present work, which sprang from interest in the so-called gas lift technique of enhanced oil recovery, where bubbles formed by injecting gas in oil extraction pipes help pumping the oil. Another motivation is the foam formation, where we study the problem of the film thickness that is formed between the free surface of a single bubble and the wall when the bubble reaches its critical size in the vertical circular tube with smooth or ribbed walls filled with a quiescent liquid of high viscosity. The foam formation is very important for enhanced oil recovery (EOR), where the foam is used for channeling oil and clogs the fractures to keep out gas, in fractured oil reservoirs. If gas mobility can be controlled, oil displacement efficiency is improved (Kovscek et al. 1995). Foam is a promising general agent for controlling gas mobility in EOR processes (Hirasaki and Lawson 1985) and in other applications, such as aquifer storage of natural gas (Witherspoon et al. 1987) and compressed air.

Scaling laws show that the volume of the bubbles generated by injecting a high flow rate of gas into a very viscous liquid increases as the power 3/4 of the flow rate and is independent of the diameter of the injection orifice. The simplest way to control the size of the bubbles in a given liquid is, therefore, to act on the flow rate of gas. This possibility, however, is limited in the application at hand, because the flow rate of gas to be injected in the confined space of an extraction pipe is often determined by other requirements of the gas lift technique. The limitation poses a problem to control the size of the bubbles and brings to the front elements of the generation process such as the viscous drag of the bubbles and the shear stress in the vicinity of the walls, which are disregarded an inviscid analyses but offer a clue to the solution of the size-control problem. In this chapter, we first neglect viscous effects and assume the flow to be irrotational so that a velocity potential exists; in second case we solve the Stokes equations for the liquid and the evolution equation for the surface of a bubble. The shape of the tube in the vicinity of the injection orifice, or the use of properly shaped injection nozzles, may cause substantial distortion of the growing and shape of the bubbles and modify their volume at detachment. In our analysis, a constant flow rate of gas is injected through a circular orifice at the horizontal base of a container filled with non-viscous and very viscous liquid, and the space where the bubbles grow can be partially confined by surrounding the orifice with a vertical cylindrical wall or an inverted vertical cone. The extent of the confinement can be gradually increased by decreasing the radius of the cylinder or the angle of the cone, which allows quantifying the effect of the wall on the evolution and size of the bubbles. This size is determined numerically and experimentally, and scaling laws that are extensions of well-known laws for unconfined liquids are proposed and validated.

To reach our goal the structure of the chapter is as follows. In the next section we formulate the problem in terms of dimensionless equations for the motion of the non-viscous liquid and the dimensionless boundary conditions for the evolution of the free surface. In Sect. 3 we formulate the problem in terms of dimensionless equations of motion of viscous fluids and the dimensionless boundary conditions for the evolution of the free surface. In Sect. 4 we briefly explain the BEM method. The main results of the numerical solution of this problem are given in Sect. 5, discussing the evolution of the free surface during the growth of the bubble at constant gas flow rate in conical and cylindrical containers. In Sect. 6 we compare some qualitative experiments to our numerical results. Finally, Sect. 7 summarizes the main findings and limitations of this work.

2 Equations for Bubbles in Non-Viscous Liquids

Generation of bubbles by injection of a gas into a liquid at rest is an important and much studied problem. Extensive research has been summarized in a variety of models that address the many facets of the problem with different levels of detail; see Clift et al. (1978), Räbiger and Vogelpohl (1986), Tsuge (1986) and Sadhal et al. (1997) for reviews. The conceptually simplest models are based on a balance of the forces acting on a bubble of assumed shape (see Davidson and Schuler (1960), among others). These models clearly show the existence of a regime of low gas flow rate in which the effect of the inertia of the liquid is negligible and the volume of the bubbles is a constant independent of the gas flow rate, and a regime of high gas flow rate in which the effect of the surface tension is negligible and the volume of the bubbles increases as the 6/5 power of the gas flow rate and is independent of the size of the injection orifice.

The original models of Davidson and Schuler (1960) and Ramakrishna et al. (1968) which served to establish these results, have been extended to include a variety of effects such as the viscous drag of the bubbles, the flow left by the viscous wake of the preceding bubble, the momentum flux of the injected gas, and the different shapes and apparent masses of the bubble at different stages of its growth. Extensions also include a set of ad hoc criteria to account for the interference, collision and coalescence of bubbles (Zhang and Shoji 2001), which are observed to occur at high flow rates and eventually lead to non-periodic and chaotic regimes of bubble generation (Leighton et al. 1991). More sophisticated non- spherical models (Marmur and Rubin 1976) postulate equations of motion for each element of the bubble surface, whose shape changes continuously during the growth and detachment. These models rely on varying degrees of solutions for the potential flow of the liquid (Wraith and Kakutani 1974). Oguz and Prosperetti (1993) numerically computed this flow using a boundary element method and described in full detail the growth and detachment of a single bubble at the end of a tube in different cases of interest, finding good agreement with high speed video visualizations (see also Oguz and Zeng 1997).

This section focuses on time periodic bubbling regimes featuring coalescence of two or more bubbles in a strictly inviscid liquid. Though the bubble generation process ceases to be periodic when the flow rate is increased to sufficiently high values, these more complex regimes will not be discussed here. Instead, the purpose of the work is to examine to what extent coalescence at moderate gas flow rates can be described in the framework of potential flow theory. In this respect, the work is an extension of those of Oguz and Prosperetti (1993) and Oguz and Zeng (1997) to include bubble coalescence. The main result is that potential flow computations suffice to describe many aspects of coalescence, without resorting to any wake effect or other effects related to the viscosity of the liquid.

Attention will be restricted to the simplest case of injection of a constant flow rate of a gas through a single circular orifice at the bottom of an inviscid liquid at rest (see Fig. 2). The gas will be treated as incompressible, with a density negligibly small compared to the density of the liquid. The only parameters of the problem are

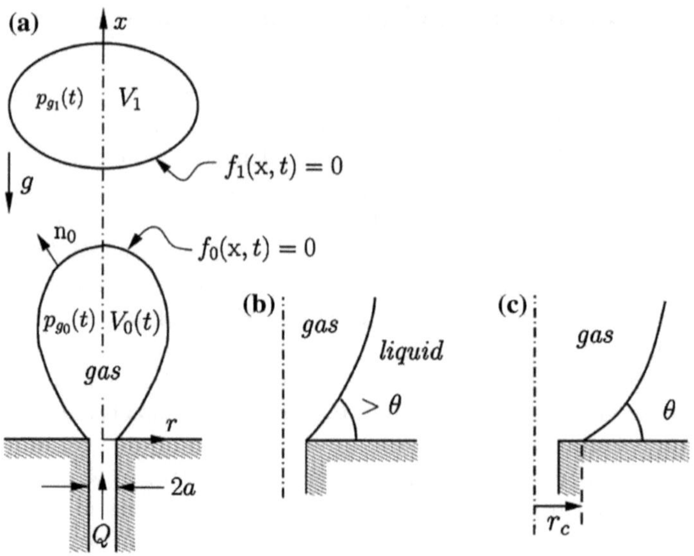

Fig. 2 Definition sketch (**a**), with details of the contact line attached to the edge of the orifice (**b**) and spreading on the horizontal bottom (**c**)

then the radius of the orifice, a, the density of the liquid, ρ, the liquid–gas surface tension, σ, the contact angle of the surface with the bottom, θ, the gas flow rate, Q (volume of gas injected per unit time), and the acceleration due to gravity, g. The dimensional parameters can be grouped into a Bond number and a Weber number:

$$B = \frac{\rho g a^2}{\sigma} \qquad We = \frac{\rho Q^2}{\sigma a^3} \tag{1}$$

The flow induced in the liquid by the train of bubbles released from the orifice is irrotational if the viscosity of the liquid is neglected. The velocity potential, φ, such that $v = \nabla\varphi$, satisfies the Laplace equation

$$\nabla^2\varphi = 0 \tag{2}$$

in the liquid, to be solved with the conditions

$$\frac{Df_i}{Dt} = 0, \tag{3}$$

$$\frac{D\varphi}{Dt} = \frac{1}{2}\,|\nabla\varphi|^2 - p_{gi} - Bx + \nabla \cdot n_i, \tag{4}$$

at the surfaces of the bubbles; and $\frac{\partial \varphi}{\partial x} = 0$ at the bottom ($x = 0$) and $\nabla\varphi \to 0$ at infinity. Here $f_i(x,t) = 0$ is the equation of the surface of the i-th bubble, with $i = 0$ denoting the bubble growing at the orifice and $i = 1, 2, \ldots$. denoting the bubbles detached previously. These surfaces are to be found as part of the solution. Distances and times are non-dimensionalized with the radius of the orifice a and the capillary time $(\rho a^3/\sigma)^{1/2}$. In Eq. (4) x is the dimensionless height above the bottom, $\frac{D}{Dt} = \frac{\partial}{\partial t} + \boldsymbol{v} \cdot \nabla$ is the material derivative at points of the bubble surfaces, $n_i = \nabla f_i/|\nabla f_i|$, and p_{gi} is the gas pressure in the i-th bubble referred to the pressure of the liquid at the bottom far from the orifice and scaled with a factor a/σ. These pressures are functions of time which are determined by the conditions that the volume of the growing bubble ($i = 0$) increases at a constant rate equal to the volume of gas injected per unit time (Q), and the volumes of the detached bubbles ($i = 1, 2, \ldots$) do not change with time.

An additional condition is needed at the contact line of the growing bubble with the solid. Here the contact line will be taken to coincide with the edge of the orifice when the angle of the liquid–gas surface with the horizontal is larger than the contact angle (i.e. when $-\boldsymbol{n}_{x0} < \cos\theta$, where $\boldsymbol{n}_{x0}$ is the vertical component of the unit normal $\boldsymbol{n}_0$ to the attached bubble, see Fig. 2), and to spread away from the orifice with the liquid–gas surface making a constant contact angle with the solid ($-\boldsymbol{n}_{x0} = \cos\theta$) otherwise. The two possibilities are sketched in Figs. 2b and c. The contact angle θ is a third parameter of the problem, along with B and *We* defined in (1). Time periodic, axisymmetric solutions of the problem have been computed numerically using a standard boundary element method to solve the Laplace equation and a second order Runge–Kutta method to advance the material nodes at the surfaces of the bubbles and the velocity potential at them according to (3) and (4), with $p_{gi}(t)$ determined at each time step. The implementation follows that of Oguz and Prosperetti (1993).

The final volume of the bubbles is shown in Fig. 3 as a function of the dimensionless flow rate $We^{1/2}$ for two different values of the Bond number, $B = 0.1$ and $B = 1$, which correspond to orifices of radii $a = 0.85$ and 2.68 mm, respectively, in pure water. The contact angle was taken as $\theta = 45°$ though results for other values of θ are qualitatively similar.

Numerical computations have been carried out of the axisymmetric, irrotational, time periodic flow induced in a quiescent strictly inviscid liquid by the growth, detachment and coalescence of bubbles (see Figs. 4 and 5) due to the injection of a constant gas flow rate through a horizontal submerged orifice. The results show that this simple potential flow formulation may qualitatively describe many aspects of the well-known transition from quasi-static generation of independent, constant volume bubbles at low Weber numbers to inertia and buoyancy controlled growth and interaction of bubbles at moderately high Weber numbers.

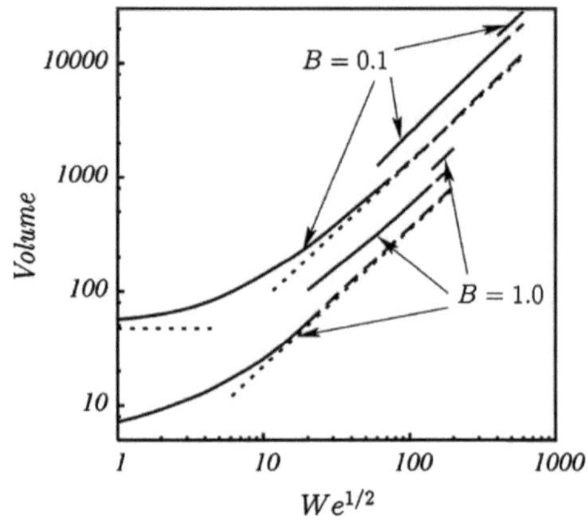

Fig. 3 Volume of the bubbles scaled with a^3 as a function of the dimensionless gas flow rate $We^{1/2}$ for $B = 0.1$ (*upper set of curves*) and $B = 1$(*lower set of curves*). The *solid curves* of each set give the final volume of the bubble. The *lower dashed curves* give the volume of the first detached bubble of a compound bubble, and the *intermediate dashed curves* give the volume of the first detached couple when double coalescence occurs. The *dotted horizontal line* is the dimensionless volume 47.497 computed in (Longuet-Higgins et al. 1991) for quasi-static detachment at $B = 0.1$. For comparison, notice that the Fritz's dimensionless volume for $B = 0.1$ is $V_F = \frac{2\pi}{B} = 62.83$. The *dotted lines* at the right correspond to volumes proportional to $We^{3/5}$ (Corchero et al. 2006)

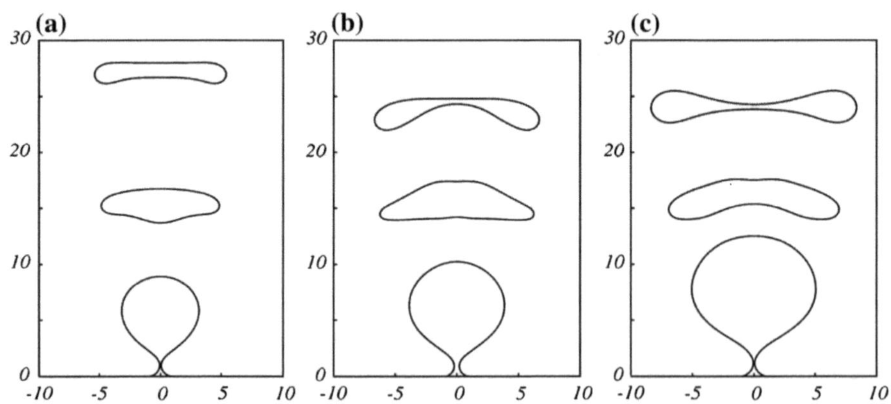

Fig. 4 Periodic generation of single bubbles for $B = 0.1$ and $We^{1/2} = 10$ (**a**), 20 (**b**), and 40 (**c**)

3 Equations for Bubbles Growing in Non-confined and Confined Viscous Liquid

The case of bubble generation in very viscous liquids is of interest in connection with polymer melts (Bird et al. 1987) and molten glasses and magmas (Sahagian 1985; Manga and Stone 1994), for example, but it has been comparatively less studied. Using a balance of buoyancy and viscous forces on the surface of each bubble,

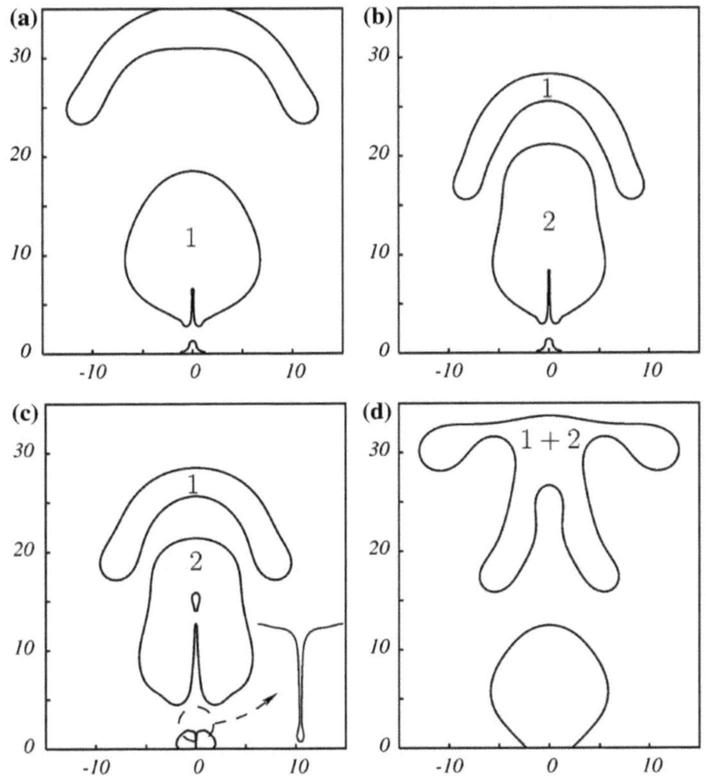

Fig. 5 Four snapshots of the generation of a (*double*) compound bubble for $B = 0.1$ and $We^{1/2} = 100$. **a** $t = 13.77$, immediately after the detachment of the leading bubble; **b** $t = 25.34$, immediately after the detachment of the trailing bubble; **c** $t = 25.64$, immediately after breakup of the thin upward jet; **d** $t = 33.68$, immediately after coalescence of the two bubbles. Times are non-dimensionalized with the capillary time $(\rho a^3/\sigma)^{1/2}$ and measured from the detachment of the bubble preceding bubble 1 in (**a**). The period of the process is 25.34. Notice the weeping in (**c**) and the displacement of the contact line away from the orifice in (**c**) and (**d**)

Davidson and Schuler (1960) proposed that the volume of the bubbles injected in a very viscous quiescent liquid increases as the 3/4 power of the gas flow rate and is independent of the radius of the injection orifice. This estimate is intended to apply for high gas flow rates, for which the effect of the surface tension acting across the contact line of the attached bubble with the solid surface of the orifice is negligible. At very small flow rates, on the other hand, viscous forces are negligible during most of the growth of the bubble, whose shape is determined by a hydrostatic balance of buoyancy and surface tension. Longuet-Higgins et al. (1991) computed the equilibrium shapes of attached bubbles and the volume at which equilibrium state is not possible anymore and the bubble should detach. In orders of magnitude, the volume of the bubble at detachment, V, is given in this small-flow-rate regime by the hydrostatic balance $\rho g V \sim \sigma a$, or $V/a^3 \sim 1/B$ in dimensionless terms.

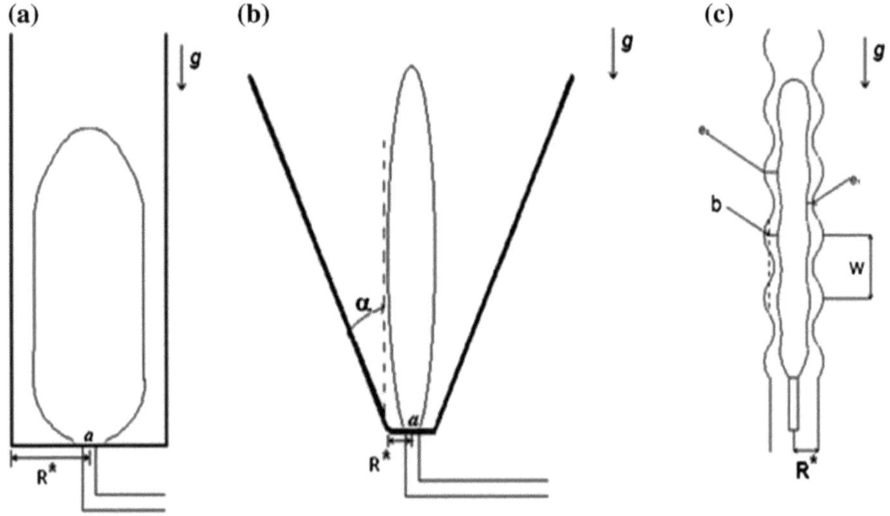

Fig. 6 Two-dimensional projections of the reservoirs symmetrical to the gas injection needles. Three different geometries are considered for bubbles growth in viscous fluids

Here ρ is the density of the liquid, σ is the liquid–gas surface tension, a is the radius of the injection orifice, g is the acceleration due to gravity, and B is de Bond Number (1).

A constant flow rate Q of an incompressible gas of negligible density and viscosity is injected into a liquid of density ρ and viscosity μ initially at rest in a reservoir under the action of the gravity. The gas is injected through a circular orifice of radius at the center of the base of radius R* of the reservoir. The lateral wall of the reservoir may be cylindrical, cylindrical corrugated or conical, making an angle θ to the vertical, as sketched in Fig. 6. The gas accumulates in a bubble attached to the base of the reservoir. The volume of this bubble increases with time until it detaches and begins to ascend in the liquid, being replaced by a new attached bubble. The effect of the inertia is assumed to be negligible in the motion induced in the liquid by the growth and displacement of the bubbles. A sufficient condition for the effect of the inertia to be negligible is that $Re = \frac{\rho Q}{\mu R_b} \ll 1$ (Wong et al. 1998; Higuera 2005; Ajaev and Homsy 2006). Here R_b is the characteristic radius of the detaching bubble or of its upper cap.

The model used here is valid to understand the bubble formation in a very viscous liquid in confined axisymmetric geometries (López-Villa et al. 2011). The equations of continuity and Stokes have the following dimensionless forms, respectively

$$\nabla \cdot \mathbf{v} = 0, \tag{5}$$

$$0 = -\nabla p + \nabla^2 \mathbf{v} - B\mathbf{i} \tag{6}$$

Here distances and times are scaled with the radius of the orifice a and the viscous time $\mu a/\sigma$, respectively, p is the pressure, v is the velocity field: $\boldsymbol{v} = \boldsymbol{v}^*\sigma/\mu$, $\mathbf{i}$ is the normal vector pointing in the upward direction (x is the vertical coordinate).

When the bubble grows the dimensionless gas flow rate Q is constant and the capillary number is a dimensionless flow rate

$$Ca = \frac{\mu Q}{\sigma a^2}. \tag{7}$$

When a bubble is formed in the liquid, a free surface of the form $f(\boldsymbol{x}, t) > 0$ exists. Equations (5) and (6) must be solved with the boundary conditions

$$\frac{Df_i}{Dt} = 0, \tag{8}$$

$$-p\boldsymbol{n}_i + \tau' \cdot \boldsymbol{n}_i = (\nabla \cdot \boldsymbol{n}_i - p_{gi})\boldsymbol{n}_i, \tag{9}$$

on the surfaces of the i-th bubbles and

$$\boldsymbol{v} = 0 \tag{10}$$

on the inner cylinder's surface ($r = R^*$), and at infinity, because the fluid does not move there. Moreover, the pressure far from the bubble must satisfy

$$p + Bx = constant. \tag{11}$$

The uniform pressure of the gas in the bubble, $p_g(t)$, is determined using the conditions that the volume V of the bubble increases linearly with time at a rate equal to Q for the attached bubble. In dimensionless variables,

$$\frac{dV}{dt} = Ca. \tag{12}$$

In above equations $\boldsymbol{n} = \nabla f / |\nabla f|$ is a unit vector normal to the surface of the bubble, $\tau' = \nabla \boldsymbol{v} + (\nabla \boldsymbol{v})^{\mathrm{T}}$ is the dimensionless viscous stress tensor, x and r are distances along the axis of the reservoir and normal to it. The condition (8) shows that the surfaces of the bubbles are fluid surfaces which separate the liquid from the gas, and therefore there is no mass exchange through them. The condition (9), in turn, expresses the balance of the stress acting on surfaces of the bubbles.

If the gas pressure p_{gi} is known in each bubble, the Eqs. (5) and (6) together with the boundary conditions determine velocity fields and fluid pressure, and in particular, the velocity on the surface of each bubble. To determine the pressures and complete the formulation of the problem conditions to be used, the volumes of the released bubbles ($i > 1$) are assumed constant and equal to the volumes at the instant of their release.

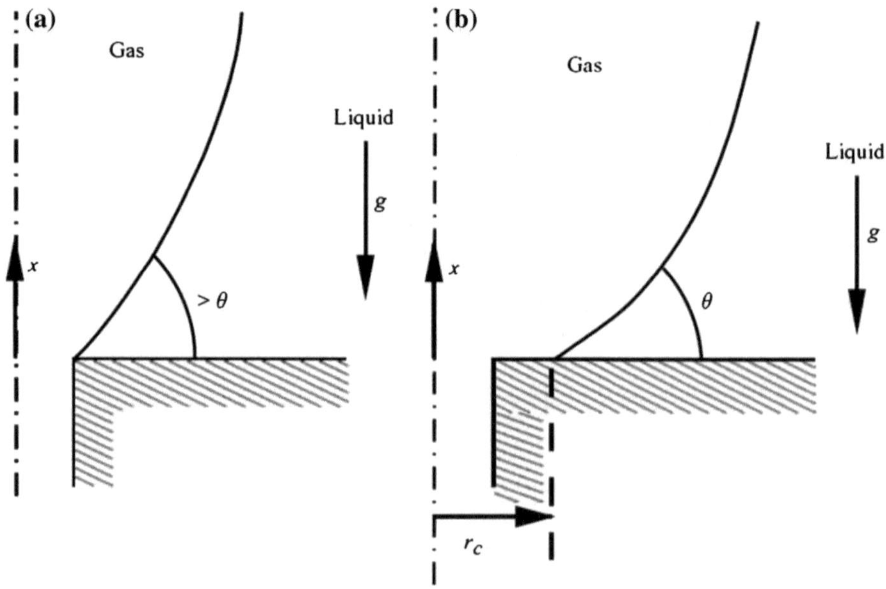

Fig. 7 Details of the contact line with two possibilities of line of contact of the bubble adhesion

An additional condition is needed at the contact line of the growing bubble with the solid. Here the contact line will be taken to coincide with the edge of the orifice when the angle of the liquid–gas surface with the horizontal bottom is larger than the contact angle (i.e. when $-\boldsymbol{n}_{x0} < \cos\theta$, where $\boldsymbol{n}_{x0}$ is the vertical component of the unit normal vector $\boldsymbol{n}_0$ to the attached bubble), and to spread away from the orifice, with the liquid–gas surface making a constant contact angle with the solid bottom ($-\boldsymbol{n}_{x0} < \cos\theta$) otherwise (Higuera 2005); see Fig. 7.

The line of contact of the bubble adhered ($i = 1$) with the reservoir base is a circle which may coincide with the edge of the orifice or move to a position $r > 1$ to be determined. Figures 7a and b illustrate both possibilities. In the first case, the radius of the contact line coincides with the hole edge, $r = 1$. In the second case, the angle which the bubble surface makes with the base of the reservoir must match the contact angle θ, which is a property of the liquid and the material of the shell.

Then the problem contains five dimensionless parameters which are the Bond number, B, the capillary number Ca, the dimensionless radius of the base of the reservoir $R = R^*/a$, semiangle of the conical base α and the contact angle θ of the liquid with the base (Fig. 7).

The set of equations given above will satisfy the outflow boundary conditions at infinity, the non-slip conditions on walls, and the quasi-static pressure balance. The evolution of the free surface (bubble shapes) is given by the solution of Eq. (8), under a fourth order Runge-Kutta scheme, which is attained after solving the hydrodynamic problem, imposed by Eqs. (5) and (6), by using the BEM method (López-Villa et al. 2011). The reservoir configurations are shown in Fig. 6.

4 Method for Numerical Solution

We seek for axisymmetric solutions of the Eqs. (5)–(12). We use a standard boundary elements method (Pozrikidis 1992, 2002) to solve the Stokes Eqs. (5) and (6) with the boundary conditions (9)–(11), and a fourth order Runge-Kutta method to calculate the evolution of free surface f given by (8).

4.1 History

After three decades of development, the boundary element method (BEM) has found a firm footing in the arena of numerical methods for partial differential equations. Compared to more popular numerical methods, such as the Finite Element Method (FEM) and the Finite Difference Method (FDM), both of which can be classified as the domain methods, the BEM distinguishes itself as a boundary method, meaning that the numerical discretization is conducted at reduced spatial dimension. For example, for problems in three spatial dimensions, the discretization is performed on the bounding surface only; and in two spatial dimensions, the discretization is on the boundary contour only. This reduced dimension leads to linear systems, less computer memory requirements, and more efficient computation. These advantages are most notorious when the domain is unbounded. Unbounded domain needs to be truncated and approximated in domain methods. The BEM, on the other hand, automatically models the behavior at infinity without the need of deploying a mesh to approximate it. In the modern day industrial settings, mesh preparation is the most intensive labor and the most costly portion in numerical modeling, particularly for the FEM. Without the need of dealing with the interior mesh, the BEM is more cost effective in mesh preparation. For problems involving moving boundaries, the adjustment of the mesh is much easier with the BEM. With these advantages, the BEM is indeed an essential part in the repertoire of the modern day computational tools (Alexander and Daisy 2005).

One can view BEM as the numerical implementation of boundary integral equations based on Green's formula, in which the piecewise element concept of the FEM is utilized for the discretization.

4.2 BEM in Axisymmetric Domains

In this section we describe the standard boundary elements method for three-dimensional flow in an axisymmetric domain. Our goal is to reduce the boundary integral equation to a one-dimensional equation, or a system of one-dimensional equations, over the trace of the boundaries in an azimuthal plane. First, the Green's functions of Stokes flow represent solutions of the continuity equation $\nabla \cdot v = 0$ and the singularly forced Stokes equation

$$- \nabla P + \mu \nabla^2 \boldsymbol{v} + g\delta\left(\boldsymbol{x} - \boldsymbol{x}_0\right) = 0, \tag{13}$$

where $\delta(\boldsymbol{x}-\boldsymbol{x}_0)$ is Dirac's delta function in three dimensions, g is an arbitrary constant, $\boldsymbol{x}_0$ is an arbitrary point. Introducing the Green's function G for three dimensions, we write the solution of (13) in the form

$$\boldsymbol{u}_i = \frac{1}{8\pi\mu}\boldsymbol{G}_{ij}\left(\boldsymbol{x} - \boldsymbol{x}_0\right)\boldsymbol{g}_i, \tag{14}$$

here $\boldsymbol{x}$ is the observation or field point. Physically, (14) expresses the velocity field due to a concentrated point force of strength g placed at the point $\boldsymbol{x}_0$, and may be identified with the flow produced by the slow settling of a small particle. In the literature of boundary integral methods, the Green's function may appear under the names fundamental solution or propagator (Pozrikidis 1992).

It is convenient to classify the Green's functions into three categories depending on the topology of the domain of flow. First, we have the free-space Green's function for infinite unbounded flow; second, the Green's functions for infinite or semi-infinite flow that are bounded by a solid surface; and third, the Green's function for internal flow are completely confined by solid surfaces. The Green's functions in the second and third categories are required to vanish over the internal or external boundaries of the flow. As the observation point $\boldsymbol{x}$ approaches the pole $\boldsymbol{x}_0$ all Green's functions exhibit singular behavior and, to leading order, behave like the free-space Green's function. The Green's functions for infinite unbounded or bounded flow are required to decay at infinity at a rate equal to or lower than that of the free-space Green's function.

$$\frac{\partial \boldsymbol{G}_{ij}\left(\boldsymbol{x} - \boldsymbol{x}_0\right)}{\partial x_i} = 0, \tag{15}$$

for (14).

Integrating (15) over a volume of fluid that is bounded by the surface D and using the divergence theorem, we find

$$\int_D \boldsymbol{G}_{ij}\left(\boldsymbol{x} - \boldsymbol{x}_0\right)\boldsymbol{n}_i\left(\boldsymbol{x}\right) d\boldsymbol{S}\left(\boldsymbol{x}\right) = 0, \tag{16}$$

independently of whether the pole $\boldsymbol{x}_0$ is located inside, right on, or outside D.

The vorticity, pressure, and stress fields associated with the flow (15) may be presented in the corresponding forms:

$$\omega_i = \frac{1}{8\pi\mu}\Omega_{ij}\left(\boldsymbol{x} - \boldsymbol{x}_0\right) g_j, \tag{17}$$

$$P = \frac{1}{8\pi\mu}\boldsymbol{p}_j\left(\boldsymbol{x}-\boldsymbol{x}_0\right)g_j, \tag{18}$$

$$\tau_{ij} = \frac{1}{8\pi\mu}\boldsymbol{T}_{ijk}\left(\boldsymbol{x}-\boldsymbol{x}_0\right)g_j, \tag{19}$$

where $\boldsymbol{\Omega}, \boldsymbol{p}$, and $\boldsymbol{T}$ are the vorticity tensor, pressure vector, and stress tensor associated with the Green's function The stress tensor $\boldsymbol{T}$, in particular, is defined as

$$\boldsymbol{T}_{ijk}\left(\boldsymbol{x}-\boldsymbol{x}_0\right) = -\delta_{ik}\boldsymbol{p}_j\left(\boldsymbol{x}-\boldsymbol{x}_0\right) + \frac{\partial \boldsymbol{G}_{ij}\left(\boldsymbol{x}-\boldsymbol{x}_0\right)}{\partial x_k} + \frac{\partial \boldsymbol{G}_{ij}\left(\boldsymbol{x}-\boldsymbol{x}_0\right)}{\partial x_i}. \tag{20}$$

It will be noted that $\boldsymbol{T}_{ijk} = \boldsymbol{T}_{kji}$ as required by the symmetry of the stress tensor $\boldsymbol{\tau}$. When the domain of flow is infinite, we require that all $\boldsymbol{\Omega}, \boldsymbol{p}$, and vanish as the observation point is moved to infinity.

Considering first axisymmetric flow with no swirling motion, we observe that in cylindrical coordinates, none of the boundary variables is a function of the azimuthal angle. This reduces the number of variable.

In the problem of bubble generation, the contours are the surfaces of the bubbles and the solid surfaces of the reservoir, which are all surfaces of revolution. The unknowns involved in the formulation of the boundary element are the velocities of fluid particle that define surfaces of the bubbles and the stresses on the vessel wall.

We introduce the dimensionless driving pressure $P = p + Bx$, which allows us to write the Stokes Eq. (6) as

$$0 = -\nabla P + \nabla^2 \boldsymbol{v} = \nabla \cdot \tau, \tag{21}$$

where $\tau = -P\boldsymbol{I} + \tau'$ is a modified stress tensor, with fluid pressure replaced by the driving pressure. The stress of liquid on the surface of the i-th bubble, given by left hand side of (9), is then

$$-p\boldsymbol{n}_i + \tau' \cdot \boldsymbol{n}_i = -P\boldsymbol{n}_i + \tau' \cdot \boldsymbol{n}_i + Bx\boldsymbol{n}_i. \tag{22}$$

We will use the notation $\boldsymbol{f} = \tau \cdot \boldsymbol{n}$ for the modified stress on the limiting contour of the liquid, where $\boldsymbol{n}$ is the normal to the contour directed towards the liquid. With this notation $-p\boldsymbol{n}_i + \tau' \cdot \boldsymbol{n}_i = \boldsymbol{f} + Bx\boldsymbol{n}_i$ and the boundary condition (9) on the bubble, the i-th surface takes the form

$$\boldsymbol{f} = \left(\nabla \cdot \boldsymbol{n}_i - Bx - p_{gi}\right)\boldsymbol{n}_i. \tag{23}$$

Additionally, we use the Green's functions for axisymmetric flow, which are the solutions of Stokes equations in unlimited space (14) with the stresses concentrated on a circumference of radius $\boldsymbol{r}_0$ centered at point $\boldsymbol{x}_0$ on the axis of symmetry. These forces can be either directed along the axis of symmetry or be in perpendicular to it, which gives rise to two distinct solutions whose velocity and motion pressure distributions denoted as $\boldsymbol{G}^x(\boldsymbol{x}, \boldsymbol{x}_0)$ and $\boldsymbol{P}^x(\boldsymbol{x}, \boldsymbol{x}_0)$ for axial force, and $\boldsymbol{G}^r(\boldsymbol{x}, \boldsymbol{x}_0)$ and

$\boldsymbol{P}^r(\boldsymbol{x}, \boldsymbol{x}_0)$ for a radial force. Here, $\boldsymbol{x} = (x, r_0)$ and $\boldsymbol{x} = (x, r)$ (a generic point) in cylindrical coordinates defined above. The equations to be solved to determine the Green's functions are

$$\nabla \cdot \boldsymbol{G}^x = 0, \qquad 0 = -\nabla P^x + \nabla^2 \boldsymbol{G}^x + 8\pi\delta(x, x_0)\, \boldsymbol{e}_x, \tag{24}$$

and

$$\nabla \cdot \boldsymbol{G}^r = 0, \qquad 0 = -\nabla P^r + \nabla^2 \boldsymbol{G}^r + 8\pi\delta(x, x_0)\, \boldsymbol{e}_r, \tag{25}$$

with the conditions $(\boldsymbol{G}^x, \boldsymbol{P}^x) \rightarrow 0$ and $(\boldsymbol{G}^r, \boldsymbol{P}^r) \rightarrow 0$ at infinity. In these equations $\boldsymbol{e}_x, \boldsymbol{e}_r$, are the unit vectors parallel and perpendicular to the axis of symmetry, respectively, δ is the Dirac function, and 8π factor is introduced by convention. We will also use the notation T^r and T^x for stress tensor of the solutions (24) and (25). These solutions are known and are given in Appendix A.

Given that $\nabla \cdot \boldsymbol{v} = \nabla \cdot \boldsymbol{G}^x = 0$ verifies that

$$\nabla \cdot \left(\boldsymbol{G}^x \cdot \tau - \boldsymbol{v} \cdot T^x\right) = \boldsymbol{G}^x \cdot (\nabla \cdot \tau) - \mathbf{v} \cdot \left(\nabla \cdot T^x\right).$$

Using (22) and (24) (i.e. $\nabla \cdot \tau = 0$ and $\nabla \cdot T^x = 8\pi\delta(x, x_0)\, \boldsymbol{e}_x$) in the second term of the above equality, and by integrating the result over the volume occupied by liquid and using the Gauss theorem for flow in an axisymmetric domain and transforming the integral of the first member in a surface integral. Then in a line integral on the meridional section of the bubbles and the walls tube denoted as C, is obtained

$$-\int_C \boldsymbol{G}^x(\boldsymbol{x}, \boldsymbol{x}_0) \cdot \boldsymbol{f}(\boldsymbol{x})\, r(\boldsymbol{x})\, dl(\boldsymbol{x}) + \int_C \boldsymbol{v}(\boldsymbol{x}) \cdot \boldsymbol{T}^x(\boldsymbol{x}, \boldsymbol{x}_0)\, \boldsymbol{n}(\boldsymbol{x})\, r(\boldsymbol{x})\, dl(\boldsymbol{x}) = 8\pi r_0 v_x(\boldsymbol{x}_0), \tag{26}$$

where $\boldsymbol{f}(\boldsymbol{x})$, $r(\boldsymbol{x})$ and $dl(\boldsymbol{x})$ are defined in a point $\boldsymbol{x}$ in the contour C, which are, respectively, the distance from this point to the axis of symmetry, and the arc element on the boundary. Similarly,

$$-\int_C \boldsymbol{G}^r(\boldsymbol{x}, \boldsymbol{x}_0) \cdot \boldsymbol{f}(\boldsymbol{x})\, r(\boldsymbol{x})\, \boldsymbol{dl}(\boldsymbol{x}) + \int_C \boldsymbol{v}(\boldsymbol{x}) \cdot \boldsymbol{T}^r(\boldsymbol{x}, \boldsymbol{x}_0)\, \boldsymbol{n}(\boldsymbol{x})\, r(\boldsymbol{x})\, \boldsymbol{dl}(\boldsymbol{x}) = 8\pi r_0 v_r(\boldsymbol{x}_0), \tag{27}$$

In the derivation of (26) and (27) it is assumed that the circumference on which the force is applied is concentrated in the volume occupied by liquid. If not, the second members of these equations are null. The second integral on the left hand side of (26) and (27) diverge when the point tends $\boldsymbol{x}_0$ to the contour C. A detailed calculation, deforming the contour in the vicinity of $\boldsymbol{x}_0$ (see, for example, Pozrikidis 1992) shows that for $\boldsymbol{x}_0 \in C$,

$$v_j(\boldsymbol{x}_0) = -\frac{1}{4\pi}\int_C G_k^j(\boldsymbol{x},\boldsymbol{x}_0)\, f_k(\boldsymbol{x})\,\frac{r}{r_0}dl + \frac{1}{4\pi}\int_C^{PV} v_k(\boldsymbol{x})\, T_{kl}^j(\boldsymbol{x},\boldsymbol{x}_0)\, n_l(\boldsymbol{x})\,\frac{r}{r_0}dl, \tag{28}$$

where PV indicates the principal value of the integral and subscripts notation was used with or $(i, j, k) = x$ or r, for compactness.

Equation (28) is a ratio between speeds and stresses on the contour surface C domain occupied by the liquid. If stresses $\boldsymbol{f}$ were known at all points of C, this equation lets calculate the velocity of the liquid in C. Similarly, if in $\boldsymbol{v}$ C is known the equation allows compute $\boldsymbol{f}$ (except for an undetermined constant P; see Pozrikidis 2002).

In the problem of bubble generation, $\boldsymbol{v} = 0$ on the solid surfaces wetted by the liquid. The velocity of liquid on the bubbles is unknown, however, the modified stress can be obtained by using Eq. (23). Moreover, suppose all bubble stresses are known at a certain instant. The stress given by Eq. (23) would then be known if p_{gi} were known. In this case, the solution of (28) determines the velocity of the liquid on the surfaces of the bubbles and stresses on the solid surfaces. With p_{gi} unknown, the stress on the bubble i is the sum of a known stress $(\nabla \cdot \boldsymbol{n}_i - B_x)\boldsymbol{n}_i$, and a normal uniform stress, $-p_{gi}\boldsymbol{n}_i$.

Given the linearity of the Eq. (28), the velocity on the surfaces of bubbles and stress on solid surfaces are

$$\boldsymbol{v} = \boldsymbol{v}^0 + \sum\nolimits_i \boldsymbol{v}^i(p_{gi}) \quad \text{and} \quad \boldsymbol{f} = \boldsymbol{f}^0 + \sum\nolimits_i \boldsymbol{f}^i(p_{gi}) \tag{29}$$

where $\boldsymbol{v}^0$ and $\boldsymbol{f}^0$ are the velocity and the stress calculated for all $p_{gi} = 0$ in (18), and $\boldsymbol{v}^i$ and $\boldsymbol{f}^i$ are the velocity and the stress calculated for $\boldsymbol{f} = 0$ on all surfaces of the bubbles except the bubble i, where $\boldsymbol{f} = -\boldsymbol{n}_i$.

Equation (28) provide the solution of (27) in terms of the pressures of the gas in the bubbles, p_{gi}. We now need to establish equations to calculate these pressures. These equations express the conditions (12) and the volumes of the bonded bubbles released $(i > 1)$ are constant and the volume of the bubble grows linearly with time. In terms of the velocities of the liquid (28), the rate of change of volume of the bubble j is

$$\frac{d\boldsymbol{V}_j}{dt} = 2\pi\int_{C_j} \boldsymbol{v}\cdot\boldsymbol{n}_j r\,dl = a_{j0} + \sum\nolimits_i a_{ji}p_{gi}, \tag{30}$$

where C_j is the contour of the bubble, and a_{j0} and a_{ji} are easily calculated from velocities $\boldsymbol{v}^0$ and $\boldsymbol{v}^i$. Thus, the linear equations are obtained

$$a_{10} + \sum\nolimits_i a_{1i}p_{gi} = Ca \quad \text{and} \quad a_{j0} + \sum\nolimits_i a_{ji}p_{gi} = 0 \quad \text{for} \quad j > 1, \tag{31}$$

which determine p_{gi} and complete the solution (28) given the surfaces of the bubbles.

To complete the formulation of the problem, we must consider the conditions (8), expressing the surface of each bubble moving with the local velocity the liquid. The position $\boldsymbol{x}(t)$ of a fluid particle on the surface of a bubble satisfies

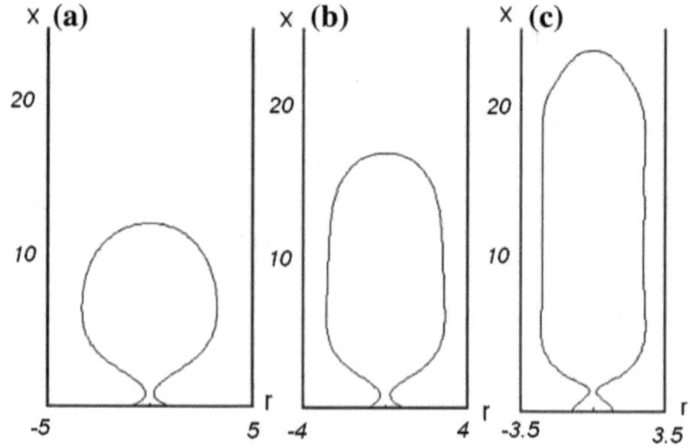

Fig. 8 Meridional sections of bubbles which are about to detach from the base of a cylindrical reservoir for $B = 0.2$, $Ca = 10$, and $R = 5$ (**a**), 4 (**b**) and 3.5 (**c**)

$$\frac{d\boldsymbol{x}}{dt} = \boldsymbol{v}(\boldsymbol{x}, t), \tag{32}$$

where $\boldsymbol{v}(\boldsymbol{x}, t)$ is the solution of (28)–(31) at the point $\boldsymbol{x}$ on the surface in the time t.

To solve numerically the integral Eq. (28), the contours of the bubbles and solid surfaces must be discretized. This is done using N_i nodes distributed uniformly on the bubble boundary i and N_d nodes on the base of the reservoir, distributed non-uniformly with space increasing with distance from the axis of symmetry. The integrals in (28) are calculated using a Gaussian integration with six points allocated in the interval between each pair of nodes. The surface is discretized with a finite number of nodes that move as material particles. Numerical tests conducted with different numbers of nodes show that 120 nodes give sufficient resolution.

The value $\theta = 45°$ has been used for the contact angle in the computations that follow. Numerical computations with other values θ of show that the effect of the contact angle on the volume of the bubbles is small as far as $\theta < 90°$.

5 Numerical Results

From the numerical solutions a set of important results are achieved. These results are compared with experiments qualitatively, where the effects of reservoir geometry and film thickness is studied.

5.1 *Cylindrical Reservoir*

Some numerical computations have been carried out to study the growth and detachment of a bubble in a cylindrical reservoir and validate it with experiments. Figure 8

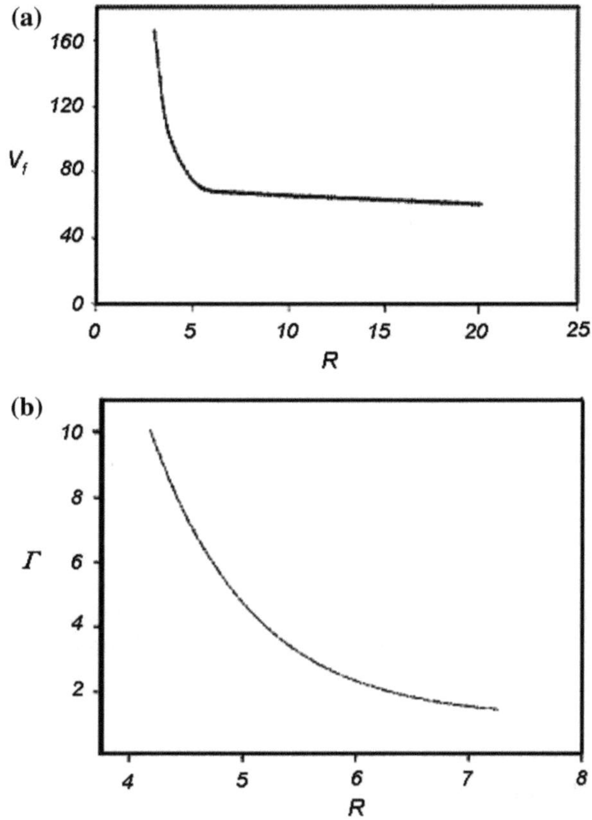

Fig. 9 Plots of **a** volume V_f and **b** aspect ratio Γ of a bubble at detachment from the base of the cylindrical reservoir as functions of it R for $B = 0.2$, and $Ca = 10$

shows a bubble which is about to detach from the base of the reservoir for $B = 0.2$, $Ca = 10$, and the three values $R = 5, 4$ and 3.5 of the dimensionless radius of the reservoir. Figure 9 shows plots of $V_{f,}$ volume of detachment, and the aspect ratio Γ of the bubble (defined as the ratio of the height to the maximum diameter of the bubble) as functions of R.

The finite radius of the reservoir affects only the high-flow-rate regime for the values of B and R used here. The decrease of Γ with increasing R in Fig. 9b is in qualitative agreement with the estimate $L_f/R \sim Ca/\left(BR^4\right)$ for columnar bubbles. The decrease of V_f in Fig. 9a also agrees with the previous estimates, according to which the ratio of the volume of a columnar bubble to the volume of a bubble detaching in an infinite reservoir is of order $(Ca/B)^{1/4}/R$ for Ca/B large compared to R^4. Figure 9 shows V_f as a function of Ca for B = 0.2 and three values of R. The nearly linear increase of V_f agrees with the estimate $V_f \sim Ca/\mathrm{B}$. Notice, in comparison, that $V_f \sim (Ca/\mathrm{B})^{3/4}$ for a bubble in an infinite reservoir.

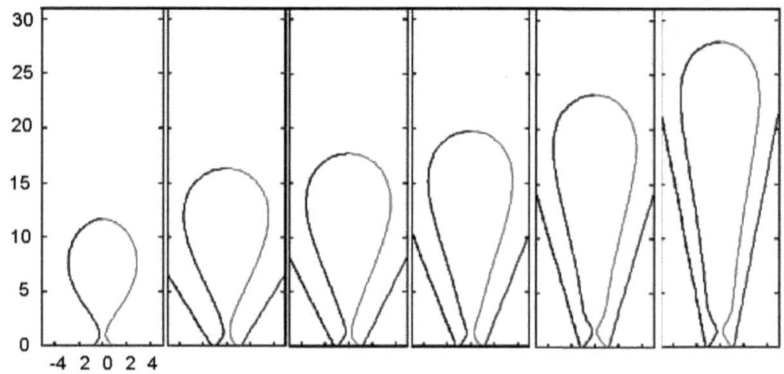

Fig. 10 Meridional sections of bubbles growing in cones which are about to detach from the base of a conical reservoir for $B = 0.2$, $Ca = 50$, $R = 1.2$, and various values of α

Numerical computations also show (results not displayed) that the center of mass of a columnar attached bubble rises linearly with time during the growth of the bubble, and that the velocity of the center of mass is nearly constant, except in the early stages of the process, when the bubble is still small compared to the radius of the tube.

5.2 Conical Reservoirs

Figure 10 shows the shape of a bubble which is about to detach from the base of a conical reservoir for $B = 0.2$, $Ca = 50$, R $=$ 1.2, and various values of the semi-angle of the cone, α, and Fig. 11 shows the volume V_f and the aspect ratio Γ of the detaching bubble as functions of α and different values of B and Ca. As it can be seen, the volume of the bubble always increases as the angle of the cone decreases, the effect being more pronounced for small values of the Bond number, for which the bubble is larger and therefore more easily affected by the wall of the reservoir. Figure 11a displays an important result of this work, namely, that at low Bond numbers and high capillary numbers, the volume of the bubbles can be easily controlled through the angle of the cone without having to change the flow rate. This is a desirable feature in some applications.

These results can be rationalized by means of a straightforward extension of the estimations of the previous section for the high-flow-rate regime in cylindrical reservoirs. Figure 10 shows that the bubbles in conical reservoirs are columnar for moderately small values of α, with a cap that increases linearly with its height above the bottom of the reservoir. (See also Fig. 11b; the bubble is slender for smaller than 30°).

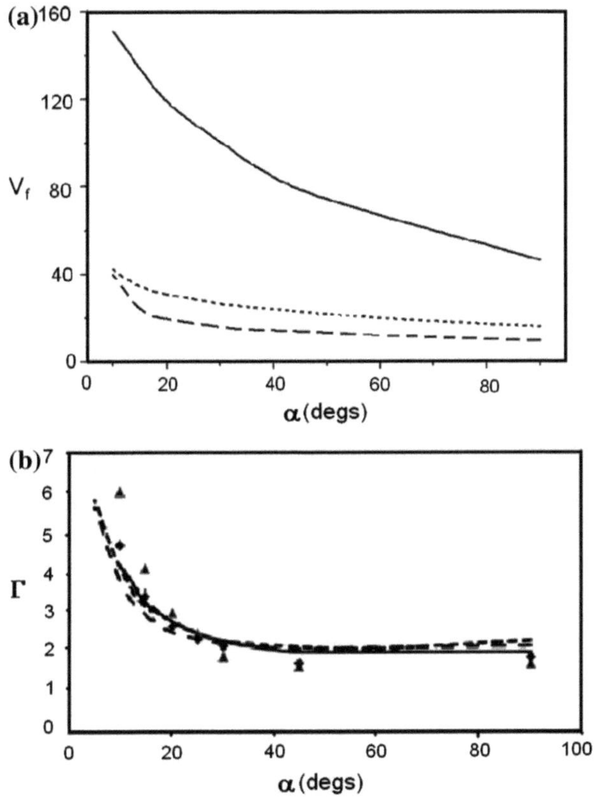

Fig. 11 Plots of **a** volume V_f and **b** aspect ratio Γ of a bubble at detachment from the base of a conical reservoir as functions of α, for $R = 1.2$ and $(B, Ca) = (0.2, 10)$(*solid*), $(2, 10)$ (*dashed*), and $(2, 20)$ (*dotted*). Symbols in panel (b) show the values of Γ measured experimentally for $B = 0.0176$ (▲) and $B = 0.15$ (◇), with $Ca = 50.78$ and $R = 1.2$

5.3 Corrugated Pipes

For the corrugated cylinder case, the numerical study was done to understand how corrugations affect the bubble shape. We tried different number of nodes distributed along the tube walls, and found that as in the case of the smooth walls, the number of nodes has no importance on qualitative behavior.

Figure 12 shows differences between bubbles formed in tubes with several corrugation wavelengths in the viscous case, and in Fig. 13 the cases in the inviscid approximation are shown with different amplitudes of corrugation. Figure 13 shows that the film thickness is small in the case of a small capillary number.

In Figs. 14 and 15 we observe the bubbles detachment process and a possible application to understand the foams formed in viscous fluids in porous media (Kovscek et al. 1995; Hirasaki et al. 2006).

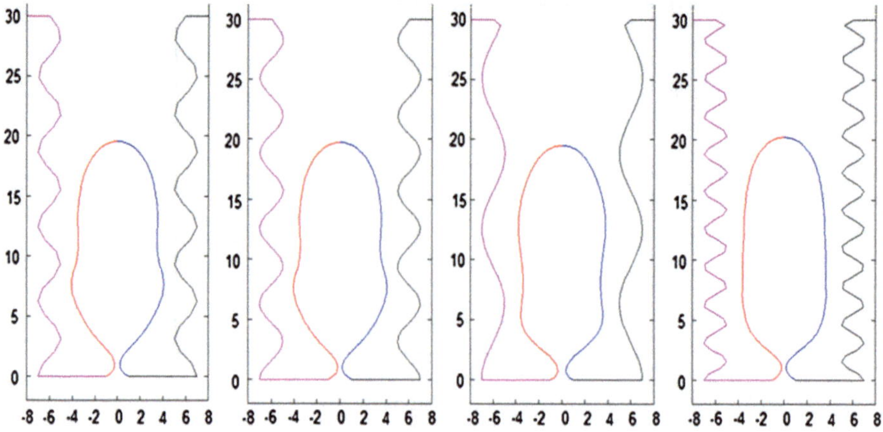

Fig. 12 Bubble shapes in corrugated cylinders of same amplitude but different wavelengths of the corrugations. The conditions are for viscous liquid and $Ca = 25$, $B = 0.2$

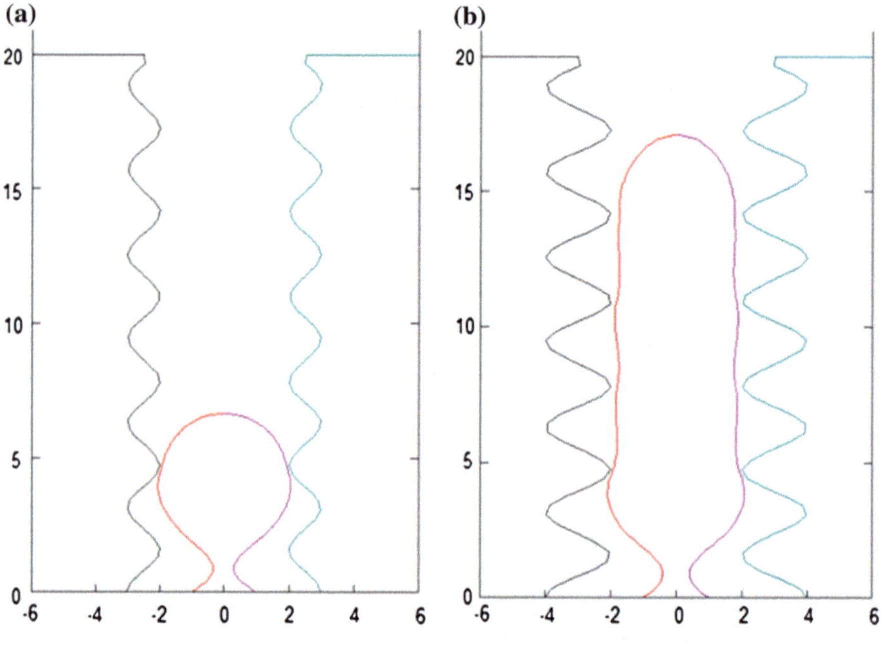

Fig. 13 Bubble shapes in corrugated cylinders of same wavelenghts but different amplitudes. **a** Bubbles growing in nearly inviscid liquid, i.e., $Ca = 0.1$ and $B = 0.2$. In **b** $Ca = 1$ and $B = 0.2$. Notice that the thickness of the film in both cases is very thin

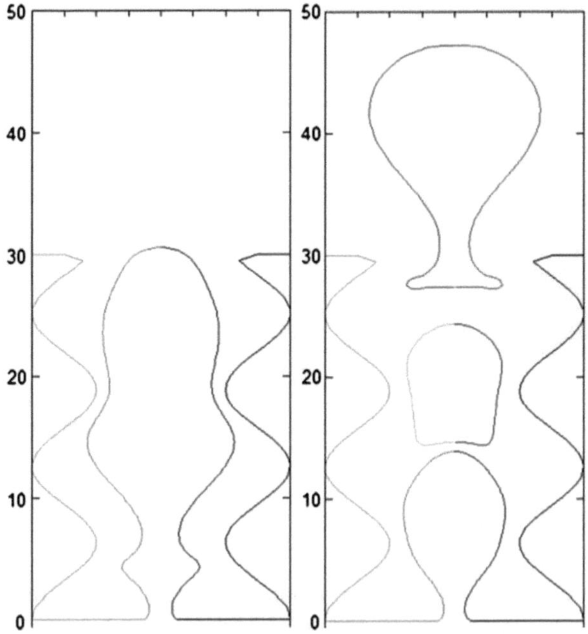

Fig. 14 Details of the bubble detachment process and the effect of the tube wall

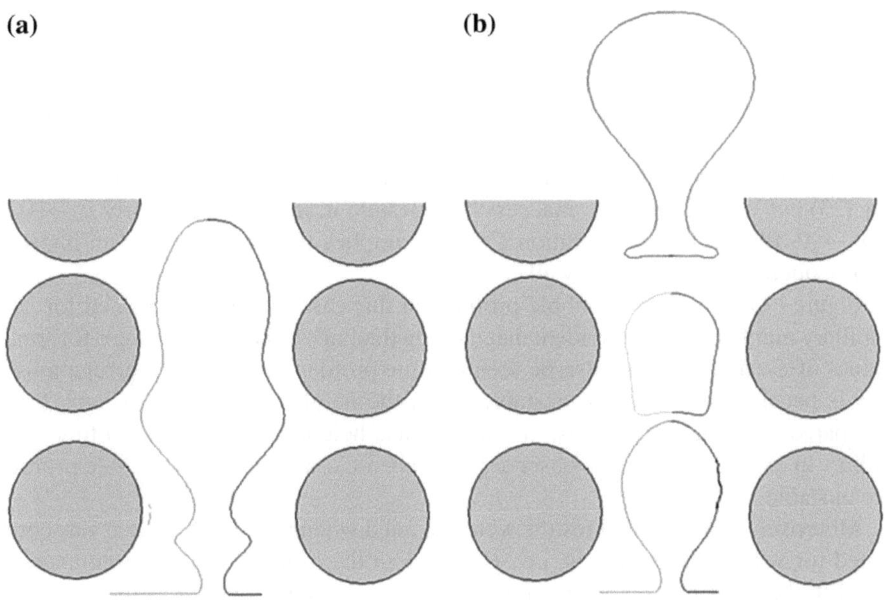

Fig. 15 Bubble shapes in the corrugated pipe compared to a bubble growing in a porous medium

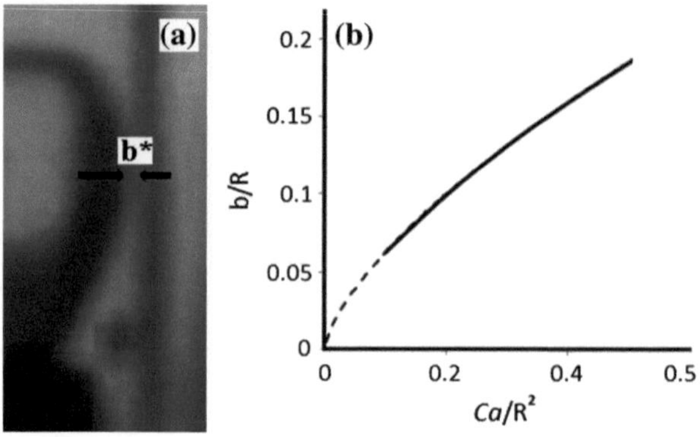

Fig. 16 Bubble in a vertical tube (**a**). Notice the existence of a film of thickness b*. **b** Plot of the scaled dimensionless film thickness b/R as a function of the scaled capillary number Ca/R^2

5.4 The Film Thickness

In Fig. 16a it is shown that during the bubble growth in the pipe there is an annular film of thickness b*. In a classical study of the lubrication theory, the dimensionless thickness of the film scales as

$$\frac{b}{R} \sim \left(\frac{Ca}{R^2}\right)^{\frac{2}{3}}, \qquad (33)$$

which is valid whenever $Ca \to 0$ and R and the Bond number are also small. In Fig. 16b a plot obtained from the numerical computations that obeys the relation (33) is given. In such a plot the continuous curve was obtained through our numerical solution. Meanwhile, the dashed part of the curve only shows the trend given by Eq. (33) but was numerically inaccessible. Despite it, in this case, clearly $b \to 0$ if $Ca \to 0$. Physically the condition $Ca \to 0$ implies that the bubble in an inviscid liquid touches the inner solid wall.

Figure 17a shows some bubble profiles: in this case they were obtained for low capillary numbers and it is evident that the film thickness tends towards zero for small values of *Ca* and *R*. It can also be seen that the profiles show some "corrugations", this is because they become unstable when the height of the tubes is very large compared with the tube radius, *i.e.* in this case height of the tube is 30 times the radius. In experiments it was observed that when $Ca \ll 1$ the small bubble profiles are unstable.

Moreover, very different results were obtained when the film thickness was computed for very viscous liquids, i.e., $Ca \gg 1$ in the limit of low Bond number. In Fig. 17b it is possible to notice that the film thickness tend towards a constant value when the capillary number increases. In Fig. 18 it is sketched how $b \to constant$ for $Ca \gg 1$. In dimensional terms the actual thickness of the annular film, $b^* \to 1.5a^*$,

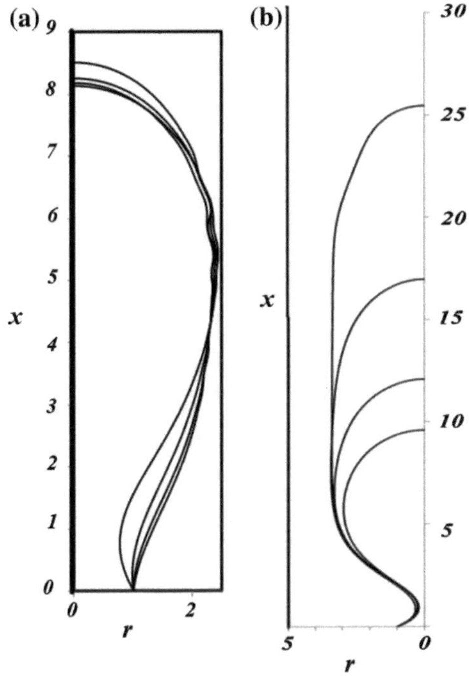

Fig. 17 Bubble shapes in cylinders filled with **a** liquid of low viscosity and **b** with very viscous liquid. In **a** the height and the film thickness between the bubble and the wall diminishes when $Ca \rightarrow 0$ ($Ca = 0.4, 0.3, 0.2$ and 0.1). In **b** the film thickness $b \rightarrow constant$, for $Ca \gg 1$. The larger bubble corresponds to $Ca = 35$; other cases are $Ca = 20$, 10 and 5. The dimensionless pipe radius was $R = 5$

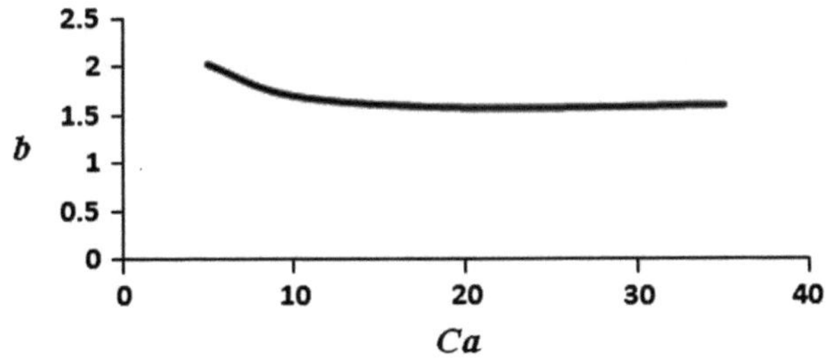

Fig. 18 Plot of the thickness of the annular film, b, as a function of the capillary number, Ca. Notice that $b \rightarrow constant$ for $Ca \gg 1$

i.e., the lower value of b^* is 1.5 times the radius of the gas injection orifice. Physically, this condition is attained in very viscous liquids or at very large gas flow rates.

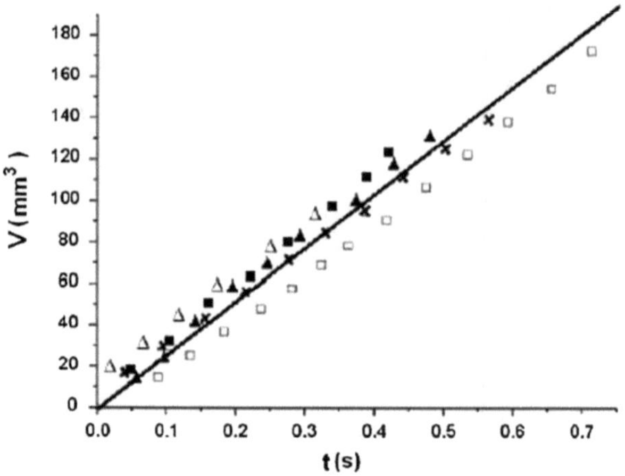

Fig. 19 Volume of a bubble attached to the base of a conical reservoir filled with glycerine as a function of time during the growth of the bubble for B = 0.0176, R = 1.2, and different values of the semi-angle of the cone. $\alpha = 10°$ (□), 20° (x), 25° (▲), 30° (■) and 90° (△)

6 Experiments

A series of experiments have been carried out to study the growth and detachment of bubbles in very viscous liquids. Glycerine and a silicone oil have been used in different experiments. The properties of glycerine at 25° are: density $\rho = 1260\,\text{kg/m}^3$, viscosity $\mu = 7.9 \times 10^{-1}\,\text{N s/m}^2$, and surface tension $\sigma = 6.3 \times 10^{-2}\,\text{N/m}$. The properties of the silicone oil at the same temperature are: density $\rho = 971\,\text{kg/m}^3$, viscosity $\mu = 9.71 \times 10^{-1}\,\text{N s/m}^2$. In each experiment, a large open container with an horizontal bottom where a circular orifice of radius $a = 0.3\,\text{mm}$ has been drilled was filled with the chosen liquid to a height of 100 mm.

A glass tube of inner radius $R^* = 3.2\,\text{mm}$ was set vertically and concentrically with the orifice to form a cylindrical reservoir. Conical reservoirs of various angles were formed by carefully inserting cones made of acetate sheet concentrically with the orifice. Air was pumped through a capillary tube 40 cm long and 0.6 mm of inner diameter which ends at the orifice in the bottom of the container. We found in a previous work (Corchero et al. 2006) that a length of 40 cm suffices to make the pressure drop in the air line it is large compared to the pressure variations in the bubble during the growth process and therefore ensure a constant flow rate in our experiments, which is one of the premises of the numerical work. To check that the flow rate is constant, the evolution of the attached bubble was video recorded; the contour of the bubble was extracted from the video images using a standard algorithm (Russ 2002) implemented in a home made code; and the volume of the bubble, V(t), and the height of its center of mass, $x_{CM}(t)$, were computed assuming that the bubble is axisymmetric. Some sample plots of V as a function of time for a bubble growing in glycerine within conical reservoirs of various angles are shown

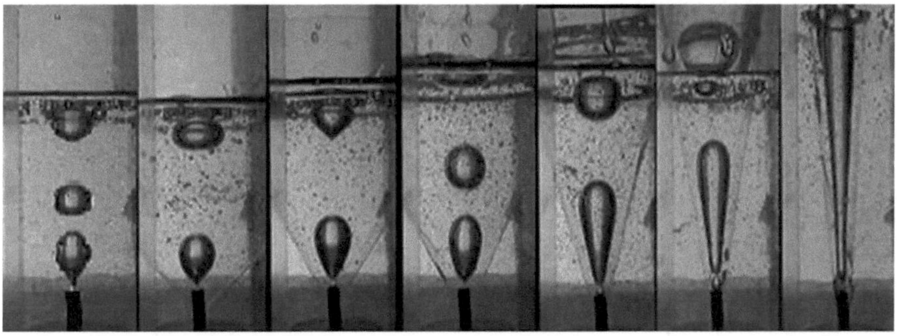

Fig. 20 Growth of bubbles at different angles of inclination α

in Fig. 19. The approximate linear variation of V with time shows that the flow rate is nearly constant and independent of the angle of the cone. The value of the flow rate determined by fitting a straight line to the experimental data of Fig. 19 is $Q = 364.5\,\mathrm{mm}^3/s$. The same procedure was used to measure the flow rate of air injected into silicone oil and in cylindrical containers. The flow rate was found to be nearly constant in all cases.

Figure 20 shows the shapes of bubbles in glycerine which are about to detach from the injection orifice in conical reservoirs of various angles. Here $\alpha = 90°$ corresponds to a bubble detaching in an infinite reservoir, and the shape of the bubbles begin to differ significantly from this case when α becomes smaller than about 30°. Coalescence between previously detached bubbles can be seen in some of the images. The presence of the conical wall of the reservoir increases the drag of the ascending bubbles, decreasing their velocity and apparently promoting coalescence. We plan to analyze this important aspect of the generation of bubbles in a future work. The gas flow rate in this sequence of experiments is that measured from Fig. 19. Values of the dimensionless parameters are $Bo = 0.0176$, $Ca = 50.78$ and $R = 1.2$.

The aspect ratio Γ of the bubbles in Fig. 20 and others was also extracted from the images and is included in Fig. 12b (triangles and diamonds), where it is compared to numerical results obtained for similar values of the dimensionless parameters. The comparison is reasonably good, though the experimental values of Γ increase with decreasing somewhat faster than the numerical values, and become larger than them for small values of α. We think that the difference is due to the vertical momentum injected with the gas, which was not taken into account in the numerical computations. The evolution of the center of mass of the bubble is shown in Fig. 21 and compared to numerical results. The nearly linear increase of x_{CM} with time is to be compared to the $x_{CM} \propto t^{1/3}$ evolution expected for a round bubble growing in an infinite reservoir (Davidson and Schuler 1960). The difference clearly shows the effect of the conical wall.

Only silicone oil was used in experiments with cylindrical reservoirs because glycerine tends to produce small bubbles that linger in the reservoir for a long time and interfere with the observation of the bubble attached to the orifice. Figure 22

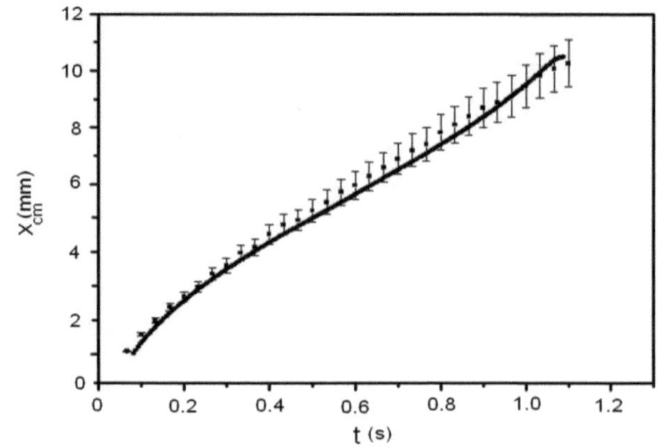

Fig. 21 Height of the center of mass of a bubble attached to the base of a conical reservoir as a function of time during the growth of the bubble for Bo = 0.04, $Ca = 70.23$, R = 1.2 and = 15°. Symbols are experimental results and the *solid curve* shows the results of the numerical computation

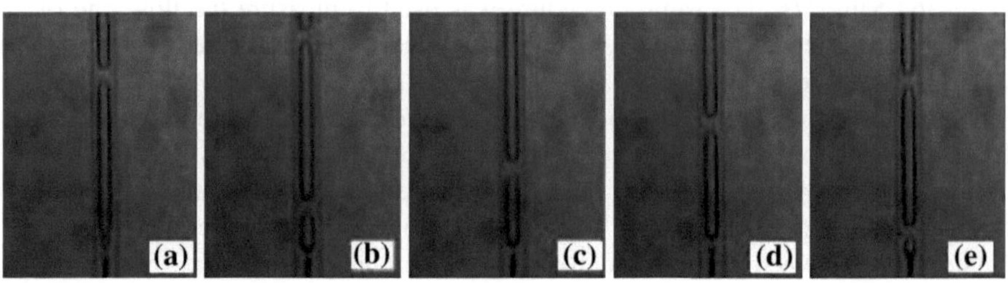

Fig. 22 Five equispaced images spanning the period of growth of a bubble attached to the base of a cylindrical reservoir for B = 0.04, $Ca = 209.94$ and R = 10.66. The period of bubbling is 1.33 s

shows five images equispaced in time that span the cycle of growth and detachment of a bubble. The flow rate of gas measured from the video record is Q = 419.59 mm^3/s in this experiment, and the period of bubbling is 1.33 s. Values of the dimensionless parameters are Bo = 0.04, Ca = 209.94 and R = 10.66. Figure 23 shows profiles bubbles formed into tubes with periodic corrugations c = 7.33, dimensionless radius $R^* = 3.7$, Bo = 0.2 to different capillary number.

7 Conclusions

The growth of a bubble due to the injection of a constant flow rate of a gas through an orifice in the horizontal base of a container filled with a very viscous liquid has been investigated numerically and experimentally in conditions in which nearby

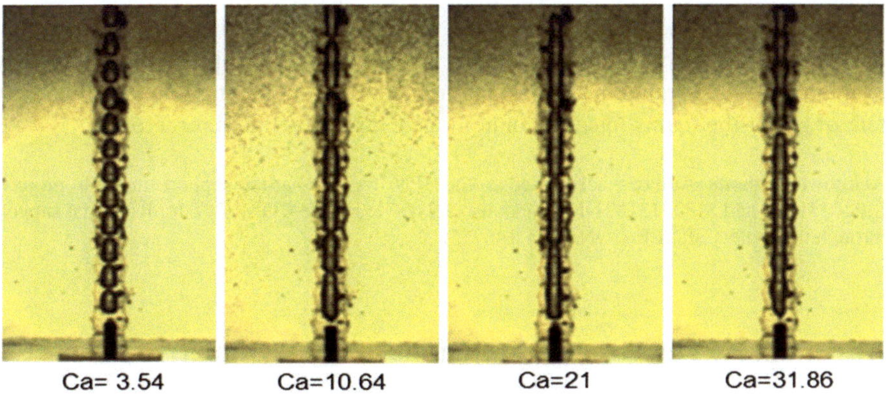

Fig. 23 dimension of profiles bubbles formed into tubes with periodic corrugations c = 7.33, dimensionless radius $R^* = 3.7$, Bo = 0.2 to different capillary number

solid walls partially confine the space where the bubble is allowed to grow. Conical, cylindrical and cylindrical with corrugations walls coaxial with the injection orifice have been used to allow easy control of the extent of the confinement by simply changing the angle of the cone or the radius of the cylinder or frequency in the corrugated case.

Numerical solutions of the Stokes equations for the liquid and the evolution equation for the free surface of the bubble show that the wall near the injection orifice may have an important effect on the shape of the bubble and its volume at detachment. Computations for small Bond numbers (Bo = 0.2) and moderately large capillary numbers (of the order of 10) made with BEM show that vertically elongated bubbles with volumes significantly larger than those of the round bubbles generated in the absence of walls are obtained when the radius of the cylindrical wall is smaller than about six times the radius of the orifice, or when the semi- angle of the cone is smaller than about 30°. The computed distributions of forces on the surface of the bubble and the wall suggest that buoyancy, viscous drag and viscous friction with the wall all play a role in the dynamics of the bubbles. A brief explanation of the foundations of the BEM in axi-symmetric domains was given in order to understand the esence of the method.

Experiments have been carried out with two different viscous liquids that have allowed to explore wide ranges of the Bond and capillary numbers keeping the effect of the inertia of the liquid small. Good qualitative agreement has been found between numerical and experimental results. The known scaling law for the volume of a bubble at detachment from the bottom of an unconfined liquid has been extended to take into account the presence of conical or cylindrical walls. For a conical reservoir, the semi-angle of the cone appears as an extra factor $\alpha^{-1/4}$ multiplying the standard $(Ca/Bo)^{3/4}$ scaling. For a cylindrical reservoir, the exponent may change from 3/4 to 1 when the radius of the cylinder decreases.

The results of the work may have a bearing on the methods of enhanced oil recovery, where properly shape injection nozzles may allow optimizing the volume of the bubbles generated in oil production pipes without having to change the flow rate of gas or the foams injection in homogeneous or fractured reservoirs.

Acknowledgments Authors acknowledge the IPN for its partial support through projects SIP20131821 and SIP20131821-IPN, and also acknowledge the CONACyT for its partial support through the project SENER-CONACyT 146735.

Appendix A. The Green's Functions for Axisymmetric Flow

This Appendix lists Green functions for axisymmetric flow generated by a ring force of unit strength located at (x_0, r_0) and pointing in the direction $\mathbf{e}_k$ with $k = r, x$. Defining the following quantities in cylindrical coordinares

$$
\begin{aligned}
Z &= x - x_0 \\
L &= \sqrt{Z^2 + (r + r_0)} \\
D &= \sqrt{Z^2 + \sqrt{Z^2 + (r + r_0)}} \\
S &= \sqrt{Z^2 + r^2 + r_0^2} \\
m &= \frac{2\,(r r_0)^{\frac{1}{2}}}{L}
\end{aligned}
$$

and elliptic integrals

$$
\begin{aligned}
K\,(m) &= \int_0^{\frac{\pi}{2}} \frac{d\theta}{\sqrt{1 - m^2 sen^2\theta}} \\
E\,(m) &= \int_0^{\frac{\pi}{2}} \sqrt{1 - m^2 sen^2\theta d\theta},
\end{aligned}
$$

we have

$$
\begin{aligned}
G_x^x &= 4\frac{r}{L}\left(K + E\frac{Z^2}{D^2}\right) \\
G_r^x &= 2\frac{Z}{L}\left(K - E\frac{S^2 - r^2}{D^2}\right) \\
G_x^r &= 2\frac{rZ}{r_0 L}\left(-K + E\frac{S^2 - 2r_0^2}{D^2}\right)
\end{aligned}
$$

$$G^r_r = 2\frac{1}{r_0 L}\left[-k\left(S^2+Z^2\right) - E\left(L^2+\frac{Z^2S^2}{D^2}\right)\right]$$

$$T^x_{x\,x} = 8\frac{rZ^3}{D^2}\left(K - E\frac{4S^2}{D^2}\right)$$

$$T^x_{x\,r} = T^x_{r\,x} = -4\frac{Z^2}{D^2L}\left[K\frac{S^2-2r^2}{L^2} - E\left(1+\frac{8r_0^2\left(2r_0^2-s^2\right)}{D^2L^2}\right)\right]$$

$$T^x_{r\,r} = -4\frac{rZ}{L}\left[K\left(\frac{1}{r^2}+\frac{2Z^2}{D^2L^2}\right) - \frac{E}{D^2}\left(6-S^2\left(\frac{1}{r^2}+\frac{8Z^2}{D^2L^2}\right)\right)\right]$$

$$T^r_{x\,r} = -4\frac{rZ}{r_0D^2L}\left[K\frac{2r_0^2-S^2}{L^2} - E\left(1+\frac{8r_0^2\left(2r^2-S^2\right)}{D^2L^2}\right)\right]$$

$$T^r_{x\,r} = T^r_{r\,x} = -4\frac{Z}{r_0}\left[K\left(\frac{Z^2S^2}{D^2L^2}-2\right) + \frac{E}{D^2}\left(2S^2 - Z^2 - \frac{16r^2r_0^2Z^2}{D^2L^2}\right)\right]$$

$$T^r_{r\,r} = -4\frac{r}{L}\left[\frac{K}{r_0}\left(\frac{Z^2\left(S^2-2r_0^2\right)}{D^2L^2} - \frac{r^2-r_0^2-2Z^2}{r^2}\right)\right.$$
$$\left. + \frac{E}{D^2}\left(\frac{8r_0Z^2\left(S^2-2r^2\right)}{D^2L^2} + \frac{r^2\left(r^2+r_0^2\right)-S^2\left(r_0^2+2Z^2\right)}{r^2r_0}\right)\right]$$

References

Ajaev VS, Homsy GM (2006) Modeling shapes and dynamics of confined bubbles. Ann Rev Fluid Mech 38:277

Bird RB, Armstrong RC, Hassager O (1987) Dynamics of polymeric liquids. Wiley, New York

Cheng Alexander HD, Cheng Daisy T (2005) Heritage and early history of the boundary element method. Eng Anal Boundary Element 29:268–302

Clift R, Grace JR, Weber ME (1978) Bubbles, drops, and particles. Academic Press, New York

Corchero G, Medina A, Higuera FJ (2006) Effect of wetting conditions and flow rate on bubble formation at orifices submerged in water. Colloids Surf A Physicochem Eng Aspects 290:41

Costabel Martin, (1986), Principles of boundary element methods. Technische Hochschule Darmstadt, lectures given at the first graduate summer course in computational physics "Finite Elements in Physics" Lausanne, pp 1–10

Davidson JF, Schuler BG (1960) Bubble formation at an orifice in an inviscid liquid. Trans Inst Chem Eng 38:335–342

Higuera FJ (2005) Injection and coalescence of bubbles in a very viscous liquid. J Fluid Mech 530:369

Hirasaki GJ, Lawson JB (1985) Mechanisms of foam flow in porous media: apparent viscosity in smooth capiliaries. Sot Pet Eng J 25(2):176–190

Hirasaki GJ, Miller CA, Pope GA (2006) Surfactant based enhanced oil recovery and foam mobility control. In: 3rd annual and final technical report for DOE project (DE-FC26-03NT15406), February

Kovscek AR, Tretheway DC, Persoff P, Radke CJ (1995) Foam flow through a transparent rough-walled rock fracture. J Pet Sci Eng 13:75–86

Leighton TG, Fagan KJ, Field JE (1991) Acoustic and photographic studies of injected bubbles. Eur J Phys 12:77
Longuet-Higgins M, Kerman BR, Lunde K (1991) The release of air bubbles form an underwater nozzle. J Fluid Mech 230:365–390
López-Villa A, Medina A, Higuera FJ (2011) Bubble growth by injection of gas into viscous liquids in cylindrical and conical tubes. Phys Fluids 23:102102
Manga M, Stone HA (1994) Interactions between bubbles in magmas and lavas: effects of the deformation. J Volcanol Res 63:269–281
Marmur A, Rubin E (1976) A theoretical model for bubble formation at an orifice submerged in an inviscid liquid. Chem Eng Sci 31:453–463
Oguz HN, Prosperetti A (1993) Dynamics of bubble growth and detachment from a needle. J Fluid Mech 257:111
Oguz HN, Zeng J (1997) Axisymmetric and three-dimensional boundary integral simulations of bubble growth from an underwater orifice. Eng Anal Boundary Element 19:319
Pozrikidis C (1992) Boundary integral and singularity methods for linearized viscous flows. Cambridge University Press, Cambridge
Pozrikidis C (2002) A practical guide to boundary element methods. Chapman and Hall/CRC, New York
Räbiger N, Vogelpohl A (1986) Bubble formation and its movement in Newtonian and non-newtonian liquids. In: Cheremisinoff NJ (ed) Encyclopedia of fluid mechanics, vol. 3, chap. 4. Gulf Publishing Company, Houston
Ramakrishna S, Kumar R, Kuloor NR (1968) Studies in bubble formation—I: Bubble formation under constant flow conditions. Chem Eng Sci 24:731
Russ JC (2002) The image processing handbook. CRC Press, Boca Raton
Sadhal SS, Ayyaswamy PS, Chung JN (1997) Transport phenomena with drops and bubbles, chap. 7. Springer-Verlag, Berlin
Sahagian DL (1985) Bubble migration and coalescence during solidification of basaltic lava flows. J Geol 93:205–211
Tsuge H (1986) Hydrodynamics of bubble formation from submerged orifices. In: Cheremisinoff NP (ed) Encyclopedia of fluid mechanics, vol. 3. Gulf Publishing Company, Houston (chap. 9)
Witherspoon PA, Radke CJ, Shikari Y, Pruess K, Persoff P, Benson SM, Wu YS (1987) Feasibility analysis and development of foam-protected underground natural gas storage facility. In: Proceedings of the operating section, American Gas Association, Paper #87-DT-110, pp 539–549
Wong H, Rumschitzki D, Maldarelli C (1998) Theory and experiment on the low-Reynolds-number expansion and contraction of a bubble pinned at a submerged tube tip. J Fluid Mech 356:93
Wraith AE, Kakutani T (1974) The pressure beneath a growing rising bubble. Chem Eng Sci 29:1
Zhang L, Shoji M (2001) Aperiodic bubble formation from a submerged orifice. Chem Eng Sci 5371

Effect of Surface Contamination on the Drag of a Bubble Rising in a Line

Jorge Ramírez-Muñoz, Sergio Baz-Rodríguez, Alberto Soria, Elizabeth Salinas-Rodríguez and Sergio Martínez-Delgadillo

Abstract The presence of surfactants critically increases the drag on bubbles rising in contaminated water compared with bubbles rising in pure water. This is explained by the Marangoni effect, occurring when the surface tension forces existing on the surface generate tangential shear stresses on the surface bubble. This mechanism has been studied by considering stagnant cap hypothesis to simulate the increase in the drag as a function of surface contamination. In this work, the steady drag for contaminated spherical bubbles was obtained numerically for $0.1 \leq Re \leq 200$ by using Comsol Multiphysics® 3.5a assuming the stagnant cap hypothesis. The numerical values of the vorticity, flow velocity and pressure fields as function of the angle of superficial contamination and *Re* were examined. The agreement of the numerical results with reported drag values for clean and partially contaminated bubbles, as well as rigid spheres was proved. By using an appropriate normalization of

J. Ramírez-Muñoz (✉)
Energy Department, Universidad Autónoma Metropolitana-Azcapotzalco, Avenida San Pablo 180, Col. Reynosa Tamaulipas, 02200 México D.F., Mexico
e-mail: jrm@correo.azc.uam.mx

S. Baz-Rodríguez
Faculty of Chemical Engineering, Universidad Autónoma de Yucatán, Campus de Ciencias Exactas e Ingenierías, Periférico Norte. Km. 33.5, Tablaje Catastral 13615, Col. Chuburná de Hidalgo Inn, Mérida, Mexico
e-mail: sergio.baz@uady.mx

A. Soria · E. Salinas-Rodríguez
Process and Hydraulics Engineering Department, Universidad Autónoma Metropolitana-Iztapalapa, San Rafael Atlixco #186, Col. Vicentina, 09340 México, D.F., Mexico
e-mail: asor@xanum.uam.mx

E. Salinas-Rodríguez
e-mail: sabe@xanum.uam.mx

S. Martínez-Delgadillo
Basic Sciences Department, Universidad Autónoma Metropolitana-Azcapotzalco, Avenida San Pablo 180, Col. Reynosa Tamaulipas, 02200 México D.F., Mexico
e-mail: samdmas@gmail.com

J. Klapp and A. Medina (eds.), *Experimental and Computational Fluid Mechanics*, Environmental Science and Engineering, DOI: 10.1007/978-3-319-00116-6_3,

the numerical data, a simple drag correlation for contaminated bubbles as a function of the spherical angle of the stiff surface zone was obtained.

1 Introduction

The description of the motion of gas–liquid flows requires information regarding the drag forces between the phases. Therefore, the determination of the drag on bubbles rising in a liquid is important for understanding the basic bubble behavior in relevant industrial applications. In the case of ascending air bubbles in water, the viscosity of the gas is negligible compared to that of the fluid, so there is little viscous resistance to the internal gas circulation, therefore, the drag and in consequence the terminal bubble rise velocity are very sensitive to the presence of surfactants. Consequently, a slight contamination of the water may significantly affect the free circulation of liquid on the bubble surface (Levich 1962; Clift et al. 1987; Cuenot et al. 1997). Besides, the surface contamination also affects the rate of coalescence of bubbles. Thus, it influences strongly the interfacial area and the residence time of the dispersed phase.

It is well-known that the presence of surfactants critically decreases the rising velocity of bubbles in contaminated water compared with bubbles rising in pure water. Thus, while a spherical bubble rising in a contaminated liquid flow behaves like a rigid sphere, a clean spherical bubble of the same size in an uncontaminated liquid behaves differently. The reason for this retardation is that adsorbed surfactant is swept at the rear surface of the contaminated bubble driving to a surface tension gradient. This creates a Marangoni effect, i.e. a tangential shear stress appears on the bubble surface, which balance the surface tension force, increasing the drag coefficient and reducing the rising velocity of the bubble (Cuenot et al. 1997). At some point, a limiting value of the rise velocity is reached, no longer affected by a further increase in concentration. In this limit, both the rate of kinetic (adsorption–desorption) or diffusive transport of surfactant to the bubble surface and surface diffusion are slow relative to surface convection. Therefore, the surfactant is collected in a stagnant cap region at the back end of the bubble while the front end is stress free and mobile (Alves et al. 2005).

Considering the mechanism that governs the phenomenon, in the literature, most mathematical models propose the stagnant cap hypothesis to simulate the increase in drag coefficient as a function of surface contamination (Cuenot et al. 1997; Alves et al. 2005; Palaparthi et al. 2006), which was first proposed by Frumkin and Levich (1947). It is valid when surface diffusion is extremely weak compared to advection and part of the bubble surface (and in certain cases the entire interface) tends to become stagnant (Cuenot et al. 1997). The stagnant cap regime has been successful in explaining experimental observations of bubbles rising in liquids contaminated with surfactants. For instance, by successive adjustments of the contaminated angle in the numerical computations, it is possible to properly reproduce the experimental rise velocities of bubbles.

In this chapter, the steady drag for spherical bubbles rising in a liquid contaminated with surfactants was obtained numerically for $0.1 \leq Re \leq 200$ assuming the stagnant cap hypothesis and by using Comsol Multiphysics® 3.5a. Different levels of surface contamination from clean bubbles up to solid sphere bubbles were considered. By using an appropriate normalization of the numerical data and incorporating well-known drag correlations for clean bubbles and solid spheres, a simple explicit drag expression for contaminated bubbles was obtained. In addition, the structure of the flow around the bubble as a function of *Re* and the angle of contamination were examined. *Re* is the Reynolds number defined through the bubble diameter and single bubble rise velocity.

This work is organized as follows: In Sect. 2, the problem formulation and the governing equations of the problem are presented. In Sect. 3, the numerical solution of the model is discussed in detail and numerical results supported with previously reported numerical data are presented. In Sect. 4, numerical results for drag, vorticity, velocity and pressure profiles are presented and discussed. Also, the obtained results for the drag of contaminated bubbles are compared with available data reported in the literature. Finally, in Sect. 5 some concluding remarks are provided.

2 Problem Formulation

Consider a spherical bubble of diameter $d = 2R$ (R is the bubble radius), at rest with respect to an infinite body of Newtonian and incompressible moving fluid, as shown in Fig. 1. The uniform velocity far away from the bubble is u, and the flow is assumed to be axisymmetric and steady. In order to simplify the analysis of the problem it is assumed that an amount of surfactant is collected in the stagnant cap at the back end of the bubble and it remains constant (stagnant cap model). Thus, θ is the angle between the front stagnation point on the bubble surface and a current arbitrary point on the interface, i.e. the region free of surfactants where a shear free condition, $\tau_{r\theta} = 0$, is imposed, then $180 - \theta$ is the fully contaminated region where a tangential velocity equal to zero is assumed ($u_T = 0$).

It is assumed that the bulk mass transfer toward the bubble surface can be neglected, thus, the surfactant concentration, C_b, in the liquid is constant, and the governing equations are further described by the steady incompressible Navier–Stokes and continuity equations, given by

$$\rho\left(\mathbf{u}\cdot\nabla\right)\mathbf{u} = \nabla\cdot\left[-p\mathbf{I} + \mu\left(\nabla\mathbf{u} + (\nabla\mathbf{u})^{\mathrm{T}}\right)\right] \tag{1}$$

and

$$\nabla\cdot\mathbf{u} = 0, \tag{2}$$

where $\mathbf{u}$, p, ρ and μ denotes the velocity and pressure field, the density and the dynamic viscosity of the liquid, respectively. Assuming that the bubble does not

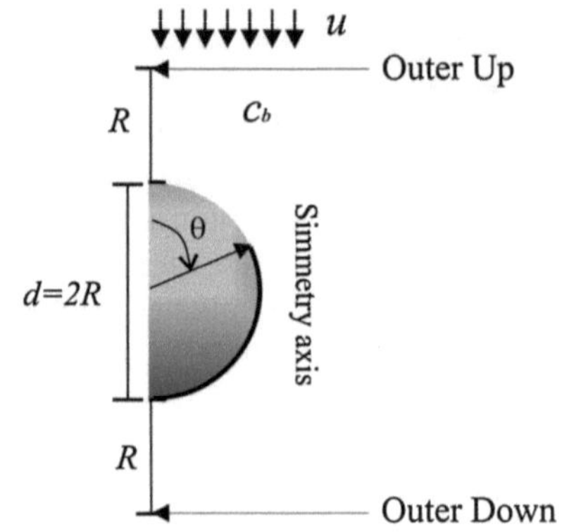

Fig. 1 Stagnant cap model scheme

deform and that no phase change occurs, the normal velocity vanishes at the interface, i.e. $\mathbf{u}_\mathrm{n} = u_\mathrm{n} = 0$.

3 Numerical Solution of the Model and Results Validation

Numerical simulations were conducted for the axisymmetric flow around a spherical bubble for Reynolds numbers ranging between 0.1 and 200. The continuity and momentum equations (Eqs. 1 and 2) were solved under the assumptions of the stagnant cap model described above (Fig. 1) using the Comsol Multiphysics® 3.5a simulation code. A 2D geometry with axial symmetry was used (see Fig. 1) and several types of meshing were included in order to obtain independent results from the numerical parameters. Furthermore, an adaptive mesh refinement scheme was useful to ensure grid independence. Different levels of surface contamination are considered based on the stagnant cap angle. Clean bubbles correspond to $\theta = 180°$, while fully contaminated bubbles to $\theta = 0°$. Two boundary conditions in the gas–liquid interface were imposed in the simulations: (1) a region free of surfactants of angle θ, where a free shear stress was imposed (slip condition), and (2) a fully contaminated region of angle $180 - \theta$, where a tangential velocity equal to zero was imposed (no slip condition).

The vertical component of the drag force on the bubble surface, F_d, was directly obtained from Comsol Multiphysics® 3.5a. The drag coefficient was calculated from its definition, given by

$$C_{\mathrm{d,CB}} = \frac{F_\mathrm{d}/4\pi d^2}{\rho U_r^2/2}, \tag{3}$$

where $C_{\mathrm{d,CB}}$, is the drag coefficient of the contaminated bubble and $U_\mathrm{r} = u - U_\mathrm{b}$, U_b is the bubble velocity, which for this particular case is equal to zero.

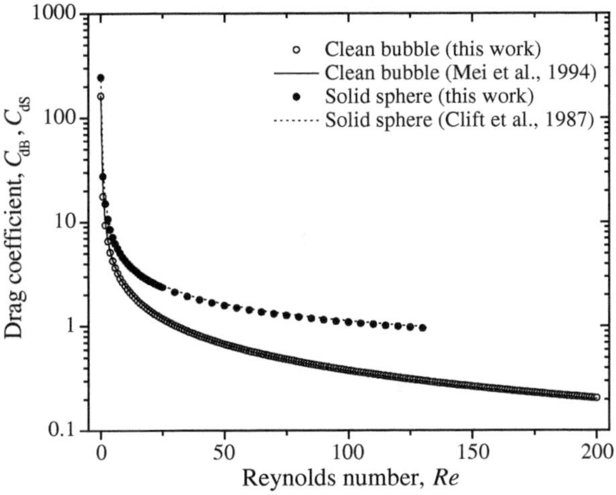

Fig. 2 Comparison of our numerical drag coefficient with reported correlations for clean bubbles and solid spheres

Figure 2 shows the variation of the numerical drag coefficient C_d as function of Re. Predictions by Mei et al. (1994) and Clift et al. (1987) for a clean bubble ($C_{d,B}$) and a rigid sphere ($C_{d,S}$), given by

$$C_{d,B} = \frac{16}{Re}\left\{1 + \left[8/Re + 1/2\left(1 + 3.315/Re^{0.5}\right)^{-1}\right]\right\} \tag{4}$$

and

$$C_{d,S} = \frac{16}{Re}\left(1 + 0.1935 Re^{0.6305}\right), \tag{5}$$

respectively, are also shown. As can be observed in Fig. 2, the numerical data recovered the values computed from both expressions. The Re range for bubbles ($Re \leq 200$) and solids spheres ($Re \leq 130$) shown in Fig. 2 were selected because for the case of bubbles or solid spheres moving at Re higher, the formed wake is no more axisymmetric (Yuan and Prosperetti 1994; Ramírez-Muñoz et al. 2007). Therefore, their paths of movement are no longer rectilinear.

4 Results and Discussions

In this section results from numerical simulations for drag, vorticity diffusion, and velocity and pressure profiles around the bubble are presented and discussed.

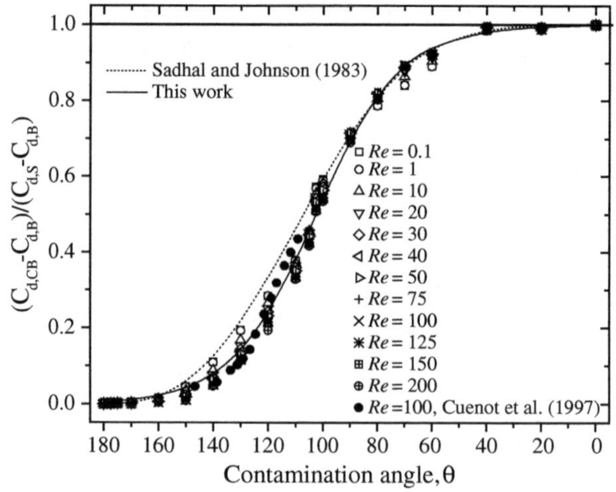

Fig. 3 Normalized drag coefficient as a function of the contamination angle θ

4.1 Drag on a Contaminated Bubble

In Fig. 3, the numerical results of the normalized drag coefficient given by

$$\frac{C_{d,CB} - C_{d,B}}{C_{d,S} - C_{d,B}}, \tag{6}$$

are shown as a function of the contaminated stagnant cap angle, $0° \leq \theta \leq 180°$, and for $0.1 \leq Re \leq 200$. Fitting of the function

$$f(\theta) = \frac{1}{1 + \exp(0.06678\ \theta - 6.84645)} \tag{7}$$

to the normalized drag coefficient defined by Eq. (6), can be appreciated as the solid line in Fig. 3, with a 0.9 % average relative error for all *Re*. These results show that the drag of a contaminated bubble is nearly the one of a clean bubble if $\theta > 160°$ and the one of a rigid sphere if $\theta < 40°$. Between these bounds, the drag is a strong function of θ, with its maximum dependency being observed in the range $90° < \theta < 130°$. These results are in agreement with those of Sadhal and Johnson (1983) in the creeping flow limit for $\theta < 90°$ and for $\theta > 160°$, see dotted line in Fig. 3. However, as it can also be observed, the normalized drag coefficient shows a slight dependence with *Re* for $90° < \theta < 160°$. Meanwhile, for $Re = 100$, the results are in agreement with the numerical data of stagnant cap model reported by Cuenot et al. (1997), see filled circles in Fig. 3.

4.2 Vorticity Diffusion Around a Contaminated Bubble

The contaminated zone of the bubble surface behaves as a vortex sheet, from the θ angle towards the rear stagnation point. In order to get insights on the effect of the surfactant on the flow structure, the numerical vorticity contours around the bubble for different contamination angles were evaluated. A shared feature to be noticed, for all contamination angles, is that vorticity diffuses in the main stream (vertical) direction, leaving a reduced vorticity zone at the rear of the bubble, which becomes shrinked as the *Re* is increased.

In Fig. 4, the vorticity contours for fully contaminated ($\theta = 0°$) and partially but strongly contaminated bubble ($\theta = 20°$) are shown. For fully contaminated bubble, the Reynolds number effect can be seen in Fig. 4–c; the stagnant surface of the bubble generates flow vorticity in the adjacent fluid. The vorticity diffusion in the vertical direction becomes more apparent as the *Re* increases. Also, a bottom zone with reduced vorticity at $Re = 0.1$, which is vertically aligned and shrinked towards the rear surface zone as *Re* increases, becomes apparent. A similar behavior is observed in the strongly contaminated bubble, with a contamination angle $\theta = 20°$, but for the front zone behavior due to the fact that the vortex sheet has now been reduced, sliding the vorticity contours towards its new position, as it can be seen in Fig. 4d–f.

For fully and strongly contaminated bubbles at same *Re* (see Fig. 4a vs d, b vs e, c vs d) the vorticity distributions are very alike and their spans are almost the same. It also should be pointed out that in Fig. 4c and f a region on the surface where the vorticity contours converge is observed. This zone is around 100° to 120°, which is close to the separation point for a solid sphere at $Re \geq 60$, located at 109.5°, as reported by Schlichting (1979). Vorticity contours convergence can already be observed in Figs. 4b and e for $Re = 20$, even when separation does not yet occurs.

On the other extreme, the vorticity behavior of the wholly clean bubble ($\theta = 180°$) can be seen in Fig. 5a–c, while the vorticity contours of a slightly contaminated bubble ($\theta = 60°$) can be observed in Figs. 5d–f, for the same Reynolds numbers as in Fig. 4.

Considering the fully clean bubble, Figs. 5a–c and the fully contaminated bubble, Figs. 4a–c, an increasing difference in vorticity span, as *Re* keeps increasing, should be pointed out. Thus the span ratio of fully contaminated to fully clean bubbles is 1.500 for $Re = 0.1$, 2.310 for $Re = 20$ and 4.192 for $Re = 100$. It can also be seen that the highest negative values of vorticity are found closer to the front for the fully contaminated bubble than for the fully clean bubble, where they locate close to the equator.

Concerning the slightly contaminated bubble ($\theta = 160°$) as compared to the fully clean bubble, its vorticity span divided by the clean bubble span is 2.773 for $Re = 0.1$, 1.527 for $Re = 20$ and 1.526 for $Re = 100$. It is remarkable that, even for the higher span ratio, at $Re = 0.1$, the vorticity maps for same *Re* look very much the same, but for a small zone where the vortex sheet is placed, giving the highest negative vorticity values.

In order to select other angles whose vorticity patterns may be interesting to be observed, let us come back to Fig. 3 where the dependence of the normalized drag

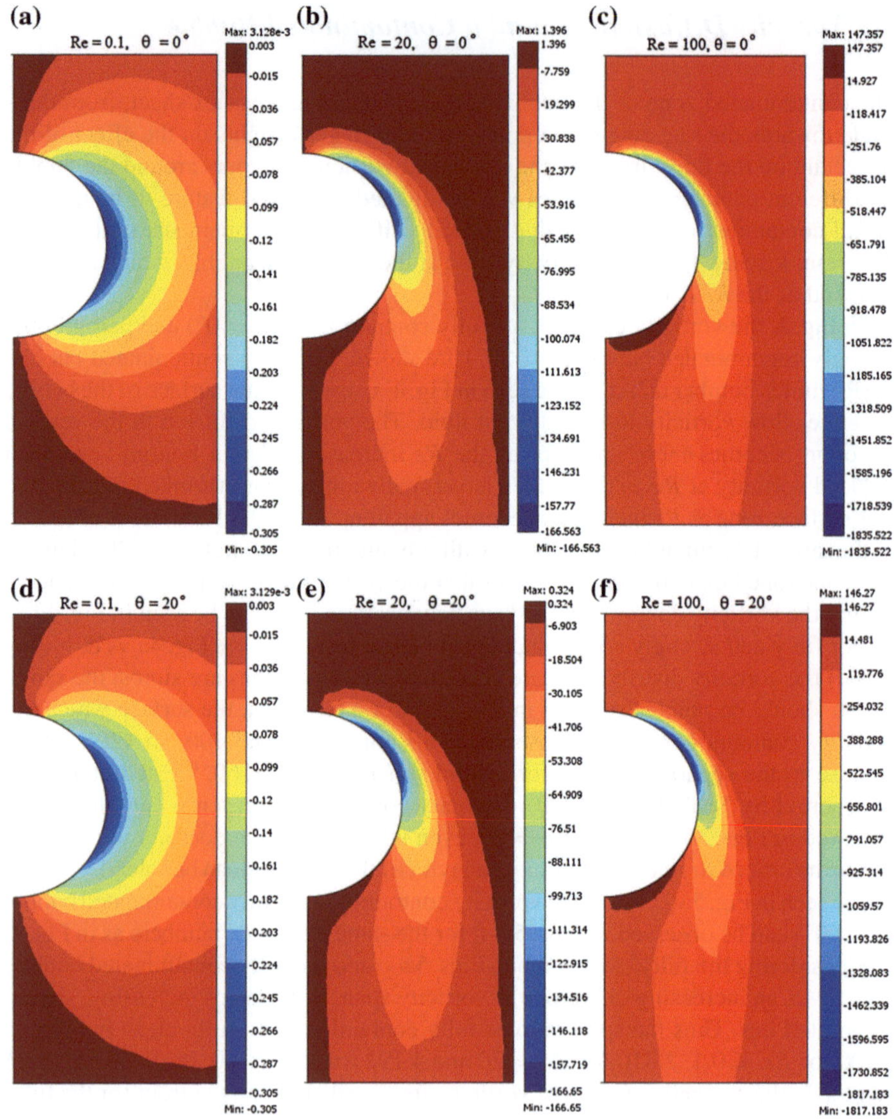

Fig. 4 Vorticity contours for $\theta = 0°$ and $\theta = 20°$, and $Re = 0.1$, 20 and 100

coefficient with respect to the *Re* appears to be lumped up to a contamination angle of $\theta = 90°$. Moreover, the higher variability of drag coefficient with respect to *Re* seems to be close to a $\theta = 120°$. Therefore we select both angles to visualize their vorticity contours in Fig. 6.

In Fig. 6, the vorticity span of the less contaminated bubble ($\theta = 120°$) divided by the span for the clean bubble gives 4.357 for $Re = 0.1$, 4.446 for $Re = 20$ and

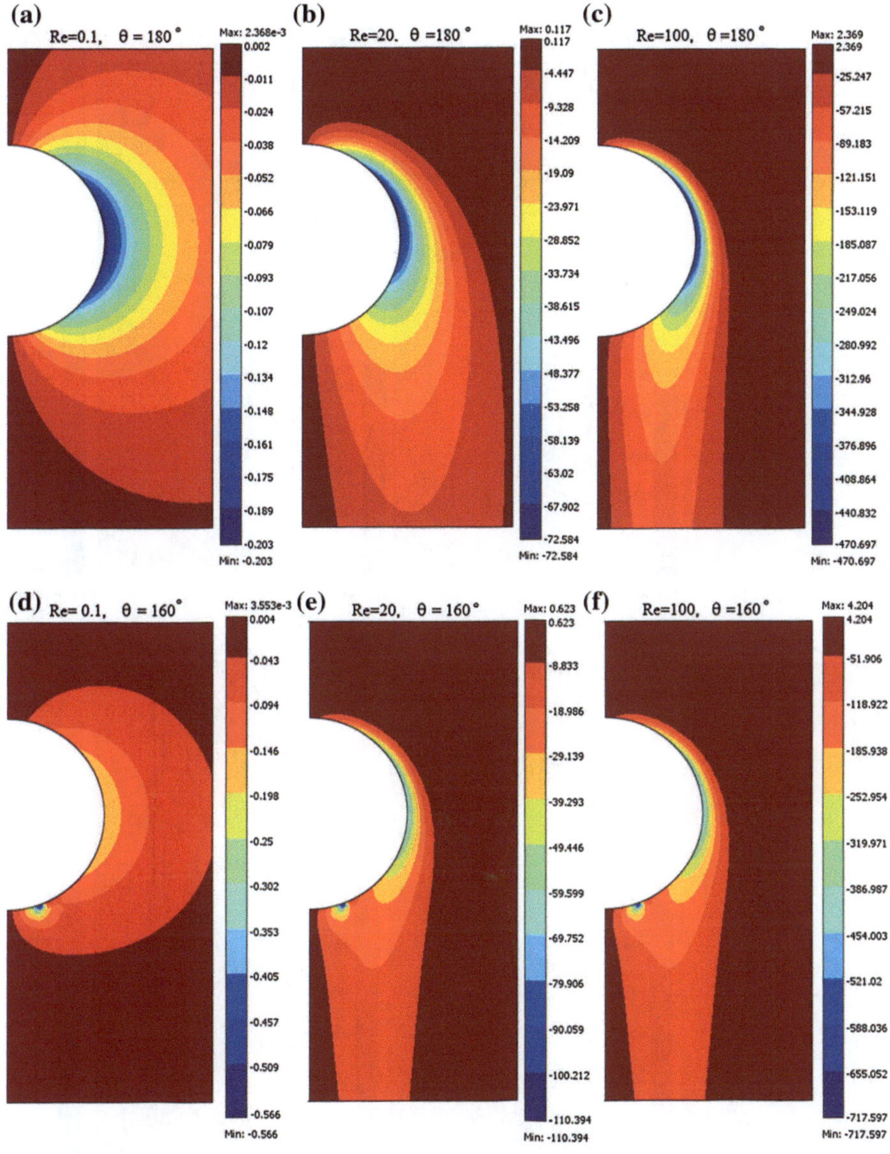

Fig. 5 Vorticity contours for $\theta = 180°$ and $\theta = 160°$, and $Re = 0.1$, 20 and 100

6.717 for $Re = 100$, meanwhile a similar comparison of $\theta = 90°$ gives 2.954 for $Re = 0.1$, 4.237 for $Re = 20$ and 9.557 for $Re = 100$. As can be seen, for Re of 0.1 and 20 the vorticity span at $\theta = 120°$ is greater than at $\theta = 90°$, however, this trend is reversed at $Re = 100$. These results indicate that a strong dependence of the vorticity distribution on the Re should be expected for $\theta = 120°$ and $Re = 100$. Meanwhile,

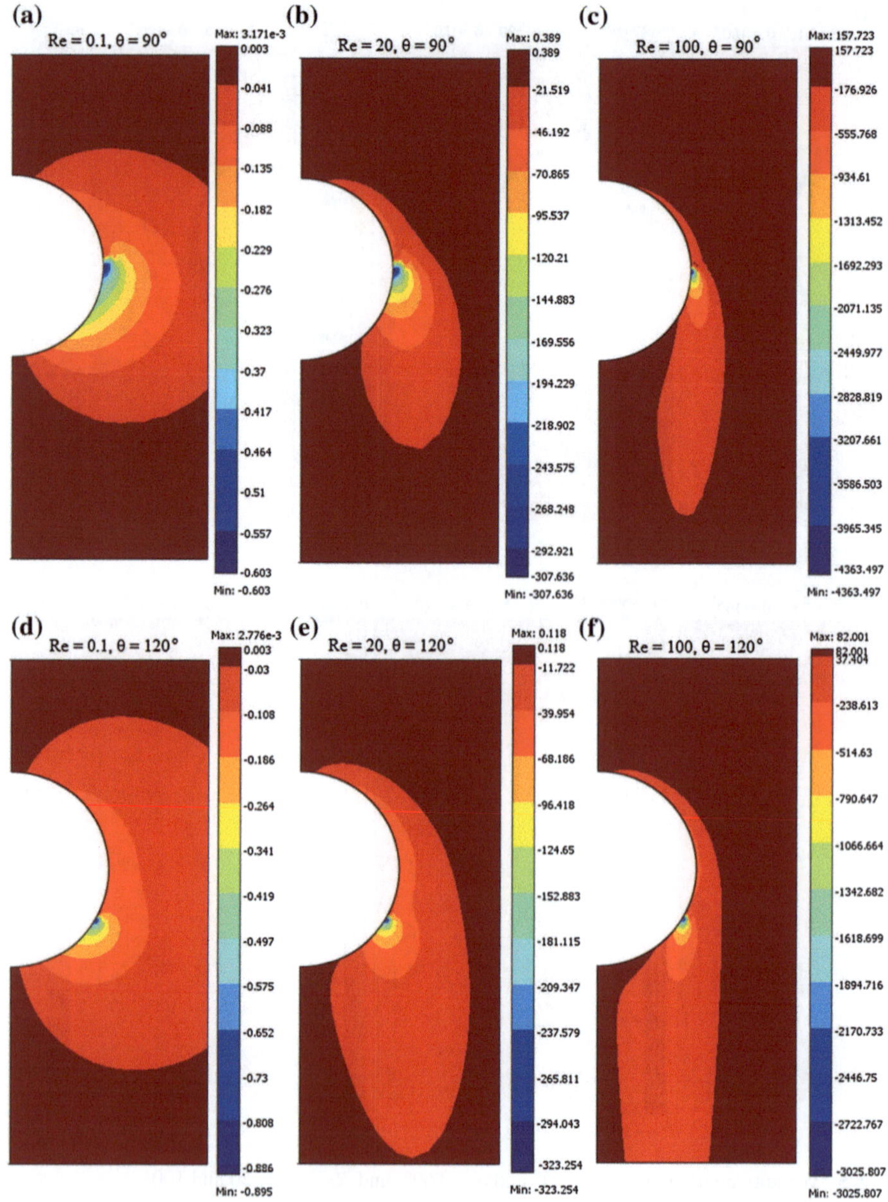

Fig. 6 Vorticity contours for $\theta = 90°$ and $\theta = 120°$, and $Re = 0.1$, 20 and 100

for $\theta = 90°$, the largest amount of vorticity is generated near—i.e. centered—to the bubble equator, as in the creeping flow regime regardless of the contamination angle (Figs. 4a, 5a, d, 6a and 6d), where the vorticity production on the bubble surface is

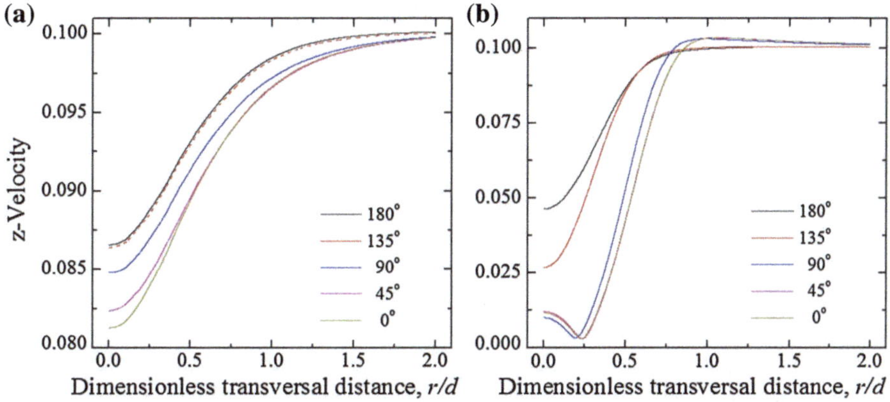

Fig. 7 z-Velocity evaluated at (**a**) outer up and (**b**) outer down positions, for $Re = 100$

symmetric—i.e. also centered—to the bubble equator. This fact could be helpful to explain why for $\theta = 90°$, the normalized bubble drag is independent of *Re*.

4.3 Velocity and Pressure Profiles Around the Bubble

Figures 7 and 8 shows the z-velocity and pressure profiles along the transversal direction to the undisturbed flow at outer up (Figs. 7a and 8a) and outer down (Figs. 7b and 8b) positions for $Re = 100$, see Fig. 1. As can be seen, the increase in the grade of contamination from clean bubble ($\theta = 180°$) to slightly contaminated bubble ($\theta = 135°$) and from strongly contaminated bubble ($\theta = 45°$) to fully contaminated bubble ($\theta = 0°$) has no important effect in both, the velocity and pressure profiles. Because the magnitude of the velocity and pressure profile can be directly associated with the drag (form and friction drag) on the bubble surface (Landau and Lifshitz 1987; Ramírez-Muñoz and Soria 2007), both fluid-dynamic variables explains the cause why the drag is not increased in the same proportion as contamination angle increased. Thus, the approach and results presented in this work can be useful for an integral understanding of the effect of contamination grade on bubble's drag. On the other hand, for $\theta = 90°$, the velocity and pressure profiles evaluated at the Outer up position differs from other contamination angles evaluated (180°, 135°, 45° and 0°); however, at the Outer down position both fluid-dynamics profiles are very alike to the fully contaminated bubble. Therefore, the wake structure downstream from the bubble surface for $\theta = 90°$ and Re = 100 tends to be similar to that of a fully contaminated bubble.

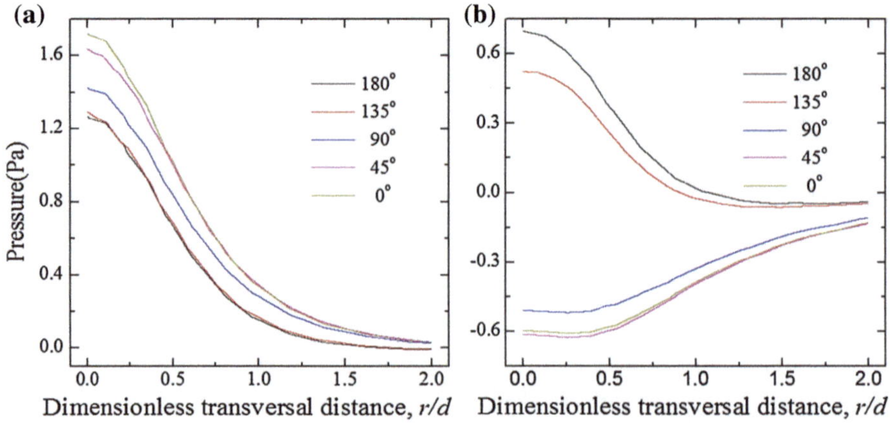

Fig. 8 Pressure profiles evaluated at (**a**) outer up and (**b**) outer down positions, for $Re = 100$

5 Conclusions

The drag, vorticity diffusion and disturbed flow around a spherical bubble rising stationary in a liquid contaminated with surfactants were investigated in this work. The model solved in the present work is based on the stagnant cap hypothesis, where the surface diffusion of surfactants is considered extremely weak, compared to advection, so steady conditions are reached. A simple explicit normalized drag correlation for contaminated bubbles as function of the stagnant cap angle, which is valid for $0.1 \leq Re \leq 200$ was obtained. Comparison of the model predictions with existing data were performed, showing very good agreement.

The simulations revealed several interesting features concerning the drag on the contaminated bubble and the flow structure around the bubble as function of the grade of contamination and Reynolds number. Among these features, the following are pointed out: (1) For $\theta > 160°$ and $\theta < 40°$, the normalized drag on the bubble remains practically constant and is independent of *Re*. (2) For $\theta > 160°$ and $\theta < 90°$ a slight dependence of the normalized drag on the *Re* can be observed. The maximum variability is achieved for $\theta = 120°$. (3) The vorticity span and its distribution around an almost fully contaminated bubble ($\theta = 20°$) are more alike to the one of the fully contaminated bubble ($\theta = 0°$), than the slightly contaminated bubble ($\theta = 160°$) with respect to the clean bubble ($\theta = 180°$). (4) The vorticity production at $\theta = 90°$ is located near to bubble equator. This is similar to the creeping flow vorticity production from a clean up to a fully contaminated bubble. This fact could be helpful to explain why for this angle of contamination the normalized bubble drag is independent of Reynolds number. (5) The velocity and pressure profiles shows that for the outer up and outer down positions for the clean bubble and the low contaminated angle bubble ($\theta = 180°$ and $\theta = 135°$) behave very similar. A similar behavior is observed as the bubble is almost fully contaminated ($\theta = 0°$ and $\theta = 45°$).

References

Alves SS, Orvalho SP, Vasconcelos JMT (2005) Effect of bubble contamination on rise velocity and mass transfer. Chem Eng Sci 60:1–9

Clift R, Grace JR, Weber ME (1987) Bubble, drops and particles. Academic Press, New York

Cuenot B, Magnaudet J, Spennato B (1997) The effects of slightly soluble surfactants on the flow around a spherical bubble. J Fluid Mech 339:25–53

Frumkin A, Levich V (1947) On surfactants an interfacial motion. Zh Fiz Khim 21:1183 (in Russian)

Landau L, Lifshitz E (1987) Fluid mechanics (Course of theoretical physics), vol 6, 2nd edn. Butterworth Heinemann, Oxford

Levich V (1962) Physicochemical hydrodynamics. Prentice Hall, Englewood Cliffs

Mei R, Klausner JF, Lawrence CJ (1994) A note on the history force on a spherical bubble at finite Reynolds number. Phys Fluids 6:418–420

Palaparthi R, Papageorgiou DT, Maldarelli C (2006) Theory and experiments on the stagnant cap regime in the motion of spherical surfactant-laden bubbles. J Fluid Mech 559:1–44

Ramírez-Muñoz J, Soria A (2007) An expression for the hydrodynamic force on a body interacting with the leading body wake. Rev Mexicana Ingen Quim 6:101–109 (in Spanish)

Ramírez-Muñoz J, Soria A, Salinas-Rodríguez E (2007) Hydrodynamic force on interactive spherical particles due to the wake effect. Int J Multiphase Flow 33:802–807

Sadhal S, Johnson R (1983) Stokes flow past bubbles and drops partially coated with thin films. J Fluid Mech 126:237–249

Schlichting H (1979) Boundary-layer theory. McGraw-Hill, New York

Yuan H, Prosperetti LA (1994) On the in-line motion of two spherical bubbles in a viscous fluid. J Fluid Mech 278:325–349

Hydrodynamic Interactions in Charged Vesicles Suspensions

C. Haro-Pérez, M. Quesada-Pérez, J. Callejas-Fernández, R. Hidalgo-Álvarez, J. Estelrich and L. F. Rojas-Ochoa

Abstract We have measured the short-time dynamics and the structure of charged liposome dispersions upon increasing the volume fraction. Although the structural properties of the suspensions suggest an interparticle potential not purely repulsive, the hydrodynamic theory of Beenakker-Mazur can explain the experimental hydrodynamic functions of the dispersions. This result suggests the generality of the theory, which up to now, only had been tested in pure repulsive systems.

1 Introduction

The complete description of colloidal interactions includes both, direct and hydrodynamic interactions. Considerable progress has been made in the description of both interactions using hard-sphere colloids suspended in apolar, nearly index-matching fluids (Segré et al. 1995; Pusey et al. 1997). Indeed, to access information on both interactions requires the measurement of the static and dynamic structure factor, which are easily accessible by static and dynamic light scattering. However, these exper-

C. Haro-Pérez (✉)
Departamento de Ciencias Básicas, Universidad Autónoma Metropolitana-Azcapotzalco, Av. San Pablo 180, 02200 México D.F., Mexico
e-mail: cehp@correo.azc.uam.mx

M. Quesada-Pérez
Escuela Politécnica Superior de Linares, Universidad de Jaén, 23700 Jaén, Spain

J. Callejas-Fernández · R. Hidalgo-Álvarez
Departamento de Física Aplicada, Universidad de Granada, E-18071 Granada, Spain

J. Estelrich
Unitat de Fisicoquímica, Facultat de Farmàcia, Universitat de Barcelona, 08028 Barcelona, Spain

L. F. Rojas-Ochoa
Departamento de Física, CINVESTAV-IPN, Av. Instituto Politécnico Nacional 2508, 07360 México D.F, Mexico

J. Klapp and A. Medina (eds.), *Experimental and Computational Fluid Mechanics*, Environmental Science and Engineering, DOI: 10.1007/978-3-319-00116-6_4,

imental methods suffer from the need of measuring single scattered light, which is difficult to achieve when the refractive index of the colloidal particles differs considerably from the one of the solvent. This, however, is the case for most charged colloidal systems, where the direct interactions are dominated by Coulomb-repulsion. Indeed, measurement on the hydrodynamics of charged particles mostly involved silica particles suspended in various organic solvent mixtures (Philipse and Vrij 1988; Härtl et al. 1999; Rojas et al. 2003; Phalakornkul et al. 1996; Riese et al. 2000) (none of them pure water), where the refractive index is nearly matched to the one of the solvent.

To gain an understanding of interplay between direct and hydrodynamic interactions in relatively dense charged colloidal systems we thus investigate the static and dynamic properties of aqueous suspensions of charged liposomes. These systems nearly fulfill conditions, index and density matching. Moreover, hydrodynamic theories have only been tested in pure repulsive systems. In fact, more than twenty years ago, Genz and Klein addressed the relevance of testing these theories in systems showing other kind of direct interactions, like attractive ones (Genz and Klein 1991). In this chapter, we will study the validity of the hydrodynamic theories in dispersions of charged liposomes, which show static structure factors that cannot be explained by a purely repulsive potential (Haro-Pérez et al. 2009).

2 Hydrodynamic Interactions

The dynamic structure factor depends on the direct interaction potential between the colloidal particles and on the indirect hydrodynamic interactions mediated by the solvent. Hydrodynamic interactions arise from the flow patterns generated in the surrounding fluid by the moving particles. The effect of direct and hydrodynamic interactions on the particle diffusion is described by the well-known expression $D_{eff}/D_0 = H(q)/S(q)$ (Pusey 1991), where $D_{eff}(q)$ and D_0 are the effective and the single particle diffusion coefficient, respectively. The hydrodynamic function, *H(q)*, contains the configuration-averaged effect of hydrodynamic interactions on the particle dynamics.

The hydrodynamic function *H(q)* is connected to the mobility tensor $\tilde{\mu}_{ij}(\{\vec{r}^N\})$ by

$$H(q) = \frac{S(q)D_{eff}(q)}{D_0} = \frac{\kappa_B T}{N D_0} \sum_{i,j=1}^{N} \langle \hat{q}.\tilde{\mu}_{ij}(\{\vec{r}^N\}).\hat{q}e^{i\hat{q}.(\vec{r}_i-\vec{r}_j)} \rangle \qquad (1)$$

where N is the particle number, $\hat{q}$ is the unit vector in the direction of the scattering vector $\vec{q}$, $D_0 = \kappa_B T(6\pi\eta a)^{-1}$ is the Stokes-Einstein free-diffusion coefficient and $\vec{r}^N$ represents the position of all colloidal particles. The brackets <..> stand for the ensemble average over all the spatial configurations of the particle systems.

Beenakker and Mazur have evaluated *H(q)* applying a renormalization method valid at arbitrary volume fractions that has been successfully employed with hard sphere systems (Beenakker and Mazur 1984). This method has also been used by Genz and Klein for dispersions of moderately charged particles (Genz and Klein 1991). The result is called the $\delta\gamma$-expansion to zero order and can be expressed in terms of *S(q)*

$$H(q) = \frac{D_S(\phi)}{D_0} + \frac{3}{2\pi}\int_0^\infty d(aq')\left(\frac{\sin(aq')}{aq'}\right)^2 \left[1+\phi S_{\gamma_0}(aq')\right]^{-1} \times \int_{-1}^{1} dx(1-x^2)(S(|q-q'|)-1) \quad (2)$$

where $D_s(\phi)$ is the self-diffusion coefficient and $x = \cos(q', q)$. The function $S_{\gamma 0}$ depends on the volume fraction ϕ through the scalars $\{\gamma_0^{(1)}\}$. An alternative hydrodynamic theory used for charged systems is the Pairwise Additivity Approximation. In our case we choose the Beenaker and Mazur formalism since the Pairwise Additivity Approximation needs as an input function in the calculations the radial distribution function which is described in the real space, and our experimental data, *S(q)* are in the reciprocal one.

3 Experimental

Liposomes are composed of egg phosphatidylcholine (PC) and phosphatidylserine (PS) at a ratio of PS/PC = 0.25. PC is a zwitterionic phospholipid, comprising a choline and a phosphate group. At the working conditions we presume PC to be uncharged. By contrast, PS is an ionic phospholipid with an expected charge of $1e^-$ per molecule, which sets the total surface charge of our liposomes, thereby controlling the long-range repulsion between liposomes. The preparation technique used (Haro-Pérez et al. 2003) leads to unilamellar vesicles that are rather monodisperse in size, which we determine by dynamic light scattering as 98 ± 5 nm. The volume fraction of the stock suspension was estimated from the lipid weight fraction, 30 μmol/mL, and the mean outer and inner vesicle radii (*a* and *b*, respectively) (Stuchly et al. 1988), considering that the thickness $\Delta = a - b$ and the density of the phospholipid shell are 4.5 nm and 1.015 g/cm^3, respectively (Huang and Mason 1978; Lasic 1991).

We perform static and dynamic light scattering using an ALV-5000 spectrophotometer with an Argon-Ion Laser (Coherent, model Innova 308) of wavelength $\lambda_0 = 514.5$ nm. Average intensities are obtained from several measurements at each angle, with different cell positions to minimize the effect of scratches on the cell surface. The intermediate scattering functions, $g_2(q,t)$, are recorded during a time around four orders of magnitude the decay time of the correlation function, which depends on density and angle, in order to obtain a good statistic. For ergodic systems, where the ensemble average of the intensity can be identified with the time average intensity, the field correlation function is obtained from the time correlation function of the intensity by the Siegert relation (Berne 1976) as

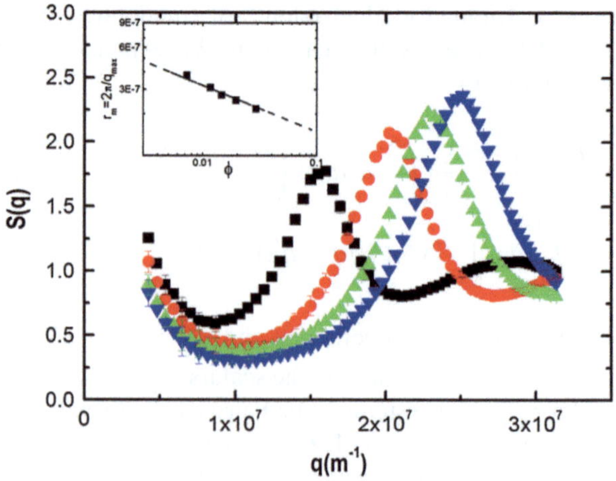

Fig. 1 Static structure factor for liposome dispersions as a function of wave vector at different volume fractions: 0.7, 1.16, 1.45, 1.93 %, (from *left* to *right*). Inset: Mean interparticle distance versus volume fraction. Experimental data (*squares*), theoretical values $0.9 \cdot n^{-1/3}$ (*dashed line*)

$$g_2(q,t) = \frac{\langle I(q,0)I(q,t)\rangle_E}{\langle I(q,0)\rangle_E^2} = 1 + |\beta g_1(q,t)|^2 \tag{3}$$

where *I(q,t)* is the scattered intensity by the sample at time *t* and at scattering vector *q* and β is an experimental constant smaller than one. The ergodicity theorem allows replacing the ensemble average of the intensity, $I^E = <..>_E$, by its time average, I^t, which is the magnitude that a correlator in a light scattering device provides.

As multiple scattering is strongly dependent on the particle concentration and particle refraction index, usually researchers eliminate it reducing the contrast between solvent and particles, reason why the medium composition has to be altered. In this case, the contrast of the system is intrinsically low since the refraction index of the liposomes, $n_l = 1.36$, is quite close to that of the water, $n_w = 1.33$ (Haro-Pérez et al. 2009).

Subsequently, several structures were formed for increasing liposome concentrations from 0.7 to 1.9 % volume fraction. The suspensions studied in this work were kept for 5 days over a bed of ion exchange resin (Amberlite NRM-150).

4 Results

The static structure factor may be calculated by $S(q) = (\rho_0/\rho)(I(q)/I_0(q)$, where $I_0(q)$ is the light intensity scattered by a sample of non-interacting particles with number density ρ_0. The results obtained are shown in Fig. 1. Obviously, this figure reveals the existence of liquid-like order in the vesicle suspension.

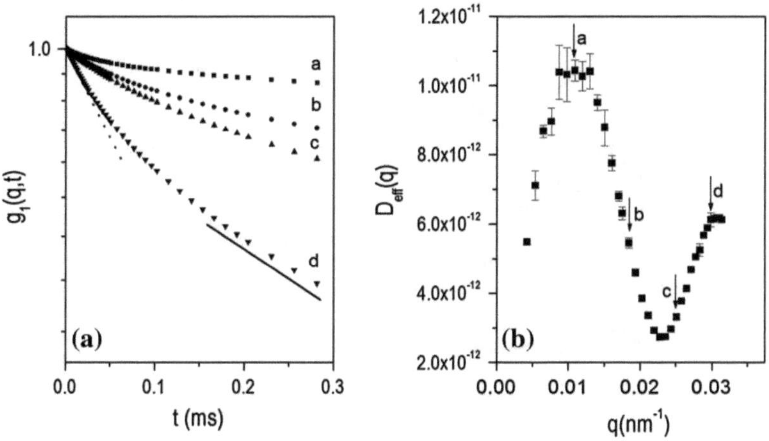

Fig. 2 **a** Field autocorrelation functions for a suspension at concentration $\phi = 1.45\%$ at the wave vectors marked with arrows in the *right side*: (a) 39°, (b) 69°, (c) 101°, (d) 133°. The *straight lines* are fits to the linear part related to the short-time dynamics (*dotted line*) and to the long-time dynamics (*solid line*). **b** Angular dependence of the effective diffusion coefficient for the same sample

The structure factor exhibits a pronounced maximum, whose position strongly depends on the particle volume fraction. The height and position of the main peak exhibit the expected behavior as the particle concentration increases, i.e., the peak height increases and their position shifts to large q-values as a function of $n^{-1/3}$, where n is the particle number density. This trend is characteristic of charged colloidal systems. However, we have seen in previous studies that the structure factor calculated theoretically by solving the Ornstein-Zernike equations along with a closure relation by assuming a purely repulsive potential can only describe the main peak of *S(q)*. The fits start to fail at lower q-values where we find an upturn of the static structure factor, being *S(q)* larger than predicted. In a similar system, we could describe the low-q regime by assuming a mixed potential that comprises a long range repulsion and a shorter range attraction (Haro-Pérez et al. 2009).

Concerning the dynamic properties, dynamic light scattering measurements were performed via the *normalized* intensity correlation function. The analysis of the curves $\sqrt{g_2(q,t) - 1}$ versus t reveals interactions between liposomes have a marked effect: $\sqrt{g_2(q,t) - 1}$ departs strongly from a simple dependence on time and its decay cannot be described by a single exponential law as can be seen in Fig. 2a). In this plot several field correlation functions measured at different wave vectors for a given sample are displayed. In all cases, we observe that the decay of $g_1(q,t)$ is linear for $t \to 0$, so the short-time effective-diffusion coefficient can be measured from their slope $-D_{eff} q^2$ (Snook and Tough 1983), $g_1(q,t) = \exp(-q^2 D_{eff}\, t)$. To obtain D_{eff}, a cumulant expansion of second order was fitted to the autocorrelation function. The second linear regime found at longer times is related to the long-time dynamics, which is not the focus of the present work. The angular dependence of the normalized short-time effective diffusion coefficient D_{eff} on the wave vector is

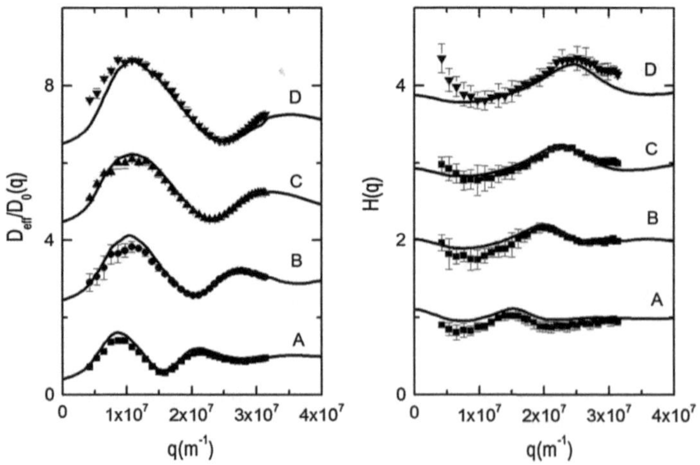

Fig. 3 *Left* Measured normalized effective diffusion coefficient at different volume fractions (a) 0.7 %, (b) 1.16 %, (c) 1.45 %, (d) 1.93 %, along with their corresponding theoretical fits according to $\delta\gamma$ expansion theory but with the experimental data of *S(q)* used as input functions in the calculation of *H(q)*. Offset by 2. *Right* Comparison of their corresponding experimental hydrodynamic function at the same volume fractions (*solid squares*) with the theoretical fits computed according to $\delta\gamma$ expansion theory but with the experimental *S(q)* used as input function in the calculation of *H(q)* (*solid line*). Offset by 1

shown in Fig. 2b). These values are obtained by fitting the initial slope of the field correlation functions, as shown in Fig. 2a).

The single particle diffusion coefficient is measured by dynamic light scattering using diluted samples of liposomes where the measurements are performed at several angles. The obtained result was $D_0 = (4.98\pm0.20)\cdot 10^{-12}$m^2/s. Similarly, we analyze the rest of the samples. The angular behavior of the normalized effective diffusion coefficient for different volume fractions is plotted in Fig. 3 (left). We can observe the inverse of the curves show surprising similarities with their corresponding *S(q)*: an atypical upturn at the collective regime, a maximum occurring at the same scattering vector q_{max} where the static structure factor *S(q)* has its first peak and the oscillatory behavior for $q > q_{max}$. The decrease at low q values of the normalized diffusion coefficient reflects that the dynamics in the collective regime slows down. Moreover, the minimum related to the mean peak of *S(q)* shifts with concentration to larger q values in the same way *S(q)* does. This link between structure and dynamics is well known (Ackerson 1976). A striking feature observed in the diffusive behavior of these dispersions is the appearance of a maximum in the collective regime, placed at the same position as the minimum of the static structure factor. This maximum occurs at a scale corresponding to 2–2.5 interparticle distance units, suggesting the existence of dynamical heterogeneities of these sizes.

Combining the effective diffusion coefficient $D_{eff}(q)$ and the measured static structure factor enables to determine the hydrodynamic function without taking recourse to any theoretical model beforehand. In Fig. 3 on the right, we plot *H(q)* versus

q obtained for different volume fractions, whose structure factors are shown in Fig. 1. All the curves show common features: An increase at low-q values followed by a minima resembling the one of $S(q)$ and the presence of a pronounced peak (concentration dependent) at the same position as the particle-particle correlation peak of the static structure factor. The height and the width of the main peak increase with ϕ, and its value is always bigger than unity. This feature indicates hydrodynamic interactions speed up the system dynamics in the vicinity of the particle correlation peak, behaviour that has been previously observed in charged systems (Härtl et al. 1999; Rojas et al. 2003). As different authors reveal, this contrasts with hard-sphere dispersions, where the peak never exceeds one and its height decreases with increasing ϕ (Segré et al. 1995).

With regard to the behaviour at low q, we observed an unusual minimum that mimics the one observed in the structure factor. For a more detailed analysis, we perform a theoretical approach to explain our experimental results. As the particles we are dealing with are charged, we tried to fit the experimental $H(q)$ according to the Modified Beenakker and Mazur Formalism where the input parameter is the structure factor. In Fig. 3, comparisons of the experimental results to model calculations are presented. Firstly we tried to calculate the $H(q)$ by assuming a pure repulsive interaction potential to generate the input function $S(q)$, data are not shown. In this case the fits are rather poor, particularly at low wave vectors. This finding is predictable, as the statics could not be explained assuming a purely repulsive interaction potential. To describe our experiments we have used a different approach, instead of using as input the theoretical curves of $S(q)$, which failed as well at low angles to describe the static experiments, we employ the experimental $S(q)$ to evaluate the hydrodynamic function. In this case, we observe that the theory (solid lines) reproduces the tendency of the experimental curves at low wave vectors although there is a little vertical displacement. The error bars are considerable since the experimental values of the hydrodynamic function are obtained from the multiplication of two experimentally determined quantities. That is the reason why we prefer to compare the theoretical fits to the experiments in terms of the effective diffusion coefficient, where we can see a better agreement between experiments and theory, see solid lines in Fig. 3 (left). As we can see, the theory starts to fail for the most concentrated sample mainly at low wave vectors. These discrepancies could be due to the proximity of a phase transition where the theory would not be valid. The novelty of our findings is that the Beenakker-Mazur formalism, in principle devised to describe hydrodynamic interactions in hard spheres systems, and later adapted to charged particles, has now been applied to systems that show hallmarks of attraction. Moreover this theory provides a qualitative description of the experimental hydrodynamic functions.

5 Conclusions

In this work ordered colloidal suspensions from moderate to large volume fractions of charged liposomes are formed and their static and dynamic properties are analyzed by light scattering techniques. The particle-particle peak of the static structure factors follows the usual $n^{-1/3}$ law, but there is an upraise at low q's that cannot be explained by a purely repulsive interaction potential. Concerning the short-time dynamics, despite the lack of understanding of the *S(q)* we can simply predict the short time dynamics by using the modified Beenaker-Mazur formalism, which describes the experimental data reasonably well. We have found the Beenaker-Mazur formalism is more general than expected and can be applied successfully to other kind of systems, not only repulsive, but also with attraction as long as the structural properties of the dispersion are well determined and used as input functions in the theory.

Acknowledgments C.H.P acknowledges financial support from Conacyt (Project 166645) and PROMEP.

References

Ackerson BJ (1976) J. Chem. Phys. 64:242–246
Beenakker CWJ, Mazur P (1984) Physica A 126:349–370
Berne BJ, Pecora R (1976) Dynamic light scattering: with applications to biology, chemistry and physics. Wiley, New York
Genz U, Klein R (1991) Physica A 171:26–42
Haro-Pérez C, Quesada-Pérez M, Callejas-Fernández J, Casals E, Estelrich J, Hidalgo-Álvarez R (2003) J. Chem. Phys. 118:5173–6167
Haro-Pérez C, Rojas-Ochoa LF, Trappe V, Castañeda-Priego R, Estelrich J, Quesada-Pérez M, Callejas-Fernández J, Hidalgo-Álvarez R (2009) Structural and functional properties of colloidal systems, vol. 146. CRC Press, Taylor and Francis Group, London, pp 77–91
Haro-Pérez C, Rojas-Ochoa LF, Castañeda-Priego R, Quesada-Pérez M, Callejas-Fernández J, Hidalgo-Álvarez R, Trappe V (2009) Phys. Rev. Lett. 102(081301):1–4
Härtl W, Beck C, Hempelmann R (1999) J. Chem. Phys. 110:7070–7072
Huang C, Mason JF (1978) Pro. Natl. Acad. Sci. U.S.A. 75:308–310
Lasic DD (1991) Liposomes: from physics to applications. Elsevier, Amsterdam
Phalakornkul JK, Gast AP, Pecora R, Nägele G, Ferrante A, Mandl-Steininger B, Klein R (1996) Phys. Rev. E 54:661–675
Philipse AP, Vrij A (1988) J. Chem. Phys. 88:6459–6470
Pusey PN, Segrè PN, Behrend OP, Meeker SP, Poon WCK (1997) Physica A 235:1–8
Pusey PN (1991) Liquids, freezing and glass transition. In: Hansen JP, Levesque D, Zinn-Justin J (eds) Volume Session L1 of Les Houches. Elsevier Science Publications, Les Houches, pp 763–942
Riese DO, Ewgdam GH, Vos WL, Sprik R, Fenistein D, Bongaerts JHH, Grübel G (2000) Phys. Rev. Lett. 85:5460–5463
Rojas LF, Vavrin R, Urban C, Kohlbrecher J, Stradner A, Scheffold F, Schurtenberger P (2003) Faraday Discuss. 123:385–400
Segré PN, Behrend OP, Pusey PN (1995) Phys. Rev. E 52:5070–5083
Snook I, Tough RJA (1983) J. Chem. Phys. 78:5825–5836
Stuchly MA, Stuchly SS, Liburdy RP, Rousseau DA (1988) Phys. Med. Biol. 33:1309–1324

Laminar-Turbulent Transition in Stratified Wakes

Patrice Meunier

Abstract This chapter presents experimental and theoretical results on the transition from a laminar to a turbulent wake in a stratified fluid. The case of a cylinder is analysed in detail at low Reynolds number since it gives rise to the famous von Karman vortex street when the Reynolds number exceeds a critical value. This value highly depends on the stratification and on the tilt angle of the cylinder. A moderate stratification tends to suppress the von Karman vortex street, in agreement with the stabilisation of shear flows at high Richardson numbers. However, it is surprising to see that a strong stratification destabilises the flow when the cylinder is tilted. This new von Karman vortex street is allowed because the vortices exhibit horizontal streamlines although the vortices are tilted. The experimental stability diagram obtained by dye visualisations are compared to numerical results. At larger Reynolds numbers, the 2D von Karman vortex street leads to a 3D instability. Shadowgraph visualisations clearly reveal that the unstable mode is similar to the mode A well known in homogeneous cylinder wakes if the cylinder is vertical. This mode seems to be more unstable for moderate stratifications and more stable for strong stratifications. When the cylinder is tilted a new unstable mode appears at moderate Froude numbers, which exhibits thin undulated dark lines. This mode is due to a Kelvin-Helmholtz instability of the critical layer which appears in each tilted vortex of the von Karman street. Finally, at high Reynolds numbers, the wake becomes turbulent in the early stages for the case of a sphere. However, the late stages of the wake exhibit once again a von Karman street of flat horizontal vortices. The size and the velocity of the wake vary algebraically with time. These scaling laws can be predicted by a simple model of turbulent diffusion in the horizontal direction and of viscous diffusion in the vertical direction.

P. Meunier (✉)
Institut de Recherche sur les Phénomènes Hors Equilibre, UMR 6594, CNRS - Aix-Marseille Université, 49 rue F. Joliot Curie, Marseille Cédex 1313384, France
e-mail: meunier@irphe.univ-mrs.fr

J. Klapp and A. Medina (eds.), *Experimental and Computational Fluid Mechanics*, Environmental Science and Engineering, DOI: 10.1007/978-3-319-00116-6_5,

1 Introduction

Despite an extensive number of studies on bluff body wakes in a homogeneous fluid, there has been very few results on wakes in a stratified fluid. The goal of this chapter is to analyse the influence of the stratification on the transition from a laminar to a turbulent wake of a bluff body.

Bluff body wakes have been a main subject of interest for engineers due to their application to terrestrial and naval vehicles, where drag reduction has been a major concern for cars, trains and boats. Bluff body wakes have also direct engineering significance for civil constructions, where the presence of alternate vortices in the wakes may cause structural vibrations, acoustic noise and even resonances, leading to catastrophic failures. Of course, the behaviour of the wake is dependent on the shape of the bluff body. However, both boundary layer separation and shedded vortices are present in the wake of a circular cylinder. This basic geometry is thus often used for fundamental studies on bluff body wakes.

The presence of a continuous stratification is characteristic of geophysical applications, since the atmosphere and oceans are stratified in density. This may have an influence on the large scale wakes of mountains and islands but also on the small scale wakes of submarines and off-shore platforms. How much mixing is caused by islands wakes in oceans? Can meteorological weather forecast simulations take into account mountain wakes? Are submarine wakes detectable? Can the off-shore platforms resist in strong storms and currents? All these questions partly motivate this study which however remains very fundamental.

2 Presentation of the Problem

The experimental set-up for the study of the stratified wake of a cylinder is presented schematically in Fig. 1. The experiments are performed in a 150 cm long, 75 cm wide and 50 cm high Plexiglas tank allowing visualisations from all sides. The tank is filled with a linearly stratified fluid up to a height $Z = 45$ cm. The density profile $\overline{\rho}(Z)$ is established by the two-tank method, using fresh water in the first tank and salt water with a density $\overline{\rho} = 1.15$kg/l in the second tank, leading to a Brunt-Väisälä frequency $N = \sqrt{-(g/\overline{\rho})(\partial\overline{\rho}/\partial Z)}$ close to 2 rad/s.

A circular cylinder of diameter D varying between 0.3 and 1 cm is towed horizontally in the stratified fluid, at a velocity U varying between 0.4 and 4 cm/s. As can be seen on Fig. 1, the cylinder axis is tilted relative to the vertical, at an angle α in the cross-stream plane. A sphere has also been used at large Reynolds number.

The tilted stratified wake of a cylinder is characterized by five non-dimensional parameters: the tilt angle α, the Reynolds number $\mathrm{Re} = UD/\nu$, the Froude number $F = U/ND$, the Schmidt number $Sc = \nu/\kappa$ (κ being the diffusivity of salt in water) and the non-dimensional stratification length L. However, the two last parameters

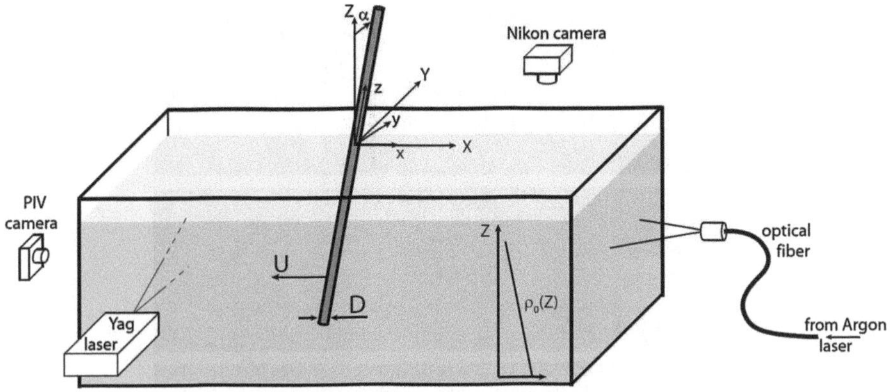

Fig. 1 Schematic of the experimental set-up: the stratified *cylinder* wake is analyzed by dye visualisations on the *right* side and by PIV measurements on the *left* side

will be assumed very large in this study, which reduces the problem to 3 main non-dimensional parameters.

In order to visualize the flow, a fluorescent dye mixture made of Fluorescein is deposited on the upstream side of the cylinder and then advected in the von Karman vortices. A laser sheet allows to reveal the 2D structure of the flow. Shadowgraph visualisations have also been done by putting a light far from the tank, whose rays are deviated inside the tank (due to the variation of the refractive index with the density) and focused inside a camera on the other side of the tank. This allows to reveal the 3D instabilities of the flow. Finally, Particle Image Velocimetry (PIV) measurements have been obtained by seeding water with small reflecting particles.

3 Two-Dimensional von Karman Vortex Street

It is well known that a cylinder wake exhibits alternate vortices above a critical Reynolds number equal to 49.9 (Williamson 1996b). This von Karman vortex street remain discernible in a turbulent wake at very large Reynolds numbers.

Figure 2 shows dye visualisations behind the cylinder tilted with a 30° angle at a moderate Reynolds number ($\mathrm{Re} = 100$). In a homogeneous fluid, it is well known that the wake is unstable at this Reynolds number and leads to a von Karman vortex street. It is indeed what is observed on Fig. 2a because the Froude number is relatively high ($F = 1.8$), i.e. the stratification relatively low.

When the Froude number is decreased to 1.3, the vortex street disappears and is replaced by a stationary recirculation bubble, even though the Reynolds number is still equal to $\mathrm{Re} = 100$. This is very clear on the visualisation of Fig.2b, where the dye rolls-up in the two counter-rotating vortices of the recirculation bubble and then stretches in a long straight line with no apparent sinusoidal perturbation.

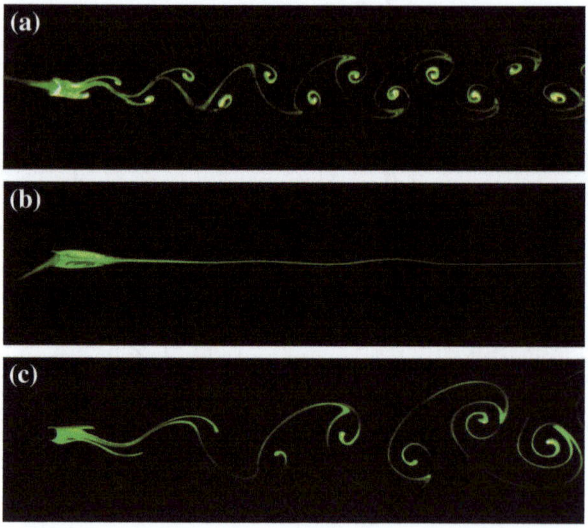

Fig. 2 Dye visualisation of the wake behind a *cylinder tilted* at a 30° angle with respect to the vertical. The pictures are obtained at the same Reynolds number Re $= 100$ and at three different Froude numbers: **a** $F = 1.8$, **b** $F = 1.3$ and **c** $F = 0.8$. The field of view is approximately 35 by 10 diameters in the cross-cut plane (x, y)

Intuitively, the presence of a stratification stabilises the flow because the restoring force tends to attenuate the vertical velocity V created by the tilted vortices (Meunier 2012a, b). This restabilising effect is very similar to the stabilisation of a vertically sheared flow which occurs when the Richardson number is larger than 1/4, as shown by Miles (1961). Indeed, the von Karman vortex street is generated by the two shear layers which detach from the cylinder and which are subject to a shear instability. This is why the wake is stabilised when the Froude number decreases.

However, it is surprising to see that when the Froude number is decreased further, the wake loses this stability and leads once again to a von Karman vortex street as for a homogeneous fluid. This is clearly shown on Fig. 2c for $F = 0.8$, where the dye rolls up again in the vortices shed on each side of the cylinder. This structure seems to be similar to the von Karman vortex street obtained at high Froude number. However, the wavelength is larger, showing that it is in fact a different unstable mode. Indeed, these vortices do not contain any vertical velocity although their axes are tilted (and parallel to the cylinder), meaning that the streamlines are horizontal ellipses.

This new unstable mode is indicated schematically at low Froude number ($F < 1.3$) in the stability diagram of Fig. 3a. It appears above $\text{Re}_c = 45$ at vanishing Froude numbers and at larger Reynolds numbers for larger Froude numbers (up to Re $= 130$). At large Froude numbers, the classical von Karman vortices appear, but they are suppressed by the moderate stratification. These two modes create a strange stability diagram which is more stable only for moderate Froude numbers with a cusp at $F = 1.3$.

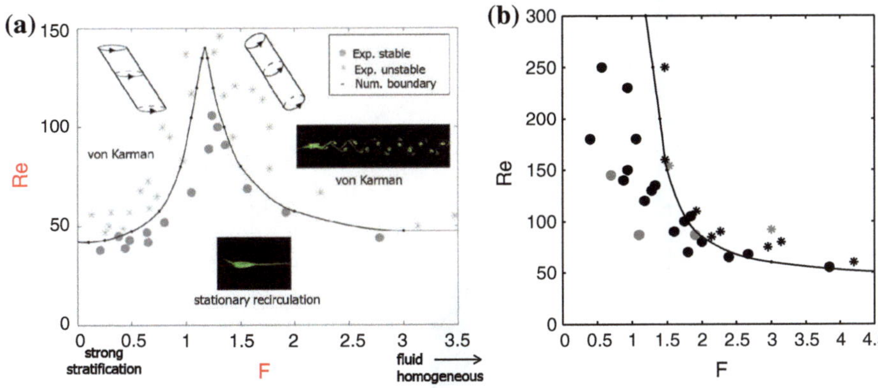

Fig. 3 Stability diagram of the wake of a *cylinder* tilted at **a** $\alpha = 30°$ and **b** $\alpha = 90°$, with respect to the vertical. Symbols correspond to stable (•) and unstable (∗) experiments. The *solid line* corresponds to numerical results. In **b** *Grey symbols* correspond to the experimental results of (Boyer et al. 1989)

However, when the tilt angle α increases, the low Froude number mode is suppressed because the streamlines become more and more elliptical. The low-Froude part of the critical Reynolds number curve is translated toward smaller Froude numbers and eventually disappears for a horizontal cylinder, as shown in Fig. 3b.

4 Three-Dimensional Instabilities

When the Reynolds number increases, the 2D von Karman vortex street becomes unstable with respect to a 3D instability. This transition has been well studied for a homogeneous cylinder wake, where the first mode (called mode A) exhibits counter-rotating vortex pairs perpendicular to the primary von Karman vortices. They are nicely visualised in Fig. 4 by shadowgraph for a weak stratification. The vortex pairs are clearly visible in the front view. The side view highlights the fact that their tails are advected by the cylinder wake. The wavelength is equal to 4 diameters, as in a homogeneous fluid (Williamson 1996a).

The stability diagram of Fig. 5 indicates that this mode becomes more unstable for moderate stratifications. This is surprising because the mode A is due to the elliptic instability of the von Karman vortices (Thompson et al. 2001), whose growth rate σ decreases with the presence of a stratification as noted by Kerswell (2002):

$$\sigma = \frac{9}{16}\varepsilon(1 - \frac{3}{4F_v^2})$$

Here, ε is the ellipticity of the streamlines and $F_v = \Omega(0)/N$ is the vortex Froude number based on the angular velocity at the center of the vortex. We would thus expect

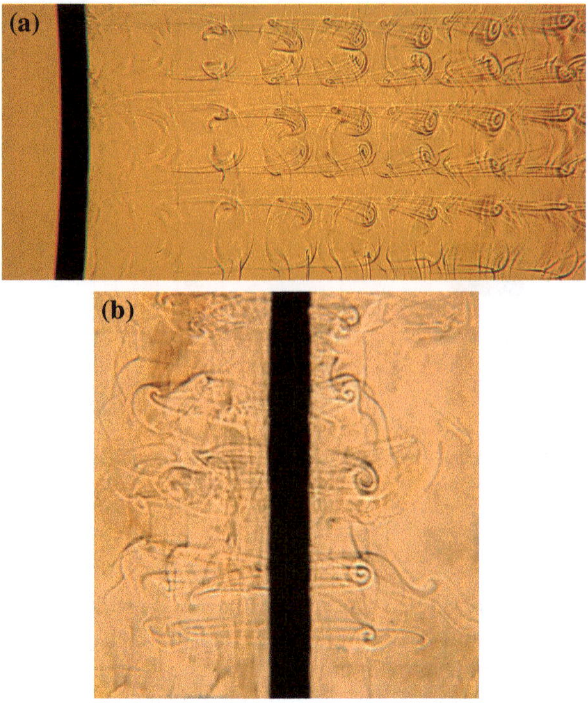

Fig. 4 *Shadowgraph* visualisations of the 3D unstable mode for a *vertical cylinder* ($\alpha = 0°$) in a side view (**a**) and in a front view (**b**). Re = 190, $F = 4$

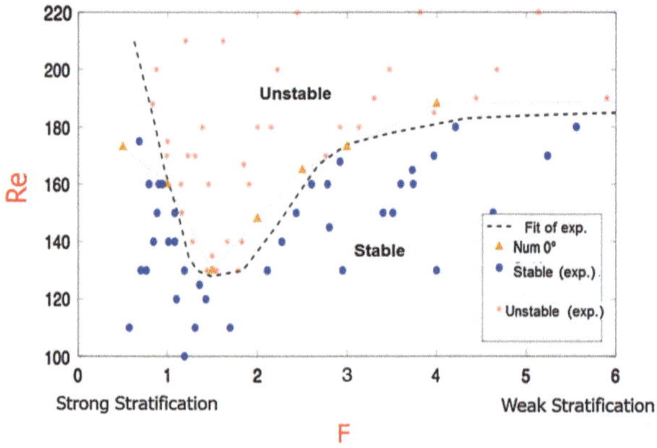

Fig. 5 Stability diagram of the mode A for a *vertical cylinder*

for weak stratification (large F_v) to have a smaller growth rate and thus a larger critical Reynolds number. However, at smaller Froude number ($F < 1.5$), there is a large increase of the critical Reynolds number, probably due to the presence of critical layers in the Kelvin modes of the vortices, as explained by Le Dizès (2008).

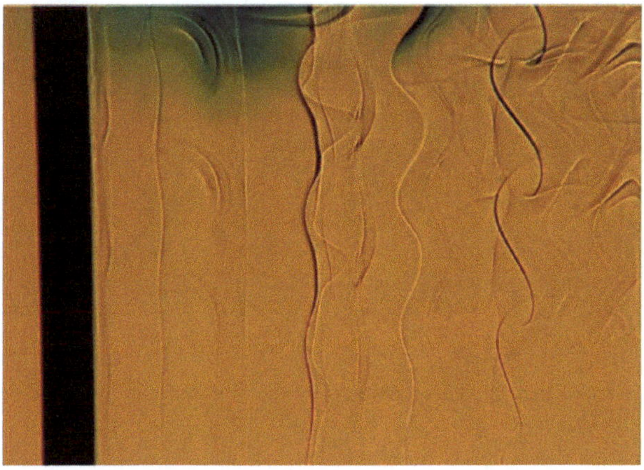

Fig. 6 *Shadowgraph* visualisations of the 3D unstable mode for a *cylinder* tilted at $\alpha = 45°$ in a side view. Re = 180, $F = 2.5$

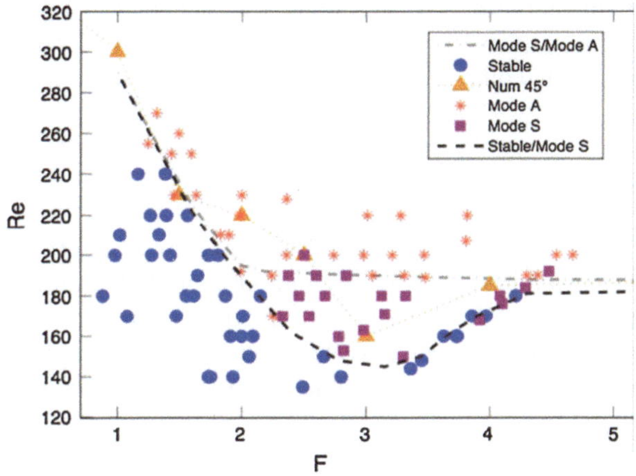

Fig. 7 Stability diagram of the mode S for a *cylinder* tilted at $\alpha = 45°$

The dynamics of the 3D wake is totally different when the cylinder is tilted with respect to the vertical. Indeed, Fig. 6 shows that thin dark and bright lines appear in the wake, and start to undulate above a critical Reynolds number, creating S shaped lines. This new unstable mode (that we called mode S), only appears for moderate Froude numbers (between 2 and 4), as can be seen in Fig. 7.

At these Froude numbers, the tilted vortices exhibit strong axial flows even in the absence of 3D instabilities. This is clearly visible in the 2D numerical simulations presented in Fig. 8a, where mushrooms of strong positive and negative axial velocity surround the vortices. This characteristic structure can be retrieved theoretically by summing the axial flow created by each vortex, as has been done in Fig. 8b. Indeed,

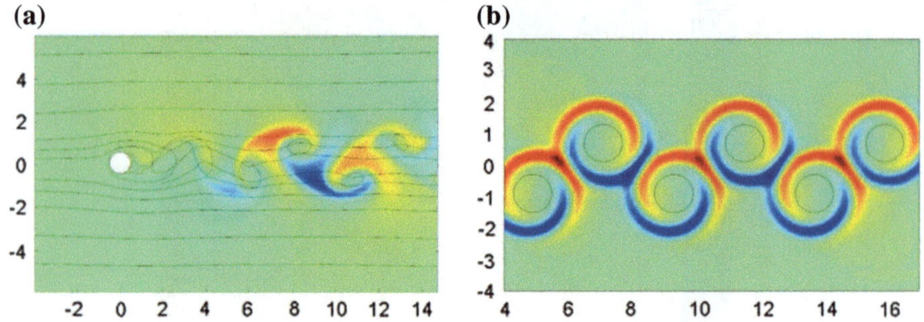

Fig. 8 Axial velocity of the 2D flow generated by the von Karman vortex street (**a**) numerically and (**b**) theoretically at $\alpha = 45°$, $F = 2.5$, $\mathrm{Re} = 180$

each tilted vortex contains a critical layer at the radius r_c where the angular velocity $\Omega(r_c)$ is equal to the buoyancy frequency N. This is due to a resonance of the stratified fluid to the periodic vertical forcing at frequency $\Omega(r)$ which is created by the tilted streamlines. At small tilt angle α, Boulanger et al. (2007) showed that the axial velocity is simply equal to

$$w(r) = \frac{r\Omega(r)^3}{\Omega(r)^2 - N^2}$$

and thus diverges at r_c. Inside the critical layer, viscous effects can be taken into account and lead to an analytic solution with an amplitude scaling as $\mathrm{Re}^{1/3}$ and a thickness scaling as $\mathrm{Re}^{-1/3}$. This structure, plotted in Fig. 8, exhibits two embedded positive and negative circular jets. When summing the critical layers of all the von Karman vortices, we recover the mushroom structures found in the exact 2D numerical simulations.

Boulanger et al. (2008) showed that this critical layer creates a strong shear which is unstable with respect to the Kelvin-Helmholtz instability. This is nicely visualised in Fig. 9, where the thin lines start to undulate leading to Kelvin-Helmholtz billows. The lines exhibit an S shape very similar to the structure of the mode S in the wake of the cylinder, which is visualised in Fig. 6. This explains the origin of this new 3D unstable mode of the cylinder wake, due to Kelvin-Helmholtz instabilities of the critical layers.

5 Turbulent Wake

When the Reynolds number increases further, the wake becomes turbulent with the presence of small-scale fluctuations, especially at large Froude numbers. However, the stratification still plays a major role at late stages when the mean velocity of

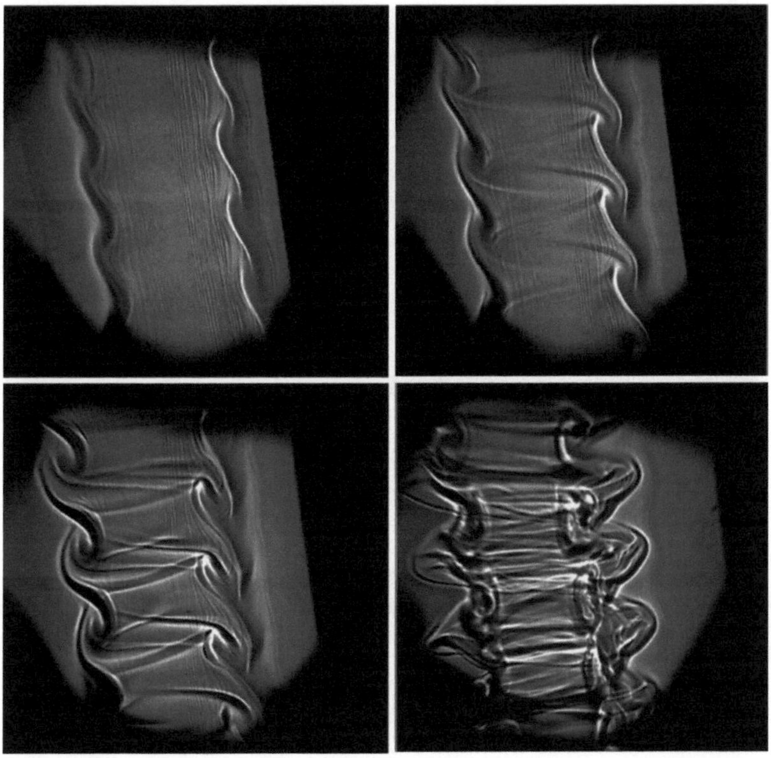

Fig. 9 Temporal development of the tilt-induced instability of a vortex in a stratified fluid tilted at $\alpha = 4°$ visualized by Shadowgraph

the wake has decayed to the characteristic value ND. This leads to the presence of horizontal vortices in order to reduce the vertical velocity, as has been well shown by Lin and Pao (1979) for a sphere wake. The stratification also prevents the extension of the wake in the vertical direction, which imposes a very small aspect ratio of the vortices. As a consequence, the velocity decays slower than in a homogeneous wake because the flow rate is constant.

This can be modeled using a turbulent diffusivity ν_T which depends on time. Indeed, assuming that the Reynolds stresses are proportional to the mean shear, it can be shown that the mean streamwise velocity satisfies a diffusion equation (see Tennekes and Lumley 1972):

$$\frac{\partial u}{\partial t} = \nu_T \left(\frac{\partial^2 u}{\partial y^2} + \frac{\partial^2 u}{\partial z^2} \right) \tag{1}$$

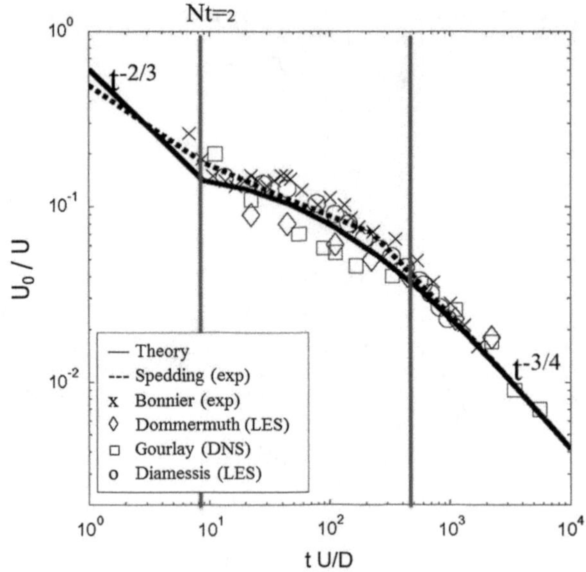

Fig. 10 Mean velocity amplitude of the stratified turbulent wake of a sphere as a function of time

In the homogeneous case, this equation has a Gaussian solution

$$u(y, z, t) = U_0(t)\exp(-\frac{y^2}{L_y^2} - \frac{z^2}{L_z^2}) \tag{2}$$

with an amplitude $U_0(t)$, a horizontal width $L_y(t)$ and a vertical height $L_z(t)$ which depend on time. If the turbulent diffusivity ν_T is proportional to $U_0 L_y$, the amplitude decreases as $t^{-2/3}$ and the width and height increase as $t^{1/3}$. This is indeed observed in homogeneous wakes and in the early stages of stratified wakes, as shown on Fig. 10 for $Nt < 2$. However, at $Nt \sim 1$, the amplitude U_0 becomes close to NL_y and the wake starts to feel the effect of the stratification. The turbulent vertical velocities are suppressed such that the vertical Reynolds stress $< u'w' >$ vanishes. The turbulent diffusive term $\nu_T \partial^2 u/\partial z^2$ must be replaced by a viscous diffusive term $\nu \partial^2 u/\partial z^2$, which is much smaller (Meunier et al. 2006). This explains why the mean profile expands in the horizontal direction but not in the vertical direction. The amplitude U_0 decreases as $t^{-1/2}$ like in a 2D homogeneous wake, leading to a slower decay of the wake. This is clearly visible in Fig. 10 for $Nt > 2$, where the initial decay is drastically slowed down when the stratified effects become important.

However, at very late stages, the viscous diffusion eventually becomes important and leads to an extension of the wake in the vertical direction. This creates a decrease as $t^{-3/4}$ of the amplitude of the wake U_0. This is visible in Fig. 10 for $Nt > 50$ and has been called the quasi-2D regime in the literature (Spedding 1997).

Taking the empirical values of the coefficients between the Reynolds stress and the mean shear, it is possible to predict the exact laws for the velocity amplitude with no fitting parameter. This is plotted in Fig. 10 and compared to all the experimental and numerical data found in the literature. There is an excellent agreement, which indicates that a simple model of turbulent diffusivity can accurately predict the mean characteristics of a stratified wake.

6 Conclusion

This chapter highlights the fact that a density stratification such as the stratification of the atmosphere or the oceans can drastically modify the structure and dynamics of a bluff body wake. At small Reynolds numbers, the transition from a stationary wake to a 2D time-periodic and then to a 3D wake are highly dependent on the Froude number and on the orientation of the bluff body, which gives a very rich dynamics of bifurcations and instabilities. At large Reynolds numbers, the stratification plays a major role in the late stages even for a weak stratification.

All these results indicate that a great care should be taken when using simple models of homogeneous wakes for geophysical applications of mountain or island wakes. The presence of resonances and waves creates a very complex and surprising structure of the wake, which has been very weakly studied in clean laboratory experiments or even in numerical simulations. How are modified the transitions for different bluff bodies? What is the structure and stability of the lee waves for a 3D bluff body? How much mixing is created by these 3D wakes? These are a few questions that will need to be answered for a better understanding and modeling of geophysical wakes.

Acknowledgments Special acknowledgements to Prof. Anne Cros for her invitation to the congress of the División de Dinámica de Fluidos. I also thank the Secretaría de Relaciones Exteriores, Dirección General de Cooperación Educativa y Cultural de México for their financial support. Finally, I would like to thank Prof. Geoff Spedding for introducing me to the study of stratified wakes.

References

Boulanger N, Meunier P, Le Dizès S (2007) Structure of a stratified tilted vortex. J Fluid Mech 583:443–458

Boulanger N, Meunier P, Le Dizès S (2008) Tilt-induced instability of a stratified vortex. J Fluid Mech 596:1–20

Boyer DL, Davies PA, Fernando HJS, Zhang X (1989) Linearly stratified flow past a horizontal circular cylinder. Philos Trans R Soc Lond Ser A 328:501

Le Dizès S (2008) Inviscid waves on a lamb-oseen vortex in a rotating stratified fluid: consequences for the elliptic instability. J Fluid Mech 597:283

Kerswell RR (2002) Elliptical instability. Ann Rev Fluid Mech 34:83–113

Lin JT, Pao YH (1979) Wakes in stratified fluids: a review. Ann Rev Fluid Mech 11:317–338
Meunier P (2012a) Stratified wake of a tilted cylinder. Part 1. Suppression of a von Karman vortex street. J Fluid Mech 699:174–197
Meunier P (2012b) Stratified wake of a tilted cylinder. Part 2. Lee internal waves. J Fluid Mech 699:198–215
Meunier P, Diamessis P, Spedding GR (2006) Self-preservation in stratified momentum wakes. Phys Fluids 18:106601
Miles JW (1961) On the stability of heterogeneous shear flows. J Fluid Mech 10(4):496–508
Spedding GR (1997) The evolution of initially turbulent bluff-body wakes at high internal Froude number. J Fluid Mech 337:283–301
Tennekes H, Lumley JL (1972) A first course in turbulence. M.I.T Press, Cambridge
Thompson M, Leweke T, Williamson C (2001) The physicsl mechanism of transition in bluff body wakes. J Fluids Struct 15:607
Williamson CHK (1996a) Three-dimensional wake transition. J Fluid Mech 328:345–407
Williamson CHK (1996b) Vortex dynamics in the cylinder wake. Ann Rev Fluid Mech 28:477–539

Flows Driven by Harmonic Forcing in Planetary Atmospheres and Cores

Michael Le Bars

Abstract It is a commonly accepted hypothesis that convective motions are responsible for most flows in planetary and stellar fluid layers, and in particular that convective motions are responsible for planetary dynamos, as it is the case on Earth today. However, the validity of the convective dynamo model can be questioned in certain planets. Besides, even in planets where the dynamo is of convective origin, additional driving mechanisms may significantly modify the organization of fluid motions in their core. The same question holds for all large-scale flows in any fluid layer of astrophysical bodies, such as atmospheres of gas giants, subsurface oceans of icy satellites, and convective/radiative zones of stars. In particular, three mechanical forcings present at the planetary scale remain largely unknown regarding their fluid mechanics and planetary consequences: libration, precession, and tidal distortions. Combining analytical studies with numerical simulations and laboratory experiments, we show here that libration and tides can drive highly energetic turbulent flows, which could for instance participate in the generation of Jupiter bands and in the generation of the past Moon magnetic field. The key point is that flows are excited by resonance mechanisms such as the elliptical instability, where the harmonic forcing only acts as a conveyor to extract energy from the huge reservoir related to the rotational dynamics of planetary systems. Even small forcing can thus have important consequences.

1 Introduction

Since the seminal works of the 1960s and 1970s (see for instance Busse 1970; Spiegel 1971, ...), most research on the domain of planetary and stellar flows has focused on convection. In planetary sciences, one of the most significant outcomes of this

M. Le Bars (✉)
Rotating & Geophysical Flows, Institut de Recherche sur les Phénomènes Hors Equilibre (IRPHE), UMR 7342, CNRS - Aix-Marseille University, 49 rue F. Joliot Curie, Marseille Cédex 13 13384 , France
e-mail: lebars@irphe.univ-mrs.fr

J. Klapp and A. Medina (eds.), *Experimental and Computational Fluid Mechanics*, Environmental Science and Engineering, DOI: 10.1007/978-3-319-00116-6_6,

research (see Glatzmaier and Roberts 1995) was to demonstrate that convective flows in a spherical shell indeed generate a dynamo, and that the present magnetic field of the Earth is most probably due to such motions driven by the solidification of its iron core. Since its formulation, this same model has been applied, with only marginal modifications, to other planetary systems. However, the validity of the convective dynamo model can be questioned in certain planets and moons, for instance in Mercury and Ganymede. More generally, it is often tacitly assumed that all motions in fluid layers of astrophysical bodies (e.g. atmospheres of gas giants, subsurface oceans of icy satellites, convective zones of stars, ...) are controlled and driven by convective effects only. For instance Jupiter's bands may be explained by the surface trace of deep convective flows (Heimpel et al. 2005). But in the view of the latest data coming from moons and planets in our solar system as well as from more exotic extrasolar ones (see for instance the fast magnetic inversions in Tau-Boo studied by Donati et al. 2008), it is now high time to re-evaluate these standard models and to explore the role of other instabilities in the organization of fluid motions at the planetary and stellar scales. In particular, three processes are generically present at the planetary or stellar scales, but remain mostly neglected when looking at their driving influence in fluid layers: precession, tidal distortion and libration.

Precession corresponds to the periodic change in the orientation of the rotational axis of a planet. The flow of a rotating viscous incompressible homogeneous fluid in a precessing container has been studied for over one century for both planetary and engineering applications. In the spheroidal geometry, the early work of Poincaré (1910) demonstrated that the flow of an inviscid fluid has a uniform vorticity and takes the form of an inclined solid body rotation called tiltover. But precession can also drive turbulence (e.g. Noir et al. 2003), whose origin remains a matter of debate and the subject of current research.

Tidal distortions come from the gravitational interactions of any planet with its neighbours. They affect all the layers of a given body, which all take an ellipsoidal shape with the long axis oriented towards the deforming body. Tidal distortions are generally time-periodic (i.e. dynamic tides), but they also have a static component in synchronized systems (i.e. static bulges). The most obvious consequences of dynamic tides are of course the oceanic flows on Earth, but they are also responsible for the intense volcanism on Io for instance, and for various flows that we will detail in Sect. 3.

The term longitudinal libration (hereafter called libration) refers to periodic variations in the rotation rate of a planet around its axis. Such oscillations are present in many bodies stacked in a spin–orbit resonance. The determination of the librational motions of a planet allows to better constrain its internal organization (see for example, Margot et al. 2007). Libration may play a fundamental role in the dynamics of planetary fluid layers, for example in the liquid core of Ganymede, in the subsurface ocean of Europa and Titan, or in the atmosphere of hot-Jupiters and in the core of super-Earths in extrasolar systems. This will be shown in Sect. 4.

From a fluid dynamics point of view, libration, precession and tides correspond to closely related mechanisms, which we generically call "harmonic forcing": they correspond to periodic perturbations of an otherwise simple solid-body rotation,

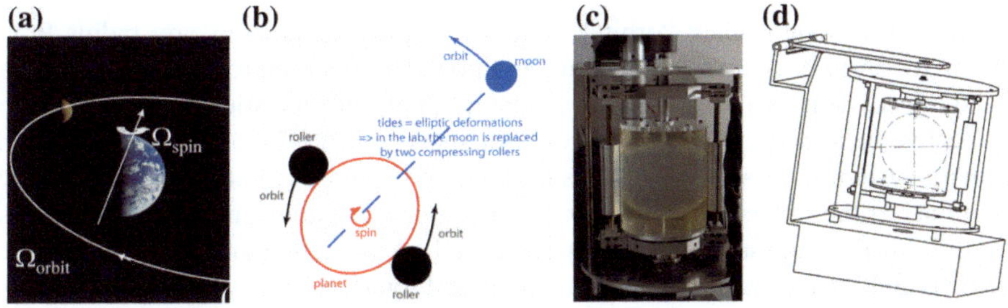

Fig. 1 Experimental set-up. The picture in (**a**) highlights the existence of two different rotation rates: the spin rate of the studied planet, and the orbital rate of the body responsible for the tidal deformation. As shown in the equatorial cross-section (**b**), this companion body is replaced in our set-up by two symmetric rollers. **c** and **d** show the experimental set-up

with an azimuthal wave number m respectively equal to 0, 1 and 2. For planetary systems, it is well known that a huge amount of energy is stored in their rotational dynamics (spin and orbit): those mechanical forcings could play the role of efficient conveyers that extract this energy and drive large scale and intense fluid motions, as initially suggested by Malkus for the Earth (1963, 1968, 1989). Note that the studies of Malkus have long been neglected by the scientific community, mainly because of a misunderstanding (e.g. Rochester et al. 1975): as later mentioned by Kerswell (1996), critiques indeed focused upon establishing the energetic irrelevance of the laminar response to mechanical forcing, rather than considering the fully turbulent case, which is significantly more energetics, thus more relevant for astrophysical bodies. Malkus' studies are now being rehabilitated, in the view of the latest results from space missions and extrasolar systems observations that highlight the need to go beyond the standard models in order to understand the variety of planetary and stellar configurations. My team "Rotating and Geophysical Flows" at IRPHE has largely contributed to this rehabilitation over the past 10 years, and I will shortly present below our main contributions in the field of tides and libration driven flows.

2 Methods

Our studies are based on combined theoretical, experimental and numerical approaches. Our purpose is to describe generic physical mechanisms and to derive and validate generic scaling laws, which are then extended towards planetary applications.

Following the first study by Malkus (1989), we have developed an original experimental set-up to study the tides driven flows in spherical geometry (Fig. 1). In this device, the container consists in a hollow sphere molded into a deformable silicone matrix and filled with water, that is set in rotation. The container is elliptically deformed by a pair of opposed rollers that mimics a tidal deformation rotating at an independent "orbital" velocity. Typical parameters are: radius R = 10 cm, rotation rates up to ±160 rpm, and tidal distorsion up to 5 mm. Our device is equipped with two measuring systems: a simple visualization by Kalliroscope (reflective flake particles) in a vertical laser sheet from the laboratory frame, and a camera system with a wireless transmission, embarked in rotation with either the container or the rollers. This visualization is especially interesting since it allows us to perform PIV measurements in an equatorial cross-section and to determine the velocity field that takes place above the imposed rotation. Two types of experiments have been performed so far with this device, as developed in the next sections:

- studies on the effects of tides for both elliptical instability (Le Bars et al. 2010) and zonal wind generation (Morize et al. 2010) in a tidally deformed sphere;
- studies of libration in a perfect sphere, in which case the rollers are removed but the spin rate is modulated sinusoidally (Sauret et al. 2010).

Numerical simulations have been performed using the commercial code COMSOL Multiphysics (e.g. Cébron et al. 2010a,b,c, 2012a). This software is capable of solving Navier–Stokes equations (and additional physics such as a thermal field or the induction equations) using a finite element method: this allows to deal with complex geometries, such as the triaxial ellipsoidal shape representative of tidally deformed and polarly flattened planets and stars. This constitutes a significant added value compared to previous numerical simulations of planetary cores and stars. Indeed, most other numerical resources use spectral methods and hence suppose an exact axisymmetry of the boundaries of the studied body around its rotation axis, which eliminates most of the interesting dynamics that we want to tackle here. All hydrodynamical and MHD aspects of the numerical simulations have been validated by direct comparison with our previous analytical and experimental results.

3 Tides Generated Flows

As for any rotating flow, fluid layers of planets and stars support oscillatory motions called "inertial waves", whose frequencies range between ± twice the spin frequency. Usually damped by viscosity, these waves can nevertheless be excited by harmonic forcings, and in particular by tides. For instance, Ogilvie and Lin (2004, 2007) showed that in stars and atmospheres of gas giants, tidal forcing might excite inertial waves that significantly alter the energy dissipation of the system. The nonlinear interaction of such forced inertial modes can then generate intense axisymmetric geostrophic jets in the bulk of the fluid, which could for instance participate in the generation of Jupiter's stripes (see e.g. our experimental study Morize et al. (2010) and Fig. 2).

Additionally, tidal forcing induces an elliptical deformation of the rotating streamlines in fluid layers that may excite a parametric resonance of inertial waves called the elliptical instability (see e.g. the review by Kerswell (2002)). Initially motivated by the aeronautical applications of the elliptic instability, our research group at IRPHE has significantly participated in its theoretical, numerical and experimental investi-

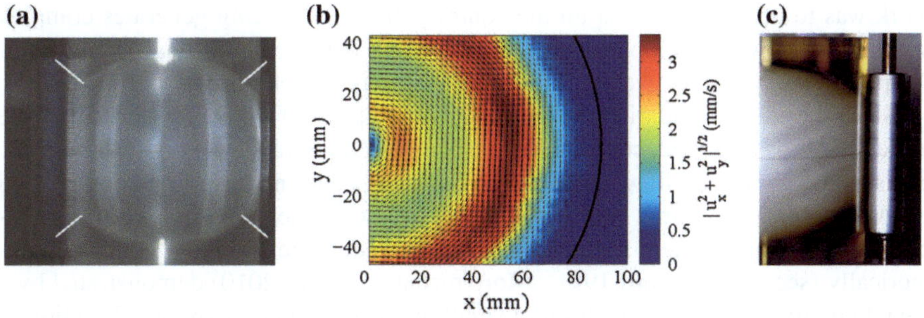

Fig. 2 **a** Axisymmetric geostrophic jet as visualized in our laboratory experiment by Kalliroscope in a meridional plane. **b** Corresponding norm of the horizontal velocity in the equatorial plane, as measured by particle imaging velocimetry (PIV) in the rotating frame. **c** Surface flow visualized by Kalliroscope in the combined presence of tides and precession: the geostrophic flows generate stripes similar to the ones observed for instance on Jupiter. See details in Morize et al. (2010)

gation in spherical containers for planetary and stellar applications (see e.g. Le Bars et al. 2007, 2010; Cébron et al. 2010a). We have also investigated the interaction between elliptic and convective instabilities (Le Bars and Le Dizès 2006; Lavorel and Le Bars 2010; Cébron et al. 2010b). Then, we have studied the response of the flow in a liquid metal when a magnetic field is imposed along the rotation axis. Experimental measurements and numerical simulations have shown the induction of a horizontal magnetic field by the instability as well as its progressive attenuation by Joule dissipation, in perfect agreement with the complete analytical resolution (Lacaze et al. 2006; Herreman et al. 2009; Cébron et al. 2012a). Besides, various chaotic behaviours with excursions and inversions of the induced magnetic field may take place depending on the relative strength of the elliptical forcing and imposed magnetic field (Herreman et al. 2010). Results from these ideal models have been applied to natural systems, showing for example that the early Earth's core was clearly unstable (Cébron et al. 2012b). We have also proposed a complex temporal evolution of binary systems, passing successively through resonance bands of the elliptical instability, separated by stability regions (Le Bars et al. 2010). Lately, we have suggested that the elliptical instability in the Moon's core, temporarily desynchronized by large meteoritic impacts, may be responsible for its early dynamo (Le Bars et al. 2011).

4 Libration Driven Flows

The determination of the librational motions of a planet allows to better constrain its internal organization (see for example, Margot et al. 2007). The main problem of these models is that they tacitly assume the complete absence of specific motions in detected fluid layers outside the viscous Ekman boundary layer. One result of our work was to demonstrate that on the contrary, libration forcing generates complex and intense flows, which participate in the energy balance and orbital dynamics of the planet, and possibly even in the generation of a magnetic field.

Since the solid boundaries surrounding a liquid layer of a planet generically have a triaxial ellipsoid shape, couplings of viscous and topographical origins have to be envisaged. Librational flows generated by viscous coupling have first been studied by Aldridge and Toomre (1969), who observed experimentally that inertial waves can be excited in a sphere at resonant libration frequencies, as latter confirmed numerically (see e.g. Rieutord 1991). More recently, Busse (2010) demonstrated by a weakly nonlinear study in the limit of small libration frequency, that the libration of a sphere generates via the nonlinearities in the Ekman layer, an axisymmetric zonal flow whose amplitude varies as the square of the libration amplitude, regardless of the Ekman number. These predictions were confirmed experimentally and numerically by our group (Sauret et al. 2010), then extended to a more generic configuration (Sauret and Le Dizès 2013). Finally, Noir et al. (2009) have demonstrated experimentally the presence of a centrifugal instability in a librating sphere under the form of Taylor-Görtler type vortices appearing in the viscous layer near the boundary. However, Calkins et al. (2010) emphasised that in the limit of small Ekman numbers relevant to planetary applications, these structures remain localised in the boundary layer and near the equator, and therefore have a very marginal role in the dynamics of the fluid layer. In the view of new numerical results in a cylindrical geometry, we have recently re-evaluated this conclusion by showing a new mechanism of inertial waves generation by the boundary layer turbulence, which could significantly alter the bulk dynamics, no matter what the libration frequency is (Sauret et al. 2012): this mechanism has also been described systematically in the spherical geometry (Sauret et al. 2013).

The libration flows generated by topographic coupling appear to have been even less studied than those with viscous coupling. Recently, Zhang et al. (2011) combined theoretical and numerical approaches, to show the persistence of a zonal wind generation in a triaxial ellipsoidal geometry. They have also suggested that no inertial wave could be forced in this case. However, 4 publications have shown that a hydrodynamic instability of elliptical type driven by librational forcing could develop in triaxial containers (see Fig. 3): the original paper by Kerswell and Malkus (1998) and subsequent contributions of our group (Herreman et al. 2009; Cébron et al. 2012c; Noir et al. 2012). Such a mechanism would be of fundamental importance at the planetary scale because it would generate space-filling turbulence in fluid layers, possibly explaining for instance the magnetic signatures of Io and Europa. This however has to be confirmed by a more systematic experimental and numerical study.

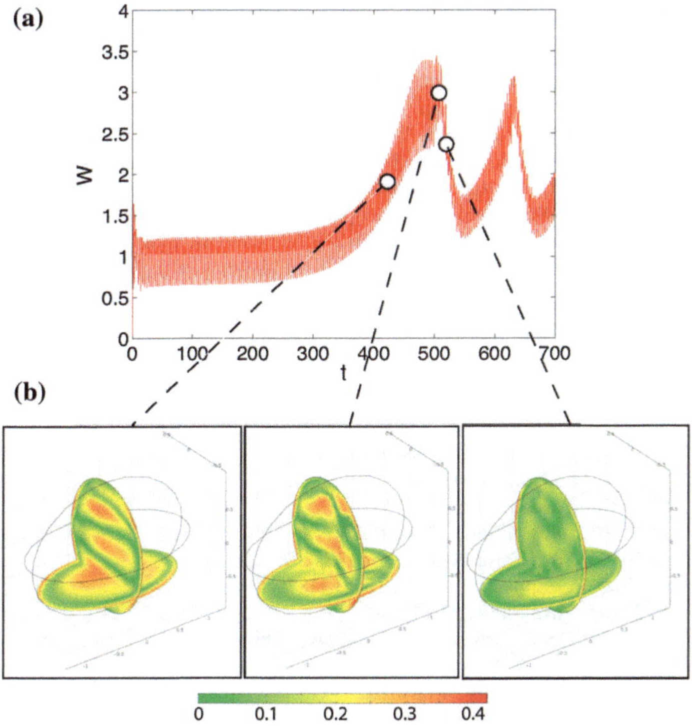

Fig. 3 **a** Time evolution of the absolute value of the axial velocity integrated over the whole container, showing the cycles related to the growth and collapse of the libration driven elliptical instability. Numerical simulations are performed at an Ekman number $E = 5 \times 10^{-4}$. **b** Norm of the velocity field in a meridional cross section and in the plane $z = -0.5$. The sequence shows, from left to right, the typical field during the exponential growth, at saturation and during the collapse. From Cébron et al. (2012c)

5 Conclusions

By combining analytical, experimental and numerical studies, we have shown that "alternative" mechanisms driven by libration, precession and tides, may participate in core and other layers fluid dynamics, where they can replace or significantly perturb usually considered convective motions. For these motions, the source of energy comes from the rotational dynamics of planetary systems (spin and orbit), and the harmonic forcing only acts as a conveyor. Hence, even very small forcing can give rise to intense flows, explaining for instance the existence of strong localized jets when a given inertial wave is resonantly excited (e.g. Jupiter's bands), or the presence of a planetary dynamo when 3D turbulence is excited by elliptical instability (e.g. Moon's dynamo). One should thus remember that on no account, planetary or stellar fluid motions systematically mean convection.

Acknowledgments The work presented here encompasses researches at IRPHE since 2005 with various co-workers that I would like to thank: colleagues at IRPHE Patrice Le Gal, Stéphane Le Dizès and Pierre Maubert; Ph.D. students and postdocs David Cébron, Guillaume Lavorel, Wietze Herreman, Laurent Lacaze, Cyprien Morize and Alban Sauret; and finally collaborators Jon Aurnou (UCLA, USA), Özgur Karatekin (ORB, Belgium), Jérôme Noir (ETH Zurich, Switzerland), M. Rieutord (LAT, France), Andreas Tilgner (University of Göttingen, Germany) and M. Wieczorek (IPGP, France). I also acknowledge continuous support since 2005 from the INSU-SEDIT then INSU-PNP programs (CNRS, France), and support from the Marie Curie International Outgoing Fellowship, which allowed me to spend the 2012-2013 academic year as a visiting researcher at UCLA, USA.

References

Aldridge KD, Toomre A (1969) J Fluid Mech 37:307–323
Busse FH (1970) J Fluid Mech 44:441
Busse FH (2010) J Fluid Mech 650:505–512
Calkins MA et al (2010) Phys Fluids 22:086602
Cébron D et al (2010a) Phys Earth Planet Inter 182:119–128
Cébron D et al (2010b) Geophys J Int 182:1311–1318
Cébron D et al (2010c) Phys Fluids 22:116601
Cébron D et al (2012a) Geophys Astrophys Fluid Dyn 106:524–546
Cébron D et al (2012b) Astron Astrophys 539:A78
Cébron D et al (2012c) Phys Fluids 24:061703
Donati J-F et al (2008) MNRAS 385:1179–1185
Glatzmaier GA, Roberts PH (1995) Nature 377:203
Heimpel M et al (2005) Nature 438:193–196
Herreman W et al (2009) Phys Fluids 21:046602
Herreman W et al (2010) J Fluid Mech 661:130–158
Kerswell RR (1996) J Fluid Mech 321:335
Kerswell RR (2002) Annu Rev Fluid Mech 34:83
Kerswell RR, Malkus WVR (1998) Geophys Res Lett 25:603
Lacaze L et al (2006) Geophys Astrophys Fluid Dyn 100:299–317
Lavorel G, Le Bars M (2010) Phys Fluids 22:114101
Le Bars M, Le Dizès S (2006) J Fluid Mech 563:189–198
Le Bars M et al (2007) J Fluid Mech 585:323
Le Bars M et al (2010) Phys Earth Planet Inter 178:48–55
Le Bars M et al (2011) Nature 479:215–218
Malkus WVR (1963) J Geophys Res 68:2871–2886
Malkus WVR (1968) Science 160:259
Malkus WVR (1989) Geophys Astrophys Fluid Dyn 48:123–134
Margot JL et al (2007) Science 316:710–714
Morize C et al (2010) Phys Rev Lett 104:214501
Noir J et al (2003) Geophys J Int 154:407
Noir J et al (2009) Phys Earth Planet Inter 173:141–152
Noir J et al (2012) Phys Earth Planet Inter 204–205:1–10
Ogilvie GI, Lin DNC (2004) ApJ 610:477
Ogilvie GI, Lin DNC (2007) ApJ 661:1180
Poincaré R (1910) Bull Astron 27:321
Rieutord M (1991) Geophys Astrophys Fluid Dyn 59:185–208
Rochester MG et al (1975) Geophys J Int 43:661–678

Sauret, Le Dizès (2013) J Fluid Mech 718:181–209
Sauret A et al (2010) J Fluid Mech 662:260–268
Sauret A et al (2012) Phys Fluids 24:026603
Sauret A et al (2013) J Fluid Mech 728:R5
Spiegel EA (1971) Annu Rev Astron Astrophys 9:323
Zhang K, Chan KH, Liao X (2011) J Fluid Mech 673:468

Numerical Simulation of Ocean Response by Offshore Wind Stress Events

F. A. Velazquez-Muñoz, J. A. Martínez and R.Durazo

Abstract In this chapter we present some of the features of the wind forced mesoscale current in the Gulf of Tehuantepec, Mexico. We use a three-dimensional numerical model with topography and a realistic wind as the only forcing. We find that starting with "at rest" conditions and horizontally uniform vertical stratification, in a few days ($\sim$20) the main features observed in regional coastal circulation are developed, which establishes the importance of local wind on coastal dynamics. Our results show that the model produces an asymmetrical ocean response on both sides of the wind jet, forming only an anticyclonic eddy on the western side of the Gulf of Tehuantepec. This response is explained by the balance of momentum in terms of external mode equations (vertically average). It was observed that the geostrophic balance dominates and is asymmetrical with respect to the axis of high wind. On the west side, the balance is part of a nearly geostrophic anticyclonic eddy. In the east, the presence of a wide platform and the shore-line, results in an altered geostrophic balance, making other terms important. The terms of lower order of magnitude also showed this asymmetry, notably on the west side where the anticyclonic eddy is formed and along the coast, influenced by the decrease in depth. This adds alternative information to the descriptions given (stationary and uniform offshore wind) of the ocean response to northerly winds and complements the explanation of the asymmetrical results. The wind used to force the model contains much of the temporal variability, therefore allowing the observation of differences in length, intensity and frequency between wind events.

F. A. Velazquez-Muñoz (✉)
Departamento de Física, Universidad de Guadalajara, Guadalajara, Jalisco, México
e-mail: federico.velazquez@red.cucei.udg.mx

J. A. Martínez · R. Durazo
Facultad de Ciencias Marinas, Universidad Autónoma de Baja California, Ensenada, Baja California, México

J. Klapp and A. Medina (eds.), *Experimental and Computational Fluid Mechanics*, Environmental Science and Engineering, DOI: 10.1007/978-3-319-00116-6_7,

1 Introduction

Most of the work in the Gulf of Tehuantepec has been focused on: the study of high winds (Chelton et al. 2000), on the generation of a large number of eddies (Palacios and Bograd 2005; Zamudio et al. 2006) and on the effects of winds over the ocean (Lavín et al. 1992; Trasviña et al. 1995; Velázquez-Muñoz et al. 2011). The wind events have been characterized in terms of the duration, its intensity and the sequential effect when two events occur during relatively short times (Velázquez-Muñoz et al. 2011). Most of the events are of short extent (1–4 days) compared to the local inertial period ($\sim$2 days), however, because of its intensity, the effect on the GT is notable (Trasviña et al. 1995; Velázquez-Muñoz et al. 2011). Wind events have been associated with the formation of eddies at the side of the wind axis (Barton et al. 1993; Trasviña et al. 1995; Trasviña and Barton 2008). From the time when the first remote measurements with satellites were made (Stumpf 1975; Stumpf and Legeckis 1977) till most recent measurements using HF radars (Velázquez-Muñoz et al. 2011; Flores-Vidal et al. 2011), it has been possible to obtain information during wind events. This has served to study the circulation for longer time periods. However, some of these measurements are limited only to the surface layer of the ocean or cover a relatively small area offshore, compared to the scale of influence of wind events.

Moreover, work done to investigate Gulf dynamics by measurements makes difficult to assess the influence of events in an isolated maner. There has been some works using numerical and analytical models (McCreary et al. 1989; Clarke 1988, Chap. 2) that study the response of the ocean to wind stress, trying to simulate the Tehuanos. While these studies provide some physical concepts that control the dynamics forced by wind, they are limited to relatively simple cases, not taking into account the influence of coastline or bathymetry variations. Works like Zamudio et al. (2006)include the heat flux at the surface, forcing at the borders and also the wind forcing, making it difficult to isolate the effect of the dynamic local wind. Other studies with numerical models like Umatani and Yamagata (1991) or, more recently Sun and Yu (2006) use monthly averages to force the model, so that wind events are considerably underestimated, or simply disappear. Even so, they find that there is a strong contribution from the GT to the seasonal variability of eastern tropical Pacific due primarily to wind events, which is even stronger than the contribution from the Gulf of Papagayo.

2 Numerical Model

In this work, we use the nonlinear, hydrostatic and three-dimensional numerical ocean model POM (Princeton Ocean Model) developed by Blumberg and Mellor (1987). This model solves the primitive equations for the momentum and thermodynamic for the conservation of temperature and salinity. The computational solution of numerical model includes the mode splitting technique that separates the problem into two sets:

the first being a set of vertically integrated equations (external mode) and the second set vertical structure equations (internal mode). The equations are solved by spatial and temporal finite difference methods using an implicit scheme in a tri-diagonal metrix wich is solved by a Gaussian elimination method. The original version of this model -coded in Fortran77- was run in a MacPro 8 cores personal computer. The model also incorporates a turbulent mixing scheme (Mellor and Yamada 1982). The vertical coordinates of the model are also called "following terrain", such that at any point of numeric domain, the same number of vertical levels exist. This feature is very useful to represent healthy Ekman layers.

For this study, we chose the simulation period from February 1st to March 18th, 2005, which coincides with measurements described in Velázquez-Muñoz et al. (2011). The considered domain extends from 104° to 82°W in longitude and 3°–19°N in latitude (Fig. 1a). In the open boundaries of the west, south and east, the Orlansky radiation conditions for sea level and velocity components for internal and external mode were applied. For temperature and salinity, an advection boundary condition was used. The initial distribution of temperature and salinity is only a function of the vertical coordinate. The bathymetry was taken from the product ETOPO2 (Geodetic Center, Boulder, Colorado), and was interpolated to a cartesian grid of 5 km horizontal resolution. The number of cells in direction (x, y) is (484, 350). In order to properly resolve the boundary layers, we used 71 sigma levels with a logarithmic distribution near the surface and bottom. Outside the boundary layers, the level distribution is linear. The Coriolis parameter varies as $f = 2\Omega \sin(\varphi)$, where Ω is the angular velocity of the Earth and φ the latitude. The coordinate system is oriented with the positive x-axis to the east and the positive y-axis northward. We use the notation (u, v) and (U, V) to refer to the horizontal three-dimensional and vertically integrated components of current velocity, respectively. Sea level is represented by η.

WE used Cross-Calibrated Multi-Platform (CCMP) of Atlas et al. (1996) for wind speed values for in the entire planet in a uniform matrix array of $0.25° \times 0.25°$ with temporal resolution of 6 h. For this study, we calculate the wind stress following Large and Pond (1981). Several features are evident in Fig. 1a: the offshore wind jet in the GT is narrow, intense and follows an inertial tray after leaving land in the Gulf of Tehuantepec. Another similar wind jet appears almost simultaneously and occurs in the Gulf of Papagayo. The intensity of the wind in the GT is only significant in a strip of about 100 km wide and is weak in adjacent regions along the coast. Towards the south, the influence of events extends for about 400 km. During the winter months, wind events are an important forcing factor on coastal dynamics in the Tropical Pacific region (Barton et al. 2009a). Figure 1b shows the magnitude of wind stress time series at the point of maximum temporal variability in front the Gulf of Tehuantepec. Throughout the 45 days selected for this study, we identify five events where the meridional component of the wind stress reaches significant values towards the South (shaded in Fig. 1b), which can be considered as northern wind events. The first and the most intense event is presented between February 2nd to 6th, reaching a value close to $1.0\,\mathrm{Nm}^{-2}(\sim 20\mathrm{ms}^{-1})$. Between February 10th and 12th, there is a second wind event with half the intensity of the previous one

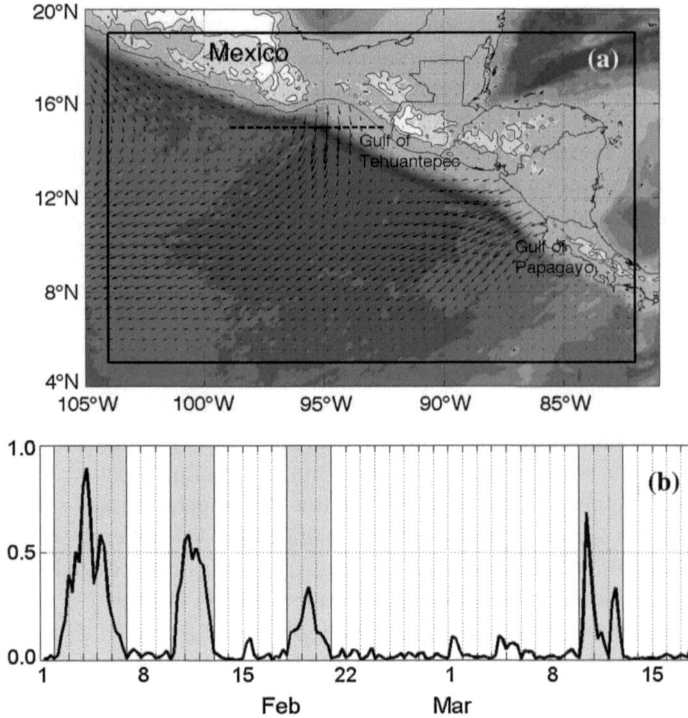

Fig. 1 (**a**) Map of the numerical model domain. *Black arrows* correspond to temporal wind average from February 1 to March 18, 2005. The time series of the magnitude of the wind stress in 95°W and 15°N are shown in (**b**), which corresponds to the maximum wind variability. We can identify four periods with strong wind (*shaded in gray*). The units are [N m^{-2}]

(0.5 Nm^{-2} or 16 ms^{-1}), while the third event is from 18th to 20th of February with 0.4 Nm^{-2}. After a period of relative calm, from February 20th to March 10th, there is a pair of short consecutive events, or pulses, which last 2 days and 1 day, respectively, reaching values of 0.7 Nm^{-2} the first and 0.3 Nm^{-2} the second.

3 Circulation Induced by the Wind

3.1 Surface Currents

Using the results of the last 25 days of simulation, temporal average surface current in Tehuantepec was calculated. Figure 2 shows the time averages of model surface currents (black arrows) and surface current measurements (gray arrows), systems registered with High Frequency Radio and reported by Velázquez-Muñoz et al. (2011). The color scale is the root mean square (rms) of the model time series currents.

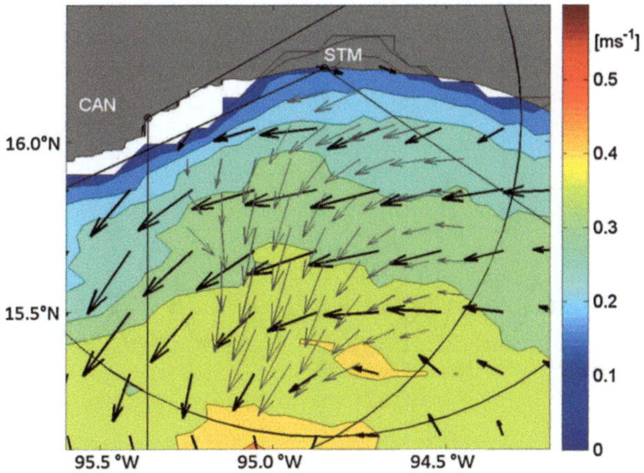

Fig. 2 rms values of surface current on numerical model (*color tones*). *Black arrows* correspond to the time average of surface current on model and the *gray arrows*, surface current reported by Velázquez-Muñoz et al. (2011). In all cases the time period is from February 20 to March 18, 2005

We must emphasize that there is a great similarity in the magnitude and direction between the average surface current model and measurements. In this part of the numerical domain, far from the borders, the sea surface dynamics consist of a coastal current entering from the east that is deflected to the south (offshore). The rms pattern of sea surface current (color in Fig. 2) is also quite similar to measurements (Fig. 2b in Velázquez-Muñoz et al. 2011) having lower values near the shoreline and increasing seaward.

3.2 Kinetic Energy and Relative Vorticity

During the northerly winds, one of the first effects that the wind produces on the ocean is through the development of current (increasing kinetic energy) on the entire surface of the numerical domain, and a way to compact the information is through the quantities integrated over a defined area. In Velázquez-Muñoz et al. (2011), it was shown that the kinetic energy integrated over the HFR coverage area is closely related to the intensity of the wind, and it is suggested that the duration and temporal interval between events can be as important or even more so, than the intensity of the event. Following the calculations of kinetic energy density and of relative vorticity presented in Velázquez-Muñoz et al. (2011), we can make a comparison between measurements and numerical model results. Taking the results of the surface current model, it is possible to calculate the density of kinetic energy (Ek^*) and relative vorticity (ω^*) for the nodes that are located within the HFR coverage area. Figure 3a shows the time series Ek^* and Fig. 3b shows ω^*. In both cases the thin line corresponds to measurements and the wide line, to model data. It can be clearly seen

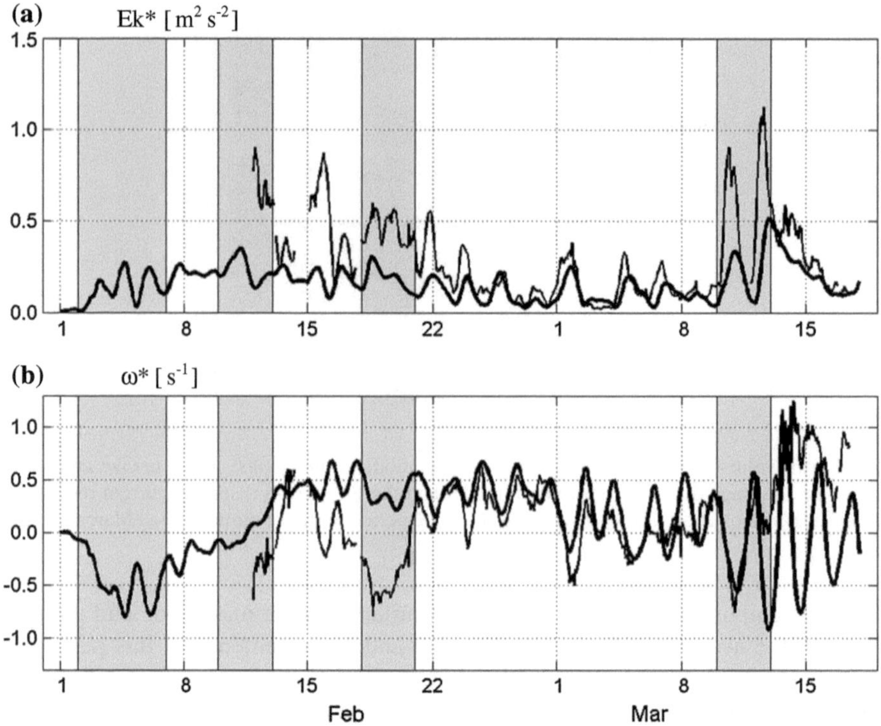

Fig. 3 Time series of density of (**a**) kinetic energy, Ek*, and (**b**) of relative vorticity, ω*, during Feb–Mar 2005. In both cases the *wide lines* show an area from the model and *thin lines* are for measurements

from the *Ek** on February 21st and ω* in the model, that observations are very similar in frequency and amplitude, although the model *Ek** is generally weaker than that obtained from measurements with HFR. From March 10th to 13th, two consecutive events occur: the first one which lasts 2 days, is much stronger than the second one, which lasts for 1 day (see Fig. 1b). Contrary to what might be expected, the increase in *Ek** is greater for the second event (weaker event) than the first one (intense event), as described by Velázquez-Muñoz et al. (2011). The model similarly reproduces the relationship described between *Ek** and the intensity of the wind, so that the relative increase in *Ek** between the first and second events is proportional to observations, although the model underestimates the *Ek**.

The relative vorticity is also satisfactorily reproduced by the model. The data shows increases in negative ω* in the three wind events where measurements were taken (Fig. 3b). For the model, the ω* is negative during the first and last wind events. During the events of February 10 and 21 the correlation between the model and the observations regarding relative vorticity cannot be appreciated well, nor in *Ek**, because the model is still in the initialization period (spinup). After day 22, the correlation improves. During the events of March 10–13, the increase in ω* shows a similar behavior to that of *Ek**, since both increases are lower during the first event than in the second event, that is more intense. The second event lasts for only 1 day and is significantly less than the first one, but the increase in *Ek** and in (negative)ω* are greater. Unlike *Ek** which is generally underestimated by the numerical model, the ω* is well represented by the model. In this case it can be assumed that the differences between observations and model results are due to the lack of other forcing factors in the model, however, the strong similarity between both variables should be noted.

3.3 Sea Level

An outstanding feature of the GT is the emergence of numerous eddies associated with strong winds, which can be identified as circular domes or depressions in the sea level. Figure 4b shows the time average for output sea level taken from the model simulation in the last 25 days and sea level anomaly from AVISO product (http://www.aviso.oceanobs.com) from February 20 to March 18. In both cases a depression is formed in front of the GT and an elevation in the west. Elevation in the pattern coincides with the forming zone of the numerous anticyclonic eddies that emerge in this area. Moreover, the corresponding sea level elevation in satellite observations is shifted westward, away from the coast and is more intense. This elevation is apparently caused by a cyclonic eddy which is already formed on February 1 with the position and size very similar to that obtained with the numerical model, and with westwards movement. Towards the south in the satellite data we can see the influence of the passage of anticyclonic eddies generated in the Gulf of Papagayo. Simultaneously, an eddy formed in Papagayo, enters the domain from the east and travels to the west as shown in Fig. 4a. This takes about two months for the eddy to cross the domain, so that by taking the time average footprint, a passing zone elevation is seen. The numerical model (Fig. 4b) also shows the formation of an eddy in the Gulf of Papagayo, but its spread is slower than in the observations (Fig. 4a). Along the coast, observations show a depression in sea level, which is not present in the model. Due to the short period of model initialization (spinup) a recurrence or residual features in the circulation or in the sea level is not expected. On the other hand, as already mentioned, the continuous passage of eddies generated south of GT, masked the comparison. The open boundary conditions, especially in the south and east do not introduce energy into the numerical domain, so that the continuous passage of coastal trapped waves from the south (Zamudio et al. 2006) and the presence of local low frequency currents, alter the statistics of the observations and are not contained in the model. It is notable that while the model generates eddies, the travel speed of these is much slower than observed and this requires further investigation.

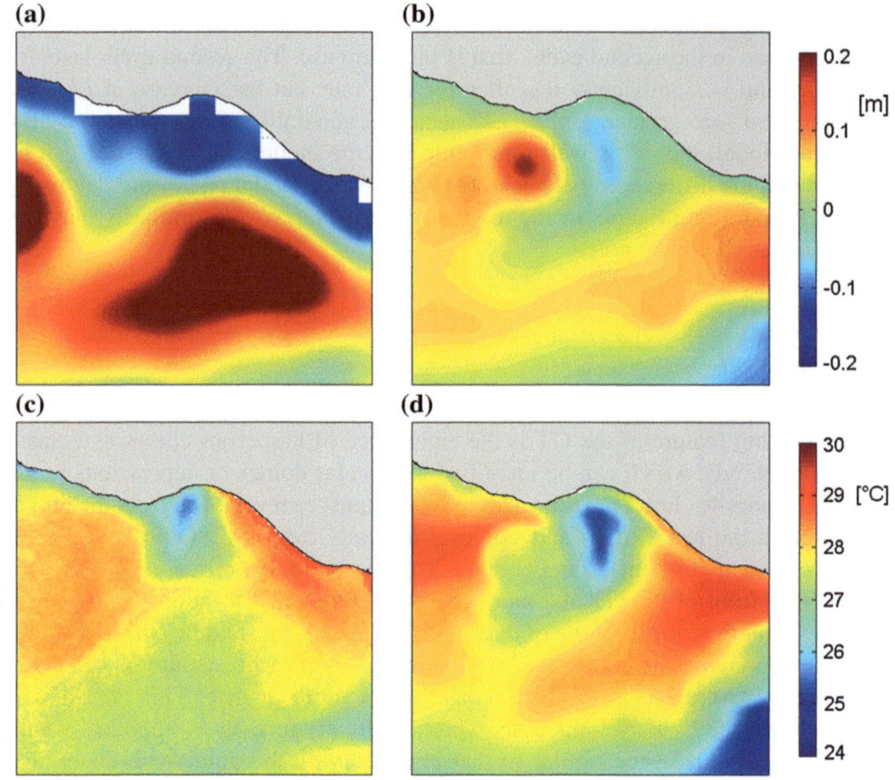

Fig. 4 Time-average of AVISO sea level product (**a**) and numerical model simulation (**b**) and time-averaged sea surface temperature of GOES product (**c**) and numerical model simulation (**d**). All averages were calculated from February 20 to March 18, 2005

3.4 Sea Surface Temperature

An obvious effect of wind stress on the GT is the cooling effect on the sea surface. The lower part of Fig. 4 shows the mean field of GOES-SST product (JPL Physical Oceanography DAAC) (Fig. 4c) and the surface temperature taken from the model (Fig. 4d), between February 20 and March 18, during the last 25 days of simulation. Both variables from the model and satellite show a sea surface cooling on the area just below the maximum wind, which for both is about 4 °C lower than the surrounding water. In both cases it is possible to observe on the east side a thin tongue of warm water extending along the coast to the northern part of the GT. The dimensions of the cold spot in the model and the observations are similar, although the model is slightly shifted eastward. The shift can be attributed to the absence of ambient large scale flows.

4 General Circulation

4.1 Surface Dynamic

The previous sections show that the numerical model reasonably reproduces the observations of surface currents, the density of kinetic energy and of relative vorticity. Temporal changes and spatial distribution of surface temperature and sea level are also well represented by the model. This section describes aspects observed in the general circulation model obtained, which have been reported by Barton et al. (1993) and Trasviña et al. (1995), among others through observations. Figure 5 shows the maps of the daily average temperature (left column in Fig. 5) and sea level (right column in Fig. 5) obtained with the numerical model for days March 3, 8 and 13, 2005. In both cases there is an overlap of black arrows showing the daily average of the surface current. The first feature to note is the presence of an anticyclonic eddy with well-defined circular shape and center in about 97°W, 14°N. It has a diameter of about 200 km moving slowly towards the southwest at $\sim$2.0 km d^{-1} on average. The propagation velocity value is less than 3.5 km d^{-1} reported by Trasviña and Barton (2008) and the 3.3 km d^{-1} reported by McCreary et al. (1989). This anticyclonic eddy can be clearly observed in the surface temperature (Fig. 5a–c) and in sea level (Fig. 5d–f). On the east side, it is possible to observe some cyclonic circulation patterns, but they fail to organize themselves to form a cyclonic eddy. As pointed out by Trasviña et al. (1995), the cyclonic eddy is formed with warm water coming from the west, outside the Gulf (Fig. 5a–c). In the center you can see an area with a cooling surface temperature, which is stretched around the eddy.

In Fig. 5a, b warm water east along the northern gulf is visible. At sea level (Fig. 5d, e) distinguishes a depression covering the central gulf, where circulation is cyclonic (which remains briefly) and extends to the east coast, causing a pressure gradient associated with the aforementioned coastal current.

After consecutive events at the end of the simulation, the characteristic surface temperature and current shows some significant changes. In Fig. 5c one may notice a cooling of $\sim$2 °C in the area of wind influence accompanied by a change in direction of the current, now towards the southwest. Associated with this decrease in surface temperature and the change in current direction, Fig. 5f shows that the depression in sea level becomes smaller.

The results shown here on cyclonic swirling development (Fig. 5a–c) can be compared with Fig. 3 of Barton et al. (1993). They show surface temperature observations and current measurements to identify an anticyclonic eddy with the same characteristics as that obtained with the model. In both cases there is the emergence of a cyclonic eddy in the east, so the question remains as to what factors are involved in the asymmetrical response in the GT and what prevents eddies developing on the east side.

It is not only the emergence of anticyclonic eddy emergency which are the main features that play an important role in asymmetrical ocean response. Barton et al. (2009a) mention the presence of a persistent coastal current entering the east side

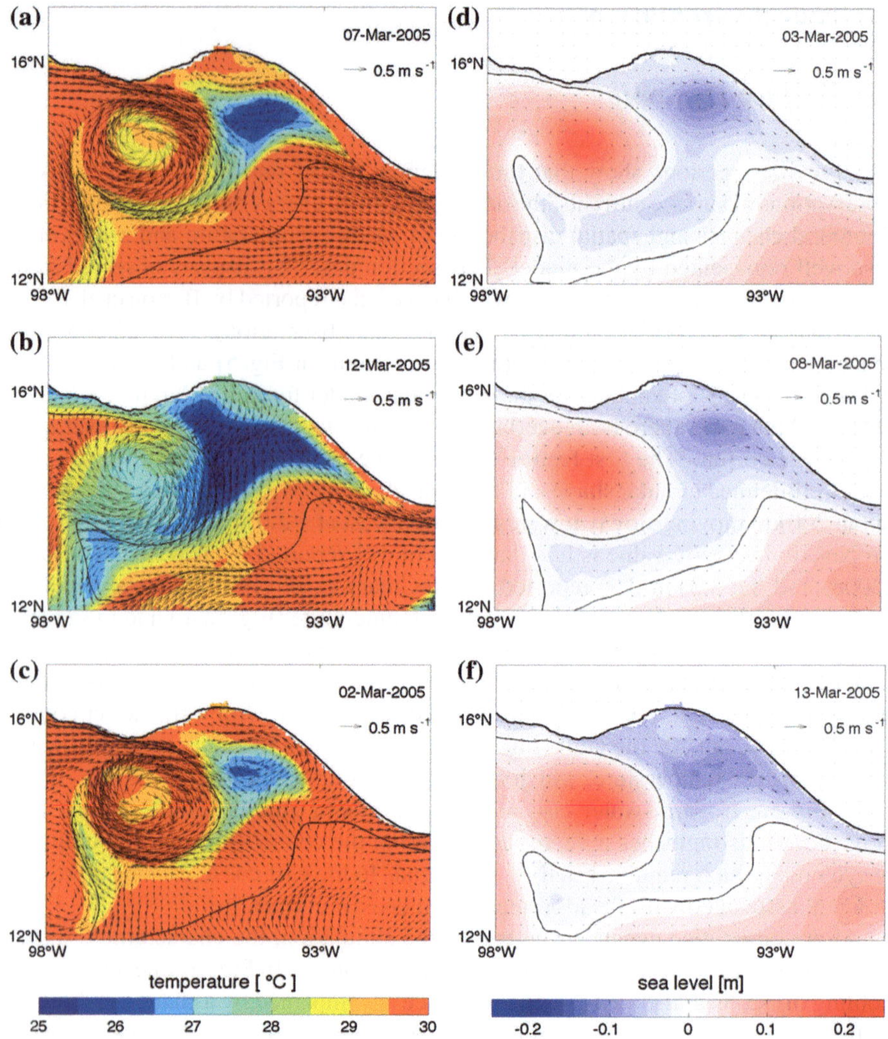

Fig. 5 Daily averages of surface temperature on the *left* (**a-c**) and sea level on the *right* side (**d-f**). In both cases the daily average of sea surface current velocity are overlap with *black arrows*

of the gulf, while on the west side, the coastal current has a strong temporal variablility. Our results show the same patterns on both sides. On the west side we find a weak current of approximately 0.20–0.25 ms^{-1} surface speed on average, while in the east it is higher, reaching values between 0.3 and 0.4 ms^{-1}. To compare the features between the data reported and the results of the model, we calculate the transport in Sverdrups ($1\,Sv = 10^6\,m^3\,s^{-1}$) produced by coastal currents generated. West transport is weak, with about 0.1 Sv, while east transport is much higher, ranging between 0.55 and 0.92 Sv. This result is very similar to that reported by

Barton et al. (2009a), where he obtained a value of 0.87 Sv with ADCP observations and 0.5–0.7 Sv through geostrophic velocity calculations. Surprisingly, the wind in the Gulf of Tehuantepec generates a stream similar to that observed, so it is very important to further explore this coastal current to clarify the factors that cause it.

4.2 Interior Dynamics

Into the ocean, the wind forcing causes upwelling of subsurface water with lower temperature and large amount of chlorophyl concentration as has been reported by several authors (Müller-Karger and Fuentes-Yaco (2000); McClain et al. (2002); Gonzalez-Silvera et al. (2004)) , whom describe the importance of wind to generate eddies that transport biological properties. It is therefore important to understand the mechanisms leading to the surface water which may be trapped by the anticyclonic eddy.

In this simulation we found that just as in the surface layer, the internal ocean responds is promptly. Figure 6 shows the vertical structure of temperature and meridional component of tridimentional velocity in a section parallel to the coast along 14.75°N. Similarly, Fig. 7 shows the temperature and the zonal component of tridimentional velocity, in this case in a section along 95°W. During the period of relative wind calm between March 3 and 8 (Figs. 6a, b) is maintaining a lifting of isotherms, so that isotherms of 22–23 °C reach the surface. Associated with this lifting, the meridional velocity component (v) in the same section parallel to the coast shows a southward flow centered near 95.5°W with opposing flows at both sizes. The flow in the west side is limited to 100 m deep, while the east side is weaker but extends from the surface to 250 m depth. It is notable that the vertical movement of the isotherms and the width of the upwelling is higher east of 96°W to the coast, while the isotherms westward almost maintained their position being narrower sinking area between 97.5 and 96°W. Similarly, the behavior of the southern current componte shows the outward flow of the coast (blue tone in Fig. 6d–f) centered at the maximum of wind at ~95°W. The sections of meridional velocity and temperature can be compared with the measurements shown in Figs. 7 and 9 of Barton et al. (2009a), who describe similar features to those shown here for measurements in winter.

The vertical section of zonal current (Fig. 7d–e) correspond to the days of relative weak wind, show that in the north, the zonal flow is westward (negative) from the subsurface layers up to 250 m depth. South of 15°N the currents is mainly toward east (positive). In temperature, for the same days of calm Fig. 7a–b shows two isotherms upwelling locations, at 13.5 and 15°N. After the two wind events from March 10 to 12, the internal conditions change significantly. It can be seen in Fig. 7c an upwelling very wide, from 13.5° to 15.5°N. At the same time, in Fig. 7f the surface current in the first 50 m has a strong zonal component moving towards the west (negative).

These results again show that the intensity of the wind events is so high that its influence is also reflected in the ocean interior quickly. This results allows us to propose that these characteristics are in temperature and current dynamics of the

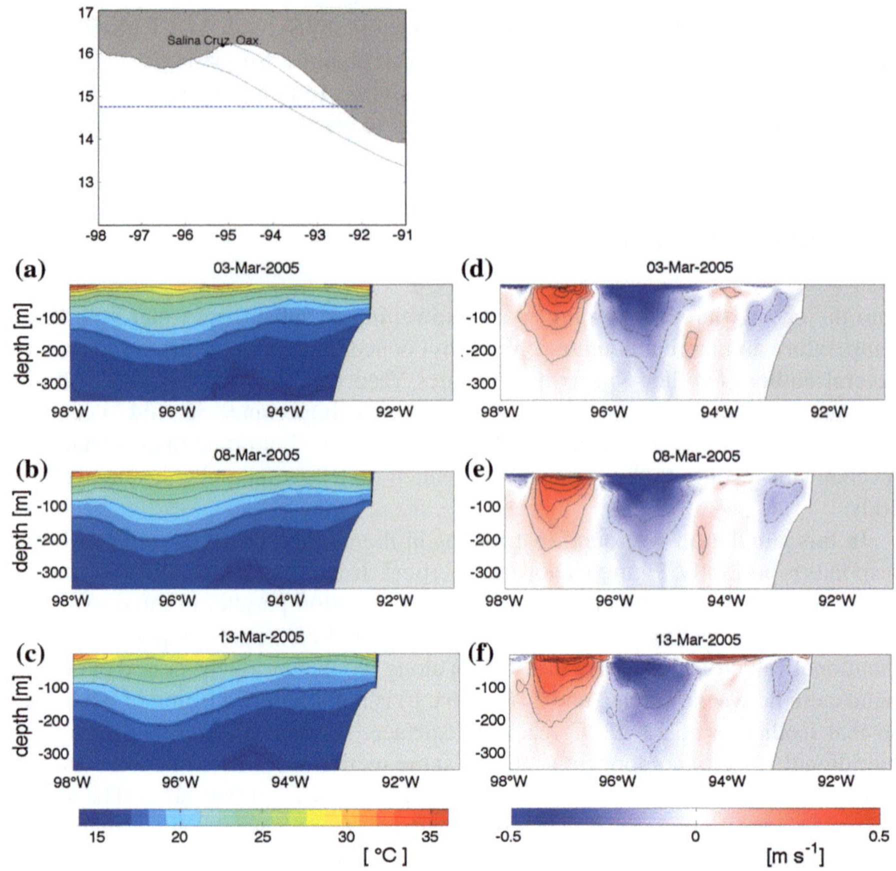

Fig. 6 Zonal section along 14.75°N of temperature (**a–c**) and meridional component of velocity (**d–f**)

gulf, at least in winter, when the main force is the wind and not only during wind events. We can assume that wind events reinforce these shape in temperature and leading upwards the isotherms and forcing the current to offshore. Once the wind event ends, the relaxation process inside the ocean should be long in time, so that the next Nortes not allow a significant change in these conditions.

5 Balance of Terms

This section presents an attempt to add new elements to the conceptual diagram of the ocean response by wind stress forcing in the region of the Gulf of Tehuantepec. The basic explanation of this ocean response to wind events in the Gulf of Tehuantepec

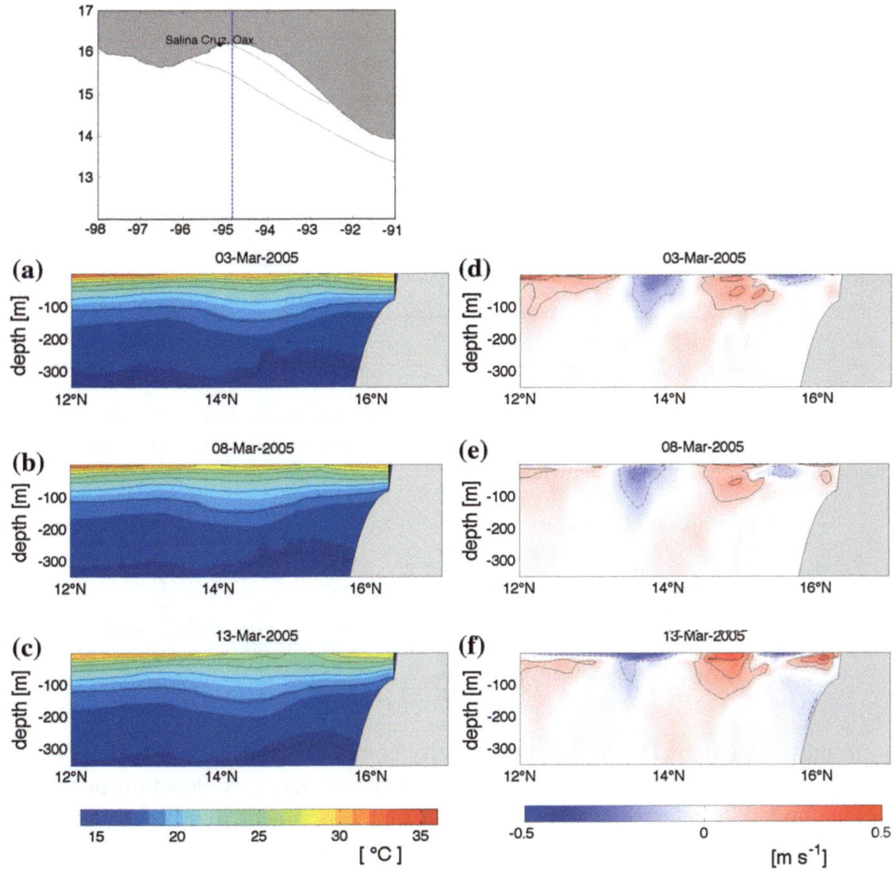

Fig. 7 Meridional section along 95.0°W of temperature (**a–c**) zonal component of velocity (**d–f**)

(Lavín et al. 1992; Willett et al. 2006) has considered only stationary and with wind direction normal to the coast (Clarke 1988; McCreary et al. 1989). It is also considered that the pressure gradient is in geostrophic balance and does not contribute directly to the wind-induced circulation.

In this section we analyze the prevailing balances temporarily averaging terms in the equations for the external mode in the model along a zonal line parallel to the coast. A robust numerical model as the used in this work allows to evaluate the balance of momentum equations terms for realistic simulations. The momentum equations in the vertical or integrated shallow water equations are

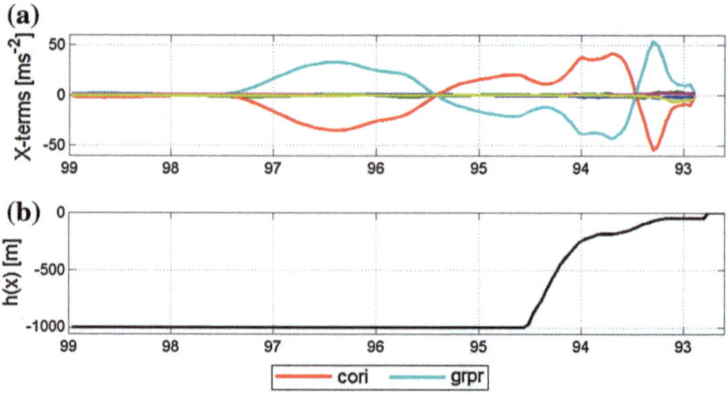

Fig. 8 (**a**) Time average of the terms in first-order zonal direction for the last 25 days of simulation. The lower order terms are shown in detail in Fig. 9. In (**b**) shows the bathymetry profile along 15°N. Horizontal axis correspond to longitude west

$$\overbrace{\frac{\partial U}{\partial t}}^{temp}+\overbrace{U\frac{\partial U}{\partial x}+V\frac{\partial U}{\partial y}}^{advc}-\overbrace{fV}^{cori}+\overbrace{g\frac{\partial \eta}{\partial x}}^{grpr}-\overbrace{\frac{\tau_x^s}{\rho h}}^{surf}+\overbrace{\frac{\tau_x^f}{\rho h}}^{botm}-\overbrace{dif_x}^{diff}=0 \tag{1}$$
$$\frac{\partial V}{\partial t}+U\frac{\partial V}{\partial x}+V\frac{\partial V}{\partial y}+fU+g\frac{\partial \eta}{\partial y}-\frac{\tau_y^s}{\rho h}+\frac{\tau_y^f}{\rho h}-dif_y=0$$

In these equations (U, V) are the velocity components and are functions of the horizontal position and time, f is the Coriolis parameter, η is elevation of the sea surface and g, acceleration of gravity. In a first experiment in which we use flat bottom (Ho = 1,000 m), the terms of geostrophic balance (*cori* and *grpr*) shows a slight asymmetry along a zonal line parallel to the coast. The advective terms are evidently asymmetrical, being important only in the side where cyclonic eddy is formed.

Once incorporated into the model topography variations, *h*(*x*,*y*), the balance of momentum terms changes considerably. It is obvious from Fig. 8 that there is an asymmetric balance. The most important terms in the simulation remain Coriolis and pressure gradient. Figure 8a shows the balance in the direction parallel to the coast (red line: Coriolis and cyan line: pressure gradient). Both terms have sign change at 95.5° and 93.5°W. West of 94.5°W is great depth (Fig. 8b), since the platform narrows north of GT. The topographic changes, coupled with the coastal current which spreads northward along the coast (Fig. 5), seem to compress the eddy formed at east of the GT, as seen in the balance of terms shown in Fig. 8a, of 94.5°W to the east. To the west, the anticyclonic eddy is limited by the presence of the coast, but the depth is still large, and seems "squeezed" to the southwest, where the depth increases rapidly. When moving into deeper water, it is likely that conservation of potential vorticity (vortex stretching) the vortex intensifies, flows whit no alterations by topographic.

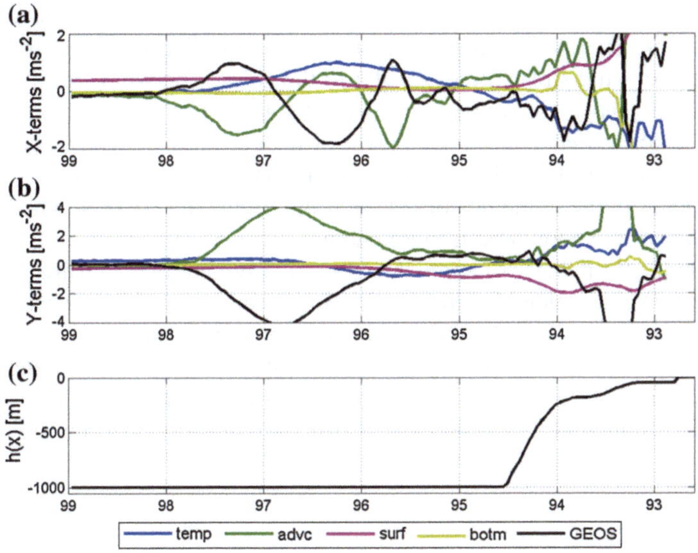

Fig. 9 Time average of the second order terms in (**a**) zonal and (**b**) meridional direction last 25 days of simulation. In (**c**) shows the bathymetry profile along 15°N. Horizontal axis correspond to longitude west

Although dominates geostrophic balance is not identical to zero, which is also shown as a single term that call GEOS (= cori + grpr) which is comparable with the other terms of second order, as shown in Fig. 9 .

When the balance is purely geostrophic, the terms of Coriolis (***cori***) and pressure gradient (***grpr***) in (1) are canceled and the other terms must be zero. But if the balance is not equal cero, the ageostrophic terms should balance each other, as has been discussed by Gan and Allen (2005). The remaining terms of (1) shown in Fig. 9 are: acceleration (***temp***); advection and horizontal diffusion (***advc***); wins stress at surface (***surf***) and bottom friction (***botm***). It also includes the term is the sum of Coriolis and the pressure gradient (***GEOS***).

In Fig. 9, the second-order terms have a very different balance on both sides of the gulf. In the part located where the cyclonic eddy emerge, between 98° and 95°W term ***advc*** balances ***GEOS*** term in meridional direction (Fig. 9b), while in the zonal direction there is a slight contribution of temporal and ***surf*** wind stress terms (Fig. 9a). On the continental shelf on the east side, all terms in zonal direction (Fig. 9a) become important, indicating a very active and complex dynamic, while in southern direction, key terms are ***advc*** and ***GEOS***, a lower ***temp*** participation terms, ***surf*** and ***botm***. In the middle, between 95.5° and 94°W, where the northerly winds acts, the terms are smaller in both directions, so we can consider that the geostrophic balance is more pure.

The time average of wind stress term (***surf***) is not significant across the Gulf of Tehuantepec, but it is important for both sides. West of 96°W wind component along shore is higher that perpendicular component, while at the 95°W both components are significant. This is an important result, especially because most of the works are focused on wind stress perpendicular to the coast. The wind stress is strongest between 93.0° and 94.5°W, where the acceleration term (***temp***) increase mainly whereas GEOS and ***advc*** terms are balanced with each other. This is also the only place where the bottom friction (***botm***) is significant. This shows that the bathymetry is fundamental in the dynamics of GT, as it highlights the friction, which does not occur to the west, where there is more depth.

In summary, the balance of terms shows a clear asymmetry in the direction parallel to the coast. We can identify three areas with different types of balance: 1. To the west the balance between ***GEOS*** and ***advc*** is moderate and have ***temp*** and ***surf*** contribution in zonal direction only; 2. The central part (between 95.5° and 94.5°W), where the terms are lower, making it more pure the geostrophic balance; 3. To the east, on the continental shelf where the coastal current form, all terms are important.

6 Conclusions

In this work we show the result of a three-dimensional numerical model with a realistic configuration to investigate the dynamics forced by the wind stress in the Gulf of Tehuantepec. Unlike previous studies in which the wind field is symmetrical and varies smoothly over time, in this study we used a forcing containing a significant temporal variability by the presence of strong wind events in direction normal to the shore. Furthermore, the model configuration includes the coastline and bathymetry variation from 50 m on the coast up to maximum depth of ~6,000 m.

The results show that from initial condition of rest, in 20 days the main features of the local circulation are developed in the GT during winter, which highlights the importance of the wind stress on local dynamics. The model response shows the formation of two coastal currents flowing into the northern part of gulf. On the East side, the poleward current is persistent, with an approximate width of 100 km and an average surface speed of 0.3–0.4 ms^{-1}. The transport for this coastal current is about 0.55–0.92 Sv, which agrees with the value previously reported by Barton and Trasviña (2009b) using data from ADCP. On the West, the coastal current has a variability associated with wind events and shows sign changes: positive during the beginning of the event and negative or very low after the events. The most important values of transport in this side reaching 0.1 Sv in the northern gulf. After February 20 can be identified an anticyclonic structure of approximately 200 km in diameter on the west side of the GT, which is formed by an incursion of warm water. This anticyclone structure moves at low speed (~2.0 km d^{-1}) compared to the observed eddies in Tehuantepec, so we can assume that current environmental conditions and ambient vorticity play an important role in the drift of the eddies.

The balance terms in momentum equations of the model external mode shows that the geostrophic terms dominate the dynamics in the GT, which have a marked asymmetry in a section along parallel to the coast. This asymmetry is influenced by the continental shelf in the eastern Gulf, where it forms a northward coastal current. In a second order, other terms are compared with the sum of geostrophic terms, that are important over Eastern continental shelf and in the west side, where anticyclonic eddy development.

These results suggest that the continental shelf prevents eddy formation in the eastern gulf. On the opposite side, where the depth rapidly decreases from de coast, the effect of wind induces an anticyclonic eddy, which has not found obstacles to develop.

Acknowledgments The author Velazquez-Muñoz thanks the committee of the XVIII Congreso de la Divisi6n de Dinámica de Fluidos for support to attend the meeting at Ensenada BC. The authors also thank Gerardo Ruiz Chavarría and Jaime Klapp for help with the manuscript preparation.

References

Atlas R, Hoffman RN, Bloom SC, Jusem JC, Ardizzone J (1996) A multiyear global surface wind velocity data set using ssm/i wind observations. Bull Amer Meteor Soc 77:869–882

Barton ED, Argote ML, Brown J, Kosro PM, Lavín M, Robles JM, Smith RL, Trasviña A, Vélez HS (1993) Supersquirt: dynamics of the gulf of Tehuantepec, Mexico. Oceanography 6:23–30

Barton ED, Lavin MF, Trasviña A (2009a) Coastal circulation and hydrography in the Gulf of Tehuantepec, Mexico, during winter. Cont Shelf Res 29:485–500

Barton ED, Lavin MF, Trasviña A (2009b) Near-shore circulation of the Gulf of Tehuantepac under offshore wind. Cont Shelf Res 29:485–500

Blumberg AF, Mellor GL (1987) A description of a three-dimensional coastal ocean circulation model. In: Heaps NS (ed) Three-dimensional coastal ocean models. American Geophysical Union, Washington, pp 1–16

Chelton DB, Michael HF, Steven KE (2000) Satellite observations of the wind jets off the pacific coast of central america. part i: case studies and statistical characteristics. Mon Wea Rev 128:1993–2018

Clarke AJ (1988) Inertial wind path and sea surface temperature patterns near the gulf of Tehuantepec and gulf of Papagayo. J Geophys Res 93(15):15 491–15 501

Flores-Vidal X, Durazo R, Flament P, Chavanne C (2011) Circulación costera en ausencia de viento en el Golfo de Tehuantepec, México: Observaciones con radares de Alta Frecuencia. Aceptado en Revista Ciencias Marinas

Gan J, Allen JS (2005) Modeling upwelling circulation off the Oregon coast, J Geophys Res 110(C10S07), doi:10.1029/2004JC002692

Gonzalez-Silvera A, Santamaria-del-Angel E, Millan-Nuñez R, Manzo-Monroy H (2004) Satellite observations of mesoscale eddies in the Gulfs of Tehuantepec and Papagayo (Eastern Tropical Pacific). Deep-Sea Res II 51:587–600

Large WG, Pond S (1981) Open ocean momentum flux measurements in moderate to strong winds. J Phys Oceanogr 11:324–336

Lavín MF, Robles JM, Argote ML, Barton ED, Smith R, Brown J, Kosro M, Trasviña A, Vélez HS, García J (1992) Física del golfo de Tehuantepec. Ciencia y Desarrollo 103:98–108

McClain CR, Christian JR, Signorini SR, Lewis MR, Asanuma I, Turk D, Dupouy-Douchement C (2002) Satellite ocean-color observations of the tropical Pacific Ocean. Deep-Sea Res II 49:2533–2560

McCreary JP, Lee HS, Enfield DB (1989) The response of the coastal ocean to strong offshore winds: with application to circulations in the Gulfs of Tehuantepec and Papagayo. J Marine Res 47:81–109

Mellor GL, Yamada T (1982) Development of a turbulence closure model for geophysical fluid problems. Rev Geophys Space Phys 20(4):851–875
Müller-Karger FE, Fuentes-Yaco C (2000) Characteristics of wind-generated rings in the eastern tropical Pacific Ocean. J Geophys Res 105(C1):1271–1284
Palacios DM, Bograd SJ (2005) A census of Tehuantepec and Papagayo eddies in the northeastern tropical Pacific. Geophys Res Lett 32
Stumpf HG (1975) Satellite Detection of Upwelling in the Gulf of Tehuantepec Mexico. J Phys Oceanogr 5: 383–388
Stumpf HG, Legeckis RV (1977) Satellite Observations of Mesoscale Eddy Dynamics in the Eastern Tropical Pacific Ocean. J Phys Oceanogr 7: 648–658
Sun F, Yu J-Y (2006) Impacts of Central Americagap winds on the SST annual cycle in the eastern Pacific warm pool. Geophys Res Lett 33(L06710)
Trasviña A, Barton ED, Brown J, Velez HS, Kosro PM, Smith RL (1995) Offshore wind forcing in the Gulf of Tehuantepec,Mexico: the asymmetric circulation. J Geophys Res 100:20,649–20,663
Trasviña A, Barton E (2008) Summer circulation in the Mexican Tropical Pacific. Deep Sea Res I Oceanogr Res Papers 5(5):587–607
Umatani S, Yamagata T (1991) Response of the Eastern Tropical Pacific to Meridional Migration of the ITCZ: The Generation of the Costa Rica Dome. J Phys Oceanogr 21, 346–363
Velázquez-Muñoz FA, Martínez JA, Chavanne C, Durazo R, Flament P (2011) Circulación costera forzada por el viento en el Golfo de Tehuantepec, México. Capítulo 1. Aceptado en Revista Ciencias Marinas
Willett CS, Leben RR, LavínMF (2006) Eddies and tropical instability waves in the eastern tropical pacific: a review. Progr Oceanogr 69(2–4):218–238 (A Rev Eastern Trop Pacific Oceanogr)
Zamudio L, Hurlburt HE, Metzger EJ, Morey SL, O'Brien JJ, Tilburg C, Zavala-Hidalgo J (2006) Interannual variability of Tehuantepec eddies. J Geophys Res 111(C05001):21

Quasi-Steady Endless Vortices with Chaotic Streamlines

Oscar Velasco Fuentes

Abstract This chapter reviews the dynamics of one or more endless vortices in an incompressible inviscid fluid. Each vortex, a thin closed tube lying on the surface of an immaterial torus, is characterised by the number of turns, p, that it makes round the torus symmetry axis and the number of turns, q, that it makes round the torus centerline. Since the vortices are assumed to be identical and evenly distributed on any meridional section of the torus, the flow evolution depends only on the vortex topology (p, q), the number of vortices (n) and the torus thickness $(r_1/r_0$, where r_0 is the centerline radius and r_1 is the cross-section radius). Numerical simulations based on the Biot-Savart law showed that a small number of vortices $(n = 1, 2, 3)$ coiled on a thin torus $(r_1/r_0 \leq 0.16)$ progressed along and rotated around the torus symmetry axis in an almost uniform manner while each vortex approximately preserved its shape. In the comoving frame the velocity field always possesses two stagnation points. The stream tube emanating from the front stagnation point and the stream tube ending at the rear stagnation point intersect along a finite number of stream lines, giving rise to a three-dimensional chaotic tangle. It was found that a single toroidal vortex $\mathcal{V}_{p,q}$ generates a larger chaotic region if it makes less coils round the symmetry axis (smaller p) or if it lies on a thicker torus (larger r_1/r_0). Similarly a set of linked ring vortices $\mathcal{V}_{11}$ generate a larger chaotic region if there are less vortices in the set (smaller n) or if they lie on a thicker torus (larger r_1/r_0).

1 Introduction

An endless vortex is a mass of fluid rotating round a closed curve. The best known example is the *smoke ring*, which is a vortex whose axis of rotation is approximately circular. Ring vortices have been observed for over 400 years in man-made situations, like the firing of cannons or the puffing of tobacco smoke, and possibly longer

O. Velasco Fuentes (✉)
CICESE, Ensenada, México
e-mail: ovelasco@cicese.mx

J. Klapp and A. Medina (eds.), *Experimental and Computational Fluid Mechanics*, Environmental Science and Engineering, DOI: 10.1007/978-3-319-00116-6_8,

in natural situations, like the exhalations of volcanoes and geysers. The earliest references to smoke rings appeared at the dawn of the seventeenth century in the writings of English dramatists, and soon afterwards the first graphical representations of them appeared in the works of Dutch painters (Velasco Fuentes 2013). Despite this ancient interest, it was seemingly not until the beginning of the nineteenth Century that it was recognised that the mysterious smoke rings are in fact vortices. At this time it was even suggested that their motion is responsible for their stability and capacity to carry fluid (B 1804): the "quick rotation of the ring, from within outwards, (...) seems, in some manner or another, as if it kept the parts together." In the following decades the interest in smoke rings increased to such an extent that Helmholtz (1858) and Rogers (1858) published, with only a few months of difference, an analytical study of the motion of *kreisförmige Wirbelfäden* (circular vortex-filaments) and an experimental study of the formation of *rotating rings*, respectively.

A few years later Kelvin (1867a) placed the ring vortex in a prominent scientific position with his hypothesis that matter consists of *vortex atoms* moving in an all-pervading ideal fluid. This turned out to be an erroneous conjecture, but it also proved to be very fruitful for fluid mechanics and mathematics. Indeed, the major advances on vortex dynamics in the following decades were made in the pursuit of the vortex atom theory (e.g., the Kelvin circulation theorem, Kelvin waves on a cylindrical vortex, the stability of a polygon of point vortices, the motion of ring vortices, etc.). And the theory of knots, now one of the most active areas of research in mathematics, received a decisive impulse from the speculations of Kelvin and Peter Guthrie Tait about the shapes that these vortex atoms could take.

One of the first things Kelvin wondered about was the motion of linked ring vortices. On 22 January 1867 he wrote to Helmholtz that he was "a good deal puzzled as to what two vortex-rings through one another would do (how each would move, and how its shape would be influenced by the other)" (Thompson 1910). Kelvin (1875) later deduced that specific configurations of individual as well as multiple endless vortices could rotate and advance uniformly along a fixed line without changing their shape. Kelvin did not give the exact configuration of these steady vortices but hypothesized that they should be thin tubular vortices uniformly coiled on an immaterial torus so that each vortex winds p times around the torus' symmetry axis and q times around the torus' centerline before closing on itself (see Fig. 1).

Building on Kelvin's hypothesis, Thomson (1883) analysed the motion of two or more toroidal vortices. He obtained an approximate analytical expression for the shape and translation speed of two steady, linked vortices of equal circulation. And by considering the limit of infinitely thin vortices lying on the surface of a torus of infinite centerline radius, Thomson (1883) obtained his celebrated result about the stability of a regular polygon of n equal point-vortices.

Almost a century later, Kida (1981) found steady vortex solutions under the local-induction approximation (LIA, an approximation that amounts to omitting distant effects when computing the vortex' self-induced velocity, which then turns out to be proportional to the local curvature of the filament). In the LIA solutions of Kida (1981) the supporting torus may have an oval cross section but steady vortices exist only when $q \geq p$. Ricca et al. (1999) studied the evolution of these vortices under

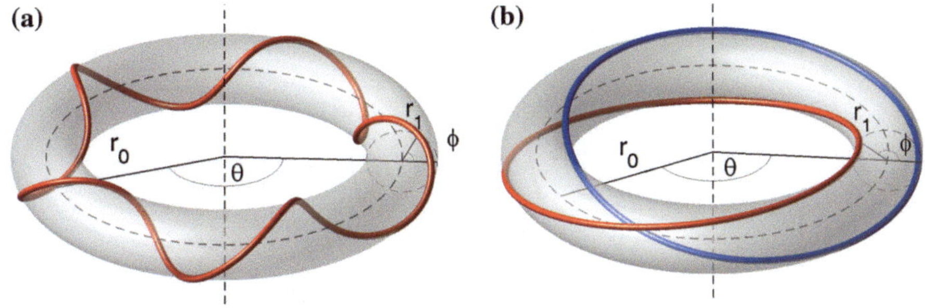

Fig. 1 Thin tubular vortices coiled on an immaterial torus represented by the *grey surface*. **a** A toroidal helix coiled once round the torus symmetry axis and five times round the torus centerline. **b** Two linked ring vortices, each one coiled once round the torus symmetry axis and once around the torus centerline. Right frame adapted from Velasco Fuentes and Romero Arteaga (2011)

both the Biot-Savart law and the LIA: the numerical simulations based on the Biot-Savart law confirmed the hypothesis of Kelvin (1875) whereas those based on LIA were consistent with the analytical results of Kida (1981).

Recently Kleckner and Irvine (2013) succeeded in generating, in a controlled and systematic way, knotted, unknotted and linked vortices in water. Their toroidal helical vortices were stable, the more complicated vortices rapidly became unstable and underwent topological changes through reconnection. The probable cause of these results is that neither the couple of ring vortices nor the trefoil-knot vortex were, in their initial condition, close to the steady solutions hypothesised by Kelvin.

Here we review published results on the dynamics of knotted and unknotted vortices (Velasco Fuentes 2010) and of linked vortices (Velasco Fuentes and Romero Arteaga 2011). Two aspects are central in this review: the vortices' steadiness, that is to say the uniformity of their motion and the constancy of their shapes; and the vortices' capacity to carry fluid, that is to say, the existence of three-dimensional islands of stability surrounding the vortices. In Sect. 2 we discuss the conservation laws discovered by Kelvin, which were the basis for his deductions. The numerical results of Sect. 3 confirm that thin tubular vortices coiled on a torus according to Kelvin's prescriptions are quasi-steady. In Sect. 4 we analyse the velocity field and the transport properties of toroidal vortices. Section 5 contains some conclusions.

2 Integrals of Motion

We assume that the vortices evolve in an inviscid, incompressible, homogeneous fluid which is unbounded and acted upon by conservative forces only. Therefore the kinetic energy, E, and the linear an angular vortex impulses, $\mathbf{I}$ and $\mathbf{A}$ respectively, are invariants of the motion. If all vorticity is concentrated on a single line vortex, these conserved quantities are defined as follows:

$$E = \frac{1}{2}\Gamma \oint \mathbf{u} \cdot \mathbf{R} \times d\mathbf{s} \tag{1}$$

$$\mathbf{I} = \frac{1}{2}\Gamma \oint \mathbf{R} \times d\mathbf{s} = \Gamma \int d\mathbf{S} \tag{2}$$

$$\mathbf{A} = -\frac{1}{2}\Gamma \oint R^2 d\mathbf{s} = \Gamma \int \mathbf{r} \times d\mathbf{S} \tag{3}$$

Here we have used standard notation: the vortex has circulation Γ, moves with velocity $\mathbf{u}$, and lies on the three-dimensional curve $\mathbf{R}(s)$; $d\mathbf{s}$ is a line element along this curve and $d\mathbf{S}$ is the surface element at the point $\mathbf{r}$ of an arbitrary surface spanning the closed curve $\mathbf{R}(s)$.

Kelvin (1869) demonstrated the conservation of linear and angular vortex impulses in the general case, and discovered their geometric meaning when the vorticity is concentrated on a set of filamentary vortices (Kelvin 1875). To achieve this, he first introduced the following definitions:

1. The resultant area of a 3D closed curve is the area of its projection on the plane that makes this projection a maximum.
2. The resultant axis of this curve is the line that passes through its centre of gravity and is perpendicular to the plane of its resultant area.
3. The areal moment of a 2D surface about any axis is equal to its area multiplied by the distance between that axis and the line passing perpendicularly through the surface's centroid.
4. The resultant areal moment of a 3D closed curve is equal to the moment, about the curve's resultant axis, of the areas of its projections on two mutually orthogonal planes that are parallel to this axis.

With these definitions Kelvin (1875) was able to spell out Eqs. (2)–(3) in the form of two theorems:

Theorem 1 *The linear impulse of a curvilinear vortex of unit circulation is equal to its resultant area.*

Theorem 2 *The angular impulse of a curvilinear vortex of unit circulation is equal to its resultant areal moment.*

These theorems, combined with the conservation of linear and angular impulses, allowed Kelvin (1875) to give the following description of the behaviour of a filamentary vortex of arbitrary shape: "the resultant area, and the resultant areal moment of the curve formed by the filament, remain constant however its curve may become contorted; and its resultant axis remains the same line in space. Hence, whatever motions and contortions the vortex filament may experience, if it has any motion of translation through space this motion must be in average along the resultant axis."

Antecedents of these results can be traced back to Helmholtz and Maxwell. The former showed that in a purely azimuthal vorticity field the sum of the projected areas of all ring elements, multiplied by their vorticity, is constant (Helmholtz 1858). The latter, writing to Tait in July 1868, stated that "two ring vortices of any form affect each others area so that the sum of the projection of the two areas on any plane remains constant" (Maxwell and Harman 1995).

2.1 Numerical Method

We compute the vortex motion with the Rosenhead-Moore approximation to the Biot-Savart law (Saffman 1995):

$$\mathbf{u}(\mathbf{x}) = -\frac{\Gamma}{4\pi}\sum_i \oint \frac{[\mathbf{x}-\mathbf{R}_i(s)] \times d\mathbf{s}}{\left(|\mathbf{x}-\mathbf{R}_i(s)|^2 + \mu^2 a^2\right)^{3/2}}, \tag{4}$$

The use of this approximation implies that the vortices are no longer infinitely thin: they now have an undeformable, circular cross-section of radius a. The value of the radius is chosen to be $a = 0.05r_0$, that of the constant μ depends on the vortex internal structure. The particular value used here, $\mu = e^{-3/4}$, corresponds to uniform vorticity on the vortex cross section (Saffman 1995).

In order to evaluate the integral on the right-hand side, we represented each vortex with a set of material markers. We chose the number of markers as $m \approx 2L/a$, where L is the vortex length, and the time step as $dt \approx a^2/\Gamma$, because preliminary tests showed that these values resulted in accurate simulations of the motion of a circular ring, i.e. the shape was preserved and the speed deviated less than 0.5 % from the analytical value. Higher spatial or temporal resolutions substantially increased the computational costs without providing major improvements in the accuracy. Since the vortices evolved without significant changes in length or shape it was not necessary to update the spatial discretization as is usually done in highly time-dependent flows (see, e.g., Baggaley and Barenghi 2011).

We used a fourth-order Runge-Kutta scheme with fixed time step to integrate the evolution equation

$$\frac{d\mathbf{x}_k}{dt} = \mathbf{u}(\mathbf{x}_k, t),$$

where $\mathbf{x}_k$ is the position of the node and $\mathbf{u}(\mathbf{x}_k, t)$ is its velocity, computed with equation (4). Note that k runs through all nodes ($j = 1, \ldots, m$) of all filaments ($i = 1, \ldots, n$). To verify the accuracy of the simulations, we monitored the evolution of the integrals of motion (1)–(3): the energy varied by less than 0.1 % of its initial value; the linear and angular impulses varied by less than 0.001 and 0.1 % of their initial magnitudes, respectively, while their directions, which initially coincided with the torus symmetry axis, deviated from this direction by angles of about 0.0001 s.

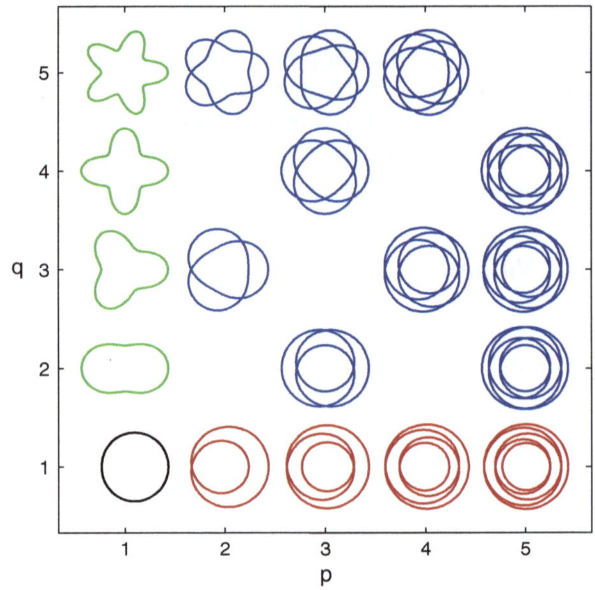

Fig. 2 Topology of toroidal vortices as a function of p and q: ring (*black*), helices (*green*), loops (*red*), and knots (*blue*)

2.2 Initial Conditions and Parameter Space

Following Kelvin (1875) we assume that a number of identical tubular vortices of small cross-section are uniformly coiled on a torus whose centerline has a radius r_0 and whose cross-section has a radius r_1. In Cartesian coordinates the centerline of the vortex is given as follows:

$$
\begin{aligned}
x &= (r_0 + r_1 \cos\phi)\cos\theta, \\
y &= (r_0 + r_1 \cos\phi)\sin\theta, \\
z &= r_1 \sin\phi.
\end{aligned}
$$

where ϕ is the angle round the torus centerline and θ is the angle round the torus symmetry axis (see Fig. 1). They are given by $\phi = qs - 2(n-i)\pi/n$ and $\theta = ps$, where n is the number of vortices, i indicates the vortex being described, p and q are co-prime integers and s is a parameter in the range $0 - 2\pi$. Therefore, before closing on itself, each vortex $\mathscr{V}_{p,q}$ makes p turns round the torus symmetry axis and q turns round the torus centerline. These numbers determine the topology of the vortex, as follows: when $p > 1$ and $q > 1$ the vortex forms a toroidal knot, when either $p = 1$ or $q = 1$ the vortex forms a toroidal unknot (see Fig. 2). In the latter situation it is useful to make a further distinction between toroidal helices ($p = 1$ and $q > 1$), and toroidal loops ($p > 1$ and $q = 1$).

When there are two or more vortices in a given configuration, all of them have the same circulation, Γ, and topology, p, q. Equal circulations are necessary for the steadiness of motion; equal topologies are necessary to avoid an intersection of the vortices, which would be in violation of the condition $\nabla \cdot \omega = 0$. Note also that the term $2(n-i)\pi/n$ ensures that the vortices are equally spaced on the section of the torus (i.e., they intersect any meridional plane on the vertices of a regular polygon inscribed on the corresponding cross-section of the torus).

Here we discuss the dynamics of a small number ($n < 5$) of toroidal vortices ($\mathscr{V}_{pq}$) coiled on thin tori ($r_1/r_0 < 0.16$). We must further set a lower bound for r_1/r_0 because of the desingularization of the Biot-Savart law, which implies that the vortices have an undeformable cross-section of radius a. Consistency then requires that the vortices are never too close to each other, i.e. their centerlines must be separated by distances about or larger than $3a$. We chose to use a value which amply satisfies this condition for $n = 2$ and narrowly does it for $n = 4$. Therefore in this study the aspect ratio of the torus will be in the range $0.1 < r_1/r_0 < 0.16$ (except for one case in Sect. 4).

3 Vortex Motion

3.1 Knotted and Unknotted Vortices

Figure 3 shows the evolution of a trefoil-knot vortex ($\mathscr{V}_{23}$). The vortex, initially coiled on a thin torus ($r_1/r_0 = 0.1$), progresses along the torus' symmetry axis (the thin line in the lateral view) while rotating around the same axis (the cross in the front view). As predicted by Kelvin (1875), all vortices $\mathscr{V}_{pq}$ coiled on thin tori are observed to progress and rotate in an approximately uniform manner. The linear speed, U, is proportional to p and is almost unaffected by the value of q (Fig. 4, left panel). This behaviour is easily explained as follows: since $r_1/r_0 = 0.1$ the progression speed behaves as if there was a single ring with circulation $p\Gamma$ instead of p loops of a filament with circulation Γ. As a matter of fact $U \approx 3/4pU_0$, where U_0 is the speed of a circular ring of strength Γ and radius r_0 (Kelvin 1867b):

$$U_0 = \frac{\Gamma}{4\pi r_0}\left[\log\frac{8r_0}{a} - \frac{1}{4}\right] \tag{5}$$

The angular speed, Ω, grows with increasing p and decreasing q (Fig. 4, right panel). Note that toroidal helices ($p = 1$) rotate in the opposite sense and at a much lower rate than toroidal loops ($q = 1$); and toroidal knots rotate in the same sense and at lower rates than toroidal loops.

3.2 Linked Vortices

Figure 5 shows the evolution of a pair of linked ring vortices $\mathscr{V}_{11}$ of equal circulation. The vortices were initially coiled on a thin torus ($r_1/r_0 = 0.1$). The lateral

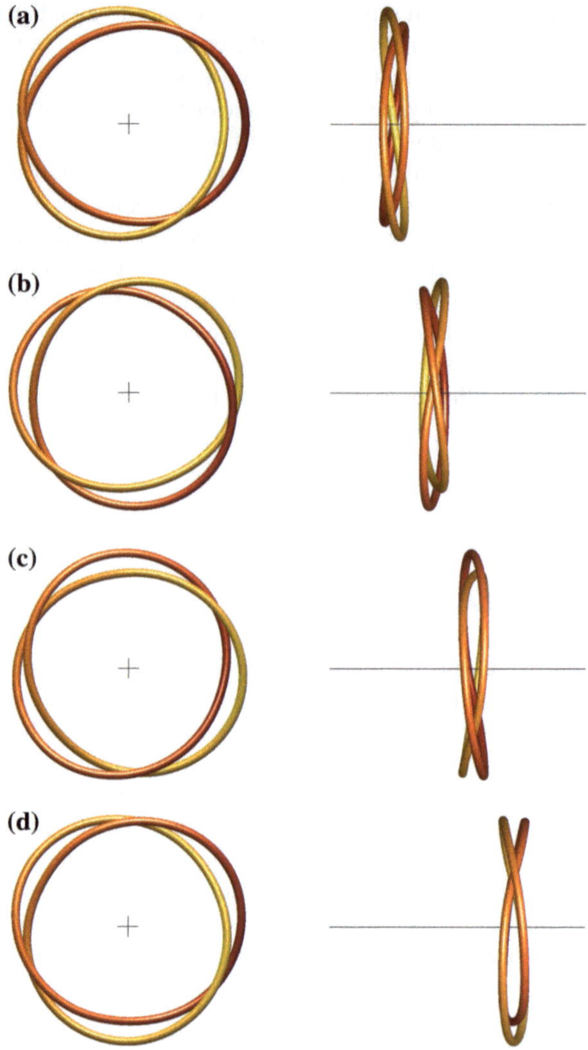

Fig. 3 Evolution of a trefoil-knot vortex $\mathscr{V}_{23}$ with $r_1/r_0 = 0.1$. The axis of the system is represented by a cross in the frontal view (*left-hand side column*) and by a straight line in the lateral view (right-hand side column). The stages depicted are (**a**) $t = 0$, (**b**) $t = 0.23T$, (**c**) $t = 0.46T$, (**d**) $t = 0.69T$, where T is the time required by a circular ring vortex of centerline radius r_0 and cross-section radius a to advance a distance equal to r_0

view shows the progression of the vortices along the torus' symmetry axis whereas the front view shows the rotation of the vortices around the same axis. This figure shows exactly one vortex rotation and since this is relatively fast, the vortices are seen to advance only a short distance during this time. They, however, continue rotating and progressing in the same way for much longer times. Figure 6, for example, shows the vortices advancing a distance equal to eight-times their diameter while

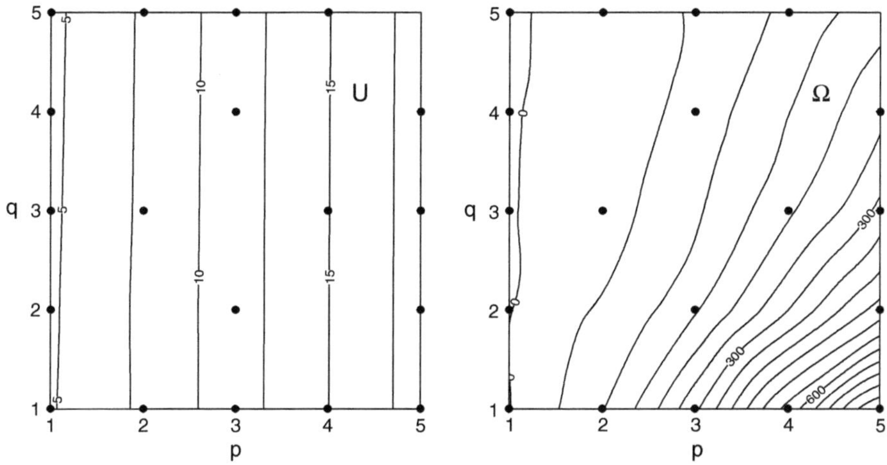

Fig. 4 Linear speed (U) and angular speed (Ω) of the vortex system as functions of p and q. The speed of linear motion along the torus axis, U, is scaled by $\Gamma/4\pi r_0$; the angular speed of rotation round the torus axis, Ω, is scaled by $\Gamma/2\pi r_0 r_1$. The *black dots* indicate the only points where toroidal vortices exist. The contours of U and Ω were drawn by interpolation to show how these speeds change in the parameter space. Taken from Velasco Fuentes (2010)

performing almost sixteen rotations around their symmetry axis. The bottom row of the same figure shows the corresponding time evolution of quantities that, theoretically, should be conserved but which are not exactly so in the numerical simulations. Instead of the instantaneous values of the energy and the linear and angular impulses, Eqs. (1)–(3), we plotted their relative change; thus Fig. 6 shows, respectively, $E(t)/E(0) - 1$, $|\mathbf{I}(t)|/|\mathbf{I}(0)| - 1$ and $|\mathbf{A}(t)|/|\mathbf{A}(0)| - 1$. In the period shown, the energy is preserved within 0.01 %, the linear impulse within 0.001 %, and the angular impulse within 0.02 %.

The progression of the vortices corresponds, because of Helmholtz (1858) vortex laws, with the advance of material elements. The vortex rotation around the symmetry axis does not match a similar motion of material elements: it is actually an azimuthal wave. To verify this, note that the hue of the colour marks fluid elements along each vortex and that, in the front view, the darker hues remain on the right-hand side and the lighter ones on the left-hand side of the vortices. The cause of the azimuthal wave is a different motion of the material elements, namely their rotation around the torus centerline. This can be qualitatively verified by close inspection of the vortices' lateral view in Fig. 6.

Hence the motion of the fluid elements that make up the vortices has two main components: (a) progression along the torus' symmetry axis, and (b) rotation around the torus' centerline. We found that these components are approximately uniform so that we characterised them by the average speeds U and Ω_c, respectively.

The linear speed U grows with the number of vortices n and decreases with the aspect ratio r_1/r_0. A simple argument accounts for this: since $r_1/r_0 << 1$ the

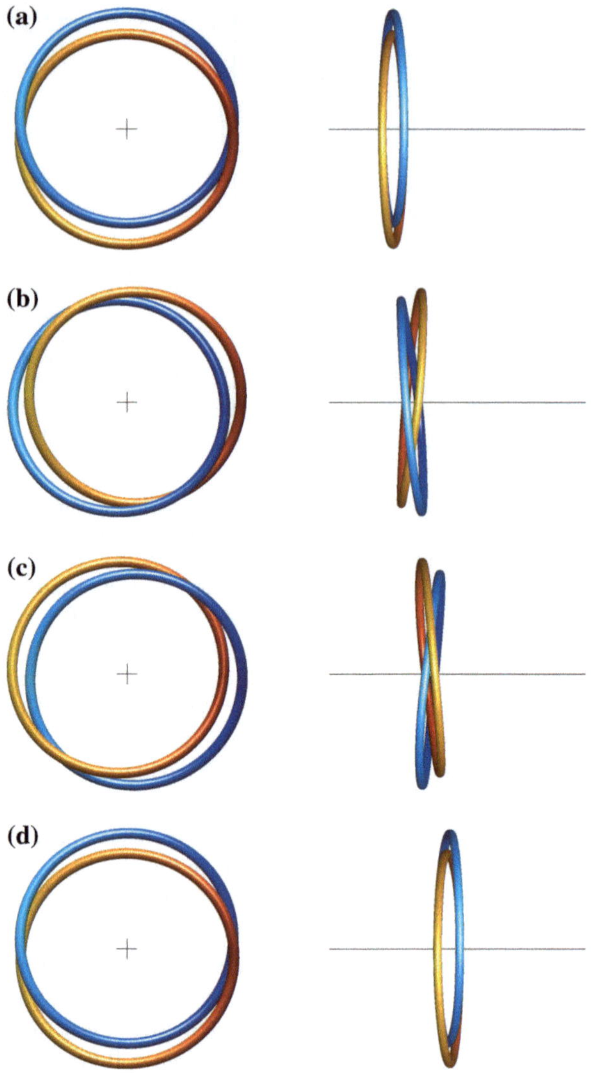

Fig. 5 Evolution of two linked ring vortices $\mathcal{V}_{11}$ with $r_1/r_0 = 0.1$. The axis of the system is represented by a cross in the frontal view (*left-hand side column*) and by a straight line in the lateral view (*right-hand side column*). The stages depicted are (**a**) $t = 0$, (**b**) $t = 0.105T$, (**c**) $t = 0.210T$, (**d**) $t = 0.315T$, where T is the time required by a circular ring vortex of centerline radius r_0 and cross-section radius a to advance a distance equal to r_0. Taken from Velasco Fuentes and Romero Arteaga (2011)

progression speed behaves as if, instead of n toroidal rings with circulation Γ, there was a single circular ring with cross-section radius r_1 and circulation $n\Gamma$. The speed of this virtual vortex is $U_0 = (n\Gamma/4\pi r_0)[\log(8r_0/r_1) - 1/4]$. Figure 7 shows that

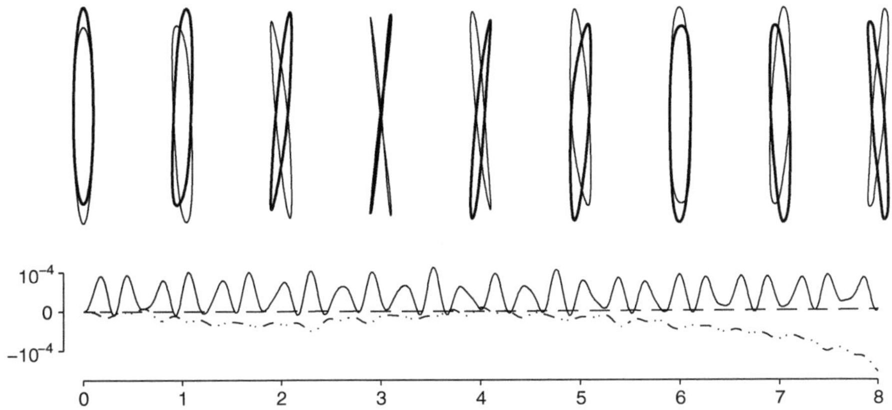

Fig. 6 *Top* Progressive motion of a pair of linked vortices (the same of Fig. 5 but for a 16-times longer period). *Bottom* Time evolution of the kinetic energy E (*continuous line*), linear impulse **I** (*dashed line*), and angular impulse **A** (*dot-dashed line*). The relative change of these quantities (see text) is shown as a function of the adimensional distance travelled by the vortices, $Z = Ut/r_0$ (where U is the speed of the vortices, t is the time and r_0 is the torus' centerline radius). Taken from Velasco Fuentes and Romero Arteaga (2011)

this is in good agreement with the speeds measured for sets of linked ring vortices, particularly when $n = 2$.

The angular speed Ω_c increases with n and decreases with r_1/r_0. This can be explained following (Thomson 1883): since the vortices are thin and $r_1/r_0 << 1$ they move on the meridional plane as if they were a set of point vortices. Indeed, in the parameter region studied here, $\Omega_c \approx 0.94\Omega_0$, where Ω_0 is the angular speed of a set of n point vortices of circulation Γ placed on the vertices of a regular polygon inscribed on a circle of radius r_1: $\Omega_0 = (n-1)\Gamma/4\pi r_1{}^2$. We argued above that the material rotation around the torus' centerline causes the azimuthal wave around the torus' symmetry axis. The close agreement, shown in Fig. 7, between Ω_0 and the angular speed of the azimuthal wave, Ω, quantitatively demonstrates the connection between these two rotations.

We applied several diagnostics to measure the deformation of the vortices throughout their evolution. The simplest one was the time evolution of the vortex length, which was observed to vary within 0.3 % of its initial value in the region of the parameter space studied here ($n = 2, 3$ and $0.1 \leq r_1/r_0 \leq 0.16$). The second diagnostic consisted in finding the torus that best fitted the vortices at every stage of the evolution. The conservation laws (2)–(3) guarantee that the fitting torus has the same symmetry axis as the initial one, therefore the former is uniquely determined by the radii $r_0(t)$ and $r_1(t)$. We found that $r_0(t)$ remained within 1 % of its initial value, whereas $r_1(t)$ remained within 5 % of its initial value. The final diagnostic was to measure the signed distance, Δr_1, from the surface of the torus to every material marker representing the vortices. The time series of histograms of Δr_1 showed that the markers remained within a distance $0.01r_0$ of the torus (as in the case $n = 2$, $r_1/r_0 = 0.1$, shown in Fig. 8).

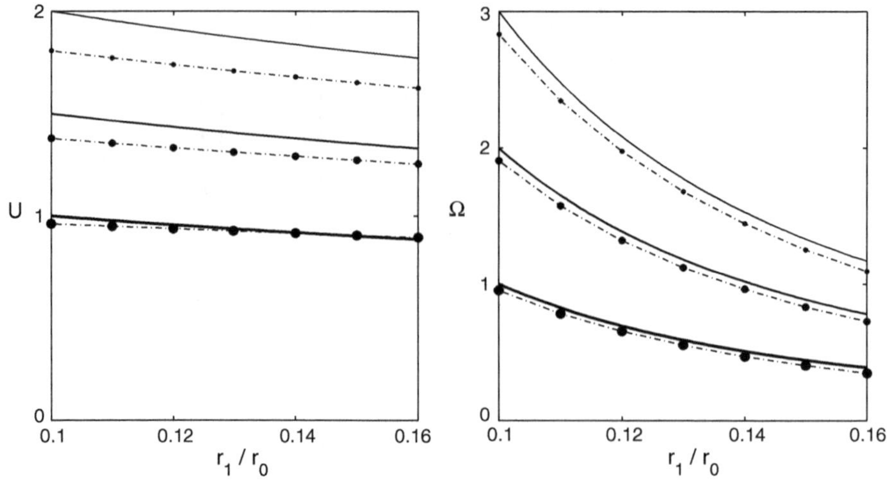

Fig. 7 Linear speed (U) and phase angular speed (Ω) of n linked vortices as functions of the torus aspect ratio (r_1/r_0), for sets of $n = 2$ (*thickest line and largest markers*), $n = 3$ and $n = 4$ (*thinnest line and smallest markers*). The continuous lines represent the analytical functions discussed in the text, the markers represent the results of the numerical simulations. The linear speeds are scaled by the speed of a circular ring vortex of circulation 2Γ, centerline radius r_0 and cross section radius $2a$, the angular speeds are scaled by the rotation speed of a pair of point vortices of circulation Γ separated by a distance $4a$. Taken from Velasco Fuentes and Romero Arteaga (2011)

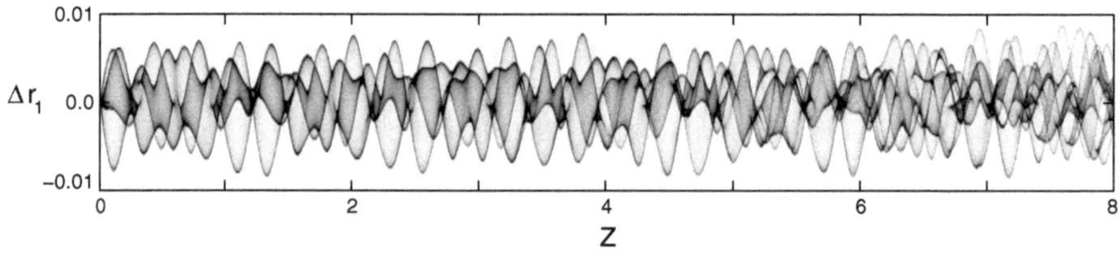

Fig. 8 Evolution of the vortex shape as illustrated by a time series of the histogram of distances from the vortex markers to the surface of the torus (see text). The signed distance, Δr_1, at which a certain percentage of the markers is located (*white*: 0 %; *black*:100 %) is shown as a function of the adimensional distance travelled by the vortices, $Z = Ut/r_0$ (where U is the speed of the vortices, t is the time and r_0 is the torus' centerline radius). The results correspond to the simulation shown in Fig. 6. Taken from Velasco Fuentes and Romero Arteaga (2011)

4 Flow geometry

The toroidal vortices discussed in the previous section very nearly keep their shape and are almost stationary when observed in a frame that translates with speed U and rotates with angular speed Ω. Hence we will use this comoving frame to analyse the geometry of the velocity field. Since the vortices lie on a thin torus the velocity field they produce may be regarded as a small perturbation of the velocity field of a circular ring vortex, at least away from the immediate vicinity of the vortices.

Hence we describe first the flow geometry of a circular ring vortex of radius r_0 and cross section r_1 in the comoving system (i.e. progressing with speed U). The flow may qualitatively change depending on the numerical value of r_1/r_0 but all values used here fall within the regime of *fat* ring vortices ($r_1/r_0 > 1/86$, see Saffman 1995 for a detailed analysis). In this regime the velocity field has two stagnation points, both lying on the ring's symmetry axis. The forward one, P, has a linear attractor and a planar repellor; the backward one, Q, has a linear repellor and a planar attractor. The two stagnation points are connected by an infinite number of streamlines starting at P and ending at Q. These lines form a surface with the shape of an oblate spheroid. This stream surface is called separatrix, because the streamlines located inside it are qualitatively different from those located outside it: the former are closed whereas the latter are open and of infinite length. From a more physical point of view, the separatrix is the surface that divides the ambient fluid from the fluid permanently carried by the vortex.

The addition of a solid body rotation, Ω, round the symmetry axis affects neither the existence nor the position of the stagnation points. The rotation transforms the plane streamlines into helical curves but it leaves the shapes of all stream surfaces unaltered. Therefore, the separatrix of a circular ring vortex in a system progressing with speed U and rotating with speed Ω is the same oblate spheroid described above.

Let us now see what happens when we substitute back the toroidal vortices in the place of the virtual ring vortex. The stagnation points survive, although somewhat displaced. The separatrix, in contrast, disappears: instead of a single surface starting at P and ending at Q, there are now two surfaces. The first one, called the unstable manifold, starts at P and ends infinitely far downstream; the second one, called the stable manifold, starts infinitely far upstream and ends at Q. These surfaces intersect along a finite number of streamlines which start at P and end at Q.

We obtained the unstable manifold by computing a set of streamlines starting on the vicinity of the front stagnation point. The starting points lay on a circle of small radius ($0.01r_0$), coaxial with the torus and centred at the stagnation point. The stable manifold could have been computed in a similar way, but this was unnecessary. Note that a time reversal in the equations of motion is equivalent to a change of sign of all vortex circulations (i.e. $\Gamma \to -\Gamma$) and this is equivalent to the transformation $(x, y, z) \to (x, -y, -z)$, because of the initial conditions described in Sect. 2.2. Therefore, to obtain the stable manifold, we rotated the unstable one by an angle π around the x axis.

In the vicinity of the vortices the flow is always very different from that of a ring vortex. In order to study the geometry of the flow in this region, we used Poincaré sections. We constructed these by numerically computing a set of streamlines that started on a radial line going from the vicinity of the torus symmetry axis to the vicinity of the vortices, and plotting every intersection of the streamlines with the meridional plane that contains the starting points.

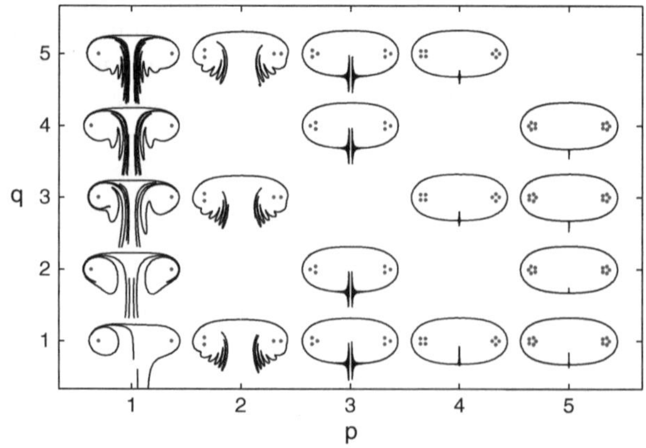

Fig. 9 The unstable manifold of toroidal vortices $\mathscr{V}_{p,q}$. All vortices have small cross-section ($a/r_0 = 0.05$) and are coiled on a thin torus ($r_1/r_0 = 0.1$); their intersections with the meridional planes $\theta = 0, \pi$ are represented by *grey dots*. Taken from Velasco Fuentes (2010)

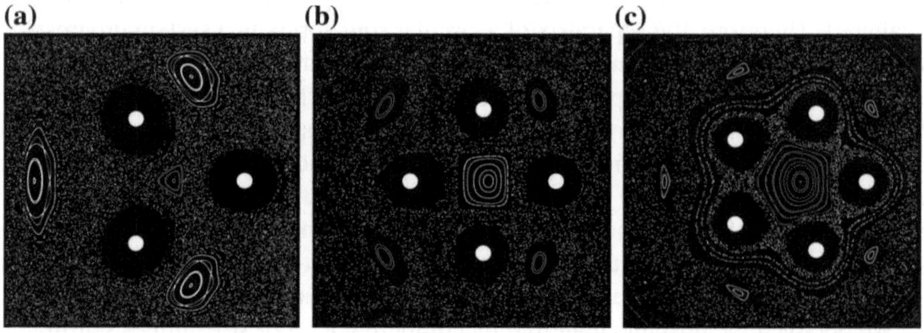

Fig. 10 Poincaré sections of streamlines induced by toroidal vortices: (**a**) $\mathscr{V}_{3,1}$, (**b**) $\mathscr{V}_{4,1}$ and (**c**) $\mathscr{V}_{5,1}$. The intersections of the vortices with the meridional plane $\theta = 0$ are represented by white circles, those of the streamlines by dots coloured according to the position of the streamline's starting point (*blue*: closer to the torus' symmetry axis; *red*: closer to torus centerline). Adapted from Velasco Fuentes (2010)

4.1 Knotted and Unknotted Vortices

Figure 9 shows meridional cross-sections of the unstable manifold for all toroidal vortices in the range $1 < p < 5$ and $1 < q < 5$. The shape of the manifolds is mainly determined by p, whereas the value of q is important only for toroidal helices ($p = 1$). Note, for example, that the manifold of the knotted vortex $\mathscr{V}_{2,5}$ is more similar to that of the unknotted vortex $\mathscr{V}_{2,1}$ than to the knotted vortex $\mathscr{V}_{3,5}$. As p grows the oscillations of the unstable manifold start closer to the backward stagnation point Q. When $p > 3$ the unstable and stable manifolds differ very little (except in the immediate neighbourhood of P and Q) from the separatrix of a fat ring vortex.

The Poincaré sections show at least $2p$ large islands of stability: p correspond to the single tube of fluid permanently trapped by the vortex, and p correspond to a single tube of irrotational fluid which runs parallel to the vortex and has approximately the same shape (see Fig. 10). When $p > 2$ there is an additional island of stability which corresponds to a tube of irrotational fluid that surrounds the torus centerline. When $p = 1, 2$ all these tubes are embedded in the unbounded chaotic sea generated by the intersections of the manifolds. When $p > 2$ the tubes are embedded in a chaotic sea that is itself bounded by a KAM-like torus.

4.2 Linked Vortices

Figure 11 shows meridional cross sections of the stable and unstable manifolds of two linked vortices; each frame corresponds to a supporting torus of a particular aspect ratio ($r_1/r_0 = 0.07, 0.1$). In both cases the red curve, which represents the unstable manifold of P, smoothly moves downstream but, as it approaches Q, it starts to oscillate about the blue curve, which represents the stable manifold of Q. Similarly, the stable manifold of Q smoothly moves upstream but as it approaches P it starts to oscillate about the unstable manifold of P. Note that when the supporting torus is thinner (frame a, $r_1/r_0 = 0.07$) the oscillations of the manifolds are of small amplitude and they start close to the opposite stagnation point. In contrast, when the supporting torus is thicker (frame b, $r_1/r_0 = 0.10$) the oscillations of the manifolds are of larger amplitude and they start closer to their own stagnation point.

The presence of this geometric structure, known as heteroclinic tangle, implies that streamlines are chaotic in this region (Wiggins 1992). It also provides a template for the wandering of streamlines around different flow regions through the following mechanism (*lobe dynamics*, for details see Rom-Kedar et al. 1990). Consider two adjacent intersections, on some meridional plane, between the unstable manifold of P and the stable manifold of Q; the two line segments bounded by these points form a closed contour which defines an area, say A_1, usually called lobe (see Fig. 11b). The streamlines passing through A_1 successively intersect the same meridional plane within the lobes $A_2, A_3, \ldots$, thus reaching at some point the interior of the so-called vortex atmosphere. This is, however, only a transient situation because the same mechanism eventually brings them out to the downstream side of the vortex.

Figure 11b shows that there are two independent sequences of lobes, the green ones and the white ones, which implies that the unstable manifold of P intersects the stable manifold of Q along four streamlines. In fact we found that manifolds always intersect along $2n$ streamlines, where n is the number of vortices. Note also that here, as in all cases we have analysed, the areas of the lobes are larger when they are closer to the torus symmetry axis. This occurs because the fluid is incompressible and the azimuthal velocity grows with the distance to the torus axis.

Analogously to the case of a single vortex $\mathcal{V}_{pq}$, the Poincaré sections show at least $2n$ large islands of stability: n correspond to the tubes of fluid permanently trapped by an individual vortex, and n correspond to tubes of irrotational fluid which run parallel to the vortices and have approximately the same shape. When $n > 2$ there is

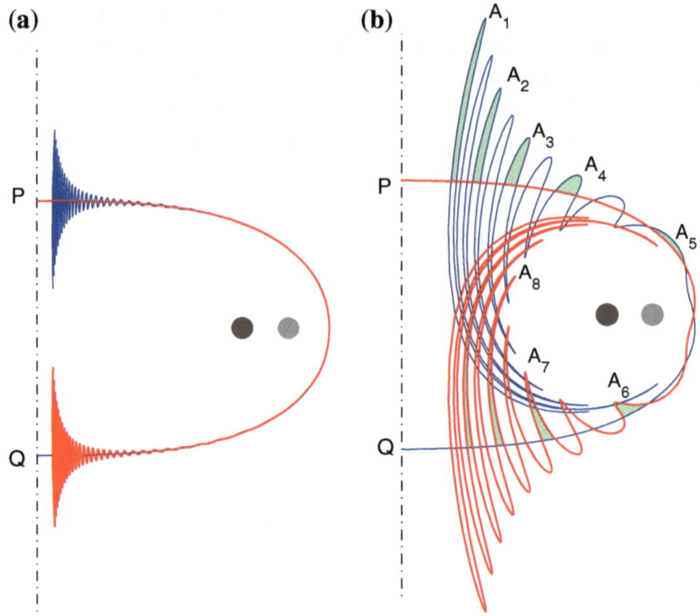

Fig. 11 Meridional cross section of the three-dimensional chaotic tangle of two linked ring vortices coiled on tori of different thickness: (**a**) $r_1/r_0 = 0.07$, (**b**) $r_1/r_0 = 0.10$. The *red* and *blue lines* represent, respectively, the unstable manifold of the front stagnation point (P), and the stable manifold of the rear stagnation point (Q); the *grey circles* represent the vortices and the *green* areas, labelled with A_i, represent successive intersections of a particular streamtube with the meridional plane (see text). Adapted from Velasco Fuentes and Romero Arteaga (2011)

an additional island of stability which corresponds to a tube of irrotational fluid that runs between the n vortices and surrounds the torus centerline.

If the number of vortices is large or the aspect ratio of the torus is small these islands of stability are embedded in a chaotic sea bounded by a nested set of KAM tori, as evidenced by the bands of differently coloured dots in Fig. 12a, c. Note that the largest KAM torus almost fills the "unperturbed" oblate spheroid. If, however, the number of vortices is small or the aspect ratio of the torus is large these islands of stability are embedded in an unbounded chaotic sea, as evidenced by the well mixed coloured dots in Fig. 12b, d.

5 Conclusions

Our numerical results confirm Kelvin (1875) deductions about knotted and linked toroidal vortices: they progress along and rotate around the torus symmetry axis with almost uniform speeds while undergoing negligible deformations. Although these results make plausible the existence of exact solutions which are both steady and stable, finding the analytical expression of such solutions is still an open problem.

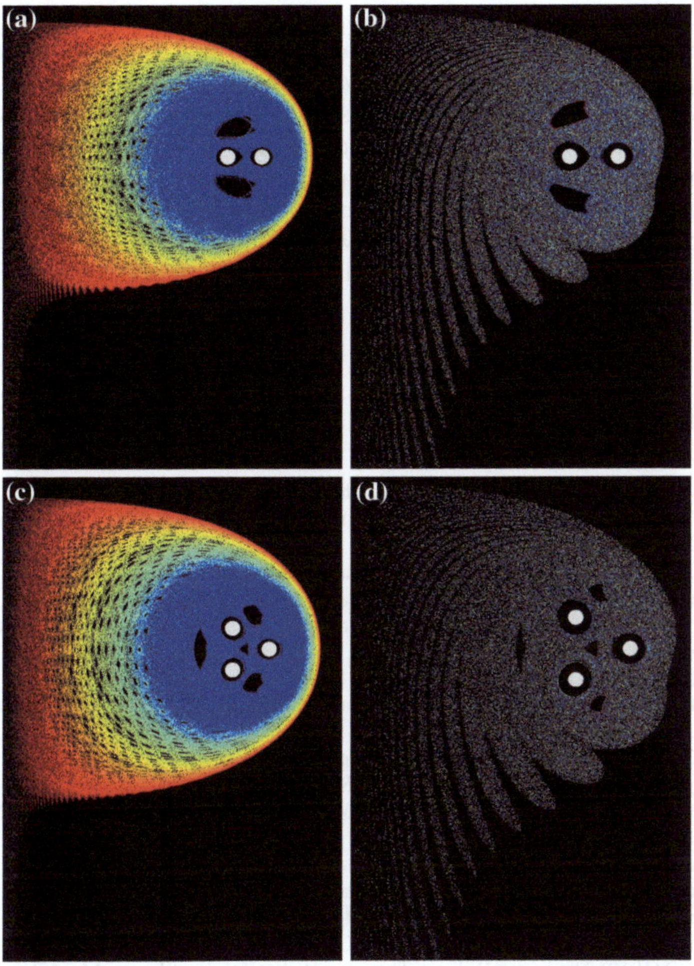

Fig. 12 Poincaré sections of streamlines in the velocity field of n linked ring vortices lying on tori of different thick nesses: **a** $n = 2$ and $r_1/r_0 = 0.07$, **b** $n = 2$ and $r_1/r_0 = 0.10$, **c** $n = 3$ and $r_1/r_0 = 0.10$, **d** $n = 3$ and $r_1/r_0 = 0.15$. The intersections of the vortices with the meridional plane $\theta = 0$ are represented by *white circles*, those of the streamlines by dots coloured according to the position of the streamline's starting point (*red*: closer to the torus' symmetry axis; *blue*: closer to the vortices). Taken from Velasco Fuentes and Romero Arteaga (2011)

The quasi-steadiness of the linked ring vortices enables us to interpret the results about the flow geometry in terms of the capacity of the vortices to carry fluid. We may thus conclude that a single toroidal vortex $\mathscr{V}_{p,q}$ carries more fluid if it makes more coils round the symmetry axis (larger p) or if it lies on a thinner torus (smaller r_1/r_0). Similarly a set linked ring vortices $\mathscr{V}_{11}$ carries more fluid if there are more vortices in the set (larger n) or if it lies on a thinner torus (smaller r_1/r_0).

Equation (4) shows that the velocity field depends on the value of μa, particularly in the neighbourhood of the vortices. This affects the self-induced velocity and, through it, the flow geometry. For if μa is smaller the vortices move faster and their stagnation points are closer to each other, and vice versa. To evaluate the extent of the modifications produced by changing the value of μa, we used thinner vortices ($a = 0.025 r_0$) with the same internal structure used above ($\mu = e^{-3/4}$) and hollow vortices ($\mu = e^{-1/2}$) with the same cross-section used above ($a = 0.05 r_0$). The thinner vortices moved with a 6 %-larger speed and the distance between their stagnation points was 8 % smaller. The hollow vortices moved with a 3 %-smaller speed and the distance between their stagnation points was 4 % larger. In neither case the chaotic tangles or the Poincaré sections exhibited significant changes with respect to those shown in the present chapter.

References

B R (1804) On phosphoric rings. J Nat Philos, Chem Arts 7:64

Baggaley AW, Barenghi CF (2011) Spectrum of turbulent Kelvin-waves cascade in superfluid helium. Phys Rev B 83:134509

Helmholtz H (1858) Über Integrale der hydrodynamischen Gleichungen, welche den Wirbelbewegungen entsprechen. Journal für die reine und angewandte Mathematik 55:25–55

Helmholtz H (1867) On integrals of the hydrodynamical equations, which express vortex-motion. Phil Mag 33:485–512

Kida S (1981) A vortex filament moving without change of form. J Fluid Mech 112:397–409

Kleckner D, Irvine WTM (2013) Creation and dynamics of knotted vortices. Nat Phys doi:10.1038/nphys2560

Maxwell JC, Harman PM (1995) The scientific letters and papers of James Clerk Maxwell, vol 2. Cambridge University Press, Cambridge, pp 1862–1873

Ricca R, Samuels D, Barenghi C (1999) Evolution of vortex knots. J Fluid Mech 391:29–44

Rogers WB (1858) On the formation of rotating rings by air and liquids under certain conditions of discharge. Am J Sci Arts 26:246–258

Rom-Kedar V, Leonard A, Wiggins S (1990) An analytical study of transport, mixing and chaos in an unsteady vortical flow. J Fluid Mech 214:347–394

Saffman P (1995) Vortex dynamics. Cambridge University Press, Cambridge

Thompson SP (1910) The life of William Thomson Baron Kelvin of Largs. McMillan, London

Thomson JJ (1883) A treatise on the motion of vortex rings. MacMillan, London

Thomson W (Lord Kelvin) (1867a) On vortex atoms. Philos Mag 34:15–24. Also in mathematical and physical papers (MPP) 4:1–12

Thomson W (Lord Kelvin) (1867b) The translatory velocity of a circular vortex ring. Philos Mag 33:511–512. Also in MPP, 4:67–68

Thomson W (Lord Kelvin) (1869) On vortex motion. Trans R Soc Edinb 25:217–260. Also in MPP, 4:13–66

Thomson W (Lord Kelvin) (1875) Vortex statics. Proc R Soc Edinburgh 9:59–73. Also in MPP, 4:115–128

Velasco Fuentes O (2010) Chaotic streamlines in the flow of knotted and unknotted vortices. Theor Comput Fluid Dyn 24:189–193

Velasco Fuentes O (2013) Early observations and experiments on ring vortices. Eur J Mech B-Fluid. http://dx.doi.org/10.1016/j.euromechflu.2013.08.008

Velasco Fuentes O, Romero Arteaga A (2011) Quasi-steady linked vortices with chaotic streamlines. J Fluid Mech 687:571–583

Wiggins S (1992) Chaotic transport in dynamical systems. Springer-Verlag, Berlin

Minimum Resistance in a Rare Medium

J. Cruz-Sampedro and M. Tetlalmatzi-Montiel

Abstract The aim of this chapter is to offer a short account of classical and recent results about Newton's problem of minimal resistance available to undergraduate students. Part of this material was presented by the first author in a lecture addressed to undergraduate students of engineering. We begin with a derivation of Newton's model for the resistance of a body moving with constant velocity in a rare medium. Then we show how to recover from Newton's geometric constructions, the corresponding solutions to Newton's aerodynamic problem for the frustum of a cone and for radially symmetric solids. Finally, we consider Newton's problem for nonsymmetric solids and describe the existence and lack of uniqueness of non-radially symmetric solutions to this minimization question.

1 Introduction

An important problem in the aerodynamic design of ships, aircrafts and rockets is the determination of shapes that experience minimal resistance during motion. In order to achieve optimal performance, nose cones of aircrafts and rockets are often designed to be conical, parabolic, elliptic, hemispheric, bi-conical, or blunted conical solids of revolution (see for example Fig. 1). This fact brings to mind the following question:

J. Cruz-Sampedro (✉)
UAM-A, Av. San Pablo 180, Col. Reynosa, 02200 Mexico, DF, Mexico
e-mail: jacs@correo.azc.uam.mx

M. Tetlalmatzi-Montiel
UAEH, Carretera Pachuca-Tulancingo km 4.5,
42090 Pachuca, HGO, Mexico
e-mail: tmontiel@uaeh.edu.mx

J. Klapp and A. Medina (eds.), *Experimental and Computational Fluid Mechanics*, Environmental Science and Engineering, DOI: 10.1007/978-3-319-00116-6_9,

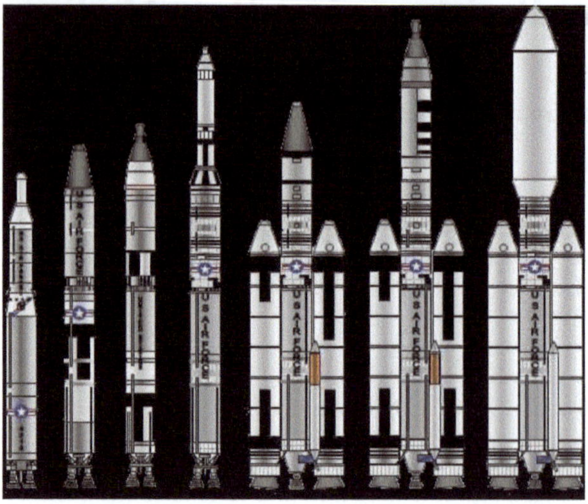

Fig. 1 The Titan rocket family displays a variety of rocket nose cones http://en.wikipedia.org/wiki/Titan_(rocket_family)

Question 1 Which of all possible solids of revolution of given base and altitude will undergo minimal resistance while moving with constant velocity in a certain medium?

This problem was already considered by Newton over 300 years ago and is referred to in the literature as Newton's Problem of Minimal Resistance or as Newton's Aerodynamical Problem (Butazzo and Kawohl 2001; Goldstine 1980; Tikhomirov 1990). In the scholium to Section VII: *Of the motion of fluids, and the resistance made to projected bodies* of his *Principia,* Newton wrote:

> ... figures may be compared together as to their resistance; and those may be found which are most apt to continue their motions in resisting mediums. ... Which Proposition I conceive may be of use in the building of ships.
>
> Principia, vol. 1, p. 328, (Newton 1934).

Newton actually solved this problem but in his note he only presents a geometric description of how to construct both the frustum and the solid of revolution of minimum resistance, without giving a single suggestion of his method of derivation. One of the goals of this chapter is to show how to recover from Newton's geometric constructions, the corresponding solutions to Newton's aerodynamic problem for the frustum of a cone and for radially symmetric solids.

Newton's problem of minimal resistance is one of the first problems of the calculus of variations (Goldstine 1980; Tikhomirov 1990). To understand its formulation the following terms need explanation:

- The class of solids where the minimizer is sought.
- The medium where the solid moves.
- The model of resistance.

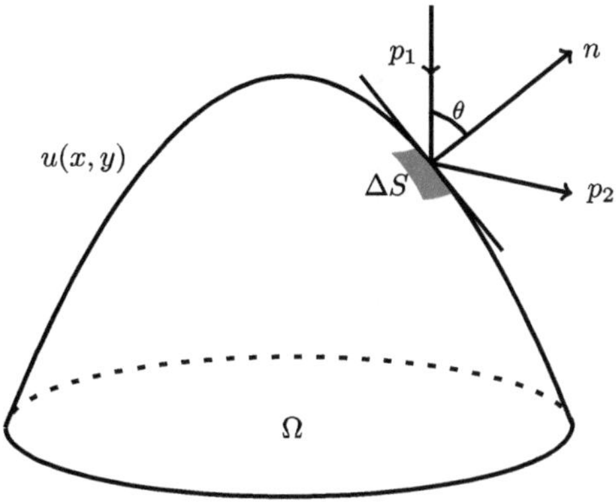

Fig. 2 Front end of a convex body with maximal horizontal cross section Ω

There are several possible classes of solids where the minimizer could be sought. Newton considered the class of truncated cones as well as the class of solids of revolution of prescribed base and altitude (Newton 1934). Recently, Butazzo et al. (1995) have studied Newton's problem for various classes of non-radially symmetric bodies. We will describe these classes with more detail in the last section of this chapter.

Concerning the medium where the solid moves, Newton calls this a *rare medium* and consists of "*equal particles freely disposed at equal distances from each other*". Each particle is assumed to be a perfectly elastic ball.

In the following section we derive Newton's model for the resistance undergone by a solid moving with constant velocity in a rare medium. It turns out that if Ω is the maximal horizontal cross section of the body whose front end is given by the graph of a nonnegative function $u(x, y)$ with $(x, y) \in \Omega$ (Fig. 2), then the resistance experienced by this solid is proportional to

$$\Phi(u) = \int_{\Omega} \frac{dx\, dy}{1 + |\nabla u|^2}. \tag{1}$$

With this terminology at hand, Newton's problem of minimum resistance can be stated as follows:

Problem 1 Minimize the functional $\Phi(u)$ over a suitable class of functions u.

In what follows we study this minimization problem. We begin with a derivation of Newton's model for the resistance. Then we analyze Newton's solution of the problem for the frustum of a cone and solve this problem using elementary methods. After this we derive the solution to the problem for radially symmetric solids from Newton's geometric construction. We conclude this chapter with a description of the recent results of Butazzo et al. about Newton's problem for non-radially symmetric solids. In particular, we show the lack of uniqueness of the solution to Newton's problem in some classes of non-radially symmetric solids.

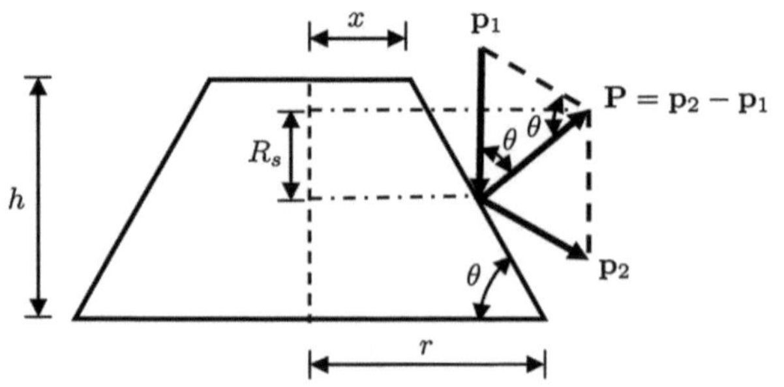

Fig. 3 Profile of the frustum of a cone hit by particles of a rare medium obeying the law of reflection

Our exposition has benefited very much from (Butazzo and Kawohl 2001; Tikhomirov 1990; Cruz-Sampedro and Tetlalmatzi-Montiel 2010; Butazzo et al. 1996) and is meant to be available to students of undergraduate level.

2 Newton's Model for the Resistance in a Rare Medium

In this section we derive a mathematical model for the resistance of an object moving in a rare medium with constant velocity. If the moving object is radially symmetric we assume that the body moves in the direction of its axis. According to Newton

> since the action of the medium upon the body is the same (...) whether the body moves in a quiescent medium, or whether the particles of the medium impinge with the same velocity upon the quiescent body, let us consider the body as if it were quiescent, and see with what force it would be impelled by the moving medium.
>
> Principia, vol. 1, p. 327, (Newton 1934).

Taking into account Newton's remark, we assume that each particle has mass m and that the density of the rare medium is a constant ρ. We assume that the object is at rest, that the particles hit it from above with constant velocity $\mathbf{v}$ or, equivalently, with momentum $\mathbf{p}_1 = m\mathbf{v}$, and that the collisions are perfectly elastic (Fig. 3). More precisely:

1. If $\mathbf{p}_1$ and $\mathbf{p}_2$ denote the momenta of the corresponding particles, before and after the collision, then $|\mathbf{p}_1| = |\mathbf{p}_2| = m|\mathbf{v}|$.
2. The collisions follow the law of reflection, i. e., the angle of incidence is equal to the angle of reflection.

2.1 *Resistance of the Frustum of a Cone*

In order to provide the reader a concrete and simple example that illustrates the general idea, before we derive a model of the resistance for a general body, we compute the resistance for the frustum of a cone.

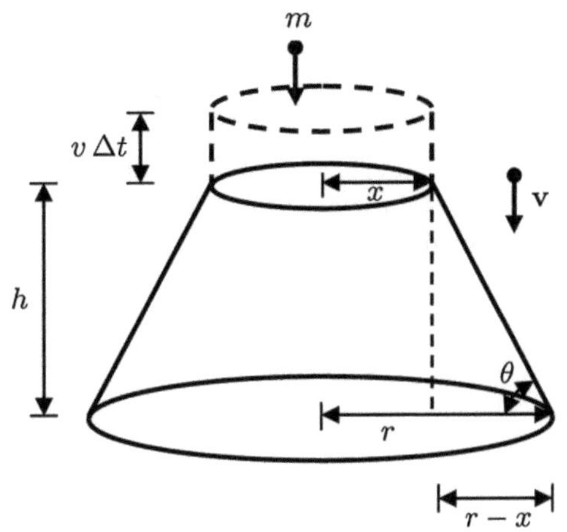

Fig. 4 Particles hitting the upper base of the frustum during the time Δt

Let us consider a frustum of a cone of altitude h, with lower base of radius $r > 0$ and upper base of radius $x \in [0, r]$ (Fig. 4). We will derive a mathematical expression for the resistance $R(x)$ encountered by this frustum when moving in a rare medium. By Newton's second and third laws, *the resistance $R(x)$ will be the vertical component of the momentum gained by the frustum from the total number of collisions received per unit time.*

Let $\mathbf{P} = \mathbf{p}_2 - \mathbf{p}_1$ denote the change of momentum of a single particle that hits the frustum (Fig. 3). If R_s denotes the vertical component of $\mathbf{P}$ for a particle hitting the side of the frustum, then

$$R_s = |\mathbf{P}| \cos\theta$$

Since $\mathbf{P}$ is perpendicular to the side of the frustum, using the sine law of trigonometry and $|\mathbf{p}_1| = |\mathbf{p}_2| = mv$, where $v = |\mathbf{v}|$, we obtain

$$\frac{|\mathbf{P}|}{\sin(\pi - 2\theta)} = \frac{mv}{\sin\theta}.$$

Thus $|\mathbf{P}| = 2mv\cos\theta$ and therefore

$$R_s = 2mv\cos^2\theta. \tag{2}$$

Note that if the particle hits the upper base of the frustum, then the vertical component R_u of the change of momentum is

$$R_u = |\mathbf{P}| = |2m\mathbf{v}| = 2mv.$$

Thus, if N_s is the number of particles that hit the side of the frustum per unit time, and N_u is the corresponding quantity for the particles that hit the upper base, then

$$\begin{aligned} R(x) &= N_u R_u + N_s R_s \\ &= 2mvN_u + 2mvN_s \cos^2\theta. \end{aligned}$$

Note next (Fig. 4) that the volume occupied by the particles that hit the upper base of the frustum during the time Δt is $\pi x^2 v \Delta t$. Since the total mass of this volume is $mN_u\Delta t$, then

$$\rho = \frac{mN_u\Delta t}{\pi x^2 v \Delta t}.$$

Hence

$$N_u = \frac{\pi\rho x^2 v}{m}.$$

Similarly, the volume filled by the particles that hit the side of the frustum during the time Δt is equal to $\pi(r^2 - x^2)v\Delta t$. Since the mass of this volume is $mN_s\Delta t$, then

$$\rho = \frac{mN_s\Delta t}{\pi(r^2 - x^2)v\Delta t}$$

and therefore

$$N_s = \frac{\pi\rho(r^2 - x^2)v}{m}.$$

Substituting N_u and N_s in the last expression for $R(x)$ we find that

$$R(x) = 2\pi\rho v^2[x^2 + (r^2 - x^2)\cos^2\theta].$$

Noting from Fig. 4 that

$$\cos\theta = (r - x)/\sqrt{(r - x)^2 + h^2},$$

we arrive at

$$R(x) = 2\pi\rho v^2\left[x^2 + (r^2 - x^2)\frac{(r - x)^2}{(r - x)^2 + h^2}\right]. \tag{3}$$

Note now that the resistance of a cylinder of radius r is

$$R_C := R(r) = 2\pi\rho v^2 r^2.$$

On the other hand, the resistance R_S of the side of a frustum of height h, lower base of radius r and upper base of radius x is

$$R_S = 2\pi\rho v^2\left[(r^2 - x^2)\frac{(r - x)^2}{(r - x)^2 + h^2}\right]. \tag{4}$$

It follows from (3) that the resistance of the cone of height $h = r$ and radius r is

$$R(0) = \pi\rho v^2 r^2 = R(r)/2 = R_C/2. \tag{5}$$

In addition, it follows from (3) that the resistance of the frustum of a cone with $h = r$ and $x = r/2$ is

$$R(r/2) = 4\pi\rho v^2 r^2/5 = 2R(r)/5.$$

As a result of these simple calculations we arrive at the following surprising conclusion:

The most aerodynamic frustum is not the cone, as one would have initially guessed, but a frustum of a cone whose top is flat.

Now we consider the northern hemisphere of a sphere of radius $x^2+y^2+z^2 = r^2$ as formed with thin frusta of height dz, lower base of radius $x + dx$, and upper base of radius x, with $0 \le x \le r$. Since $dz = -x\,dx/\sqrt{r^2 - x^2}$, after a short calculation, using (4) and ignoring second order terms, we find that the resistance R_H of this hemisphere is

$$R_H = 4\pi\rho v^2 \int_0^r \left(x - \frac{x^3}{r^2}\right) dx = \pi\rho v^2 r^2 = R(r)/2 = R_C/2,$$

which is in agreement with Newton's prediction:

> ..., a globe and a cylinder described on equal diameters move with equal velocities in the direction of the axis of the cylinder, the resistance of the globe will be but half so great as that of the cylinder.
>
> Principia, vol. 1, p. 327, (Newton 1934).

It is interesting to note that the resistance of the sphere is the same as that of a cone (5) of the same radius and altitude.

2.2 Resistance of a General Body

Now we derive a model of the resistance for a general body. To this end we suppose that the maximal horizontal cross section of the body is described by $\Omega \subseteq \mathbb{R}^2$, and that the shape of the object is given by a smooth nonnegative function $u(x, y)$ with $(x, y) \in \Omega$ (Fig. 2). In addition to the assumptions on the medium and the velocities of the particles considered above, we assume that the particles and the body interact at most once and that tangential friction, turbulence and other effects are neglected. Hence, if ΔR denotes the resistance due to the infinitesimal piece ΔS of the surface of u, using (2) we find that

$$\Delta R = 2mvN\cos^2\theta,$$

where N denotes the number of particles hitting ΔS per unit time. Since the normal to the surface u is $\mathbf{n} = (-u_x, -u_y, 1)$ we have

$$\cos\theta = 1/\sqrt{1+u_x^2+u_y^2}.$$

Moreover, since

$$\rho = mN/v\Delta x\ \Delta y$$

we find that

$$\Delta R = 2\rho v^2 \frac{\Delta x\ \Delta y}{1+u_x^2+u_y^2}.$$

Integrating over Ω we find that the resistance of the solid bounded by Ω and the surface u is given by

$$\begin{aligned} R(u) &= 2\rho v^2 \int_\Omega \frac{dx\,dy}{1+|\nabla u|^2} \\ &= 2\rho v^2 \Phi(u), \end{aligned} \tag{6}$$

where $\Phi(u)$ is as in (1).

In particular, if the body is a sphere of radius r we have $u(x, y) = \sqrt{r^2 - x^2 - y^2}$ and $\Omega = \{(x, y) : x^2 + y^2 \le r^2\}$. Thus, using (6) and polar coordinates $x = \eta\cos\theta$ and $y = \eta\sin\theta$ we obtain once again

$$\begin{aligned} R_H &= 2\rho v^2 \frac{1}{r^2} \int_\Omega (r^2 - x^2 - y^2)dx\,dy \\ &= 4\pi\rho v^2 \int_0^r \left(\eta - \frac{\eta^3}{r^2}\right) d\eta \\ &= \pi\rho v^2 r^2. \end{aligned}$$

3 Newton's Aerodynamic Frustum

As pointed out in the comment following (5), the most aerodynamic frustum of a cone will have a flat top. In 1687, in his *Philosophia Naturalis Principia Mathematica*, Newton provided the following geometric answer for the problem of the frustum of a cone:

> As if upon the circular base $CEBH$ from the centre O, with the radius OC, and the altitude OD (Fig. 5), one would construct a frustum $CBGF$ of a cone, which should meet with least resistance than any other frustum constructed with the same base and altitude and going forwards towards D in the direction of its axis: *bisect the altitude OD in Q, and produce OQ to S, so that QS may be equal to QC, and S will be the vertex of the cone whose frustum is sought.*
>
> Principia, vol. 1, p. 328, (Newton 1934).

Following Newton's directions quoted above we observe that setting $h = OD$, $r = OC$, and $p = h/r$, then $OQ = pr/2$ and

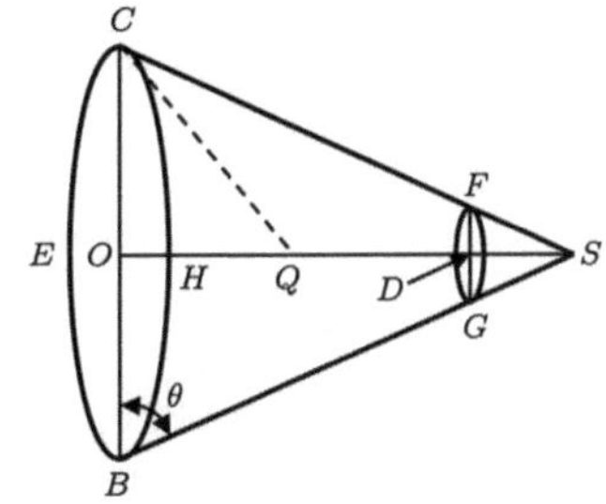

Fig. 5 Newton's construction of the frustum of a cone of minimal resistance

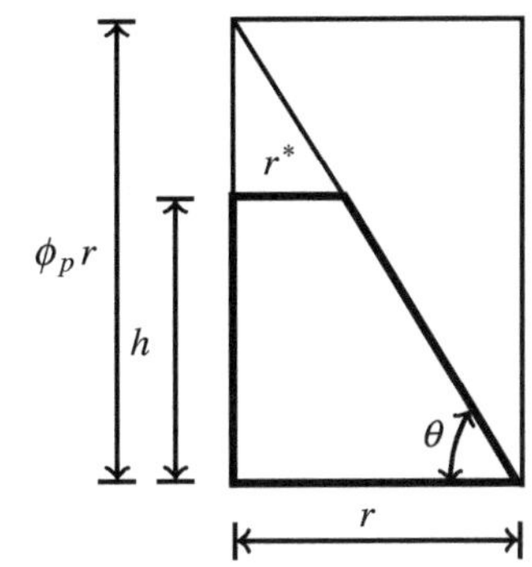

Fig. 6 A simple construction of Newton's frustum of a cone of minimal resistance

$$QS = QC = \frac{\sqrt{p^2+4}}{2}r.$$

Therefore,

$$OS = OQ + QS = \frac{p+\sqrt{p^2+4}}{2}r = \Phi_p r,$$

where Φ_p is the positive solution to $x^2 - px - 1 = 0$, and is referred to in the literature as p^{th} *order extreme mean* or POEM (Cruz-Sampedro and Tetlalmatzi-Montiel 2010).

Thus, Newton's frustum is precisely the one for which the angle θ, between the generator of the frustum and the plane of its lower base (Fig. 6), satisfies

$$\tan\theta = \Phi_p.$$

This means that Newton's frustum can be obtained with the following simple construction: *Rotate the right angled triangle of base r and height $\Phi_p r$ of Fig. 6 about its longest side. Then cut the cone so constructed at a height $h = pr$.*

In view of (3), to solve Newton's problem for the frustum of a cone we must minimize

$$\begin{aligned} S(x) &= x^2 + (r^2 - x^2)\frac{(r-x)^2}{(r-x)^2 + h^2} \\ &= r^2 - h^2 \frac{r^2 - x^2}{(r-x)^2 + h^2}. \end{aligned} \tag{7}$$

for $x \in [0, r]$. This minimization problem can be solved using the standard calculus technique but we shall do it here using the simple fact that the positive solution Φ_p to $x^2 - px - 1 = 0$ satisfies

$$\Phi_p^2 = p\Phi_p + 1 \quad \text{and} \quad 1 - \frac{p}{\Phi_p} - \frac{1}{\Phi_p^2} = 0. \tag{8}$$

Using these last properties of Φ_p and little algebra we have

$$\begin{aligned} \frac{1}{p\Phi_p} - \frac{r^2 - x^2}{(r-x)^2 + h^2} &= \frac{(1 + p\Phi_p)x^2 - 2xr + (1 - p\Phi_p)r^2 + h^2}{p\Phi_p((r-x)^2 + h^2)} \\ &= \frac{\Phi_p^2(1 + p\Phi_p)x^2 - 2x\Phi_p^2 r + \Phi_p^2(1 - p\Phi_p)r^2 + h^2\Phi_p^2}{p\Phi^3((r-x)^2 + h^2)} \\ &= \frac{\Phi_p^4 x^2 - 2x\Phi_p^2 r + (1 - p^2\Phi_p^2)r^2 + h^2\Phi_p^2}{p\Phi_p^3((r-x)^2 + h^2)} \\ &= \frac{(\Phi_p^2 x - r)^2}{p\Phi_p^3((r-x)^2 + h^2)}. \end{aligned}$$

To go from the second to the third line in this last identity we used the first equation in (8). From the third line to the fourth we used $h = pr$ in the numerator. Thus, combining this last identity with (7), and using the second equation in (8) we obtain

$$\begin{aligned} S(x) &= r^2 - \frac{h^2}{p\Phi_p} + \frac{(\Phi_p^2 x - r)^2 h^2}{p\Phi_p^3((r-x)^2 + h^2)} \\ &= \left(1 - \frac{p}{\Phi_p}\right) r^2 + \frac{pr^2}{\Phi_p^3} \frac{(\Phi_p^2 x - r)^2}{(r-x)^2 + h^2} \\ &= \frac{r^2}{\Phi_p^2} + \frac{pr^2}{\Phi_p^3} \frac{(\Phi_p^2 x - r)^2}{(r-x)^2 + h^2}. \end{aligned} \tag{9}$$

It follows that the minimum value μ_p of $S(x)$ for $x \in (-\infty, \infty)$ is attained at

$$r^* = \frac{r}{\Phi_p^2},$$

and is given by

$$\mu_p = \frac{r^2}{\Phi_p^2}.$$

Since $\Phi_p \geq 1$ we have that $0 \leq r^* \leq r$. Therefore μ_p is the minimum value of $S(x)$ for $x \in [0, r]$. In addition, if θ is the angle between the generator of Newton's frustum (Fig. 6) and the plane of its lower base, using $r^* = r/\Phi_p^2$ and the second identity in (8) yields

$$\tan\theta = \frac{h}{r - r^*} = \frac{pr}{r - (r/\Phi_p^2)} = \frac{p}{1 - (1/\Phi_p^2)} = \Phi_p.$$

Thus, we have recovered Newton's result in terms of the POEM's. An analogous calculation was done in (Cruz-Sampedro and Tetlalmatzi-Montiel 2010), using the algebraic properties of the so called *golden ratio* Φ_1, to find the most aerodynamic frustum with altitude equal to the radius of its base.

4 Newton's Aerodynamic Solid of Revolution

In this section we present the solution to Newton's aerodynamic problem for radially symmetric convex bodies. To this end we assume that u is a radial concave function, that is to say $u = u(\eta)$, with $0 \leq \eta \leq r$. First, we note that in polar coordinates $x = \eta\cos\theta$ and $y = \eta\sin\theta$ the gradient is

$$\nabla u = \frac{\partial u}{\partial \eta}\mathbf{e}_\eta + \frac{1}{\eta}\frac{\partial u}{\partial \theta}\mathbf{e}_\theta, \tag{10}$$

where $\mathbf{e}_\eta = (\cos\theta, \sin\theta)$ and $\mathbf{e}_\theta = (-\sin\theta, \cos\theta)$. Then the resistance (6) can be written in polar coordinates as

$$\begin{aligned} R(u) &= 2\rho v^2 \int_\Omega \frac{dx\,dy}{1 + |\nabla u|^2} \\ &= 4\pi\rho v^2 \int_0^r \frac{\eta\,d\eta}{1 + (u'(\eta))^2}. \end{aligned}$$

Thus, from all concave functions $u(x)$ with $u(0) = h$ and $u(r) = 0$ (Fig. 7) we must find the one that minimizes

$$\Phi(u) = \int_0^r \frac{x\,dx}{1 + (u'(x))^2}.$$

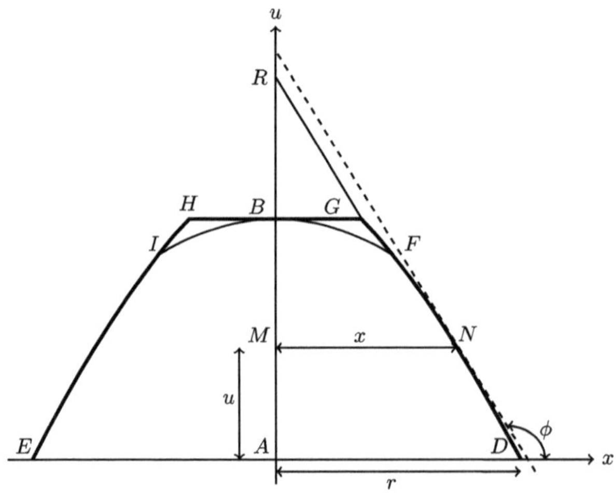

Fig. 7 Newton's geometric construction of the radially symmetric solid of minimal resistance

In the note to Section VII: *Of the motion of fluids, and the resistance made to projected bodies* of his *Principia*, Newton provided the following geometric description of the solid of revolution of least resistance

> If the figure $DNFG$ is such a curve, that if, from any point thereof, as N, the perpendicular NM let fall on the axis AB, and from the given point G there be drawn to the right line GR parallel to a right line touching the figure in N, and cutting the axis produced in R, MN becomes to GR as to GR^3 to $4BR \cdot GB^2$, the solid described by the revolution of this figure about its axis AB, moving in the before mentioned rare medium from A towards B, will be less resisted than any other circular solid whatsoever, described of the same length and breadth.
>
> Principia, vol. 1, p. 328, (Newton 1934).

In spite of being so brief, Newton's note contains essentially all the information to obtain the desired function u. In fact, setting $MN = x$ and $BG = r_0$ we see from Fig. 7 that

$$BR = -r_0 \tan\phi \quad \text{and} \quad GR = r_0(1 + \tan^2\phi)^{1/2}.$$

In addition, since $\tan\phi = u'(x)$ we have

$$R = -r_0 u'(x) \quad \text{and} \quad GR = r_0(1 + (u'(x))^2)^{1/2}.$$

Hence, Newton's statement

$$\frac{MN}{GR} = \frac{GR^3}{4BR \cdot GB^2}$$

yields

$$\frac{x}{r_0(1+(u'(x))^2)^{1/2}} = -\frac{r_0^3(1+(u'(x))^2)^{3/2}}{4r_0^3 u'(x)}$$

from which we arrive to the so called differential equation of Newton's curve

$$\frac{xu'(x)}{(1+(u'(x))^2)^2} = -\frac{r_0}{4}, \tag{11}$$

which is subject to the conditions $u(0) = h$, $u(r) = 0$, and $u'(r_0^+) = \lim_{x \downarrow r_0} u(x) = -1$. This last condition was also foreseen by Newton

> ... and FG, HI may be inclined to GH in the angles FGB, BHI of 135 degrees...
>
> Principia, vol. 1, p. 328, (Newton 1934).

This last condition can be naively explained as follows. As the altitude h of the most aerodynamic frustum goes to zero we have that $p \to 0$ and therefore $\Phi_p \to 1$ and $\theta \to 45°$. The reader familiar with the basics of the calculus of variations may verify that the Euler-Lagrange equations for the functional $\Phi(u)$ render

$$\frac{xu'(x)}{(1+(u'(x))^2)^2} = c, \qquad r_0 \le x \le r,$$

where c is a constant which, in view of the condition $u'(r_0^+) = -1$, turns out to be equal to $-r_0/4$, in agreement with (11).

Now we turn to the solution of the differential equation of Newton's curve. Setting $t = -u'$ we have

$$x = -\frac{c(1+t^2)^2}{t} \tag{12}$$

and

$$\frac{du}{dt} = \frac{du}{dx}\frac{dx}{dt} = c\left(-\frac{1}{t} + 2t + 3t^3\right).$$

Integrating this last equation with respect to t we find that

$$u = c\left(\frac{3}{4}t^4 + t^2 - \ln t\right) + d.$$

Using $u(r_0) = h$ and $u'(r_0^+) = -1$ yields $d = h + 7/16r_0$. Thus Newton's curve is given by

$$u = h - \frac{r_0}{4}\left(-\frac{7}{4} + \frac{3}{4}t^4 + t^2 - \ln t\right), \qquad x = \frac{r_0}{4}\frac{(1+t^2)^2}{t}, \qquad 1 \le t \le T,$$

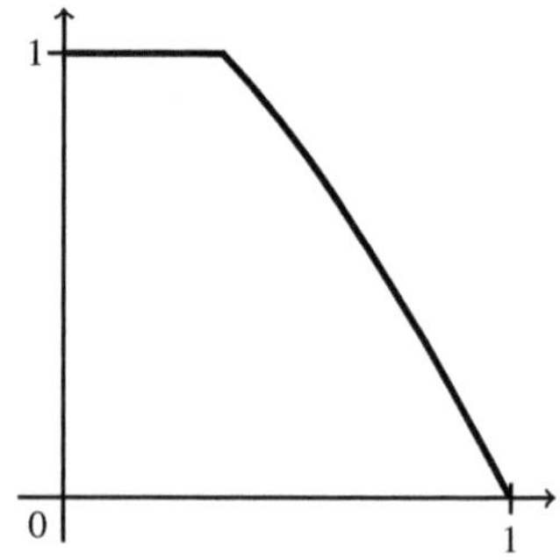

Fig. 8 Newton's curve that generates the radially symmetric solid of minimal resistance for $h = r = 1$

where T is defined by

$$r = \frac{r_0}{4}\frac{(1+T^2)^2}{T}.$$

Using this last equation and the fact that $u(r) = 0$ we obtain

$$\frac{h}{r} = \frac{T}{(1+T^2)^2}\left(-\frac{7}{4} + \frac{3}{4}T^4 + T^2 - \ln T\right) \doteq f(T).$$

Hence $T = f^{-1}(h/r)$ with $r_0 = 4rT/(1+T^2)^2$. In particular, for $h = r = 1$ we have $T \approx 1.9168$ and $r_0 \approx 0.3509$. Figure 8 shows Newton's curve in this situation.

5 Newton's Problem for Non-Symmetric Solids

In this section we describe some results concerning Newton's problem of minimal resistance for non-symmetric solids. We remark that the results presented here are fairly recent in comparison to the classical results of Newton.

First, we consider the problem of minimal resistance for the class of solids of height h, with a concave front and a convex maximal horizontal cross section Ω. The solution to this problem was given in 1995 by Butazzo et al. (1995) and is stated precisely in the following

Theorem 1 *Let $h > 0$ be given and let $\Omega \subseteq \mathbb{R}^n$ be open, convex and bounded. Then the functional*

$$\Phi(u) = \int_\Omega \frac{dx}{1+|\nabla u|^2},$$

attains its minimum in the set

$$C_h(\Omega) = \{u : \Omega \to \mathbb{R} | u \quad \textit{is concave and} \quad 0 \le u \le h\}.$$

That is to say, there exists $u_0 \in C_h(\Omega)$ such that

$$\Phi(u_0) \leq \Phi(u),$$

for all $u \in C_h(\Omega)$. Furthermore, if Du_0 is the differential of the minimizer u_0 then $|Du_0| \notin (0, 1)$.

Since the proof of this theorem is beyond the scope of the purpose of this chapter it will not be presented here. The interested reader may verify the details in Butazzo et al. (1995). There is however a number of remarks that we would like to make:

- The existence part follows from the fact that in an appropriate space the set $C_h(\Omega)$ is compact (Marcellini 1990) and the functional $\Phi(u)$ is lower semi-continuous.
- We will see below that if $\Omega \subseteq \mathbb{R}^2$ is a disk, then u_0 does not coincide with Newton's radially symmetric solution. Therefore, in general u_0 might not be unique.
- It is not known if the top of the minimizer is flat.
- It is not known if u_0 must vanish identically on the boundary of Ω.

The following result due to Butazzo et al. (1996) establishes the lack of uniqueness of Newton's aerodynamic problem for convex solids on the unit disk.

Theorem 2 *Let Ω be the unit disk in $\mathbb{R}^2$ and let u be Newton's solution to the radially symmetric problem of minimum resistance. Then u does not minimizes $\Phi(u)$ on the class of functions $C_h(\Omega)$.*

Proof Let u be Newton's radial solution to the radially symmetric problem of minimum resistance corresponding to Ω with $h = 1$. Let $r_0 \geq 0$ be as in (11) and let

$$A(r_0, 1) = \{(x, y) \in \mathbb{R}^2 : r_0^2 < x^2 + y^2 < 1\}.$$

For $t > 0$ and any infinitely differentiable function φ with support in $A(r_0, 1)$ we define

$$g(t) = \Phi(u + t\varphi).$$

A short calculation shows that

$$g'(0) = -2 \int_\Omega \frac{\nabla u \cdot \nabla \varphi}{(1 + |\nabla u|^2)^2} dx\, dy$$

and that

$$g''(0) = \int_\Omega \frac{8(\nabla u \cdot \nabla \varphi)^2 - 2|\nabla \varphi|^2(1 + |\nabla u|^2)}{(1 + |\nabla u|^2)^3} dx\, dy.$$

If φ is nonzero and radial, then in view of (10) we have $\nabla u \cdot \nabla \varphi = |\nabla u|\, |\nabla \varphi|$ and thus

$$g''(0) = 2 \int_\Omega \frac{|\nabla \varphi|^2(3|\nabla u|^2 - 1)}{(1 + |\nabla u|^2)^3} dx\, dy > 0,$$

since $|\nabla u| > 1$ on $A(r_0, 1)$. On the other hand, if

$$\varphi(r,\theta) = \eta(r)\sin(k\theta),$$

where $\eta(r)$ is an infinitely differentiable function with support in $(r_0, 1)$, then for k sufficiently large we have

$$g''(0) = 2\int_{\Omega} \frac{4u_r^2\varphi_r^2 - (1+u_r^2)(\eta_r^2\sin^2(k\theta) + k^2\eta(r)^2\cos^2(k\theta)/r^2)}{(1+|\nabla u|^2)^3}dx\,dy < 0.$$

Since $u \pm t\varphi(r,\theta) \in C_h(\Omega)$ for $t > 0$ sufficiently small and

$$g(t) = g(0) + g'(0)t + t^2 g''(0)/2 + o(t),$$

if $g'(0) = 0$ there exists $t > 0$ such that $g(t) < g(0)$, that is to say

$$\Phi(u + t\varphi(r,\theta)) < \Phi(u).$$

It follows from this result that if $\Omega \subseteq \mathbb{R}^2$ is a disk, then Newton's aerodynamic problem for the class $C_h(\Omega)$ does not have a unique solution.

The problem of proving existence of a minimizer for the functional $\Phi(u)$ has been intensively investigated in recent times by several authors for various classes of functions. For example, Butazzo et al. (1995) also prove that Φ attains its minimum in the class

$$E_h(\Omega) = \{u \in H^1_{\text{loc}}(\Omega) : 0 \le u \le h, \nabla^2 u(x) \le 0, \quad x \in \Omega\},$$

where $\nabla^2 u$ represents the Laplacian of u. Note that in view of the concavity of the functions in $C_h(\Omega)$ the class $E_h(\Omega)$ is larger than $C_h(\Omega)$.

Another class of functions which is often considered for physical reasons is

$$S_h(\Omega) = \{u : \Omega \to \mathbb{R} \mid 0 \le u \le h \text{ and every particle hits the body at most once}\}.$$

As far as we know, the existence of a minimizer for the functional Φ in the class $S_h(\Omega)$ is up to now an open problem.

We remark that the choice of appropriate classes of functions to minimize the functional $\Phi(u)$ is a delicate matter. For example,

- If u_h is the graph of a cone on the unit disk then, according to (3),

$$\Phi(u) = \frac{2\pi\rho v^2}{1+h^2}.$$

Thus $\Phi(u_h) \to 0$ as $h \to \infty$ but there is no u such that $\Phi(u) = 0$.

- Assuming the existence of $h > 0$ such that $\|u\|_\infty \leq h$ for all u is not enough either. If $u_n = h\sin^2(n|x|)$ then $\Phi(u_n) \to 0$ as $n \to \infty$ but there is no u such that $\Phi(u) = 0$.

6 Conclusion

The solutions to Newton's Problem of Minimal Resistance for the class of truncated cones and for the class of radially symmetric solids can be recovered from the corresponding geometric constructions of Newton. The minimizers for various classes of non-radially symmetric solids are not unique; in addition, the results for these classes are still far from complete. Furthermore, although other constraints for Newton's model of resistance are also considered in the literature (Miele 1965), to our knowledge rigorous results for models of resistance in other kind of fluids are still missing.

Acknowledgments Both authors thank the referee for his useful comments and remarks. The first author acknowledges Professor Abraham Medina-Ovando for his kind invitation to participate in the XIX Enzo Levi Seminar, held at Universidad Autónoma Metropolitana Azcapotzalco in Mexico City in the Spring of 2012.

References

Butazzo G, Ferone V, Kawohl B (1995) Minimum problems over sets of concave functions and related questions. Math Nachr 173:71–89

Butazzo G, Ferone V, Kawohl B (1996) A symmetry problem in the calculus of variations. Calc Var 4:593–599

Butazzo G, Kawohl B (2001) On Newton's problem of minimal resistance. Springer, Heidelberg

Cruz-Sampedro J, Tetlalmatzi-Montiel M (2010) POEM's and Newton's aerodynamic frustum. Coll Math J 41(2):145–153

Goldstine H (1980) A history of the calculus of variations. Springer, New York

Marcellini (1989) P Non-convex Integrals of the Calculus of Variations. In: Cellina A (ed) Proceedings of methods of non-convex analysis, Varenna, Lecture notes in mathematics No. 1446. Springer Verlag 1990:16–57

Miele A (1965) Theory of optimum aerodynamic shapes. Academic Press, New York

Newton I (1934) Philosophia naturalis principia mathematica. University of California Press, California (Motte's Translation)

Tikhomirov VM (1990) Stories about maxima and minima. In: Proceeding of AMS-MAA, Washington

Competitive Adsorption of Surfactants and Polymers on Colloids by Means of Mesoscopic Simulations

Armando Gama Goicochea

Abstract The study of competitive and cooperative adsorption of functionalized molecules such as polymers, rheology modifiers and surfactant molecules on colloidal particles immersed in a solvent is undertaken using coarse-grained, dissipative particle dynamics simulations. The results show that a complex picture emerges from the simulations, one where dispersants and surfactants adsorb cooperatively up to certain concentrations, on colloidal particles, but as the surfactant concentration increases it leads to dispersant desorption. The presence of rheology modifying agents in the colloidal dispersion adds complexity through the association of surfactant micelles to hydrophobic sites of these agents. Analysis of the simulation results reported here point clearly to the self-association of the hydrophobic sites along the different polymer molecules as the mechanism responsible for their competitive and cooperative adsorption.

1 Introduction

Polymer adsorption is crucial for the performance in modern applications of complex fluids, such as in stimuli—responsive systems, biological membranes, and consumer goods such as paints, cosmetics or food products. In particular, polymer adsorption on pigments surfaces remains a popular mechanism to stabilize architectural paints (Napper 1983). There are other types of polymeric molecules that can also be adsorbed on particles, such as surfactants and rheology modifying agents. These functionalized molecules have usually different lengths and interact not only with each other and the solvent, but also with themselves.

A. Gama Goicochea (✉)
Departamento de Ciencias Naturales, DCNI, Universidad Autónoma Metropolitana Unidad Cuajimalpa, Av, Pedro Antonio de los Santos 84, 11850 México, D. F., Mexico
e-mail: agama@alumni.stanford.edu; agama@correo.cua.uam.mx

J. Klapp and A. Medina (eds.), *Experimental and Computational Fluid Mechanics*, Environmental Science and Engineering, DOI: 10.1007/978-3-319-00116-6_10,

The characterization of polymer and surfactant adsorption is usually carried out through measurements of adsorption isotherms, which yield directly information about the optimal amount of polymer needed to achieve stability (Kronberg 2001). However, the simultaneous presence of more than one type of polymers in the dispersion can give rise to a complex combination of competition and synergy between polymer molecules, which leads to competitive adsorption isotherms. These types of experiments are laborious and time consuming, taking up to several weeks to complete. One attractive alternative is the use of molecular modeling using appropriately adapted algorithms for relatively complex fluids, which can then be solved highly accurately using modern computers.

This work is devoted to the presentation of coarse-grained computer simulations for the prediction and understanding of competitive adsorption isotherms of polymers and surfactants on colloidal particles. It is argued that the mesoscopic reach of the simulations carried out is especially important to obtain results that are directly comparable with scales probed with experiments on soft matter systems. This study, which is the first of its kind to the best of the author's knowledge, is a useful representation of architectural paints and coatings, as well as of other complex fluids of current academic and industrial interest.

2 Models, Methods and Systems

The force model used in the simulations presented here is a mesoscopic, coarse-grained method known as dissipative particle dynamics (DPD) (Hoogerbrugge and Koelman 1992). It involves central, pairwise forces between DPD beads, which are not physical particles but rather momentum—carrying sections of the fluid. There are three types of forces in the DPD model: a conservative force ($\boldsymbol{F}_C$), which determines the local hydrostatic pressure; a dissipative force ($\boldsymbol{F}_D$), that represents the local viscosity of the fluid, and a random force ($\boldsymbol{F}_R$), constituted by the Brownian motion of DPD beads. The latter two forces exactly balance each other by construction, as a result of the fluctuation—dissipation theorem (Groot and Warren 1997); this feature is the essence of the DPD model. The functional dependence of the forces is not specified by the DPD model, but they are usually chosen as simple as possible; the most employed ones are repulsive, linearly decaying (for $\boldsymbol{F}_C$) and short ranged. The structure of the DPD model, as well as some of its strengths and weaknesses are well known, and the reader is referred to recent reviews, like the one by Murtola et al. (Murtola et al. 2009) for details.

The systems studied are constituted by the polymeric molecules of different functionality (surfactants, dispersant polymers, rheology modifiers), the solvent (water), and the colloidal particles (pigments, fillers). The latter are typically much larger than the rest, so one can consider them as flat surfaces fixed in space, and then invest the computational cost on solving the motion of the rest of the particles. Although these polymeric molecules share the characteristic that they are amphiphilic in nature, they are usually distinguished by the role they play. Hence, surfactants are typically short molecules whose only purpose is to reduce the surface or interfacial tension. Dispersant polymers are longer and they are used to adsorb on colloids and keep them apart, hence their name. Rheology modifiers are large polymeric molecules, generally made of units of different chemical nature, with hydrophobic and hydrophilic parts. Their function is to modify the viscosity of the fluid in which they are dissolved. The polymeric molecules are constructed as DPD beads joined by freely rotating harmonic springs, and can be linear or branched; the solvent is represented by single beads and for the surfaces, an effectively exact DPD wall force is used, given by a repulsive, short range polynomial (Gama Goicochea and Alarcón 2011). For the surfactant, a non-ionic, linear, 14—bead polymer was used as a model for a nonylphenol etoxylate surfactant. The dispersant was modeled as a 48—DPD bead linear polymer, to represent a hydrophobic dispersant made of maleic anhydride and styrene. As for the rheology-modifying agent, I used a hydrophobically modified alkali-swellable emulsion (HASE) polymer, which is represented by 60 DPD units. In regards to the conservative DPD force interaction parameters, they have been chosen following the standard procedure (Groot and Warren 1997), beginning with the isothermal compressibility of water at room temperature to choose the equal—particle interaction. For different particles interaction, the Flory—Huggins parameter is used based on the chemical composition of each DPD bead. As for the choice of wall—DPD particle force, it has been chosen by fitting the interfacial tension values of the confined fluid with the wall—particle value. The interaction parameters as well as the specific bead sequence shall be omitted for brevity but may be consulted in reference (Gama Goicochea 2013), along with all simulation details.

Adsorption experiments are generally performed following a route in which the polymers to be adsorbed are added to the system and the measurements are performed when chemical, thermal and mechanical equilibrium is achieved. To properly reproduce those conditions, the simulations are carried out in the grand canonical thermodynamic ensemble, where the chemical potential, temperature and volume are kept constant as the polymer concentration is increased. The DPD method has been adapted to the grand canonical ensemble (constant chemical potential, volume and temperature) to obtain the competitive adsorption isotherms presented here. The procedure is the following: the volume of the simulation box is fixed ($L_x = L_y = 7$; $L_z = 14$ DPD dimensionless units), then a fixed number of one type of additives, say, dispersant polymers is added to it, along with a fixed number of rheology modifying agents. Then, the adsorption is monitored by adding molecules of, for example, surfactants to the box and performing the simulations until equilibrium has been achieved, while the temperature, volume and chemical potential are kept fixed. The chemical potential is fixed through the exchange of solvent particles with the virtual bulk. In the simulations, this chemical potential was fixed at $\mu = 37.7$ units so that the total average density in the simulation box was nearly equal to 3. In doing so, one ensures that the equation of motion of the DPD fluid is invariant under changes of the interaction parameters (Groot and Warren 1997). Full details of the DPD algorithm adapted to the grand canonical ensemble, as well as simulation details such as the integration algorithm, time step, simulation length, etc., can be found in (Gama Goicochea 2007).

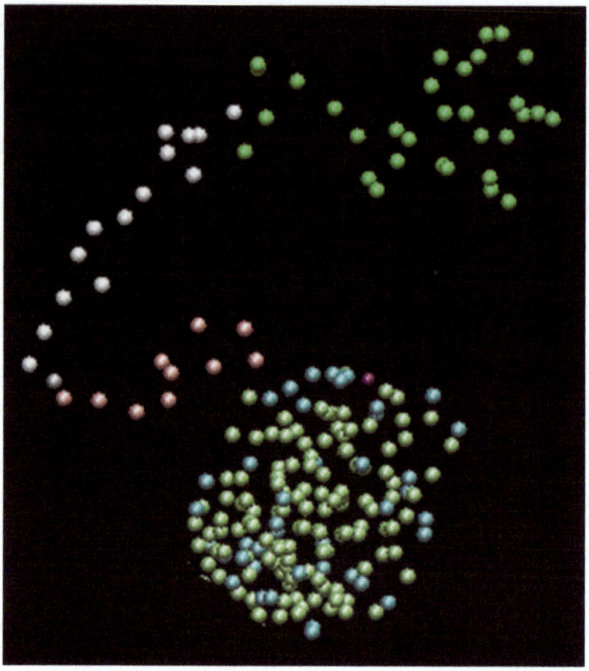

Fig. 1 Equilibrium configuration of a single linear molecule of a rheology-modifying agent (HASE) with a surfactant micelle formed at one of its hydrophobic sites. The colors represent the different chemical characteristics of the molecules (see Gama Goicochea (2013) to see the exact chemical composition and DPD mapping). The hydrophilic parts of the HASE and surfactant molecules, as well as the solvent molecules are omitted for clarity

3 Results

Let us first illustrate the capabilities of the DPD method by presenting the association of a surfactant molecules with a single HASE (rheology modifier) molecule. The system consists of 60 surfactant molecules, in addition to the HASE polymer, in solution with solvent monomers. No colloidal particles were present therefore periodic boundary conditions were used in all directions. All molecules positions are chosen at random initially and are allowed to evolve, subjected to the DPD forces. Figure 1 shows the final configuration obtained after equilibrium was reached.

As suggested by Fig. 1, HASE molecules modify the rheology of fluids by promoting the formation of surfactant micelles on specific hydrophobic sites on the HASE backbone. Self-association, and association between different HASE molecules can then be modulated through the judicious choice of surfactants, which in turn will modify the rheology of the fluid. This obviously follows from Fig. 1: when many HASE molecules are present in a solution with surfactants, they shall tend to associate as shown in Fig. 1 and therefore an association between HASE molecules will be unavoidable due to the steric interactions between those complex molecular conglomerates. Figure 1 represents a textbook example (Glass 2000) of the mechanism through which these types of molecules are thought to associate, but here it has been shown to emerge from molecular modeling.

I shall now proceed to the presentation of the adsorption isotherms, of which 2 different types were calculated. One, where the dispersant polymer concentration was fixed while the surfactant concentration was increased, and one where it was the surfactant concentration what was kept fixed while the dispersant concentration was varied. The purpose of carrying out the adsorption isotherms through these two routes is deciding which procedure leads to the optimal dispersion conditions. The fluid in all cases is confined by two different types of surfaces: one is a metal oxide, TiO_2, and the other is a silicate-based colloid with almost negligible interactions with the polymers involved in this study, whose only purpose is that of occupying space, hence its name "filler". The parameters of interaction between these surfaces and the DPD fluid have been tested and have been successfully used before, see Gama Goicochea (2007, 2013).

In the left of Fig. 2 I show the adsorption isotherm of the surfactant when the dispersant and the thickener (HASE) concentrations are fixed. It may appear that the surfactant adsorption is hardly influenced by the presence of the other types of polymers in the dispersion, for the saturation concentration of the surfactant remains almost constant. However, when the isotherm of the surfactant is obtained, at fixed dispersant concentration (without rheology modifiers), which is shown in the inset, the number of adsorbed surfactant molecules is found to increase slowly with added surfactant concentration. Hence, there is clearly an interplay between the surfactant and the dispersant, which enhances the adsorption of the surfactant by the thickener, cooperatively. While the surfactant adsorption is greatly influenced by the thickener and the filler, the dispersant is not. This conclusion is obtained from the right panel in Fig. 2.

The isotherm on the right in Fig. 2, which corresponds to that of the dispersant at fixed surfactant and rheology modifier concentrations, is not at all perturbed by these additives. When the adsorption isotherm for the dispersant only was calculated (not shown, for brevity), the same trend was obtained, namely, a constant saturation concentration, as shown on the right panel in Fig. 2. Therefore, the adsorption mechanisms that take place even if the components of the colloidal dispersion are the same, can change radically depending on the variable of control.

A clear image of the evolution of the adsorption process which may not be appreciated from the isotherms alone can perhaps be better gained from inspection of Fig. 3. In it I show snapshots obtained from the DPD simulations, after equilibrium was reached. At the lowest surfactant concentration ($[c] = 20$) all the dispersant is adsorbed on the TiO_2 surface, with the thickener almost completely extended and the surfactant associated with the dispersant. As the surfactant concentration is increased to $[c] = 40$ molecules, some of the dispersant molecules were desorbed and even migrated to the filler substrate, on the right. At the largest surfactant concentrations, the dispersant got even more desorbed, with the surfactant replacing it at the adsorption sites, on both substrates. The thickener shows self association (see the middle of the simulation box) and the dispersant prefers to associate with the surfactant and

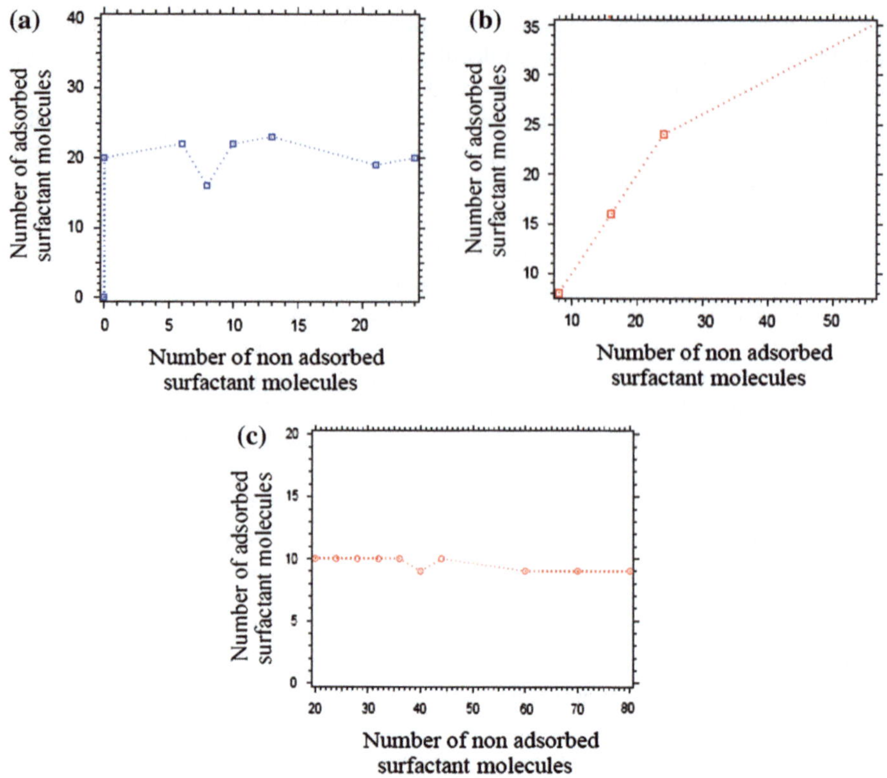

Fig. 2 Adsorption surfactant isotherm obtained for (a) fixed dispersant concentration (10 dispersant molecules, with the number of surfactant molecules varying from 20 up to 80) and (c) dispersant adsorption isotherm at a fixed surfactant concentration (10 surfactant molecules, with the number of dispersant molecules ranging from 6 up to 40). (b) The single (non competitive) isotherm for the surfactant alone. For cases (a) and (c) the system contains 6 HASE molecules and is confined by flat walls representing TiO_2 and a filler (silicate-based colloid) surfaces

the thickener rather than remain adsorbed. Evidently, at low concentrations the surfactant promotes the adsorption of the dispersant, i.e., they behave synergistically, whereas at large surfactant concentrations the opposite happens.

Precisely this type of behavior has been observed in experiments of competitive adsorption carried out with polymers and cationic, anionic and nonionic surfactants (Karlson et al. 2008) where the authors found that at low surfactant concentration, the polymer (which plays the role of the dispersant) remained adsorbed (on polystyrene and silica particles) while the surfactant formed micelles. As the concentration of the surfactant was increased, and if the polymer and the surfactant attract, they form complexes that can be desorbed. If one of them, be it the surfactant or the polymer has higher affinity for the surface, it will replace the other on the particle surface. The conclusions derived from the experimental model, water-based paint designed

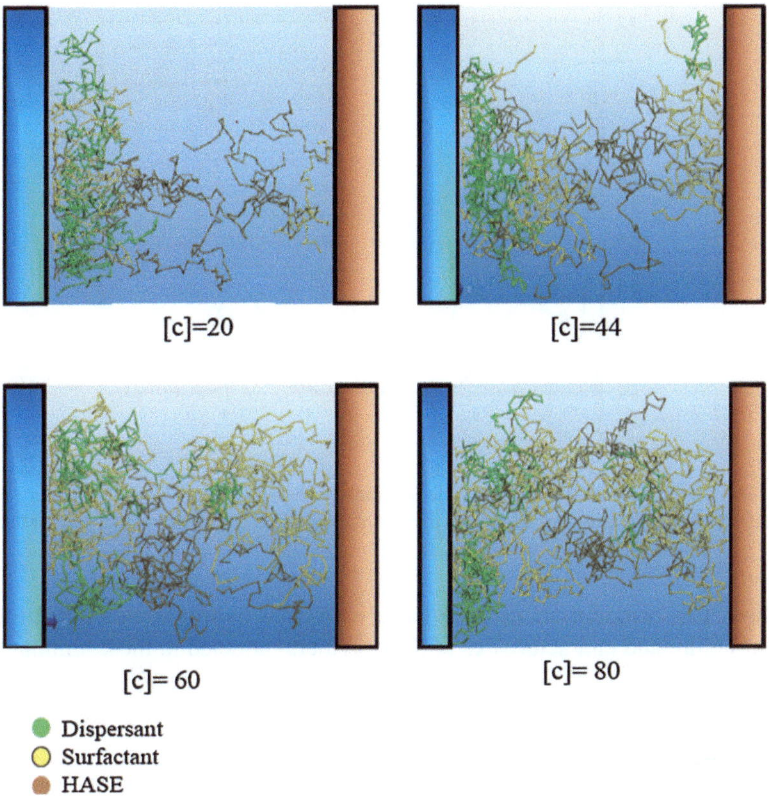

Fig. 3 Configuration of the dispersant (*green*), surfactant (*yellow*) and rheology modifier (*brown*) molecules as the surfactant concentration is increased, from 20 up to 80 molecules. For all cases the system contains 10 dispersant molecules and 6 HASE molecules and is confined by flat walls representing TiO_2 (*left*) and filler (*right*) surfaces. The solvent has been removed for clarity

by Karlson and co workers are fully supported by the results of the simulations presented in this work.

The simulations presented here give additional insight into why the phenomena presented in Figs. 2 and 3 occur. Figure 4 shows the density profiles of the hydrophobic sections of all three types of polymers in the colloidal dispersion.

The density profiles shown in Fig. 4 shows that the polymers associate because of the affinity of their hydrophobic sections, as clearly indicated by the maxima around $z = 5$ and $z = 15$ (dimensionless units). Although most of the dispersant remains adsorbed on the TiO_2 surface (on the left), some of it desorbed and formed a complex associated structure with the surfactant and the rheology modifier close to the pigment. Additionally, the surfactant formed a micelle around the hydrophobic sites of the thickener, and some dispersant molecules were completely desorbed and associated with the surfactant micelle, as shown by the structure form around $z = 15$.

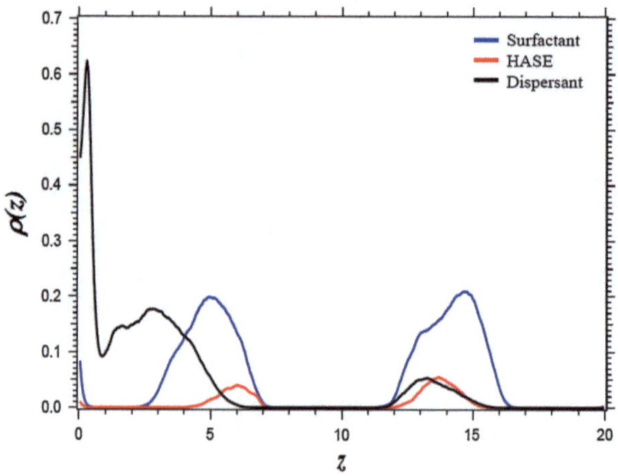

Fig. 4 Density profiles of the hydrophobic sections of the surfactant (*blue*), thickener (*red*) and dispersant (*black*). The pigment surface is the one on the *left* and the filler surface is on the *right*

Obviously this behavior arises from basic molecular hydrophobic interactions due to the structure and characteristics of the polymers modeled in these simulations.

4 Conclusions

The complex mechanisms that give rise to competitive and cooperative adsorption of polymers with different functional groups in a colloidal dispersion were studied for the first time, using mesoscopic, DPD computer simulations. Two different colloidal particles were included: a pigment (TiO_2) and a silicate-based filler. The surfactant, dispersant and rheology modifying polymers were found to associate cooperatively at low surfactant concentration, promoting the full adsorption of the dispersant which, in turn, leads to a more stable paint. This is the result of the affinity that the hydrophobic groups present in all three types of molecules have. However, as the surfactant concentration is increased, the same affinity of the hydrophobic groups makes it energetically and entropically more advantageous for some of the dispersant molecules to be desorbed, forming micelles with the thickener that eventually lead to a less stable dispersion. It was argued that these conclusions are fully supported by recent experiments on model paints. This work is expected to be useful not only to formulators and expert designers of modern water-based paints and coatings, but also to those studying smart materials and biological membranes.

Acknowledgments The author is indebted to the following individuals for enlightening discussions: F. Alarcón, M. Briseño, N. López, A. Ortega, H. Ortega, E. Pérez, and F. Zaldo. This work was sponsored in its initial phase by the Centro de Investigación en Polímeros (Grupo Comex), and afterward by PROMEP through project 47310286-912025.

References

Glass JE (2000) Associative polymers in aqueous media, ACS Symposium Series 765. Chap. 21, pp 22

Gama Goicochea A (2007) Adsorption and disjoining pressure isotherms of confined polymers using dissipative particle dynamics, Langmuir 23:11656

Gama Goicochea A (2013) Supplementary information for 'Competitive adsorption of surfactants and polymers on colloids by means of mesoscopic simulations' in https://www.researchgate.net/profile/Armando_Gama_Goicochea/

Gama Goicochea A, Alarcón F (2011) Solvation force induced by short range, exact dissipative particle dynamics effective surfaces on a simple fluid and on polymer brushes. J Chem Phys 134:014703

Groot RD, Warren PB (1997) Dissipative particle dynamics: bridging the gap between atomistic and mesoscopic simulation. J Chem Phys 107:4423

Hoogerbrugge PJ, Koelman J (1992) Simulating microscopic hydrodynamic phenomena with dissipative particle dynamics. Europhys Lett 19:155–160

Karlson L, Olsson M, Bostrom G, Picullel L (2008) Influence of added surfactant of particle flocculation in waterborne polymer-particle systems. J Coat Technol Res 5:447–454

Kronberg B (2001) Measuring adsorption. In: Krister H (ed) Handbook of applied surface and colloid chemistry 2, Chapter 22. Wiley, Chichester

Murtola T, Bunker A, Vattulainen I, Deserno M, Karttunen M (2009) Multiscale modeling of emergent materials: biological and soft matter. Phys Chem Chem Phys 11:1869–1892

Napper DN (1983) Polymeric stabilization of colloidal dispersions. Academic Press, New York

Annular Two-Phase Flow Regimen in Direct Steam Generation for a Low-Power Solar System

Iván Martínez, Rafael Almanza, María Dolores Durán and Miriam Sánchez

Abstract This study aims to quantify and to model the temperature profile around an absorber tube belonging to a parabolic trough concentrator when fluid is applied at low powers. This study was specifically developed for the Solar Power Plant of the Engineering Institute, National University of Mexico. This work presents experimental results under saturated conditions and low pressures (1.5–3 bars) using water as the thermal and working fluid for direct steam generation (DSG). The control variable was feed flow. Solar irradiance was used as the restriction variable because all experimental tests should be developed under very specific values of this variable (for example, $I > 700$ W/m^2). The objective of this experiment was to study the thermal behavior of a temperature gradient around the absorber tube under steady-state conditions and with low flow. Additionally, a theoretical analysis was carried out by means of the homogeneous heat conduction equation in the cylindrical coordinate system using only two dimensions (r, φ). The finite-difference numerical method was used with the purpose of proposing a solution and obtaining a temperature profile. The aim of this theoretical analysis was to complement the experimental tests carried out for direct steam generation (DSG) with annular two-phase flow patterns for low powers in parabolic trough concentrators with carbon steel receivers.

Nomenclature

A	area in m^2
Cp	heat capacity in kJ/(kg · K)

I. Martínez (✉) · M. D. Durán · M. Sánchez
Facultad de Ingeniería, Universidad Autónoma del Estado de México (UAEMex) Cerro de Coatepec s/n, Ciudad Universitaria, 50100 Toluca, México
e-mail: igmartinezc@uaemex.mx

R. Almanza
Instituto de Ingeniería, Universidad Nacional Autónoma de México (UNAM), Ciudad Universitaria, 04510 Coyoacán, D.F., México

J. Klapp and A. Medina (eds.), *Experimental and Computational Fluid Mechanics*, Environmental Science and Engineering, DOI: 10.1007/978-3-319-00116-6_11,

d	pipe diameter in m
G	irradiance in W/m^2
h	convective heat transfer coefficient in $W/(m^2K)$
k	thermal conductivity in $W/(m \cdot K)$
$\overline{Nu}$	overall Nusselt number
P	pressure in bar
Pr	Prandtl number
q''	heat flow in kW/m^2
$\dot{q}$	energy generation in kW/m^2
Q	volumetric flow in L/s
r	radius in m; radial coordinate
Re	Reynolds number
t	time in seconds
T	temperature in K
u	velocity in m/s

Greeks

α	absortance; void fraction
Δ	interval or difference
ε	emittance
φ	radial angle; azimuth coordinate
ν	kinematic viscosity in m^2/s
ρ	reflectance; density in kg/m^3
σ_{SB}	Stefan-Boltzman constant

Subscripts and Superscripts

a	air
atm	atmospheric
cov	convective
Fe	Iron
l	liquid
rad	radiation
sup	superficial
S	solar
T	total
int	internal
ext	external

1 Introduction

Direct steam generation (DSG) in parabolic trough concentrators is a technique being increasingly developed by many researchers world-wide. The method has great potential to improve hybrid power systems and retain competitive energy prices. In this case, the main interest is to develop a low power system (Almanza et al. 2002). During the operation of the Solar Power Plant of the Engineering Institute, National University of Mexico (UNAM), it was observed that some problems for DSG occurred under specific flow and temperature conditions. These problems corresponded to the stratified two-phase flow regimen that occurs when boiling water is transferred into the receiver tube when the thermal gradient in its periphery is increased (Almanza et al. 2002). Such problems are related to the bending of tubes owing to thermal stress, causing their permanent deformation and breaking their glass covers.

In order to find a solution to this problem, a project was carried out that included experimental tests and the production of a mathematical model to predict stratified two-phase flow pattern under transient conditions of normal solar beam irradiance to low flows (Flores 2003). This study considered flows between 1 and 2.5 L/min, with solar beam irradiance that fell on the lower part of pipe or on one side of the pipe. It is essential to understand the thermal behavior of absorber pipes under annular two-phase flow conditions with low pressures and low flows, in order to contribute to the development of low power systems like the ones proposed by Almanza et al. (2002). In order to continue the project previously described, it was considered necessary to develop experimental tests and a mathematical model for annular flow pattern during DSG. Preliminary results were generated by Martínez and Almanza (2003) but, because of high uncertainty, experimental tests required more replication and better data acquisition techniques.

These works form part of the basis required to develop a proposal for direct commercial application of DSG using parabolic trough concentrators in Mexico. For example, it is possible that in the near future the Federal Commission for Electricity (CFE) of Mexico, along with the Engineering Institute (UNAM) and CIEMAT (Centro de Investigaciones Energéticas, Medioambientales y Tecnológicas) from Spain, might build an experimental hybrid installation in Baja California State, Mexico. However, it is known that there is a high probability that annular two-phase flow exists in most of their pipes. It is hence required to experimentally explore and evaluate this idea. It is necessary to increase the mass quality of steam, from the current mixture of 40 % steam and 60 % brine found in the geothermal wells of Cerro Prieto in the Northwest of Mexico. This brine has a mass concentration of different sorts of salts, mainly NaCl, of approximately 2 %. Therefore, it is not advisable to increase this parameter too much in order to avoid the buildup of scale in the receiver pipes, which is important in maintaining the annular flow pattern along the DSG with solar energy.

A geothermal system in Cerro Prieto works well with low pressures between 10 and 16 bars. Hence, it is possible that all tests made at the Engineering Institute's solar power plant could be extended for parameters like these, with the purpose of establishing better operational conditions with a steady-state regimen for the hybrid system proposed. Therefore, annular two-phase flow is a good regime to minimize thermal gradients around absorber pipes so to avoid thermal stress.

2 Experimental Development

The main intention of this experiment was to determine the temperature gradient around the receiver tube under steady-state conditions with an annular two-phase flow pattern. In order to achieve this, it was necessary to maintain control over the feed water (the control variable), which was done by choosing three volumetric flow rates; 4, 8, and 12 L/min. Each test was replicated at least five times per value.

The temperature and pressure were measured using a high performance acquisition system registered at the beginning and at the end of the solar field. The steam volumetric flow, which contains a small quantity of liquid drops because the steam trap cannot separate them completely, was also measured. The liquid from the steam trap was quantified and it was possible to establish its effectiveness in approximately 80 % of the trials. Temperatures, as detected by an RTD (Platinum Resistance Temperature Detector) fixed on the external surface around the end of the last absorber pipe, were captured by another acquisition system.

Additional data recorded by the weather station were also taken into account in order to determine the possible influence of these parameters on the values of the process variables and the superficial absorber temperature.

Steady-state conditions were reached when the fluid was under saturated conditions that depended on pressure. The mass quality of the generated steam was hence dependent on feed flow and the irradiance level. The experiments were carried out near 800 W/m^2 of normal solar beam irradiance. Each experiment was conducted maintaining the flow with constant feeding, as for a sub-cooled liquid, so that any change in the process could cause changes in the amount of heat absorbed by the working fluid and hence be automatically controlled to be within $\pm$0.5 L/min variation. This sort of oscillation cannot be avoided and, on some occasions, the system oscillated beyond this control range. In these cases it was necessary to terminate the experiment. The process variables (pressure, flow and temperature) were registered every two seconds, and the absorber's surface temperature was measured every five seconds.

The data registered during the commencement and termination of each experimental test were not taken into account for this study, although it was noted that they were very useful for analyzing the thermal behavior of this system during transient conditions.

2.1 Description of Facilities and Equipment

All tests were carried out in the solar field of parabolic trough concentrators located at UNAM in Mexico City (19° 19' 6.9" north latitude, 99° 11' 29.7" west longitude)

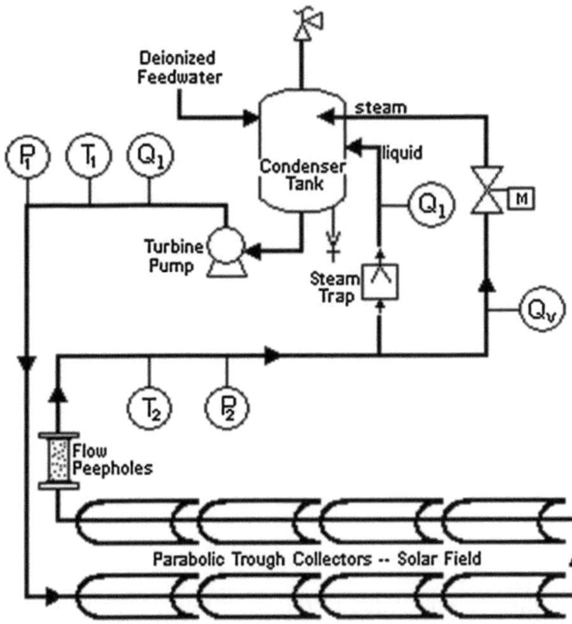

Fig. 1 Solar power plant, Engineering Institute (UNAM)

and at an altitude of 2330 ± 20m meters above sea level (as measured with a GPS). Eight modules are connected in serial mode, each one with a length of 15 m and an aperture area of 34m^2 (Fig. 1). The total aperture area of solar field is 272m^2, with mirrors whose reflectance is approximately 0.85. The absorber pipes are 25.4 mm (1") nominal diameter and are covered with a black chrome selective film with an absorptance of approximately 0.89 and emittance of approximately 0.18 at 25 °C.

The solar plant has a water deionizer system that works with ionic interchange resins, so the water conductivity is reduced to 0.39 mS/cm at 20 °C. Although the main application of this study is for geothermal wells, as a first step it would be better to start with this water quality in order to avoid damage to the experimental equipment. This water is stored within a cistern, which serves as a reservoir for a feeding tank that is connected to one condenser, and this also serves as a feed water tank. The pump is a double pass regenerative turbine with a capacity ranging from 3 to 32 L/min. It is controlled by a 3 hp motor with a current frequency converter.

At the end of the 8th module concentrator, the two-phase flow achieved can be observed through two peepholes made of borosilicate glass of 24.5 mm diameter and 30 cm length. With these elements is was possible to visually determine the flow pattern that was obtained in the experimental tests. In order to show how a two-phase flow looks through a flow-peephole a photograph is shown at Fig. 2.

After the peepholes, a steam trap serves to separate the phases and allows the flow to be measured for each phase. Both currents reach the condenser tank and a

Fig. 2 Flow peephole photograph with annular two-phase flow

Table 1 Average values of feed water

Experimental test (L/min)	Feed flow (L/min)	Velocity (m/s)
4	4.3	0.47
8	8.1	0.87
12	12.6	1.37

saturated liquid is then re-circulated to the solar field. Measuring instruments used to make recordings were:

– 8 RTDs (platinum Resistant Temperature Detector) around the absorb er pipe.
– 2 RTDs at the beginning and the end of the system.
– 2 pressure transducers at the beginning and the end of the system.
– 1 'Headland' variable area transducer for measuring flow of liquid.
– 1 'Endress & Hauser' vortex sensor transducer for measuring flow of steam.
– 9 pressure dial indicators along the tube
– 8 temperature dial indicators along the tube

In addition, data were collected from a meteorological station that registers the values of global and diffuse horizontal irradiance (using a rotating shadow band pyranometer), dry bulb temperature, and speed and direction of wind every 5 min. Direct normal irradiance was also calculated.

2.2 *Experimental Results and Discussion*

The temperature distribution around the absorber pipe when the working fluid has an annular two-phase flow pattern is the most interesting result of this work. The goal of always working with a fluid in conditions of saturation with constant feeding flow was achieved (Table 1), and the maximum pressure reached on the system was 3.5 bars. The next step will be to repeat the experiment at 10 bars. The Goebel (1997) and Herbst et al. (1996) experiments were useful for this study in order to know what happens with DSG under higher values of process conditions.

Because the experimental tests were carried out in winter, the heating tended to be from below and to one side of the receiver, between $3/4\pi$ and $7/4\pi$ radians of the circumference. As a result, it was necessary to carry out tests by heating exclusively from below part of the absorber pipe.

Owing to small instabilities during the heating process, such as normal beam irradiance changes, pressure values oscillated throughout each test. It was thus necessary to calculate average values of pressure, temperature and velocity. This allowed the average value of each process property to be estimated from simply fixing the feed. At the beginning, the feed had a specific velocity which was constant until the boiling process started. When steam appeared we started to increase the feed's velocity in order to maintain a good mass balance. The liquid phase then changed its initial velocity according to the process conditions. For example, 4 L/min flow had 27 % reduction of its speed value, for 8 L/min the reduction was 7 % and for 12 L/min there was no change because variations were minimal with respect to feed. Hence the liquid velocity diminished for the first two tests at 4 and 8 L/min, whereas for the third test, at 12 L/min, it stayed relatively constant.

Temperature and pressure behaviors of feed flow were similar to a sub-cooled liquid (97 °C and 3.5 bars). These values are important in order to reach boiling point as soon as possible in the receiver pipes. It was observed that for greater flow, the system was more stable with these parameters. It was also observed that its variation did not have great repercussions with respect to low feed flow throughout heating in the solar concentrators. Its behavior is more stable when the evaporation stage begins.

In order to verify changes in temperature and pressure during the heating and evaporation process of water in the absorber tubes, dial indicators for temperature and pressure were installed in the bridges between each module of concentrators. At the end of the last module, the variables were measured with instruments connected to a data acquisition system.

Pressure is the main parameter affecting inlet and outlet currents. The other parameter that independently affects the process is solar beam irradiance, a factor that depends on the weather. The graph of outlet pressures (Fig. 3) shows a uniform behavior; nevertheless, small variations in pressure that occurred were translated into temperature changes. The velocities reached for the generated steam are shown in Fig. 4, and the typical patterns for different feed flows can be seen under similar conditions of irradiance, which oscillated around $\pm 10\,\%$. In order to calculate the steam quality, a phase separator with a steam trap was installed and steam flow was measured with a flow meter.

The mass balance was calculated under similar values of irradiance, and the results are presented in Fig. 5. Figure 6 shows a graph depicting the relationship between liquid velocities and steam velocities; this demonstrates that a smaller feed flow generates a greater amount of steam, and therefore the liquid film is thinner and the friction factor has more influence on its velocity.

In agreement with Hahne et al. (1997), the convective heat transfer coefficient increases if the liquid film thickness diminishes. The measurements confirm this effect because external temperature is higher at 12 L/min than at 4 L/min.

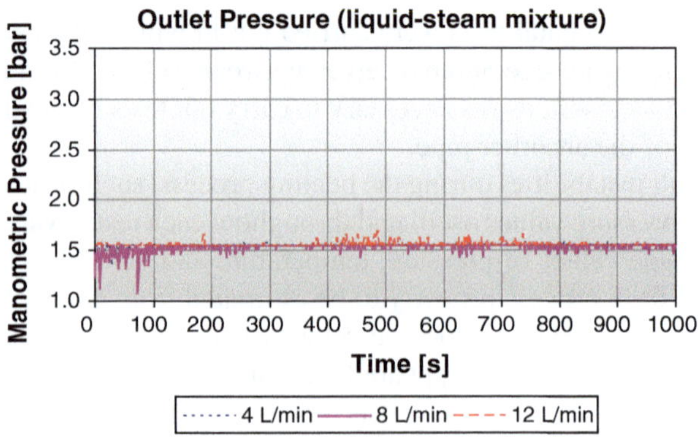

Fig. 3 Liquid-steam mixture manometric pressure

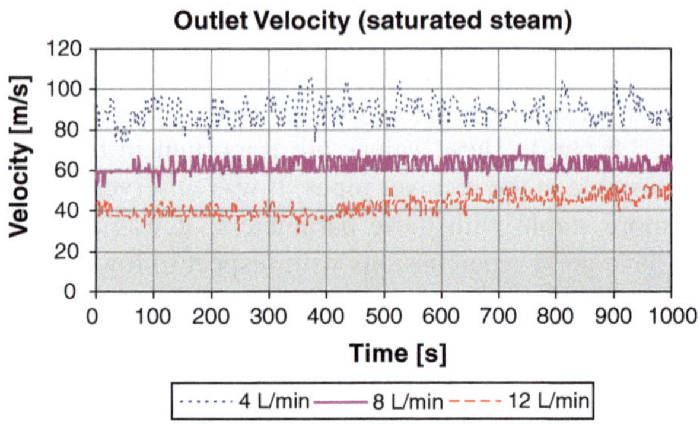

Fig. 4 Outlet steam velocity

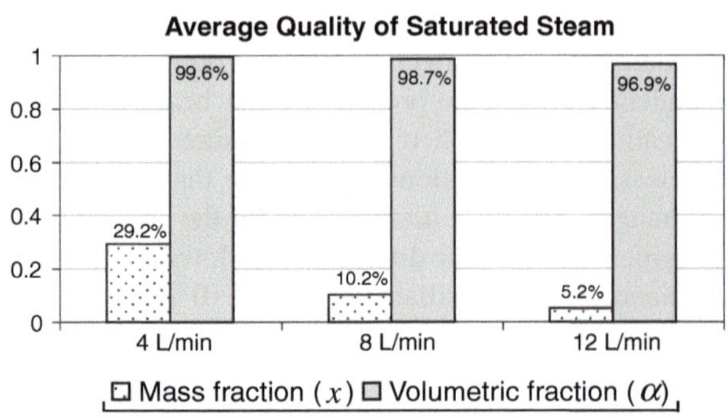

Fig. 5 Mass and volumetric fraction for outlet steam

Nevertheless this condition is directly related to the amount of normal beam irradiance that is being received in each test. Table 2 shows the average values of normal beam irradiance that were recorded during these tests. The highest temperature dif-

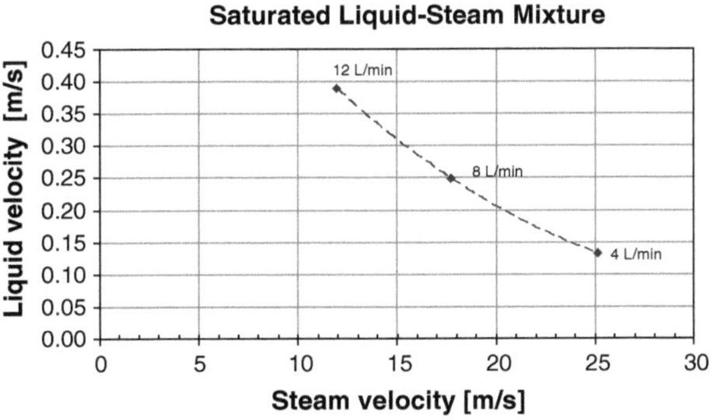

Fig. 6 Relation of velocities between phases and the feed flow

Table 2 Normal beam irradiance for experimental tests (average values)

Irradiance (W/m^2)	4 (L/min)	8 (L/min)	12 (L/min)
Global	934	895	893
Normal beam	882	793	821
Diffuse	52	102	72

ferential value that was reached with annular flow was 41 K, which persisted for only a few minutes. If an average differential of all maximums is calculated, the resulting value is 31 K.

3 Theoretical Development

Equation 1 represents the general form of the heat diffusion equation in the cylindrical coordinate system; it determines the transference velocity of energy by conduction in a unitary volume, at any point within the work media. In addition, the volumetric generation velocity of thermal energy must be equal to the change velocity of the stored thermal energy within this volume (Incropera and DeWitt 2001). Specifically, the system under study takes the form of tubes which are warmed up by concentrated solar radiation falling on half of their external circumference. Therefore heat flux, as shown in Fig. 7, has been established for very clear conditions and to enable subsequent analysis of these conditions.

$$\frac{1}{r}\frac{\partial}{\partial r}\left(kr\frac{\partial T}{\partial r}\right)+\frac{1}{r^2}\frac{\partial}{\partial \varphi}\left(k\frac{\partial T}{\partial \varphi}\right)+\frac{\partial}{\partial z}\left(k\frac{\partial T}{\partial z}\right)+\dot{q}=\rho Cp\frac{\partial T}{\partial t} \qquad (1)$$

The following conditions are proposed in order to develop a simplified analysis: Steady state $\left(\frac{\partial T}{\partial t}=0\right)$; no variation along longitudinal axis $\left(\frac{\partial T}{\partial z}=0\right)$; it does not have internal heat generation ($\dot{q}=0$).

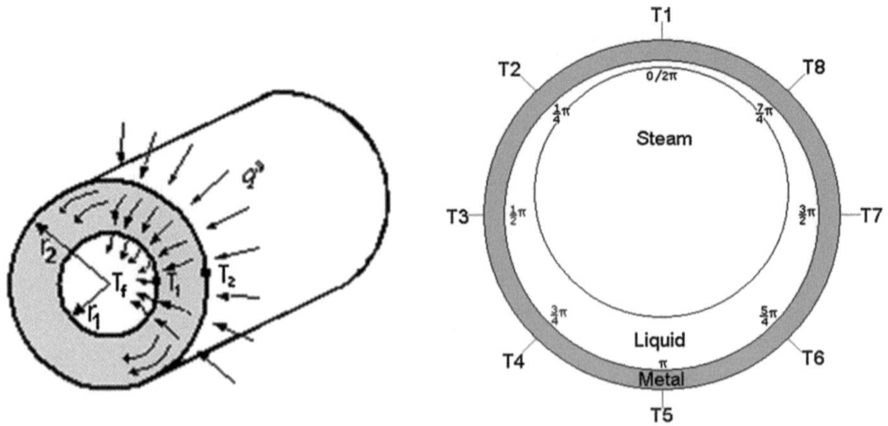

Fig. 7 Heat flux and reference point in an absorber pipe

Because of the preceding conditions, the diffusion equation is reduced to Eq. 2.

$$\frac{1}{r}\frac{\partial}{\partial r}\left(kr\frac{\partial T}{\partial r}\right)+\frac{1}{r^2}\frac{\partial}{\partial \varphi}\left(k\frac{\partial T}{\partial \varphi}\right)=0 \tag{2}$$

Numerical methods are useful for solving heat transfer problems in order to obtain reliable results, when such problems can not be handled by exact analysis because of nonlinearities, complex geometries, and complicated boundary conditions. One major approach currently used in the numerical solution of partial differential equations of heat transport is the finite-difference method. The boundary conditions that correspond to this problem are of the third kind type, since they express the heat flux between the surface of the system under study and a moving fluid that is in contact with this surface, at a precisely-known temperature. This condition must demonstrate the relationship between the heat transmitted by conduction in the system and the heat transferred by convection.

3.1 Boundary Conditions

(1) If $r = r_{\text{int}} \quad \wedge \quad 0 \leq \varphi < 2\pi$ then $-k\frac{\partial T}{\partial r} = h_l\,(T - T_l)$
(2) For concentrated irradiance on the lower part of the receiver tube:
If $r = r_{ext} \quad \wedge \quad 1/2\pi < \varphi < 3/2\pi$ then $-k\frac{\partial T}{\partial r} = q'' - h_a\,(T - T_a)$
If $r = r_{ext} \quad \wedge \quad 0 \leq \varphi \leq 1/2\pi \quad \wedge \quad 3/2\pi \leq \varphi \leq 2\pi$ then $-k\frac{\partial T}{\partial r} = h_a\,(T - T_a)$
(3) For concentrated irradiance on one side of the receiver tube:
If $r = r_{ext} \quad \wedge \quad 0 < \varphi < \pi$ then $-k\frac{\partial T}{\partial r} = q'' - h_a\,(T_a - T)$
If $r = r_{ext} \quad \wedge \quad \pi \leq \varphi \leq 2\pi$ then $-k\frac{\partial T}{\partial r} = h_a\,(T - T_a)$

3.2 Finite Differences

Different numerical methods allow the determination of the temperature only at discrete points, in contrast with an analytical solution that allows the determination of temperature at any point of interest in a specific environment. Therefore in order to obtain functions which describe the distribution of temperatures around a receiver tube, the first step is to select reference points, which usually are called "nodal points" or simply "nodes", so that a set of these points is known as a nodal grid Özişik 1993.

Each node represents a specific region whose value is a measurement of the average temperature in that region. The selection of these points depends generally on geometric convenience and the degree of precision that is desired. Figure 8 shows a proposed cross-sectional profile for this work and it displays the proposed nodal network in external circles that correspond to the walls of the absorber pipe. A third circle is also shown at the centre of Fig. 8, which has a slight displacement on the vertical axis, representing the steam-liquid interface. Therefore Fig. 8 exhibits a two-phase flow with an annular pattern. Such displacement is considered to be a phase velocity function as well as a function of liquid film thickness in the upper part of the pipe compared to the lower part.

Using this method, the heat equation is obtained by applying the law of energy conservation to a control volume around the nodal region, supposing that all the heat flow occurs towards the node (Rohsenow and Hartnett 1973). Heidemann et al. (1992) used the finite differences method in order to propose a mathematical model for steady-state and transient conditions of temperature during DSG with stratified two-phase flow.

The bi-dimensional model considers the following elements: carbon steel pipe, liquid-steam flow with annular pattern, and air atmosphere around the external surface of the receiver tube. The reference angle is in radians, and begins in the upper part of the pipe in order to agree with Flores (2003). The general exposition of the energy balance equations for each node is based on the nodes that surround it. For this reason, two sorts of equations can be written according to the location of such nodes.

For the equation that represents the nodes from 1 to 60, a radiation term is added which can be greater than or equal to zero, depending on the boundary conditions that are being considered. The energy balance begins with Node 1, expressed in Eq. 3.

$$\frac{k_{Fe}\cdot(T_2+T_{60}-2T_1)(\Delta r/2)}{r_{ext}\Delta\varphi}+\frac{k_{Fe}\cdot(T_{61}-T_1)\cdot(r_{ext}-\Delta r/2)\cdot\Delta\varphi}{\Delta r}+q''_T=0 \tag{3}$$

where,

$$q''_T = q''_{rad} + q''_{\text{cov}} - E_{\text{sup}} \Rightarrow \left\{\begin{array}{l} q''_{rad} = \alpha_S G_S + \alpha_{atm} G_{atm} \\ q''_{\text{cov}} = h_a r_1 \Delta\varphi (T_a - T_1) \\ E_{\text{sup}} = \varepsilon_{\text{sup}} \sigma T^4_{\text{sup}} \end{array}\right\} \rightarrow G_{atm} = \sigma_{SB} T^4_{atm}$$

The heat contribution by radiation only applies for those nodes in which concentrated irradiance is being considered on half their circumference, whereas for nodes

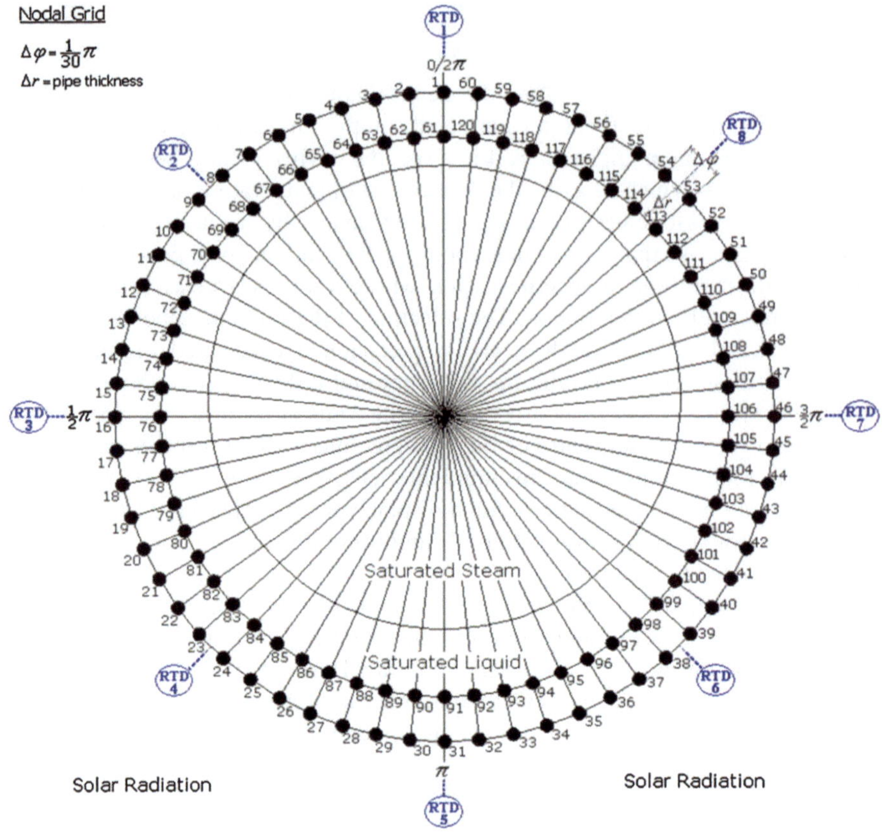

Fig. 8 An (r, ϕ) nodal network for an absorber pipe with annular two-phase flow

that are not exposed to the concentrator effect, this term is near zero. After regrouping Eq. 3, each part can be modified in order to have an equation for each node. For example, node 1's equation is shown as Eq. 4.

$$-\left[\frac{\Delta r}{r_{ext}\Delta\varphi}+\frac{(2r_{ext}-\Delta r)\Delta\varphi}{2\Delta r}+\frac{h_a r_{ext}\Delta\varphi}{k_{Fe}}\right]T_1+\frac{\Delta r}{2r_{ext}\Delta\varphi}T_2 + \frac{\Delta r}{2r_{ext}\Delta\varphi}T_{60}+\frac{(2r_{ext}-\Delta r)\Delta\varphi}{2\Delta r}T_{61}=-q''_T \quad (4)$$

where $q''_T = \frac{\varepsilon_{\sup}\sigma T_{\sup}^4-(\alpha_S G_S+\alpha_{atm}\sigma T_{atm}^4+h_a r_1\Delta\varphi T_a)}{k_{Fe}}$

In order to enumerate Eq. 4, it is necessary to know the convective heat transference coefficient of the air next to the external surface of the receiver, which is determined by applying Eqs. 5, 6, and 7 (Bejan 1995).

$$\overline{Nu} = 0.3 + \frac{0.62 R_e^{1/2} P_r^{1/3}}{\left[1 + (0.4/\Pr)^{2/3}\right]^{1/4}} \left[1 + \left(\frac{Re}{282000}\right)^{5/8}\right]^{4/5} \tag{5}$$

conditions:

$$\left\{ \begin{array}{l} Re \cdot \Pr > 0.2 \\ 7 \times 10^4 < Re < 4 \times 10^5 \end{array} \right\} \tag{6}$$

Annular two-phase flow regimen in direct steam generation.
where:

$$Re = \frac{u_a \cdot d_{ext}}{\nu_a} \text{ and } h_a = \frac{\overline{Nu} k_a}{d_{ext}} \tag{7}$$

Correlations expressed in Eqs. 5 and 6 assume the following conditions:

- Single cylinder in cross-flow
- Speed of the air-flow is uniform
- Temperature of the air-flow is equal to room temperature

For nodes that are in the internal part of the steel tube, i.e. nodes from 61 to 120, the energy balance is expressed by Eq. 8.

$$\frac{k_{Fe}(T_1 - T_{61})(r_{int} + \Delta r/2)\Delta\varphi}{\Delta r} + \frac{k_{Fe}(T_{62} + T_{120} - 2T_{61})(\Delta r/2)}{r_{int}\Delta\varphi} + h_f r_{61} \Delta\varphi (T_f - T_{61}) = 0 \tag{8}$$

After rearranging Eq. 8, we obtain the expression (Eq. 9) that we will use to calculate the temperatures in the aforementioned nodes.

$$\frac{(2r_{int} + \Delta r)\Delta\varphi}{2\Delta r} T_1 - \left[\frac{\Delta r}{r_{int}\Delta\varphi} + \frac{(2r_{int} + \Delta r)\Delta\varphi}{2\Delta r} + \frac{h_f r_{int}\Delta\varphi}{k_{Fe}}\right] T_{61} + \frac{\Delta r}{2r_{int}\Delta\varphi} T_{62} + \frac{\Delta r}{2r_{int}\Delta\varphi} T_{120} = -\frac{h_f r_{int}\Delta\varphi}{k_{Fe}} T_f \tag{9}$$

The heat convective coefficient of heat transfer for two-phase flow was calculated by means of a self-developed algorithm that is described by Martinez (2005) and is supported from different proposed correlations by authors such as Goebel (1997), Gungor and Winterton (1986), and Kattan et al. (1998).

Once all equations for each node have been considered, it is possible to solve them by means of any algebraic method. For this work it is possible to assume the following values are constants: the pressure of the system for the two-phase mixture, the cross-sectional area of the receiver, and the mass flow of the mixture. One assumes that the convective coefficient of heat transfer is based on the liquid film thickness, and it can be calculated according to Luninski et al. (1983).

Table 3 Data used for the simulation of temperature profile

Symbol	Quantity	Unit	Description
Ta	349.5	(K)	Air film temperature
Tf	406.15	(K)	Temperature of the saturated fluid
ha	8.05	(W/m^2K)	Air convective heat transfer coefficient
hf	6,768.1	(W/m^2K)	Fluid convective heat transfer coefficient
kFe	50.89	($W/m \cdot K$)	Iron thermal conductivity
r_{ext}	0.0167	(m)	External pipe radius
r_{int}	0.0131	(m)	Internal pipe radius
Δr	0.0036	(m)	Pipe thickness
$\Delta\phi$	0.1047	(rad)	Angular separation between nodes
G_{dir}	550	(W/m^2)	Useful solar beam irradiance
G_{tub}	27,614	(W/m^2)	Concentrated solar beam irradiance

3.3 Results and Discussion

In order to obtain a theoretical temperature profile as function of r and ϕ, it is necessary to adjust some process parameters that are expressed in Table 3. We know that in annular flow conditions the liquid film thickness at the top of the pipe is thinner than at the bottom. It is possible, therefore, that the value of the two-phase heat transfer convective coefficient changes as a function of the film thickness.

It is a known fact that heat losses always exist, and it is therefore necessary to determine a thermal efficiency to quantify these losses. Nevertheless, it is possible to assume that heat losses only exist when the receiver temperature is lower than the working fluid. When the calculated values of temperature are compared with the experimental data, it is necessary to take this assumption into account.

In calculating the effect of heat transfer during the liquid's boiling process, we can assume the liquid-steam interface can be almost homogeneous and have a variable distance with respect to the centre of the tube, the reason being that the thickness of the annular liquid film will be based on slip ratio and the wet angle calculated from 0 radians (top part) to π radians (bottom part) in such a way that the pattern is symmetrical in the interval from π to 2π radians.

In order to know what happens when the convective heat transfer coefficient is considered constant (theoretical A) or variable with respect to the liquid film thickness (theoretical B), a relative error factor was calculated and results were compared with experimental data. Table 4 shows a maximum value of 18.7 % for option A and 5.3 % for option B. Figure 9 shows the comparison and data behaviour around the external surface of the absorber pipe. The calculated data of liquid film thickness were 1.06mm at the top and 2.66mm at the bottom. These values accorded with flow peephole observations, where it was possible to see a double thickness of liquid film at the lower part compared to the upper part, and the estimated thickness observed correlated approximately with the values already given above.

Table 4 Relative error between theoretical model and experimental temperature

RTD No.	Node No.	ϕ angle (rad)	Temperature (°C)			Relative error	
			Experimental	Theoretical A	Theoretical B	A (%)	B(%)
1.	1	0.00	107.0	108.0	104.5	1.0	2.4
2.	8	0.79	117.6	126.6	120.1	7.1	2.1
3.	16	1.57	118.8	131.6	123.6	9.7	3.9
4.	23	2.36	125.4	131.6	123.7	4.8	1.4
5.	31	3.14	114.2	129.1	122.0	11.6	6.4
6.	39	3.93	107.2	111.1	107.0	3.5	0.2
7.	46	4.71	99.3	104.7	102.4	5.2	3.0
8.	54	5.50	97.0	104.1	102.0	6.8	4.8

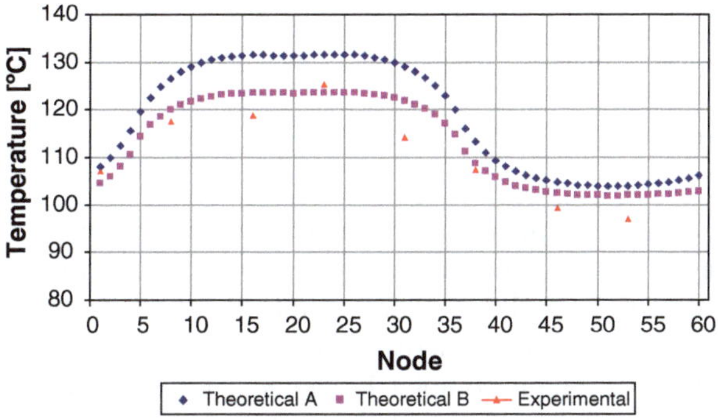

Fig. 9 Temperature profile around absorber pipe

According to the graph in Fig. 9, it is possible to observe the tendency of data to demonstrate periodicity. This is because temperature analysis is done in radial and angular directions, and the angular direction should exhibit a periodic behavior. This aspect is very important because is possible to propose an analytical solution in trigonometrical terms that could be like Eq. 10. All constants (*A, B, C* and *D*) could be obtained from boundary conditions and specific experimental values, but its particular solution is a subject for other work.

$$T(r,\varphi) = \left(A \cdot r^{\nu} + B \cdot r^{-\nu}\right) \cdot (C \cdot \sin(\nu \cdot \varphi) + D \cdot \cos(\nu \cdot \varphi)) \qquad (10)$$

In any sort of mathematical model, the convective heat transfer coefficient is a very important parameter that should be calculated by applying the best correlation that satisfies the specific process parameters required. To define an acceptable value for this coefficient, we have taken the model proposed by Goebel (1997), which is based on experimental data. Another theoretical analysis related to two-phase flow with DSG was developed by Zarza (2003); he emphasizes the importance of the convective coefficient.

Where this kind of analysis is applied to brine, for example, when analyzing a fluid in a working environment, it could be necessary to define a maximum salt concentration of 5 %, as in, for example, geothermal brine. This is because the heat transfer properties of the working fluid are not affected considerably, and therefore it is possible to consider it as a pure substance. Such an approximation should be acceptable in order to obtain preliminary results.

On the other hand, different types of methods can be applied to solve the homogeneous heat conduction equation in the cylindrical coordinate system for an absorber tube that is analyzed in two or more dimensions. For example, in the separation of variables in the cylindrical coordinate system method, where an exact will be required whose results will have to be very similar on order to separate the variables. However the choice of methods depends on the approaches that are needed to implement each one of them.

4 Conclusions

It has been shown that for annular two-phase flow within a pipe (i.e. an absorber tube) warmed by concentrated solar energy, the temperature differential registered between the hottest point and the coldest point over the external wall of the pipe will increase if the feeding flow increases too. This situation even happens when the internal wall of the pipe is completely wetted. This result verifies that the heat transference from the pipe to the liquid phase of the fluid is not constant and could depend on the liquid film thickness.

In order to be able to raise an experimental correlation for the change of heat transfer coefficient with respect to the film thickness, it is necessary to carry out further experiments with reliable and specialized devices.

The finite-differences method is useful in order to obtain a preliminary approximation for predicting the temperature values of an absorber tube with annular two-phase flow. Because the maximum relative error value was almost 12 % and the minimum was approximately 1 %, these results do not change significantly if the number of nodes in the network is increased. Nevertheless an improvement is obtained because of the error reduction.

In order to obtain preliminary comparative results to determine the viability of a hybrid power project, it is recommended to implement a theoretical analysis of this type. Results will depend on the boundary conditions and approximations that are considered in order to calculate thermal parameters such as the convective heat transfer coefficient. This is why it is very important to verify the theoretical values with experimental results.

To calculate a theoretical value for the convective heat transfer coefficient, some acceptable correlations may be used whose application depends on the operating conditions of the system. In the case of this work with low power and process conditions that are not extreme, is possible to obtain a very good approximation of the temperature profile around the receiver tube.

With respect to the comparison of experimental and theoretical data it was possible to prove that the convective heat transfer coefficient changes as a function of liquid film thickness. The theoretical temperature could correlate better with the experimental data if the convective heat transfer coefficient is considered a variable instead of a constant.

As a final suggestion, if low power solar systems with lower flows are necessary, it might be possible to operate them with an annular two-phase flow and a liquid-steam separator before the last module, in order to obtain steam of high mass quality without completely evaporating the feed fluid. This aspect will be tested in the near future.

Acknowledgments This work has been financed by DGAPA (Dirección General de Apoyo al Personal Académico) and CONACYT (Consejo Nacional de Ciencia y Tecnología) who awarded a fellowship to the corresponding author. Ceferino Figueroa, Cuauhtémoc Salazar and Lauro Santiago helped to carry out the experimental work, and also installed and calibrated measuring equipments.

References

Almanza R, Jiménez G, Lentz A, Valdés A, Soria A (2002) DSG under two-phase and stratified flow in a steel receiver of a parabolic trough collector. J Solar Energy Eng 124:140–144 (ASME)

Bejan A (1995) Convection heat transfer, 2nd edn. Wiley, New York

Flores V (2003) Transitory behaviour of a bimetallic receiver during direct steam generation, using solar parabolic trough concentrators. Doctoral Thesis, National Autonomous University of Mexico, pp 30–42 (In Spanish)

Goebel O (1997) Modelling of two phase stratified and annular flow in heated horizontal tubes. In: Mayinger F, Lehner M (eds) Convective flow and pool boiling. Taylor & Francis, New York, pp 303–310

Gungor K, Winterton R (1986) A general correlation for flow boiling in tubes and annuli. Int J Heat Mass Transf 29(2):351–358

Hahne E, Herrmann J, Rheinlander J (1997) The effect of tilt on flow patterns of water/steam flow through heated tubes. In: Giot M, Mayinger F, Celata GP (eds) Experimental heat transfer, fluid mechanics and thermodynamics. Edizionni ETS, Washington D.C., pp 925–934

Heidemann W, Spindler K, Hahne E (1992) Steady-state and transient temperature field in the absorber tube of a direct steam generating solar collector. Int J Heat Mass Transf 35(3):649–657

Herbst O, Fechner A, Köhler W, Goebel O, Oberfe B (1996) Heat transfer behaviour of an absorber tube with direct steam generation by water injection. In: Chen J (ed) Convective flow and pool boiling. Taylor & Francis, New York, pp 365–370

Incropera F, DeWitt D (2001) Fundamentals of heat and mass transfer, 5th edn. Wiley, New York, pp 68–72, 185–189, 196–198

Kattan N, Thome J, Favrat D (1998) Flow boiling in horizontal tubes: part 3—development of a new heat transfer model based on flow pattern. J Heat Transf 120:156–165 (Trans ASME)

Lentz A, Almanza R, Ruiz V (2002) A hybrid geothermal-solar system to increase the quantity of steam in Cerro Prieto, Baja California, México. In: Proceedings of the 11th solarPACES international symposium, Zurich, Switzerland, pp 553–556

Luninski Y, Barnea D, Taitel Y (1983) Film thickness in horizontal annular flow. Can J Chem Eng 61:621–626

Martinez I (2005) Theoretical and experimental analysis of annular two phase flow in horizontal pipes for a solar-geothermal system. Doctoral thesis, National University of Mexico (UNAM), pp 68–82 (In Spanish)

Martínez I, Almanza R (2003) Theoretical-experimental analysis of annular two-phase flow in horizontal pipes. In: Memories of XXVII solar energy national week, solar energy national association (ANES), Chihuahua, Chihuahua, México, pp 165–168

Özişik N (1993) Heat conduction, 2nd edn. Wiley, New York

Rohsenow WM, Hartnett JP (1973) Handbook of heat transfer. McGraw-Hill Book Co., USA, pp (4–22)–(4–23)

Zarza E (2003) Direct steam generation with solar parabolic trough collectors, direct solar steam project (DISS). Doctoral Thesis, Seville University, pp 139–149 (In Spanish)

Random Forces on Obstacles in Channels with Grains: A Mechanical Analogy of Crowd Disasters

A. Medina, A. López-Villa and G. J. Gutiérrez

Abstract In this work we have studied experimentally, through sensitive force measurements, the fluctuating forces on a rigid obstacle located at the middle of a horizontal channel, when a two-dimensional (2D) granular forced flow is induced by a wall-piston which moves with constant velocity along the channel. In this piston-like system, the force measured show strong fluctuations which are much larger than the average force, giving rise to intermittent behavior. Two different initial packing factors were employed, showing different flow characteristics. In the well-ordered high packing system, the force on the obstacle is very high, producing force peaks at time intervals almost constants. The main frequency is well correlated with the residence time of each row of grains. However, in case of initially disordered, loose packing systems, there is an initial relaxation time where force on the obstacle is extremely low, thus allowing the free flow of several grain rows. The temporal force traces have a $1/f^{\alpha}$ character and depend on the initial arrangement of grains.

1 Introduction

The nature of the forces on objects embedded inside granular flows has been a topic scarcely studied Atkinson et al. (1983); Tuzun and Nedderman (1985); Wieghardt (1974, 1975); Albert et al. (1999, 2000). However, this phenomenon frequently occurs, for instance, during the gravity induced granular flows in silos where fixed

A. Medina (✉) · A. López-Villa · G. J. Gutiérrez
SEPI ESIME Azcapotzalco, Instituto Politécnico Nacional, Av. de las Granjas 682, Col. Santa Catarina, 02250 Azcapotzalco, D.F., México
e-mail: amedinao@ipn.mx

A. López-Villa
e-mail: abelvilla77@hotmail.com

G. J. Gutiérrez
e-mail: julygp04@yahoo.com.mx

J. Klapp and A. Medina (eds.), *Experimental and Computational Fluid Mechanics*, Environmental Science and Engineering, DOI: 10.1007/978-3-319-00116-6_12,

inserts are commonly used to improve the flow rate or to add mechanical strength to the silo structure Tuzun and Nedderman (1985); Johanson and Kleysteuber (1966); Tuzun and Nedderman (1985). Following this line, Tuzun and Nedderman Tuzun and Nedderman (1985) in an experimental work, measured the normal and shear loads on big obstacles and analyzed the temporal fluctuations of these quantities. The main result in such conditions was that the force signals evidenced the existence of temporal force oscillations of nearly constant amplitude and frequency. In another experiments, some researchers have studied the fundamental problem of the drag force on solid objects. In particular, rigid cylinders fixed to load cells were introduced, to different depths, within a granular flow originated during the axisymmetric rotation of buckets of sand Wieghardt (1974, 1975); Albert et al. (1999, 2000). In such experimental configuration the existence of periodic and stepped fluctuations in the drag force were detected. Moreover, for a wide range of velocities, little Wieghardt (1974, 1975) or none Albert et al. (1999, 2000) dependence with respect to the magnitude of the velocity was noted.

The aim of this work is to study experimentally the behavior of the instantaneous force and its time-averaged properties on an obstacle due to the forced motion of a nearly two-dimensional (2D) granular flow composed of spherical, big rigid grains, in order to determine the role of a fundamental quantity of the granular material: the granular packing.

Two well defined and very different grain arrangements have been induced in a horizontal container in order to seek its influence on the force fluctuations. We have used a detailed approach to determine, through an idealized 2D system, the force in two types of initial packing: a hexagonal 2D monodisperse arrangement whose packing factor is $\eta = \pi/(2 \cdot 3^{1/2}) \approx 0.9069$ and a loose uncorrelated packing whose value was 0.81 Meakin and Jullien (1991). In this study we treated cases of relatively low velocity, 20–50 mm/s. We have examined the random behavior of the force acting on the obstacle and its relation to the collective dynamics of grains, like the formation of force chains during motion.

This mechanical study can be assumed as a mechanical analogy of the problem of crowd disasters, because people in a crowd stampede, move and push towards the exit causing a domino effect Helbing et al. (2000). A similar effect can be induced by a collective of grains which move and interact among them and against a fixed obstacle located at the middle of a channel. Thus, this study can be used to understand the collective motion of pedestrians under panic conditions, which occurs very frequently.

This work is structured as follows: in the next section we explain why the mechanical problem of interaction of grains is associated with the crowd disasters. After, in Sect. 3, we describe the experimental setup and the measurement procedure. In Sect. 4, we analyze the temporal properties of the time series of the force for several velocities and packings of grains. There, we detect interesting properties of the force fluctuations and the character of the noise associated to the time series. Finally, in Sect. 5 we discuss the main results and their implications on the problem of how improving crowd safety.

Fig. 1 Muslims circled the Kaaba, Islam's holiest site (taken from http://www1.whdh.com/news/articles/world/BO38831/)

2 Crowd Disasters

Due to growing populations in the urban centers, the frequency of programmed events like sporting event egress and ingress, religious events, food distribution and entertainment events, among others, is increasing. Similarly, the occurrence of sudden incidents as stampedes and failures of crowd control, fires and protests is also increasing.

These events tend to involve a large number of people. For instance, see Fig. 1. They also often occur in times of mass panic (e.g. as a result of a fire or explosion); as people try to get away crowd forces can reach levels that almost impossible to resist or control. Virtually all crowd deaths are due to compressive asphyxia and not the "trampling" reported by the news media. Evidence of bent steel railings after several fatal crowd incidents show that forces of more than 4,500 N (1,000 lbs.) occurred. Forces are due to pushing, and the domino effect of people leaning against each other.

Compressive asphyxia has occurred from people being stacked up vertically, one on top of the other, or horizontal pushing and leaning forces. In the Ibrox Park soccer stadium incident, police reported that the pile of bodies was 3 m (10 ft) high. At this height, people on the bottom would experience chest pressures of 3600–4000 N (800–900 lbs.), assuming half the weight of those above was concentrated in the upper body area.

Horizontal forces sufficient to cause compressive asphyxia would be more dynamic as people push off against each other to obtain breathing space. In the Cincinnati rock concert incident, a line of bodies was found approximately 9 m (30 ft) from a wall near the entrance. This indicates that crowd pressures probably came from both directions as rear ranks pressed forward and front ranks pushed off the wall.

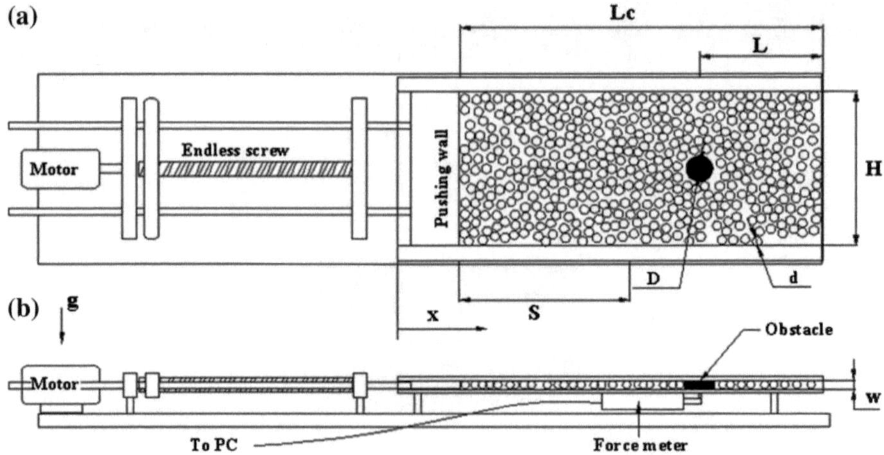

Fig. 2 Experimental setup for the force measurements in a rectangular and horizontal channel of dimensions: length L_c, width H and thick w. The obstacle has a characteristic length D (diameter for the circular obstacle and side for the square obstacle) and the granular material is made up of bearing balls of diameter d. In **a** we depict the top view where the obstacle position is shown. In **b** we depict the containers lateral view and we show the location of the force meter. $\boldsymbol{g}$ and x indicate the gravity field and the direction of the induced flow, respectively

3 Experimental Setup and Force Measurements

Our experimental set-up consists of a piston-like system where a high-torque motor coupled to an endless screw moves a slim Plexiglas plate 8 mm thick with a constant velocity, along a Plexiglas channel filled with grains. This set-up configuration is shown in Fig. 2. The grains used in this work were bearing balls with a diameter, $d = 7.85 \pm 0.05$ mm, Young modulus E $= 2 \times 10^{11}$Pa and friction coefficient $\mu \approx 0.3$. The channel has the following dimensions: $w = 0.8$ mm thick, $L_c = 450$ mm length and variable wide, so that, $H = 143$, 165 and 195 mm width. The obstacles included two different shapes: circle and square, the circular obstacles has a diameter $D = 8, 16$ and 24 mm, while the square obstacle has a side $D = 16$ and 24 mm. The motion of the plate along the channel generates a 2D granular flow: all the grains are moved by the plate and many of them nucleates force chains which reach and, consequently, push the rigid circular obstacle located just at the middle of the channel (Fig. 2a). In order to obtain reproducible force measurements, the obstacle, which was attached just at its center to a rigid harm and this to the force sensor, does not touch neither the upper nor the lower confining walls (Fig. 2b). During the motion we measured the force, $F(t) = F_x(t)$, on the obstacle transmitted by the continuous pushing of the grains along the direction of the main flow each 2 ms (i.e., at a measurement frequency, $f_m = 500$ Hz) by using a force meter with a dynamic range between 0 and 50 N with a sensitivity of 0.001 N. In Fig. 2b we show the way

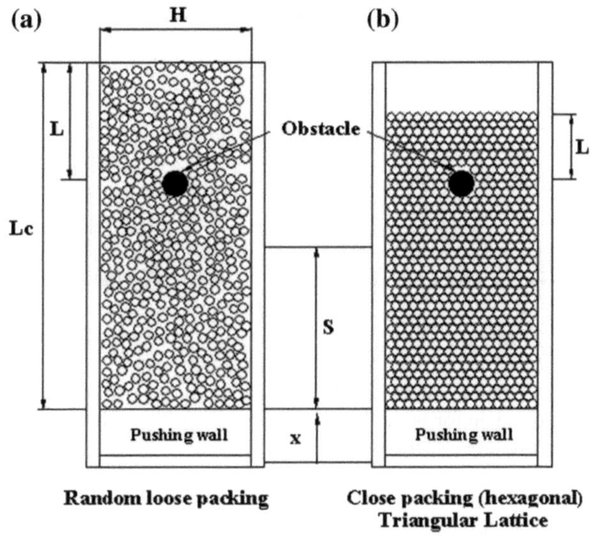

Fig. 3 **a** Schematic view of random loose packing which was obtained by raining the grains from the top and **b** the ordered (hexagonal) packing in the rectangular channel

in which the force meter was placed. The experimental error of this measurement procedure is of around 1 %.

We have treated cases where the initial packing was hexagonal and other ones where the initial packing was disordered as shown in Fig. 3. This latter packing, called random loose packing, was obtained by raining the grains from the top and after tilting the container to the horizontal position. It is also important to note that along the container's width, we can have several arrays of exactly 18, 21 and 24 grains that allow us to build the hexagonal packing. We made experiments with several velocities and we have not observed appreciable dependence on this quantity despite of the forces generated by the ubiquitous chains of grains from the pushing plate to the frontal area of this obstacle.

The velocities of the piston-plate used in the experiments, v, range between $20 \leq v \leq 50$ mm/s. A couple of representative force measurements are shown in Fig. 4, where it is evident the strong force fluctuations (peaks) which are of one order of magnitude larger than the mean value for tight packing (Fig. 4a) and less than 8 times the mean value for loose packing (Fig. 4b). Another difference occurring between the temporal trace for the tight packing and that occurring for the lose packing is that the strong pulses appear more evenly distributed in the initially ordered cases.

We have observed appreciable dependence between the quantity of material after the obstacle so-call L/d for both types of packing where is more strong for the hexagonal packing, add to above a large influence of the parameters H/d and D/d.

In the following Sect. 4 we will analyze with more detail these facts.

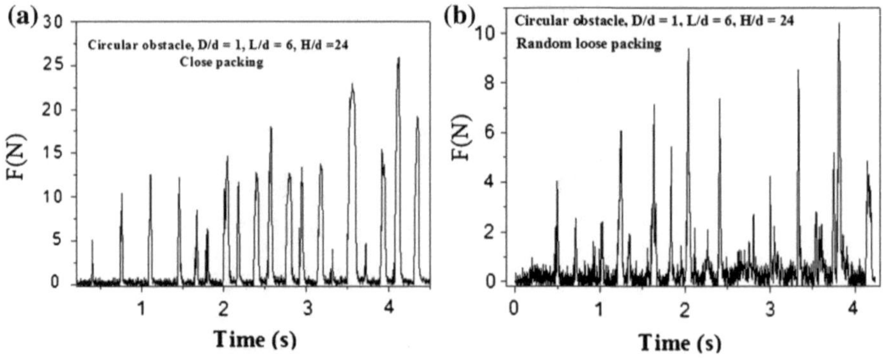

Fig. 4 Temporal measurements of the effective force along the *x*-direction. In **a** we show the force variation on the obstacle for initial hexagonal packing while in **b** we show this quantity but considering initial random loose packing

4 Analysis of Results

We analyzed several cases from both shape obstacle, the initial packing is hexagonal (close packing) with a packing factor $\eta_1 = 0.9083$, and the random loose packing with $\eta_2 = 0.814$. We vary the speed of granular material and change the parameter *L/d* and *H/d*. The representatives cases are from the circular obstacle with $\eta_1, v =$ 37.5 mm/s (Case 1), and $\eta_2, v =$ 38 mm/s (Case 2), hexagonal and loose packing respectively, $L/d = 6$, $H/d = 24$ and $D/d = 1$.

The time series for the force on the obstacle for Case 1 is shown in Fig. 4a and for Case 2 is shown in Fig. 4b. These plots show the influence of the packing in the force on the object: the close packing transmits to the obstacle, almost immediately, the strong fluctuating forces nucleated in the grain chains and the occurrence of these strong peaks is very frequent and almost equidistant. Conversely, in systems with initial loose packing the nucleation of the force chains which produces strong fluctuations takes a relaxation time and are not so frequent as in Case 1. Moreover, the force chains in the tight packing are more robust than in case of loose packing and therefore the intensity of the force fluctuations are larger in the first case. The initial packing factor is then extremely important in determining the force induced by the granular flow. In Case 2, the flow does not show any big resistance in the first second, indicating that more than 4 rows of grains can move around the obstacle without any big problem. This relaxation time is used to rearrange the grains upstream increasing its packing factor, increasing then the flow resistance. In Fig. 5 we show, for Cases 1 and 2, the root mean square (rms) for the force fluctuations, defined by

$$\frac{\sqrt{\langle (F(t) - \langle F(t) \rangle)^2 \rangle}}{\langle F(t) \rangle}, \tag{1}$$

where <*X*> represents the time moving average of *X(t)*, i.e.,

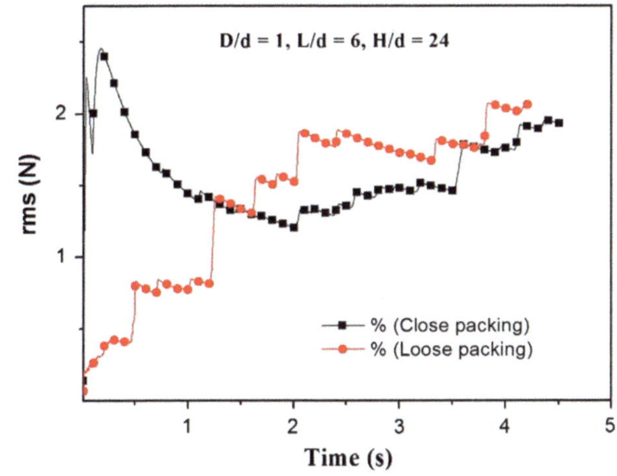

Fig. 5 Normalized rms force fluctuations for the systems here considered: system initially ordered with $v = 37.5$ mm/s (line-square) and system initially disordered with $v = 38$ mm/s (line-circle)

$$<X>= 1/t \int_0^t X(s)ds.$$

This quantity allows us to understand the role of the packing in the occurrence of the peaks of force respect to the mean, i.e., intermittency. Fig5, in fact, shows that the nucleation of force chains in Case 1 is immediate and after these occur in a random manner. It is easy to note that each force peak (a sudden increase in the fluctuation) is reached after a nonlinear decrease of the intensity of the fluctuation which indicates that formation of force chains is neither cyclic nor monotonic in time. In the same Fig. 5, are shown the fluctuations for Case 2, where initially the intensity of the fluctuations are lower than those of Case 1 and they increase suddenly with time.

The increase in the level of the fluctuations in this latter case is sudden because the chains of force are weaker and they break near instantaneously. In Fig. 6 we show the rms of the force for Case 1 but changing the quantity *L/d*. It determines that a certain quantity of grains affects the subsequent history of the force chains.

We show in Fig. 7 the behavior of the average force on the circular obstacle and square obstacle, too we have observed that the shape of the obstacle produce a significant change on the average force due to the instabilities of the chain force and that is produced by small perturbations on the contact point of material with the obstacle. This plot also indicates that apparently the mean force is larger on circular obstacles than on angulated square obstacles.

The noisy behavior of the time series of the force has been also studied. In this case we obtained the power spectra, *S(f)*, for the traces of Fig. 4. In both cases we found that in the UV-range, $f >> 1$, the spectra obey the relation

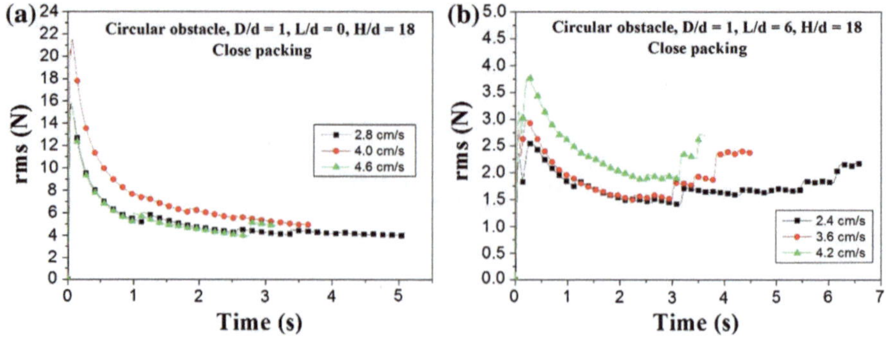

Fig. 6 Normalized rms force fluctuations for several velocities in the system with circular obstacle and D/d = 1 and H/d = 18: **a** L/d = 0 and **b** L/d = 6

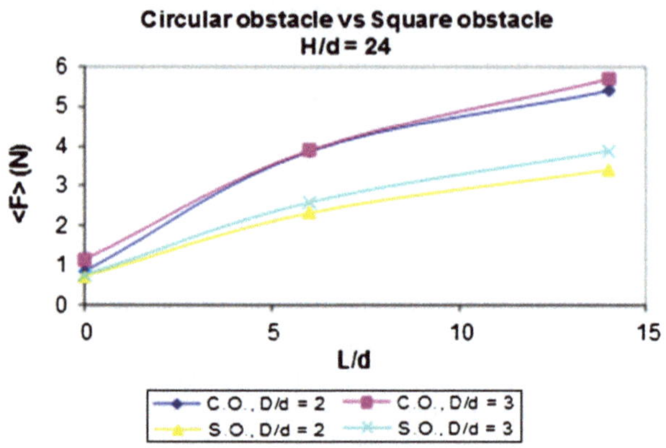

Fig. 7 Comparison of the mean force for circular and square obstacles

$$S(f) \sim f^{-\alpha},$$

where the exponent α depends on the initial packing. The plots for Cases 1 and 2 are shown in Fig. 8. In Fig. 8a we have plotted the power spectrum of the tight packing and its best fit which yields a very accurate value for the exponent $\alpha = 2.23 \pm 0.1$. This value of α implies that successive points in the time series are strongly correlated. Similarly, in Fig. 8b we show the power spectrum and the best fit where $\alpha = 1.45 \pm 0.1$. In this case we have time series which obeys fractional Brownian motion which implies that successive points in the time series are weakly correlated.

One of the most important parameters to be obtained from the present measurements is related with the maximum peak of the power spectrum of the measured force. There is a characteristic frequency dictated by the way in which a layer of grains arrives at the obstacle and is defined by $f_g = v/d$. In the present experiments we have for Case 1, $f_g = 2.7$ Hz. For the case of loose packing, the characteristic

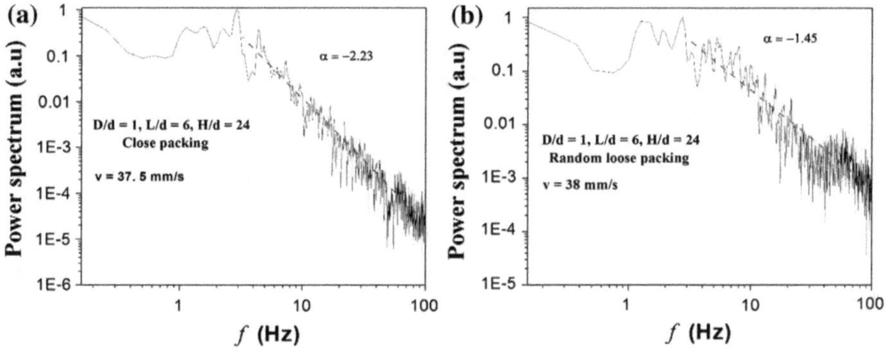

Fig. 8 **a** Power spectrum (in arbitrary units, a.u.), *S(f)*, for the time series of the force in a system with tight packing and **b** system with loose packing. The straight line was obtained by using a least squares fit for the noisy part. Speeds were very similar

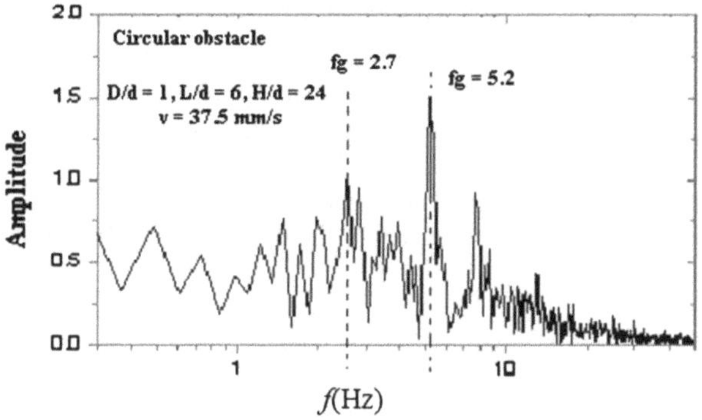

Fig. 9 The characteristics frequencies for a close packing

frequency is $f_g = 6.1$Hz, respectively. The occurrence of this frequency for the first case is shown in Fig. 9, indicating that the peaks in the force are strongly related with this characteristic frequency, f_g.

5 Conclusions

In this work we studied experimentally the force produced by a granular flow on an obstacle in a finite horizontal channel, for different initial packing. There are four main immediate results depending on the initial configuration: (a) the force fluctuations in the case of initial ordered systems are distributed through the overall measurement interval. Moreover, the force fluctuations are about twice the force fluctuations in initially disordered systems. In this latter case, the flow resistance is very low at the beginning, thus allowing several rows of grains to flow around the obstacle, without any problem. (b) The time-averaged force apparently does not depend on the velocity, c) the time interval between consecutive force peaks is well correlated with the transit time of the flow needed to move the distance given by a particle diameter. The resulting Strouhal number defined by St $= f_{peak}/f_g$ is then of order unity and (c) The power spectra, in initially tight packing systems, of the time series for the force are strongly correlated and systems with initial loose packing the time series are weakly correlated.

Finally, this study could be connected to the collective motion of pedestrian under panic conditions, for instance, fire which advance in a corridor at constant velocity. There, people in a crowd stampede move and push towards the exit at constant velocity Helbing et al. (2000). So, it is possible that forces on fixed obstacles originated in escape panics have, in a first approximation, the same nature as that found here and many of our results, adequately scaled, could be useful to guide the theoretical efforts. We have shown how initial order (high packing) produces strong flow resistance, generating forces, due to the domino effect, one order of magnitude larger than the average. For loose packing the level of forces is weaker. Finally, this type of studies can be of interest to propose better models in order to introduce public officials to the best crowd safety standards, laws and techniques for preventing common safety missteps and haphazard event planning in the communities.

Acknowledgments One of us (A. M.) acknowledges Prof. H. J. Herrmann for his useful comments about this paper. This work has been supported by the IPN through a SIP project.

References

Atkinson TD, Butcher JC, Izard MJ, Nedderman RM (1983) Chem Eng Sci 38:91
Tuzun U, Nedderman RM (1985) Chem Eng Sci 40:337
Wieghardt K (1974) Mech Res Comm 1:3
Wieghardt K (1975) Annual Rev Fluid Mech 7:89
Albert R, Pfieifer M, Barabási A-L, Schieffer P (1999) Phys Rev Lett 82:205
Albert I, Tegzes P, Albert R, Sample JG, Pfieifer M, Barabási A-L, Vicsek T, Schieffer P (2000) Phys Rev Lett 84:5122
Johanson JR, Kleysteuber WK (1966) Chem Eng Prog 62:79
Tuzun U, Nedderman RM (1985) Chem Eng Sci 40:325
Meakin P, Jullien R (1991) This value was obtained through actual measurements of the packing and agrees with the predicted value of η for random deposition in 2D narrow channels. Europhys Lett 15:851
Davies JC (1973) Statistics and data analysis in geology. John Wiley & Sons, New York
Helbing D, Farkas I, Vicsek T (2000) Nature 407:487

Part II
Multiphase Flow and Granular Media

Experimental Study of the Growth of Bubbles in Corrugated Tubes

U. Romero, A. López-Villa, A. Medina and G. Domínguez Zacarías

Abstract In this chapter we study some experiments with different liquids to validate theoretical results on the growth of bubbles in corrugated pipes. These experiments were performed using a high range of capillary numbers *Ca*, as well as different values of the Bond number *Bo*, the purpose is to approach the cases of inviscid and viscous limits in liquids. The experiments are done in the tubes with different diameters and different lengths with periodic corrugations and different amplitudes. We also characterized the effect of the corrugated walls on the shape and size jets of bubbles for constant flow rate.

1 Introduction

In literature there are different works studying bubbles, most of the extensive research carried out on the generation of bubbles by injection of gas into a liquid at rest has been devoted to the important case of liquids of small viscosity, for which the flow induced by the expansion and rise of the bubbles is dominated by inertial effects; see Kumar and Kuloor (1970); Oguz and Prosperetti (1993); Clift et al. (1978);

U. Romero (✉) · A. López-Villa · A. Medina
ESIME Azcapotzalco, Instituto Politécnico Nacional, Av. de las Granjas No. 682, Col. Sta. Catarina, 02550 México, D.F., México
e-mail: bich14nos@hotmail.com

A. López-Villa
e-mail: abelvilla77@hotmail.com

A. Medina
e-mail: abraham_medina_ovando@yahoo.com

G. Domínguez Zacarías
Cordinación Tecnológica de Ingeniería de Yacimientos,
Eje Central Lázaro Cárdenas Norte 152 Col. San Bartolo Atepehuacán Del.,
C.P. 07730 Gustavo A. Madero, D.F., México

J. Klapp and A. Medina (eds.), *Experimental and Computational Fluid Mechanics*, Environmental Science and Engineering, DOI: 10.1007/978-3-319-00116-6_13,

Räbiger and Vogelpohl (1986); Longuet-Higgins et al. (1991). Applications include direct-contact operations in chemical, metallurgical, and biomedical systems, among many others. The opposite case of bubble generation in very viscous liquids is of interest in connection with polymer melts (Bird et al. 1987; Davidson and Schuler 1960; Doshi et al. 2003; Higuera 2005) and molten glasses and magmas (Manga and Stone 1993, 1994).

Results may be expressed in terms of two dimensionless numbers, the Bond number and the Capillary number. Bond number is a measure of the intensity of flotation forces respect to the surface tension

$$Bo = \frac{\rho g a^2}{\sigma}, \tag{1}$$

where ρ is the density, g is the acceleration of gravity, a is the inner radius of the capillary and σ the surface tension.

Meanwhile the capillary number is a measure of the competition between viscous forces caused by the air inlet in fluid and the surface tension forces that keep the bubble adhered with the mouth of the air injection tube where

$$Ca = \frac{\mu Q}{\sigma a^2}, \tag{2}$$

where μ is the viscosity, Q the gas flow rate which is constant.

When expressing the results in terms of these two parameters we can have a wide variety of physical conditions under which bubbles grow.

2 Experiments

It is considered the simple case of incompressible gas (air) when it is injected at constant flow rate with viscosity and density negligible into a reservoir filled of quiescent and incompressible liquid of density ρ and viscosity μ.

In experiments the air is injected through a capillary tube with a 2 mm inner radius and 40 cm length, fed by a pump from a tank and a programmable syringe pump, that maintained a constant flow, has already been shown in a previous work (Corchero et al. 2006). The purpose of using two coupled pumps is to prevent the entry of liquid into the capillary tube when liquid is injected and to measure very accurately the flow can be performed in each test. The aquarium pump passes through a bypass valve its output to the syringe pump, then it joins the capillary tube as shown in Fig. 1. The bubbles are generated by injecting controlled air into a tube with periodic corrugations filled of viscous and non viscous liquids.

The process of growth and detachment was video recorded. Subsequently, each video was digitized to have photos of events every 1/30 s. The experiments were performed by using transparent plastic tubes with different periodic corrugations

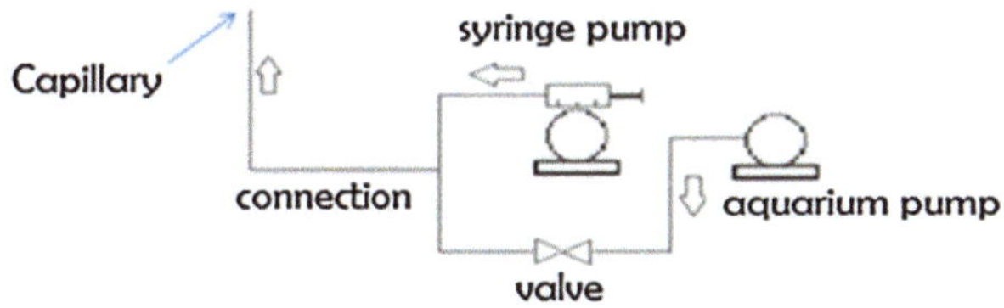

Fig. 1 Schematic array of the programmable syringe pump and the aquarium pump

and different diameters. Which were coaxially connected on the capillary pipes, perfectly upright and centered.

The liquids level in the container was always =100 mm, and its dimensions of length, width and height were 10 × 10 × 30 cm respectively.

The liquids used for the experiments were glycerin and honey for the viscous case and water for the near inviscid case.

At a temperature of about 25 °C the glycerin properties used are density $\rho = 1260\,\mathrm{kg/m^3}$, $\mu = 1.2$ Pas and surface tension $\sigma = 6\,\mathrm{mN/m}$. On the other hand, the honey properties depend on the type of honey at use, in this case $\rho = 1413\,\mathrm{kg/m^3}$, $\mu = 10$ Pas and surface tension $\sigma = 33$ N/m.

The same experiments were performed with water at a temperature of about 25 °C, the properties of water at that temperature are $\rho = 998\,\mathrm{kg/m^3}$, $\mu = 1.002$ mPa and surface tension $\sigma = 72.8\,\mathrm{mN/m}$.

3 Results and Discussions

The corrugated tube walls have a great friction which causes the bubbles to growth slowly, ip to they reach a volume that allows the buoyance force equals to the force exerted by the viscous drag of the walls.

In this work it can be seen as in other earlier works that the volume bubble increases as the pipe radius, *R* is reduced, where *R* is the mean distance of the corrugate tube walls from the center and *R/a* is the dimensionless radius. The ratio increases non linearly with the capillary number growth and is faster than that found in a semi infinite or conical vessel (Ortiz et al. 2009; López-Villa et al. 2011).

In experiments with glycerin when $R/a = 3.7$ and the corrugation wavelength is $c/a = 7.33$ it was observed that the film thickness in between the bubble and the tube wall increases with the volume of the bubble, i. e., the capillary number increase, as shown in Fig. 2.

When the Bond number is constant and the Capillary number changes, by varying the flow provided by a programmable syringe pump, we found the behavior shown in Fig. 3. The volume grows in a near linear form with the increase of the capillary number (the flow rate increase), the behavior is shown for $R/a = 3.7$ and $Bo = 0.2$.

Figure 4 shows the bubble profiles with the values used in Fig. 3.

Others experiments were bubble performed at constant flow rate and diameter, with variations only in the wavelength of the corrugations, values were $R/a = 4.5$,

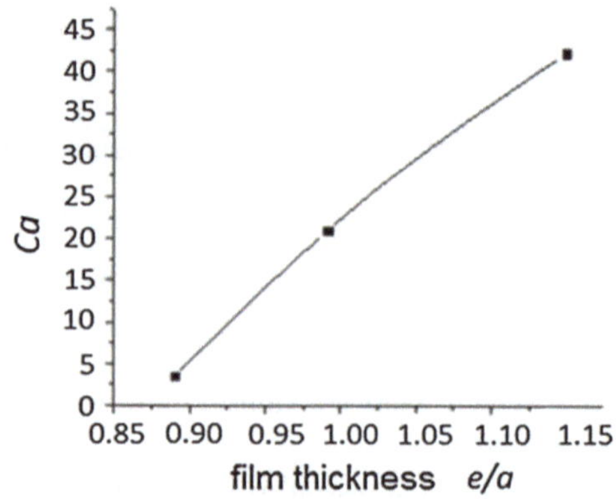

Fig. 2 plot of the capillary number against the dimensionless film thickness for in dimensionless radius $R = 3.7$ of corrugated tube and a wavelength $c/a = 7.33$ and a Bond number $Bo = 0.2$

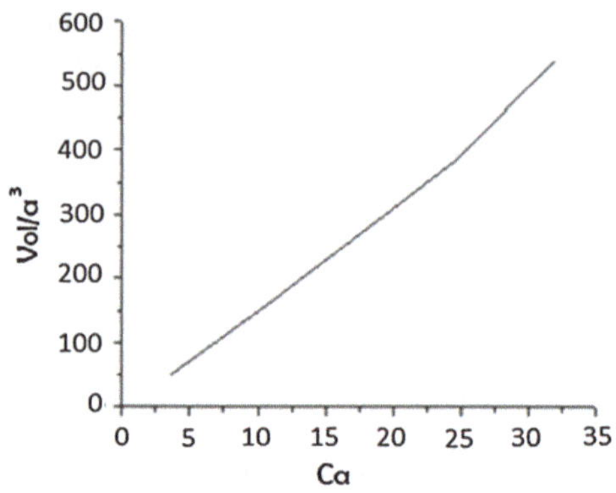

Fig. 3 Plot of the dimensionless bubble volume versus Ca, with $R/a = 3.7$ and $Bo = 0.2$

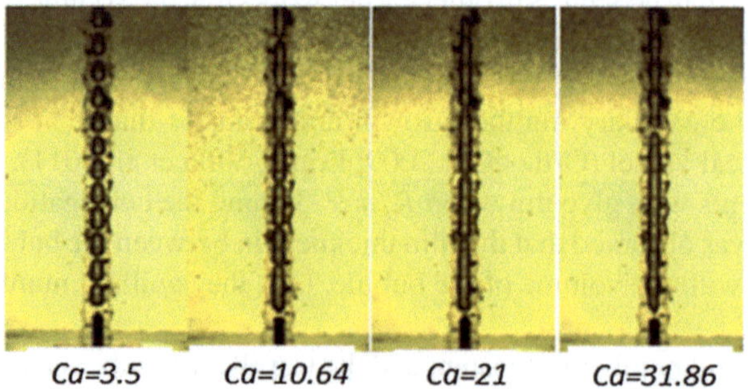

Fig. 4 Profiles of bubbles growing into tubes with wavelength $c/a = 7.33$, dimensionless radius $R/a = 3.7$ and $Bo = 0.2$ with different capillary numbers

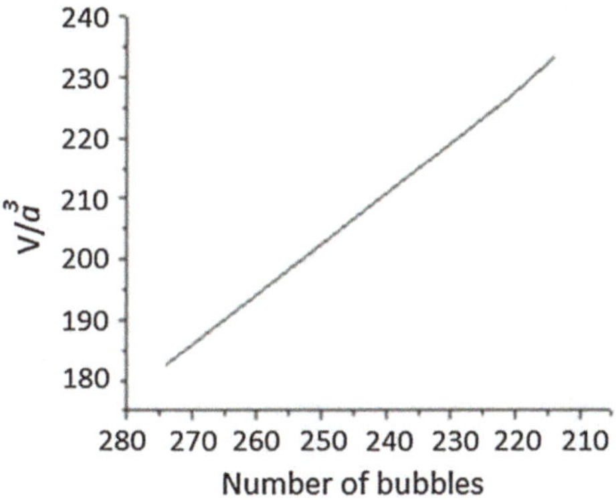

Fig. 5 Plot of the bubble volume versus the number of bubbles generated each minute, with $R/a = 3.7$, $Bo = 0.2$ and $Ca = 10.64$

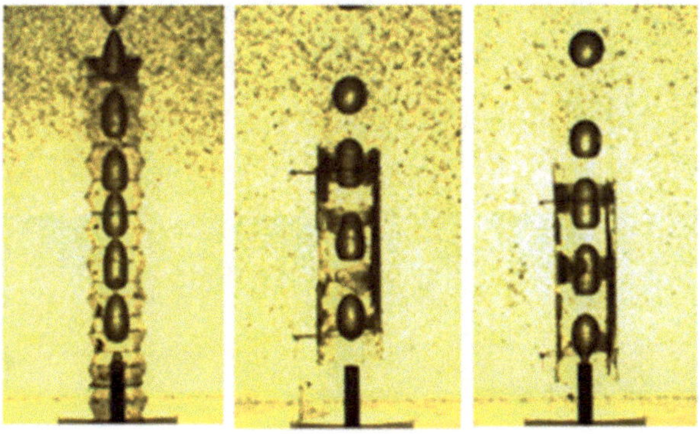

Fig. 6 Bubble profiles growing in tubes with $c/a = 6.4$, 11.0 and 15.1, dimensionless radius $R/a = 4.5$, $Bo = 0.2$, $Ca = 10.64$. Number of bubbles/min $= 274, 221, 214$, $\mathrm{Vol}/a^3 = 182.5, 226.6, 233.3$

flow rate $25.28\,\mathrm{cm}^3/\mathrm{min}$, $Bo = 0.2$, $Ca = 10.64$ and $c/a = 6.4$, 11.0 and 15.1. In such experiments it was found that the bubbles reached a maximum volume at a function of the wavelength. When this distance is large, the bubble attains a volume greater than the reached when the wavelength is shorter. The bubble generation rate is a function of the period of corrugation thus if we know the flow rate and know this rate, we can know the volume of the bubbles. As a consequence the rate at which bubbles are generated decreases linearly with the increasing volume of the bubbles, this is shown in Figs. 5 and 6. We can observe the bubbles profiles with different wavelength, the same used in the plot of the Fig. 5.

We observed different bubbles profiles growing in tubes with corrugated walls at constant flow rate, in which $Ca = 10.64$, $Bo = 0.2$, and the dimensionless radii are $R/a = 3.7$, 4.5 and 5.5 (see Fig. 7). In experiments with honey we also observed dimensionless profiles similar to those observed with glycerin, but with flow rate

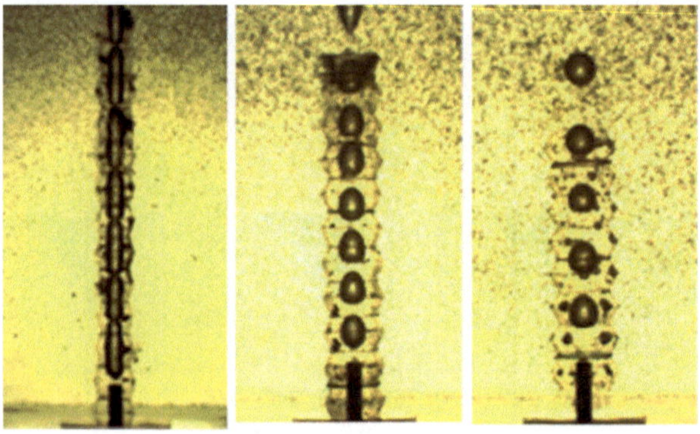

Fig. 7 Bubble profiles growing in tubes, are maintained at a constant flow rate of 25.28 cm^3/min with dimensionless radii $R/a = 3.7, 4.5, 5.5$, $Bo = 0.2$ and $Ca = 10.64$

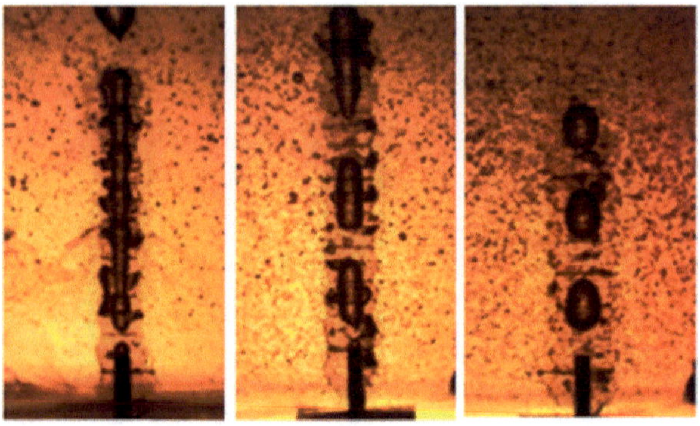

Fig. 8 Bubble profiles in tubes growing at constant flow rate of 25.28 cm^3/min with dimensionless radii $R/a = 3.7, 4.5, 5.5, 0.42$ and $Bo = 0.42$ and $Ca = 127.72$

of magnitude 25.28 cm^3/min (equal to that used with the glycerine) are bubbles of larger volume obtained for $Ca = 127.72$, $Bo = 0.42$, and dimensionless radii are $R/a = 3.7, 4.5$ and 5.5, respectively, see Fig. 8.

Experiments made with water at a flow rate of 25.28 cm^3/min show a different behavior to that of highly viscous liquids, experiments were performed with $Ca =$ 0006, $Bo = 0.134$, and the dimensionless radius $R/a = 3.7, 4.5$ and 5.5 as shown in Fig. 9.

The bubble volume is approximately constant for a wide range of capillary number (small capillary number), which has not been possible to identify with precision (see López-Villa et al. 2011). On the other hand an increasing flow, the bubbles are obtained have non-symmetrical shape see Fig. 9.

In previous works, such as Higuera (2005) and Corchero et al. (2006) we have observed the same results with a small flow.

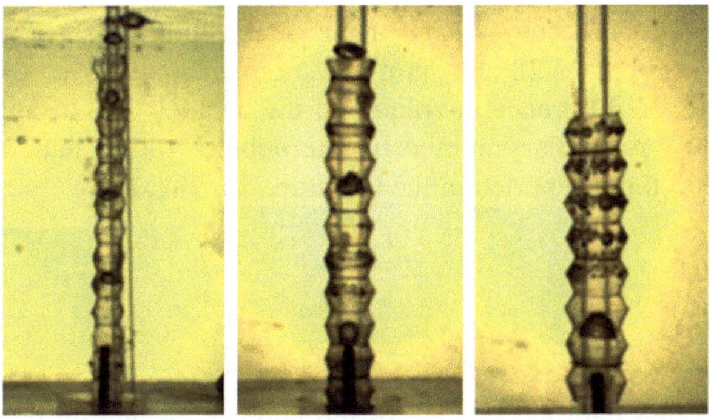

Fig. 9 Bubble profiles in tubes growing at constant flow rate of 25.28 cm^3/min with dimensionless radii $R/a = 3.7, 4.5, 5.5$, $Bo = 0.134$ and $Ca = 0.006$

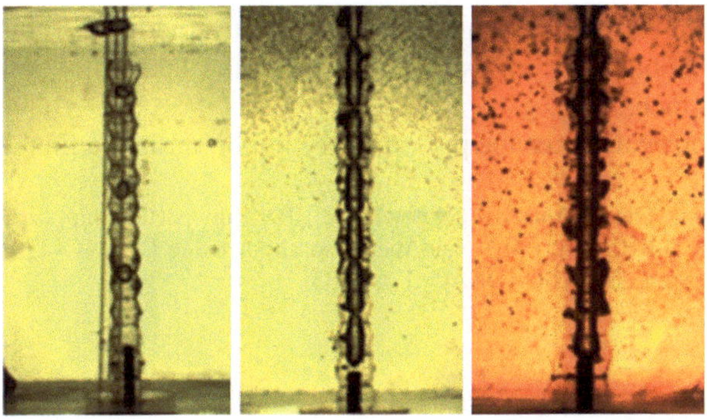

Fig. 10 Bubble profiles in water, honey and glycerin, growing in tubes with $R/a = 3.7, c/a = 7.33$, at 25.28 cm^3/min, $Bo = 0.134, 0.2, 0.42$, $Ca = 0.006, 10.64, 127.72$, dimensionless volume $V/a^3 = 61.3$, 158 and 1336 respectively

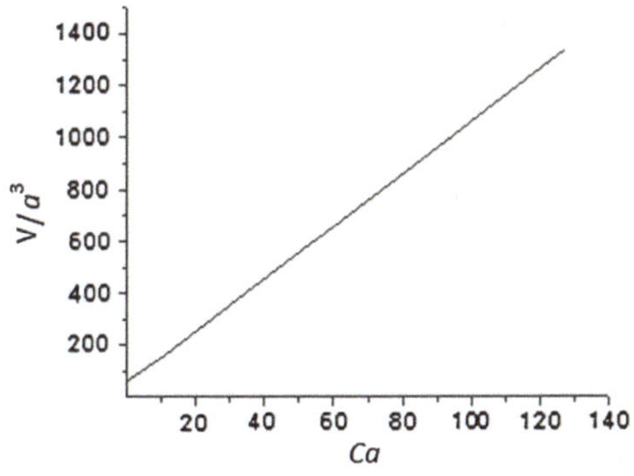

Fig. 11 Dimensionless Plot volume as a function of capillary number, *Ca*, with $R/a = 3.7$

We also performed a comparison of the three fluids used before, maintaining a constant flow rate of 25.28 cm^3/min and a dimensionless radius $R/a = 3.7$ in order to observe the difference in volume of the bubbles with a same flow rate see Fig. 10. We plot the capillary number vs the bubble volume and we compare the graph obtained to those reported in the literature, see Fig. 11.

4 Conclusions

We have found that the final volume of the bubbles depend on the dimensionless numbers *Bo* and *Ca*, therefore the increased viscosity creates a bubble of a large volume and on the other hand a greater film thickness formed between the wall and the bubble.

When Bond and Capillary numbers are kept and the radius of the tube is decreased, the volume of bubbles increases, it is similar to the fact when the wavelength of corrugations is increasing.

Thus, the corrugations of the pipe are very important in the final bubble shape.

Acknowledgments These Authors acknowledge IPN for partial support through projects SIP20131821 and SIP20131821-IPN, and they also aknowledge CONACyT for the partial support through the project SENER-CONACyT 146735.

References

Bird A, Armstrong RC, Hassager U (1987) Dynamics of polymeric liquids. Wiley, New York

Clift R, Grace JR, Weber ME (1978) Bubbles, drops, and particles. Academic Press, New York

Corchero G, Medina A, Higuera FJ (2006) Effect of wetting conditions and flow rate on bubble formation at orifices submerged in water. Colloids Surf A 290:41

Davidson JF, Schuler BOG (1960) Bubble formation at an orifice in a viscous liquid. Trans Inst Chem Engrs 38:144–154

Doshi P, Cohen I, Zhang WW, Siegel M, Howell P, Basaran OA, Nagel SR (2003) Persistence of memory in drop breakup: the breakdown of universality. Science 302:1185–1188

Higuera FJ (2005) Injection and coalescence of bubbles in a very viscous liquid. J Fluid Mech 530:369

Kumar R, Kuloor NR (1970) The formation of bubbles and drops. Adv Chem Eng 8:255–368

Longuet-Higgins MS, Kerman BR, Lunde K (1991) The release of air bubbles from an underwater nozzle. J Fluid Mech 230:365–390

López-Villa A, Medina A, Higuera FJ (2011) Bubble growth by injection of gas into viscous liquids in cylindrical and conical tubes. Phys Fluids 23:102102

Manga M, Stone HA (1993) Buoyancy-driven interactions between two deformable viscous drops. J Fluid Mech 256:647–683

Manga M, Stone HA (1994) Interactions between bubbles in magmas and lavas: effects of the deformation. J Vulcanol Res 63:269–281

Oguz HN, Prosperetti A (1993) Dynamics of bubble growth and detachment from a needle. J Fluid Mech 219:143

Ortiz A, López-Villa A (2009) Formación de burbujas en líquidos viscosos contenidos en conos y cilindros. Revista Mexicana de Física 55(3):166–179

Räbiger N, Vogelpohl A (1986) Bubble formation and its movement in newtonian and non-newtonian liquids. In: Cheremisinoff NP (ed) Encyclopedia of fluid mechanics, vol 3, Chap. 4, Gulf Publishing Company, Houston

The Cooling of a Granular Material in a Rotating Horizontal Cylinder

Ever Góngora-Leyva, Gerardo Ruiz-Chavarría, Ángel Columbié-Navarro and Yoalbis Retirado-Mediacejas

Abstract In this chapter we investigate the cooling of the lateritic ore moving inside a rotating horizontal cylinder. The mineral, which has been previously milled, forms a granular bed. Due to rotation, the granular material undergoes discrete avalanching, whose characteristics depends either upon the angular velocity and the filling degree. The ore enters into the cooler at a temperature of approximately 750 °C and after moving along the cylinder its temperature decreases. The goal is to reduce the ore temperature to a value around 170 °C, so its metallic properties are preserved. To model this process a system of two differential equations and an algebraic equation depending only on the axial coordinate is solved. Otherwise we present data of the cooling in a cylinder 30 m long, a diameter d = 3 m and rotating with an angular velocity of 6.24 rpm. We show that numerical solution exhibits a partial agreement with experimental data.

1 Introduction

During the production of nickel, the ore is extracted from open-pit mines. After extraction the ore is milled to convert it in a granular media (the mean diameter of grains is 0.074 mm). In the next step the lateritic ore enters in a reduction oven where its temperature rises. After, the ore passes through the cooler before its metallurgical treatment. The cooler is a system consisting of a hollow steel cylinder and a basin

E. Gongora-Leyva (✉) · Á. Columbié-Navarro · Y. Retirado-Mediacejas
Centro de Estudio de Energía y Tecnología de Avanzada de Moa, Instituto Superior Minero Metalúrgico de Moa, Holguín, Cuba
e-mail: egongora@ismm.edu.cu

G. Ruiz-Chavarría
Departamento de Física, Facultad de Ciencias, Universidad Nacional Autónoma de México, Mexico, México
e-mail: gruiz@unam.mx

J. Klapp and A. Medina (eds.), *Experimental and Computational Fluid Mechanics*, Environmental Science and Engineering, DOI: 10.1007/978-3-319-00116-6_14,

filled with water (hereafter referred as the pool). The cylinder is 30 m long, has a diameter of 3 m and the shell thickness is 16 mm. In standard operating conditions the cylinder rotates at an angular velocity of 6.24 rpm, whereas the filling degree lies in the range 15–25 %. The ore enters into the horizontal cylinder through a conical cover, and then it is continually stirred and shifted from the entrance toward the exit, with the aid of internal scrapers. The mean speed of the granular media is in the range 0.01–0.017 m/s, so the residence time lies between 30 and 50 min. An additional feature of the scrapers is to avoid the ore to adhere to the solid walls, a fact that hinders the heat transfer to the wall. Concerning the pool, the water is continuously renewed with a flow rate varying from 15 to 100 m^3/h. The liquid moves in the opposite direction to the displacement of the granular material. It enters at ambient temperature at the end of the cooler and at the far side the water has attained a temperature lying in the range 60–85 °C. Besides from enhancing heat transfer, the water has also the role to make floating the cylinder, reducing the amount of energy to maintain it in rotation (Góngora-Leyva et al. 2012).

A frequent problem in the coolers is the inability to lower the ore temperature from 750 °C to a value around 170 °C, which is the design temperature (Valle-Matos et al. 2000a,b). The final temperature of the granular media depends on variables like the water flow rate, the ore mass flow and the angular velocity of the cylinder. The latter factor is closely related to avalanching because this phenomenon produces a mixing of material inside the cylinder.

Góngora-Leyva et al. (2007, 2009) deduced a model based on the energy balance for a volume element, from which a system of partial differential equations is derived. The system is supplemented with other equations that relate the heat transfer coefficients with some physical parameters. Following the way of some previous works, we assume that only the axial coordinate is relevant, then the system is transformed into one of ordinary differential equations. This is the case of the works by Wang et al. (2010) and Xiaodong et al. (2012) where the cooling process of ash is investigated. Their simplified model predicts and explains some important features of the system they studied.

The chapter is organized as follow: in Sect. 2 some details of the system under study are given, in particular, the features of granular and liquid flows. In Sect. 3 the system of equation for heat transfer is outlined. In Sect. 4 some data obtained for a cooler 30 m long are present and a comparison with numerical solution is made. Finally, the conclusions are drawn in Sect. 5.

2 Description of the Cooling System

The Fig. 1 shows the cross section of the system under study, formed by the reduced lateritic ore (1), the rotating horizontal cylinder (2), the pool (3) and the surrounding air (4). The ore is discharged into the cooler at a temperature of 750 °C. For this temperature the modes of heat transfer involved in the ore cooling are conduction, convection and radiation.

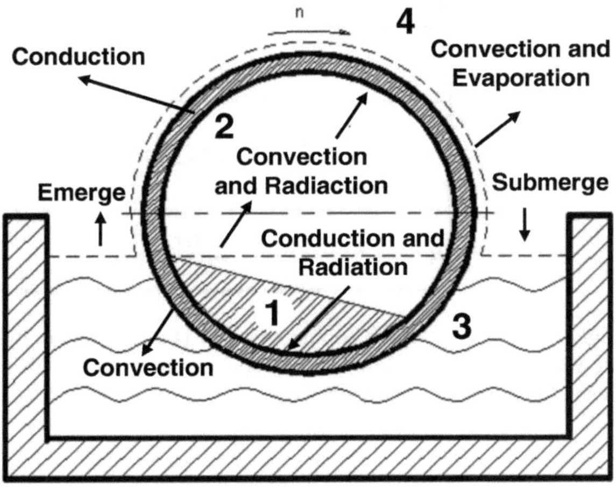

Fig. 1 Cross section of the cooler. The reduced ore (*1*) partially fills the rotating cylinder (*2*). Under standard operating conditions the filling degree is the range from 15 to 25 % and the angular velocity is 6.24 rpm. Otherwise, the cylinder is partially submerged in a pool (*3*) and it is contact with the surrounding air

In the system under study two flows exist. The first one is the motion of the granular bed, which is a mixing of a translation in the axial direction and a succession of avalanches induced by the rotation. The avalanching is characterized by a critical angle (θ_c) and a stop angle (θ_s), which determines the position of the granular bed inside the cylinder. According to a work Liu et al. (2010) in a system with low rotation and with irregular grains (they used rice grains), the value of θ_s is dependent on the filling degree, while the critical angle is nearly constant. Due to the difficulties to make measurement in the plant, the evolution of the granular bed has been investigated in a small rotating cylinder, its diameter being 30 cm and in which the front and back ends are covered by Plexiglas plates. Two different amounts of reduced ore were put inside the channel, which correspond to filling degrees of 10 and 20 %. Evolution of surface of the bed has been recorded with a video camera. The images have been digitized and the angle of the surface of the granular bed is measured. The results show that $\theta_c \approx 45°$ for a filling degree of 10 %, while $\theta_c \approx 35°$ for a filling degree of 20 %. For both cases the stop angle θ_s is less than 10°.

The second flow is the motion of the water from B to A (see Fig. 2). The liquid is injected into the pool at ambient temperature in the region where the ore leaves the cooler. As the water moves the temperature increases attaining a value between 60 and 85 °C at point A. Finally the water leaves the basin. The water motion is a forced convection used to flush thermal energy outside the system.

3 A Model for the Cooling Process

In order to construct a model for describing the ore cooling we start from the heat equation:

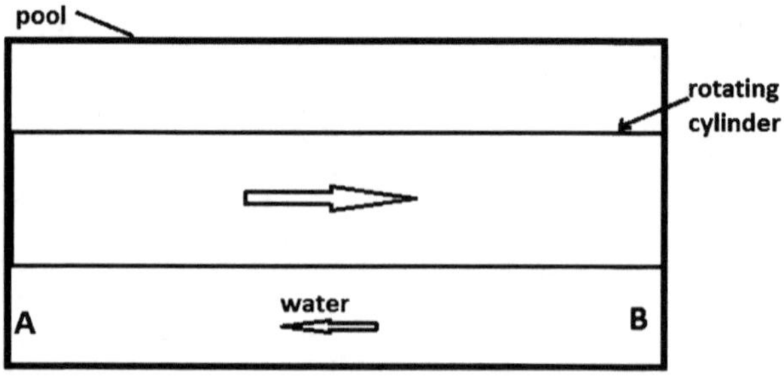

Fig. 2 *Upper view* of the cooling system. The lateritic ore moves from *left* to the *right*, whereas the water in the basin flows in the opposite direction. For this cooling system the water flow rate lies in the range 15–100 m^3/h

$$\frac{\partial T}{\partial t} + \vec{u} \cdot \nabla T = \chi \nabla^2 T + \dot{q} \tag{1}$$

where $\vec{u}$ is the velocity of the fluid (or granular media), T is the temperature and $\dot{q}$ stands for heat sources and sinks.

To derive the model some assumptions must be made. First, we assume that the process is stationary, so the time derivative vanishes. This assumption is well verified because most of time the operating conditions of the cooler remain unchanged. Second, we assume (Wang et al. 2010) that temperatures of the granular bed, the cylinder and the water have a weak dependence on the radial and angular coordinates. This implies that solution depends only on the axial coordinate. Measurement in plant show that this assumption is reasonably fulfilled in certain cases. Third, we assume that heat transfer between two adjacent control volumes along the axial direction is negligible and consequently the laplacian in Eq. 1 also vanishes. These assumptions lead to a system of two differential equations and an algebraic equation:

$$\dot{m}_o C_o \frac{dT}{dx} = \alpha_{wco} A_{wco} (T - T_w) + \alpha_{wnco} A_{wnco} (T - T_w) \tag{2}$$

$$\alpha_{wco} A_{wco} (T - T_w) + \alpha_{wnco} A_{wnco} (T - T_w) = \alpha_l A_c \left(T_w - T_l\right) \tag{3}$$

$$\dot{m}_l C_l \frac{dT_l}{dx} = \alpha_{air} A_l (T_l - T_{air}) + \alpha_l A_c (T_w - T_l) + Q_{evp} \tag{4}$$

In Eqs. 2–4 subscripts *w, o, l* and *air* refer respectively to the solid wall (cylinder), the granular bed (ore), the liquid (water) and the air, T is the temperature, $\dot{m}$ is the mass flow (of the ore and the water), A_{wco} is the arc length where the ore and the cylinder cross section are in contact, A_{wnco} is the arc length of the cylinder cross section not in contact with the ore, A_c is arc length where water is in contact with cylinder cross section, α represents the coefficient of the heat transfer and Q_{evp} is the heat transferred by evaporation. The variable without subscript T is the ore temperature.

Table 1 Temperature of the ore at x = 30 m (exit of the cooler)

	$\dot{m}_o = 20$ experiment	$\dot{m}_o = 20$ model	$\dot{m}_o = 34$ experiment	$\dot{m}_o = 34$ model
$\dot{m}_l = 50$	144 °C	149 °C	214 °C	250 °C
$\dot{m}_l = 100$	132 °C	148 °C	188 °C	248 °C

The temperature is a decreasing function of the water flow rate and an increasing function of the ore mass flow. The agreement between numerical solution and measurements in plant is good only for the case $\dot{m}_o = 20$ and $\dot{m}_l = 50$

The coefficients of heat transfer were taken from a work of Góngora-Leyva (2013). As already stated in the introduction, convection, radiation and conduction are taken into account. The conduction is the dominant mode for heat transfer of the ore in contact with the cylinder, convection occurs in the gases that are in contact with the ore and radiation is present because the high temperature of the ore. The heat absorbed by the cylinder is transferred to the outer wall by conduction. For the heat transfer between the cylinder and the water we consider that the dominant mode is convection and that boiling doesn't exist. We assume that the cylinder is always covered with a thin layer of water in the region that is not inside the pool and that the liquid layer rotates at the same angular velocity as the cylinder. Then, the convection in water is the same as in a Couette flow (Incropera and Dewitt 1999). Finally, the heat transfer from the water to environment involves convection and evaporation of the water.

In order to obtain a solution of Eqs. 2–4 we need to impose two boundary conditions. The first one is the ore temperature at the entrance of the cooler (x = 0), which has been set to 750 °C. The second condition is the temperature of the water at x = 30, which is set to 30 °C.

The solution of the mathematical model (Eq. 2–4) was calculated with a fourth order Runge-Kutta method. The domain we use is 30 m long and the step is $\Delta x = 0.01$, then the solution is calculated in 3,000 points. The numerical code was implemented in the free software Scilab. The numerical data were calculated for values of ore mass flow and water flow rate respectively in the ranges 15–44 tons/h and 40–100 m^3/h.

4 Numerical Solution and Measurements in Plant

In Table 1 we present data (taken in plant) of the ore temperature at the exit of the cooler (x = 30) for two different values of ore mass flow (20 and 34 tons/h) and two values of water flow rate (50 and 100 m^3/h). We also include the results of the numerical solution. It must be stressed that final temperature is a decreasing function of the water flow rate and an increasing function of the ore mass flow. In fact, if water flow rate is kept constant we see that a change in the ore mass flow leads to important changes in the final ore temperature. The data of table I show that when

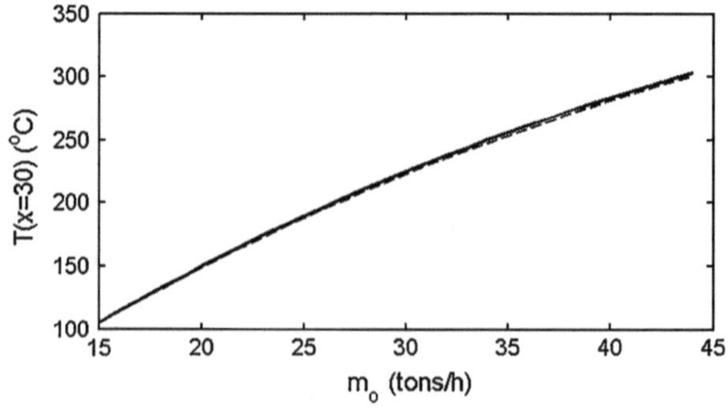

Fig. 3 Final temperature of the ore versus ore mass flow. Continuous curve corresponds to a water flow rate of 50 m^3/h whereas dashed line corresponds to a water flow rate of 100 m^3/h. Both curves are very similar, indicating that, according to the numerical model, the water flow rate has a weak influence in the cooling process

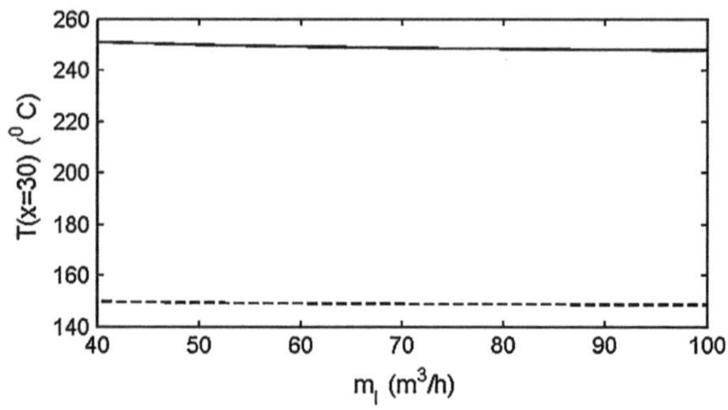

Fig. 4 Curves of final temperature of the ore T(30) versus water flow rate. **a** $\dot{m}_o = 34$ tons/h (*continuous line*) and **b** $\dot{m}_o = 20$ tons/h (*dotted line*). An increase in the water flow rate leads to a small decrease in the final temperature

$\dot{m}_o$ changes from 20 to 34 tons/h the increase of the final ore temperature is greater than 50 °C. The comparison between measurement and numerical solution show an agreement only for the case $\dot{m}_o = 20$ and $\dot{m}_l = 50$. In the remaining cases the model overestimates final ore temperature for more than 10 °C.

The Fig. 3 shows the curves of final temperature of the ore as a function of $\dot{m}_o$ in two cases, $\dot{m}_l = 50\,m^3/h$ (continuous lines) and $\dot{m}_l = 100\,m^3/h$ (dotted line). Both curves are very similar, indicating that an increase in the water flow rate has a weak influence in the cooling of the lateritic ore.

In Fig. 4 we present the curves of the final temperature T(30) versus water flow rate $\dot{m}_l$. The continuous line corresponds to an ore mass flow $\dot{m}_o = 34$ tons/h, while the dashed line corresponds to an ore mass flow $\dot{m}_o = 20$ tons/h. In both cases the increase of the water flow rate from 50 to 100 m^3/h leads (according to the numerical model) to a change of the final temperature T(30) of only few degrees.

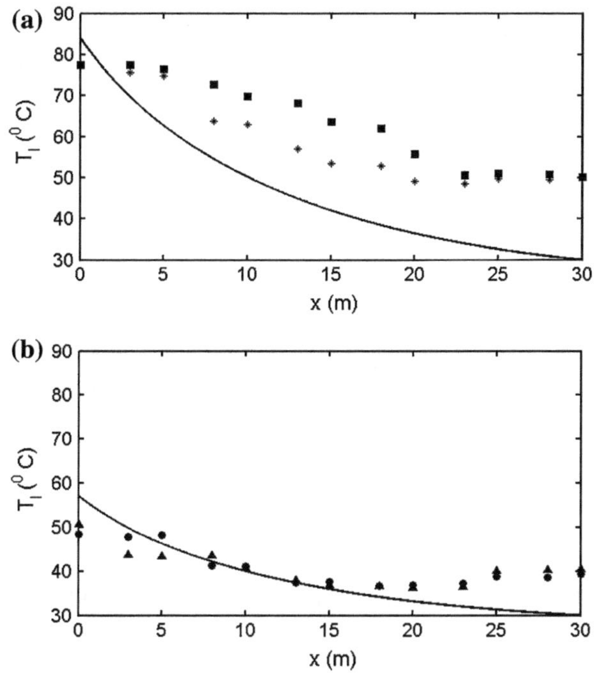

Fig. 5 Data of water temperature T_l versus x. **a** $\dot{m}_o = 34$ tons/h and $\dot{m}_l = 50\,m^3/h$, (□) measurements in the *left side*, (*) measurements in the *right side*, (*continuous line*) data obtained from numerical solution, **b** $\dot{m}_o = 34$ tons/h and $\dot{m}_l = 100\,m^3/h$, (*o*) measurement in the *left side*, (Δ) measurement in the right side, (*continuous line*) data obtained from numerical solution

Some measurements of the water temperature were made at different points along the pool, using PT-100 probes. Temperatures were recorded to the left and to the right of the cylinder (see Fig. 1). We have encountered a difference in data obtained along right and left sides, showing that there is a dependence of the cooling on the angular coordinate. This effect is related to the fact that surface of granular bed does not remain horizontal and also to the existence of discrete avalanches. However differences are only of few degrees and consequently one dimensional model can retain some important features of the cooling process. In Fig. 5 we present the data of water temperature T_l versus x and we include the prediction of the numerical solution. The Fig. 5a shows the temperature profile for an ore mass flow of 34 tons/h and a water flow rate of $50\,m^3/h$. There is an important difference between data obtained from measurement and the prediction of numerical solution, however, prediction of temperature of water at $x = 0$ is close to the experimental value. With respect (Fig. 5b), the ore mass flow and the water flow rate are respectively 34 tons/h and $100\,m^3/h$. The agreement between experimental data and numerical solution is good for $5 < x < 15$. In this case, the temperature of water at $x = 30$ is different from the assumed value $T_a = 30$ °C.

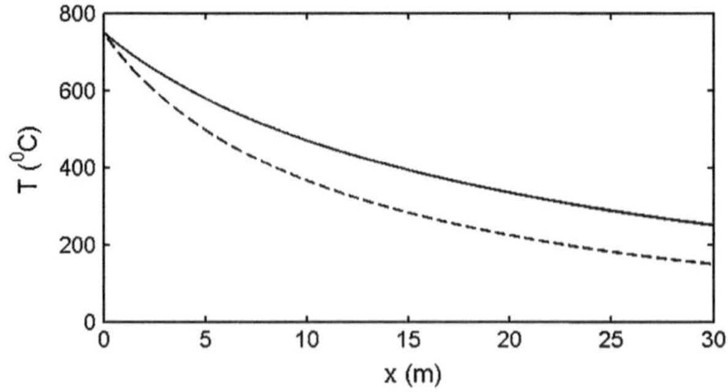

Fig. 6 Temperature profile of the granular bed, $\dot{m}_l = 50\,\mathrm{m}^3/\mathrm{h}$ **a** $\dot{m}_o = 34\,\mathrm{tons/h}$ (*continuous line*) and **b** $\dot{m}_o = 20\,\mathrm{tons/h}$ (*dashed line*). A reduction in the ore mass flow allows to a reduction of the temperature of the ore at the exit of the cooler

Finally, in Fig. 6 we present the ore temperature versus x for water a flow rate equal to 50 m^3/h and two different values of ore mass flow, namely $\dot{m}_o = 34$ tons/h and $\dot{m}_o = 20$ tons/h. In both cases the temperature at x = 0 is 750 °C. But the final temperature is different, for the first case T(30) = 250 °C and for the second case T(30) = 149 °C.

5 Conclusions

In this work we have presented data measured in situ and results of a model for the process of the cooling of lateritic ore occurring during the production of nickel. This simplified model is based on a system of two ordinary differential equations and an algebraic equation. Despite its simplicity the model allows to make some predictions for improving the efficiency of the cooler system. As a result of the numerical simulation we have found that an increase in the mineral flow and a decrease of the water flow rate lead to an increase of the temperature of the mineral at the cooler outlet. Therefore it is recommended to work with flows of mineral of 20 t/h and a water flow rate around $50\,\mathrm{m}^3$/h. On the other side It was found in experimental data that the temperature of the cooler (the rotating horizontal cylinder) have a dependence on the angular coordinate, then an improvement of this model could be the modeling of the cooling process with a system of partial differential equations dependent on two coordinates (x, θ).

Acknowledgments Ever Gongora Leyva thanks to SEP for the grant inside the "Programa de Cooperacion de Movilidad Estudiantil de Educación Superior México-Cuba".

References

Góngora-Leyva E (2013) Ph D Thesis (in preparation). Instituto Superior Minero Metalúrgico de Moa.

Góngora-Leyva E, Guzmán-del Río D, Columbié-Navarro AO, Marrero-Ramírez S, Retirado-Mediaceja Y (2007) Modelo matemático multivariable para un proceso de enfriamiento industrial de sólidos en cilindros rotatorios horizontales. Revista Energética 28(2):15–25

Góngora-Leyva E, Lamorú-Urgellés M, Columbié-Navarro AO, Retirado-Mediaceja Y, Legrá-Legrá A, Spencer-Rodríguez Y (2009) Coeficientes de transferencia de calor en enfriadores de mineral laterítico a escala piloto. Minería y Geología 25(3):1–18

Góngora-Leyva E, Palacio-Rodríguez A, Matos-Casals D, Rodríguez-Moreno JA (2012) Evaluación del proceso de enfriamiento del mineral laterítico reducido en la empresa Comandante Ernesto Che Guevara (Parte 1) Minería y Geología 28(3):50–69

Incropera F, Dewitt D (1999) Fundamentos de transferencia de calor Prentice Hall, Mexico, pp 912

Liu X, Zhou S, Specht E (2010) Avalanche time of granular flows in rotary kilns. Chem Eng Technol 33(6):1029–1033

Valle-Matos M, García-Pérez M, Rabell-González D, Morales-Morales Y (2000a) Evaluación de los enfriadores de mineral de la empresa Comandante Ernesto Che Guevara de Moa (1). Tecnología Química 18(1):70–77

Valle-Matos M, García-Pérez M, Rabell-González D, Morales-Morales Y (2000b) Evaluación de los enfriadores de mineral de la empresa Comandante Ernesto Che Guevara de Moa (2). Tecnología Química 20(2):10–15

Wang W, Si HR, Yang XD et al (2010) Heat-transfer model of the rotary ash cooler used in circulating fluidized-bed boilers. Energy Fuels 24:570–2575

Xiaodong S, Yang H, Wu Y, Wang W, Zhang H, Lu J (2012) Heat transfer in the rotary ash cooler with residual char combustion considered. Fuel Process Technol 94:61–66

Experimental Study of Mass Flow Rate in a Silo Under the Wall Width Influence

D. A. Serrano, D. Cabrera, G. J. Gutiérrez and A. Medina

Abstract The mass flow rate, $\dot{m}$, associated with the lateral outflow of dry, cohesionless granular material through circular orifices of diameter D made in vertical walls of silos was measured experimentally in order to determine also the influence of the wall thickness of the silo, w. Geometrical arguments, based on the outflow happening, are given in order to have a general correlation for $\dot{m}$ embracing both quantities, D and w. The angle of repose appears to be an important characterization factor in these kinds of flows.

1 Introduction

Since the first studies by Hagen (1852), it has been well known that the mass flow rate, $\dot{m}$, of the gravity flow of dry granular material emerging from the bottom exit of a silo scales essentially as $\dot{m} \sim \rho g^{1/2} D^{5/2}$ where $\dot{m}$ is the mass flow rate (grams/second), ρ is the bulk density, g is the acceleration due to gravity and D is the diameter of the circular orifice. This result contrasts with that occurring in liquids where the mass flow rate depends on the level of filling above the orifice. Moreover, in granular material a continuous flow occurs provided $D > 6d_g$, where d_g is the grain's diameter. After the fundamental study of Hagen was established, many researchers have proven the validity of his law, and slight modifications have also been proposed in order to improve the agreement with the experimental data. See, for instance Beverloo et al. (1961) and Ahn et al. (2008).

Despite the enormous utility of Hagen's law only a few studies have conducted to tests its validity for the flow of grains from exit holes located in the vertical wall of a silo Bagrintsev and Koshkovskii (1977) and Sheldon and Durian (2010). The aim of this work is to study experimentally the mass flow rate when the exit holes

D. A. Serrano (✉) · D. Cabrera · G. J. Gutiérrez · A. Medina
SEPI ESIME Azcapotzalco, Instituto Politécnico Nacional Av. de las Granjas 682, Col.,
02250 Azcapotzalco DF, Santa Catarina, México
e-mail: arman2390@hotmail.com

J. Klapp and A. Medina (eds.), *Experimental and Computational Fluid Mechanics*, Environmental Science and Engineering, DOI: 10.1007/978-3-319-00116-6_15,

are located in the vertical walls of the silos for different orifice sizes, D, and several wall thicknesses, w. As can be seen afterwards, the thickness of the wall can be used to control the grains dosage but this flow can be arrested if w overcomes a critical value that depends also on the angle of repose the granular material.

In order to reach our goal, the division of this work is as follows: in next section we give a short review of experimental studies of the mass flow rate where the orifices were made in lateral walls. After, in Sect. 3, we report new experiments where the influence of D and w was studied. In Sect. 4 we propose, on the basis of our experimental results, a correlation that embraces both changes in D and w. In Sect. 5 we discuss some new results based on this new correlation and finally, in Sect. 6, we give the main conclusions of this work.

2 Previous Work

To our knowledge, Bagrintsev and Koshkovskii (1977) were the first researchers who studied experimentally the problem of the gravity driven lateral outflow of granular material in vertical cylindrical silos with vertical walls. They used oval and circular exit holes made in transparent plastic walls and in relation with the wall thickness uniquely observed that "the outflow capacity decreases as wall thickness increases". Later, other authors did experiments in silos with rectangular exit holes 6–9 and circular exit holes. Bagrintsev and Koshkovskii (1977) and Choudary and Kesava (2006) and Kumar and Kesava Rao (2006) have found that apparently the better correlation among $\dot{m}$ and D was of the form $\dot{m} \sim \rho g^{1/2} D^{7/2}$, meanwhile Davies and Foye (1991) and Sheldon and Durian (2010) reported measurements of the mass flow rate in silos with lateral orifices that follows a relation of the type $\dot{m} \sim \rho g^{1/2} D^{5/2}$ (it is important to comment that in some studies Bagrintsev and Koshkovskii (1977) and Davies and Foye (1991) D is essentially a hydraulic diameter). Nevertheless, none of the referred works have analyzed systematically the effect of the wall thickness on $\dot{m}$.

In order to reach a better understanding of the behavior of the mass flow rate in lateral circular exit holes we performed a series of experiments, using circular orifices, where we mainly analyzed the influence of D and the wall thickness, w, on such a quantity. Clearly, wall thickness will be important for the occurrence of the flow because if the silo wall is very thick the outflow of granular material will be arrested. Detailed experiments given in the next section allow quantify these and other facts.

3 Mass Flow Rate Measurements

A. Bottom exit holes

It is well known that the wall thickness does not affect substantially the value of $\dot{m}_0$ the mass flow rate when the exit hole is located at the bottom of a silo. In Fig. 1 we

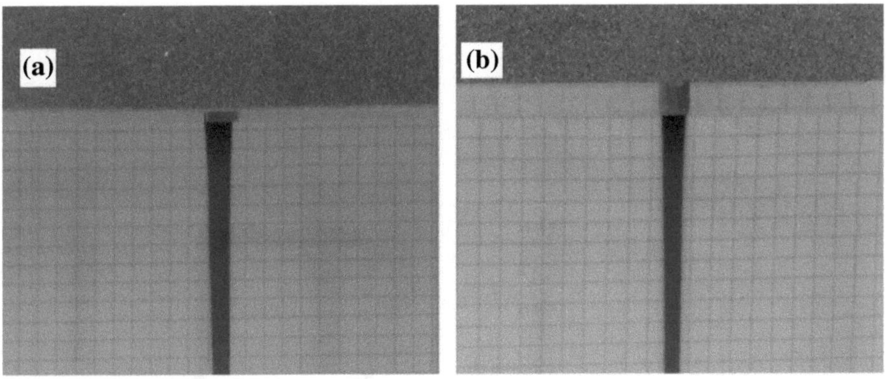

Fig. 1 Snapshots of the outflow of granular material: **a** thin bottom wall and **b** thick bottom wall

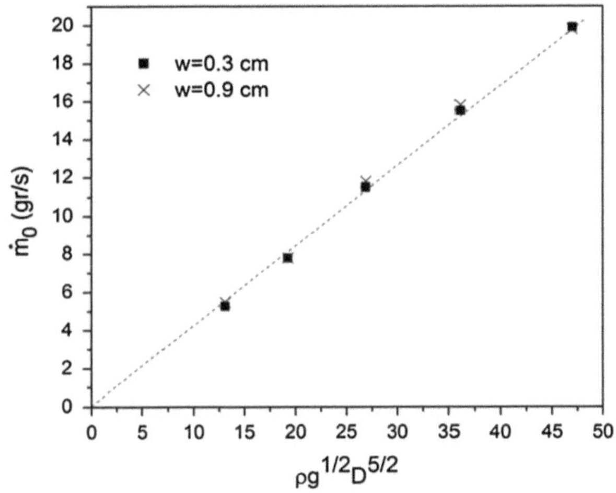

Fig. 2 Plot of the mass flow rate through horizontal orifices, $\dot{m}_0$, as a function of $\rho g^{1/2} D^{5/2}$. Data fit the Hagen's law $\dot{m}_0 = 0.48 \rho g^{1/2} D^{5/2}$. Error bars are of 4 %

show a jet of sand (mean diameter $d_g = 0.3$ mm and bulk density $\rho = 1.5$ gr/cm^3) that crosses through circular orifices in an acrylic-made box with a slim wall (Fig. 1a) and a bold wall (Fig. 1b).

In Fig. 2 we show the experimental plot of $\dot{m}_0$, as a function of $\rho g^{1/2} D^{5/2}$, for two different wall thicknesses: $w = 0.3$ cm and $w = 0.9$ cm.

In this figure we observe that both cases fit very well the Hagen's law, and thus the effect of w does not is appreciated. The relation that follows the straight line is

$$\dot{m}_0 = a \rho g^{1/2} D^{5/2}, \tag{1}$$

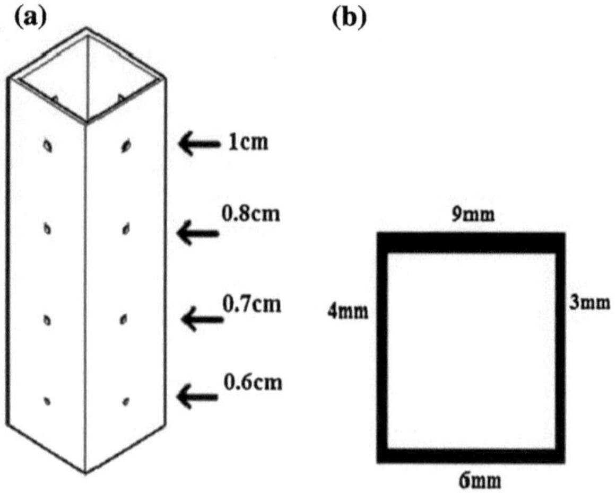

Fig. 3 Depict of the silo used in experiments. **a** a schematic of the silo where the staggered exit holes are shown. Orifices of equal diameter are at the same height. **b** Top view of the silo showing the four different wall thickness

where the dimensionless constant has the value $a = 0.48$. Afterwards we will discuss more about our measurement procedure of $\dot{m}$ and its accuracy degree.

B. Lateral exit holes

In order to reveal the influence of the wall thickness on the mass flow rate in silos with vertical walls, experiments were made using an acrylic box, 50 cm height and $10 \times 10\,\text{cm}^2$ inner cross-section, as the one shown in Fig. 3a the positions of the staggered orifices of different diameter D are sketched. As is sketched, orifices were made at the middle part of each wall. Diameters of the exit holes were: $D = 0.6$ cm at a height $H = 5$ cm from the bottom, $D = 0.7$ cm at $H = 15$ cm, $D = 0.8$ cm at $H = 25$ cm, and $D = 1.0$ at $H = 35$ cm. The circular orifices were made in each wall of the silo; in experiments four different wall thicknesses were used: $w = 0.3$, 0.4, 0.6 and 0.9 cm. In Fig. 3b we show a top view of the silo with the four different wall thicknesses. Thus, an exit hole of a given diameter is at the same height H in each wall.

A schematic of the experimental procedure employed to get the mass flow rates is shown in Fig. 4: a reservoir attached to a force sensor model Pasco CI-6537 with a resolution of 0.03 N was located close to the wall and data of weights were acquired each 1 s for each orifice in each wall, i.e., in each experimental run only one orifice was opened.

To understand our measurement procedure is crucial to quantify the involved forces in the fall of sand. In quantitative terms, suppose that the sand is dropping at a steady rate of $\dot{m}$ gr/s and that it takes t_1 seconds for the sand particles to hit the reservoir. The velocity of sand hitting the compartment would be essentially $v = gt$ (in the simplest case we have neglected the initial velocity of grains just at the exit of

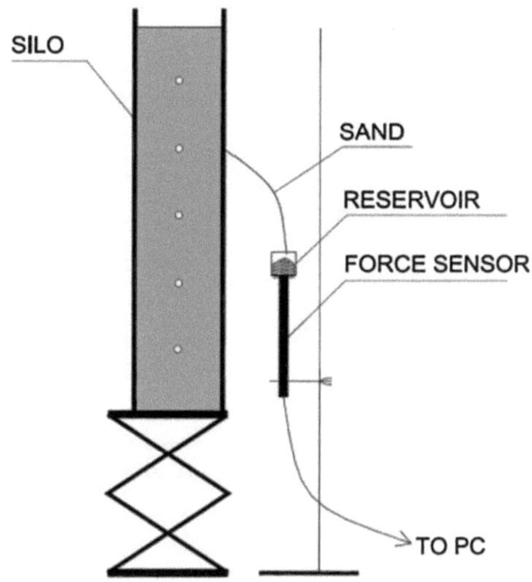

Fig. 4 Schematic of the experimental setup to measure the mass flow rate of granular material through circular orifices in vertical walls

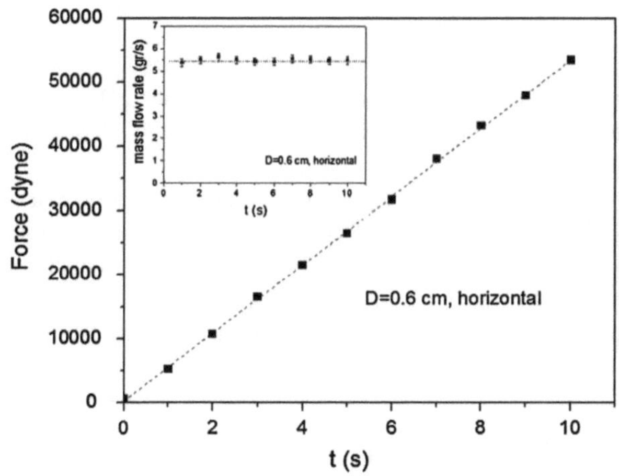

Fig. 5 Plot of the weight of the granular material in the reservoir of the apparatus depicted in Fig. 4, F, as a function of time. In the inset is the plot of $\dot{m}_0$ as a function of time. Error bars are of 4 %. Notice that the several measurements of $\dot{m}_0$ are around a constant (horizontal line)

the hole). In an infinitesimal time, dt, the change in momentum experienced by the reservoir would be $v\dot{m}dt$, so that the impulsive force experienced by the reservoir would be $F = v\dot{m}dt/dt = \dot{m}v = \dot{m}gt_1$. This force corresponds to a stepwise increase in the weight in the reservoir and for a time of measurement t_m, longer than t_1, it will occur a succession points as that shown in plot of Fig. 5. Finally, we can plot $\dot{m}$ as a function of time. See inset of Fig. 5. There, is appreciated that $\dot{m}$ is near a constant. Thus, actually the time-average of $\dot{m}$ is the quantity that we report as the mass flow rate. For example, the mean value of this quantity was computed from

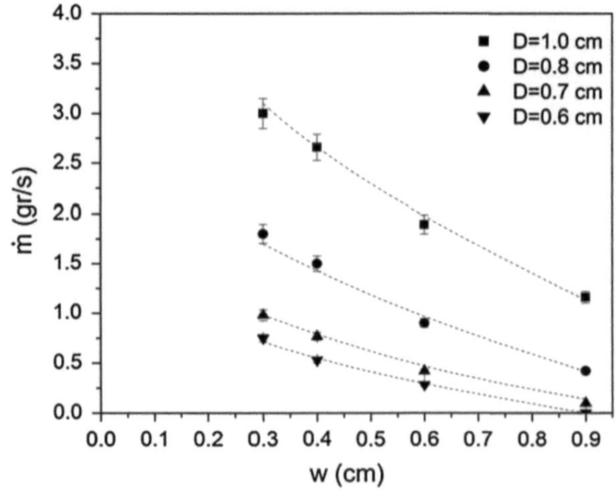

Fig. 6 Plot of the mass flow rate, $\dot{m}$, as a function of the wall thickness w. Each fit to experimental data corresponding to different diameters was made by using the relation (2) which is discussed afterwards. Error bars are of 4 %)

ten independent measurements for each couple of values (D, w). It is important to comment that no dependence on the level of filling of the silos was detected in our mass flow rate measurements.

In Fig. 6 we show the plot of $\dot{m}$ *vs* w for different diameters. It is noticed that for a fixed value of D there is a strong dependence of the mass flow rate on the wall thickness w. In some cases, for instance, when $D = 0.6$ cm in each wall, the outflow is arrested if $w \simeq 0.9$. In Fig. 6 there is a set of non linear fits which will be discussed in the next section on the basis of a simple geometrical correlation.

4 A Geometrical Correlation

From the results reported in plot of Fig. 6 it is clear that the wall thickness w also modulates $\dot{m}$. As an illustration, in Fig. 7a we show visually such a behavior: in a thin wall the outflow is strongest than in a thick wall, meanwhile in Fig. 7b it is observed that the flow is strongest as the height increases because the diameter of the staggered holes also increases.

All these results allow us to propose the next model. As can be seen in Fig. 8, we show schematically the lateral view of the zone where there is a hole of size D in a vertical wall of a silo Fig. 8a. In Fig. 8c is more evident that always there is a natural angle of wall, α, which can be defined as $\alpha = \arctan(D/w)$. In this same figure is observed that if there is no flow, due to the wall thickness is wide enough, the granular material maintained there will attains its angle of repose, θ_r. Thus, an outflow is occurs as long as $(\alpha - \theta_r)$, i.e., (see Fig. 8a and Fig. 8b) and conversely the

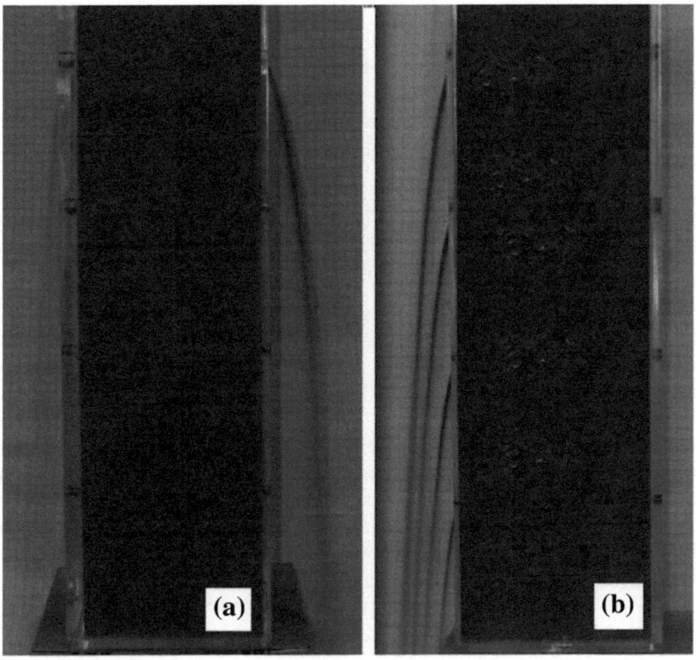

Fig. 7 Snapshot of the outflow of sand: in **a** we show the effect of the wall thickness on the mass flow rate because the two holes have the same diameter but in the thinner wall the flow is strongest. **b** in the staggered holes the mass flow rate increases due to the diameter of the holes in a same wall increase

outflow should be arrested if $\theta_r \geq \alpha$. Consequently, the mass flow rate itself must be proportional to $(\alpha - \theta_r)$, i.e., $\dot{m} \sim (\alpha - \theta_r)$.

Another important feature to get a general relation for $\dot{m}$ is that the mass flow rate through vertical holes is a fraction of $\dot{m}_0$(the mass flow rate through bottom holes). Consequently, the mass flow rate dependence on D and w would be a relation of the form

$$\dot{m} = c\dot{m}_0 \left[\arctan\left(\frac{D}{w}\right) - \theta_r \right], \tag{2}$$

where c is a dimensionless fitting parameter.

In order to show if the Eq. (2) is a correct correlation, in Fig. 9 we have plotted $\dot{m}$ as a function of $\dot{m}_0[\alpha - \theta_r]$, where $\dot{m}_0$ is given by Eq. (1) and also was measured. Here, we used the experimentally measured angle of repose $\theta_r = 33° = 0.57$ rad. This angle was measured by using the circular heap method (Wieghardt and Ann, 1975).

The direct comparison between $\dot{m}$ and $\dot{m}_0[\alpha - \theta_r]$ yields a linear relation of the form

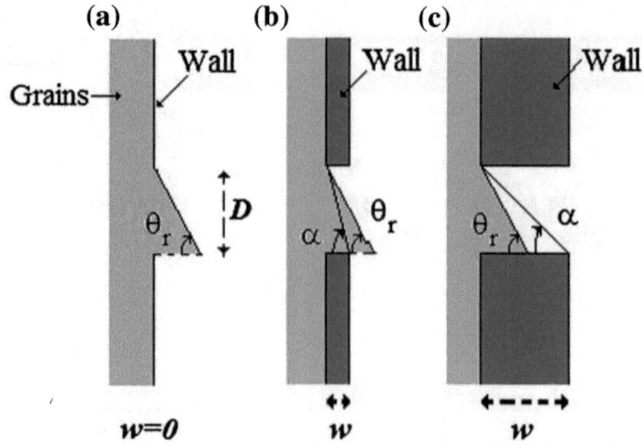

Fig. 8 Schematic lateral view of the exit hole in a vertical silo. In **a** the wall thickness is ideally null, meanwhile in **b** the wall thickness satisfies that $\alpha > \theta_r$ and a granular outflow proportional to $(\alpha - \theta_r)$ occurs. Finally, in **c** the wall thickness produces the condition $\alpha < \theta_r$ and consequently there is no flow. The granular flow is arrested in such a form that the slope of the granular material in the exit hole is featured by the angle of repose

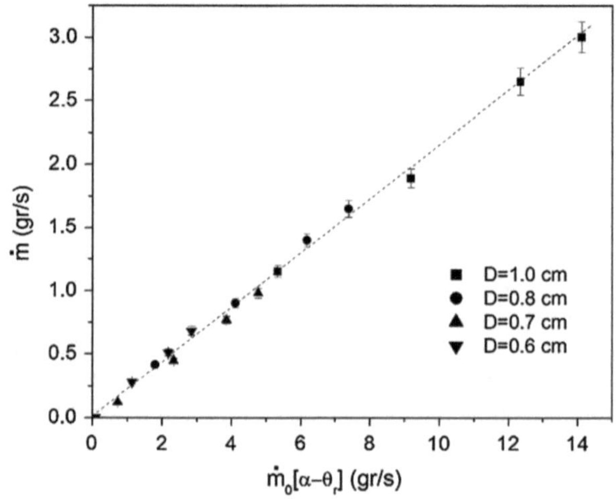

Fig. 9 Plot of the mass flow rate $\dot{m}$, as a function of $\dot{m}_0[\alpha - \theta_r]$. Data fit the straight line and produces the relation $\dot{m} = 0.23\dot{m}_0[\alpha - \theta_r] = 0.23\dot{m}_0[arctan(D/w) - \theta_r]$. Error bars are of 4 %

$$\dot{m} = 0.23\dot{m}_0 \left[\arctan\left(\frac{D}{w}\right) - \theta_r \right], \tag{3}$$

and the different sets of data (taken from Fig. 6) fit very well this straight line (in our experiments the value of c was $c = 0.23$, but, as in the case of the Hagen's law, this proportionality factor must be determined depending on the different materials, grains and silo walls, involved in the flow).

Giving all these results, we can conclude that Eq. (2) is a universal curve and that, essentially, in this work we would need of a unique experiment to fix $\dot{m}_0$ and then determine the main parameter of such equation (c), because the other point is the value $\dot{m}_0[\alpha - \theta_r] = 0$. Giving two points, always it is possible to depict the corresponding straight line, as the one shown in Fig. 9.

5 Other Results

The results of the previous section give us the confidence to explore another important consequences of Eq. (2). The first one is that Eq. (2) allows us to determine the critical value, w_c, for which the outflow will be arrested, as was discussed in Sect. 1.3. Graphically, the outflow will be arrested when $\dot{m}_0[\alpha - \theta_r] = 0$ or if $\alpha = \theta_r$. See Fig. 9. Equation (2) also allows us to find the critical value of w for the arrest, as a function of D and θ_r, this is

$$w_c = \frac{D}{tan\theta_r}, \tag{4}$$

By employing the four values of D used in the experiments and bearing in mind that $\theta_r = 33°$ we have verified that this relation predicts the precise thickness wall for which the outflow will be arrested.

A second consequence takes in to account that $\dot{m}_0 = a\rho g^{1/2}D^{5/2}$, and by using it in Eq. (2), we have that

$$\dot{m}_0 = c'\rho g^{1/2}D^{5/2}[\arctan(D/w) - \theta_r] \tag{5}$$

where the dimensionless parameter c' here takes the value $c' = ac = 0.11$. The Eq.(5) is the explicit formula of the mass flow rate with simultaneous dependence on D and w. An expansion in series of $(D/w) < 1$ transforms Eq. (5) into

$$\dot{m} = c'\rho g^{1/2}\left[\frac{1}{w}D^{\frac{7}{2}} - \theta_r D^{\frac{5}{2}} + O(D^{\frac{11}{2}})\right]. \tag{6}$$

In Eq. (6) the term proportional to $D^{7/2}$ will be dominant if the wall thickness satisfies the condition

$$\frac{D}{\theta_r} > w, \tag{7}$$

this condition is commonly fulfilled when the wall thickness is small respect to the hole diameter. Due to it, experiments of Bagrintsev and Koshkovskii (1977) and Choudary and Kesava (2006) Kumar and Kesava Rao (2006) where were used very thin walls, have shown that apparently the better correlation among $\dot{m}$ and D takes the form $\dot{m} = c'\rho g^{1/2} D^{7/2}/w$.

6 Conclusion

In this work we studied experimentally the problem of the mass flow rate of granular material through circular orifices in vertical walls of silos. Specifically, we have studied the dependence of $\dot{m}$ on D and w by using well characterized beach sand. To our knowledge, this is the first time that a systematic study of the effect of the wall thickness on the mass flow rate has been done. Our results show that Eq. (2), based in geometrical arguments, describes very well the influence of w and D on the mass flow rate, which can be considered a general formula including both quantities. In such an equation the role of the angle of repose is fundamental to describe the occurrence of the outflow of grains and its arrest. We gave also evidence that the use of Eq. (2) only requires to measure $\dot{m}_0$(if α and θ_r are known) to depict the corresponding straight line. Moreover, when the wall thickness is small respect to the hole diameters, the explicit dependence on D yields that $\dot{m}_0 = \rho g^{1/2} D^{(7/2)}/w$; this later result has been reported previously in some experiments of vertical orifices Bagrintsev and Koshkovskii (1977), Choudary and Kesava (2006) and Kumar and Kesava Rao (2006) and is obtained straightforward from our model. Finally, some studies have shown that the discharge rates from bottom exit decrease with the increase in particle size Hersam (1914). In such a case, instead the term $D^{5/2}$ in Eq. (1), it has been introduced the annular zone effect through the term $(D - kd_g)^{5/2}$ (where k is a constant); however, this assumption and the existence of many different orifice shapes requires the realization of more studies in order to have a general formula valid in a wide range of practical configurations. Work along these lines is in progress.

References

Ahn H, Basaranoglu Z, Yilmaz M, Bugutekin A, Zafer GlM (2008) Powder Tech 186:65–71
Bagrintsev II, Koshkovskii SS (1977) J Chem Petroleum Eng 6:13
Beverloo WA, Leniger HA, Van de Velde J (1961) The flow of granular solids through orifices. J Chem Eng Sci 15:260–269
Choudary S, Kesava Rao K (2006) Experiments on the discharge of granular materials through vertical and horizontal orifices of a vertical tube. Indian Acad Sci Project Report
Davies CE, Foye J (1991) Trans Inst Chem Eng 69:369
Hagen GHL (1852) Bericht uber die zur Bekanntmachung geeigneten Verhandlungen der Koniglich Preussischen Akademie der Wissenschaften zu Berlin 35

Hersam EA (1914) J Franklin Inst 177:419
Kumar SA, Kesava Rao K (2006) Experiments on the discharge of granular materials through orifices in the wall of a vertical tube. Indian Acad Sci Project Report
Sheldon HG, Durian DJ (2010) Granular Matter 12:579
Wieghardt K (1975) Ann Rev Fluid Mech 7:89

Surface Tension and Interfacial Tension Measurements in Water-Surfactant-Oil Systems Using Pendant Drop Technique

A. H. Cortés-Estrada, L. A. Ibarra-Bracamontes, A. Aguilar-Corona, G. Viramontes-Gamboa and G. Carbajal-De la Torre

Abstract The study and modeling of physical properties such as surface tension and interfacial tension are important factors in the formation and stability of fluid systems such as emulsions. The present work shows the experimental results for surface tension and interfacial tension measurements in water and/or oils systems in the presence of surfactants, using the pendant drop technique. Distilled water and straight-chain alkanes such as hexane, dodecane and hexadecane were used. The surfactants employed were sodium dodecyl sulphate (SDS) and sorbitan monooleate (SPAN 80), which are hydrophilic and lipophilic surfactants respectively. Some results show the dependence of surface tension or interfacial tension with respect to the surfactant concentration, other results were obtained by varying the temperature in a range from 20 to 60 °C.

1 Introduction

Research in surface physics is of relevance to many scientific areas such as Chemical Engineering, Materials Sciences, Physics, Electronics, among others. Surface Tension is one of the most widely used thermophysic property for surface characterization. Moreover, many industrial applications require (as part of their technological developments) the surface characterization of water–oil interfaces in the presence of other additives; this is the case of some processes in the Paint, Agrochemical, Oil and Cosmetics industries.

Several investigations have been carried out in the field of surface and interfacial tension of fluid systems. It is worth mentioning the work of Goebel and

A. H. Cortés-Estrada · L. A. Ibarra-Bracamontes (✉) · A. Aguilar-Corona
G. Viramontes-Gamboa · G. Carbajal-De la Torre
Universidad Michoacana De San Nicolás de Hidalgo, Av. Francisco J. Mujica s/n,Ciudad Universitaria, C.P. 58030 Morelia, MICH, México
e-mail: laibarrab@gmail.com

J. Klapp and A. Medina (eds.), *Experimental and Computational Fluid Mechanics*, Environmental Science and Engineering, DOI: 10.1007/978-3-319-00116-6_16,

Lunkenheimer (1997), where surface tension between water and a series of n-alkanes (from pentane to hexadecane) were presented for a single temperature (22 °C). In Zeppieri et al. (2001), surface tension between water and different alcanes (from hexane to dodecane) for a range of temperatures between 10 – 60 °C were reported. In turn, Rolo et al. (2002) reported surface tension for different kind of oils and binary mixtures of them.

The objective of this work is to analyse the experimental results obtained from surface tension measurements on different alcanes such as hexane, dodecane and hexadecane at different temperatures, as well as interfacial tension results between the aforementioned alcanes and water for different surfactant concentrations. The main purpose is to provide information that contribute to the study of emulsion type fluid systems, in particular, the surfactants SDS and Span 80 were selected due to their wide applicability in emulsions.

2 Methodology

In order to obtain surface and interfacial tension measurements an optical tensiometer was used. Image analysis of an emerging (pending) droplet was carried out so the determination of the surface parameters was achieved through the geometry of the drop and its relationship with surface.

The pendant drop technique allows for the determination of the surface or interfacial tension from distinct characteristics of the droplet profile. This calculation requires the solution of the Young-Laplace equation, which for the case of a pending drop can be expressed as:

$$2H = -\frac{\Delta \rho g}{\sigma} y + \frac{2}{R_0}$$

where y is the vertical coordinate, $\Delta \rho$ is the density difference between both fluid phases, R_0 is the curvature radius of the drop bottom, H refers to its average curvature (which is a function of y) (Zeppieri et al. 2009). Figure 1 shows a sketch of the experimental setup used to obtain the measurements under controlled temperature conditions.

Different configurations for the dispersed phase injection may be used depending on the type of measurement, as shown in Fig. 2. For surface tension measurements a straight needle was used, while a curved "U" shaped immersed needle was selected for the interfacial tension measurements.

Once the droplet is fully developed for each case, Image Analysis was carried over in order to determine the curvature parameters, which complemented with the density differences determine the interfacial and surface tension. The fluid phases were deionized water, three types of oils: hexane, dodecane and hexadecane. The selected surfactants were: Sodium Dodecil Sulfate (SDS), which is a hydrophilic (affinity for water, i.e. soluble in water) surfactant; Sorbitan Monooleate (Span 80), which is a lipophilic (affinity for oil, i.e. soluble in oil) surfactant.

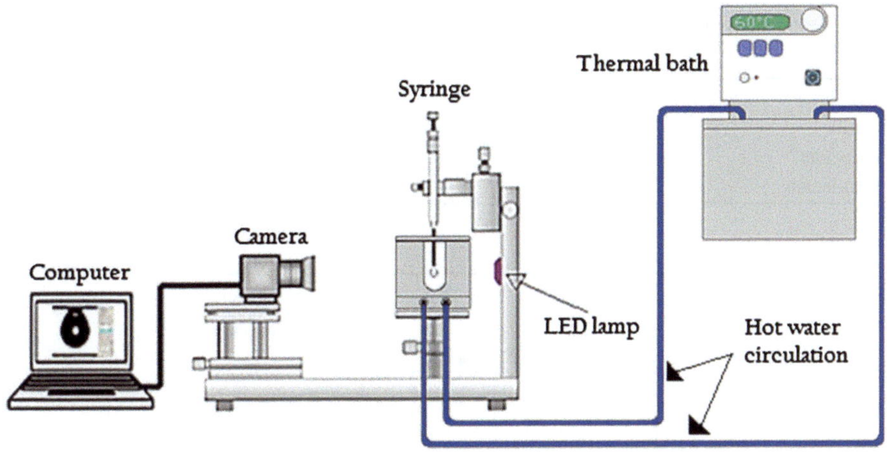

Fig. 1 Experimental setup for the surface and interfacial tension measurement using the pending drop technique

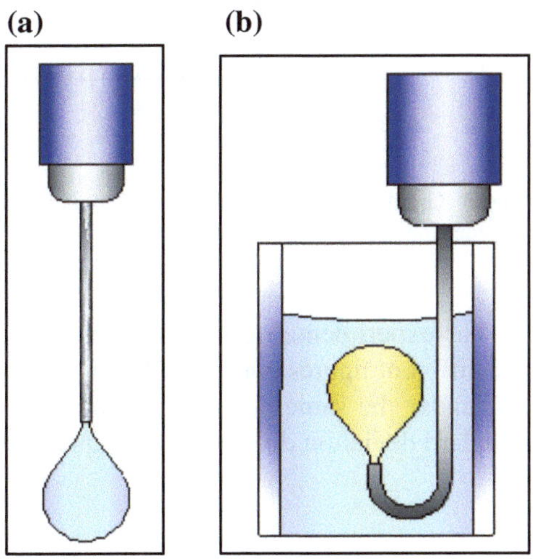

Fig. 2 Sketch of the drop injection depending on the type of measurement: **a** surface tension, **b** interfacial tension

Next, Table 1 shows the corresponding values for density ρ, surface tension σ and interfacial tension between water–oil γ reported in the literature for the different systems studied here Pardo et al. (2001); Vargaftik et al. (1983); Giner et al. (2007); Rolo et al. (2002) and Zeppieri et al. (2001).

For the experiments under different controlled temperatures, the fluid densities were selected from Pardo et al. (2001), as well as the corresponding values of the surface and interfacial tension were compared from Giner et al. (2007); Rolo et al. (2002); Zeppieri et al. (2009); Goebel and Lunkenheimer (1997).

Table 1 Physical properties for the selected fluids at 20 °C

Fluid	ρ(gr/cm^3)	σ (mN/m)	γ (mN/m)
Water	0.99823	71.8	–
Hexane	0.6616	18.40	50.80
Dodecane	0.7680	24.47	52.87
Hexadecane	0.7750	28.12	53.10

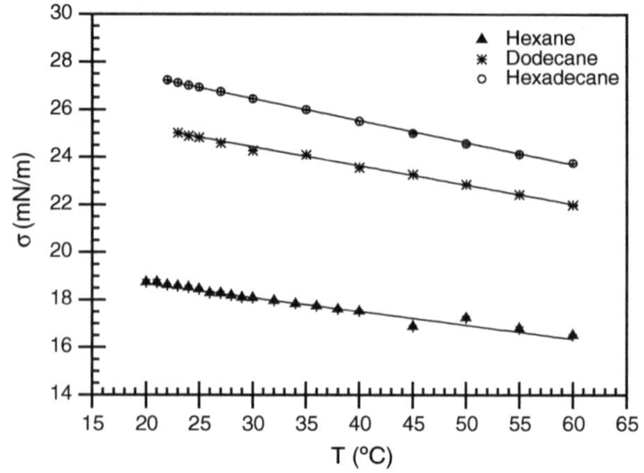

Fig. 3 Graph of surface tension (**σ**) versus temperature (**T**) for the oils: hexane, dodecane and hexadecane

The experimental results are presented in the next section. The values shown in the different figures correspond to averages of five independent experiments carried out under the same conditions. The error bars were determined from the sample standard deviation divided by the square root of the sample size.

3 Results

The first measurements correspond to surface tension in the range of temperature from 20 to 60 °C. Figure 3 shows the graph of surface tension versus temperature for three different types of oil, such as hexane, dodecane, and hexadecane.

In Fig. 3 it can be seen that as the temperature increases, the surface tension decreases. The rate of change of surface tension with respect to the temperature is different for the three different types of oils that were studied. In the case of hexane a rate of change of -0.058 mN/m °C was obtained, while for dodecane the rate of change was -0.081 mN/m °C, and for hexadecane the rate of change was -0.092 mN/m °C. These results indicate that as the linear chain length increases, a larger slope in the surface tension is obtained as a function of temperature.

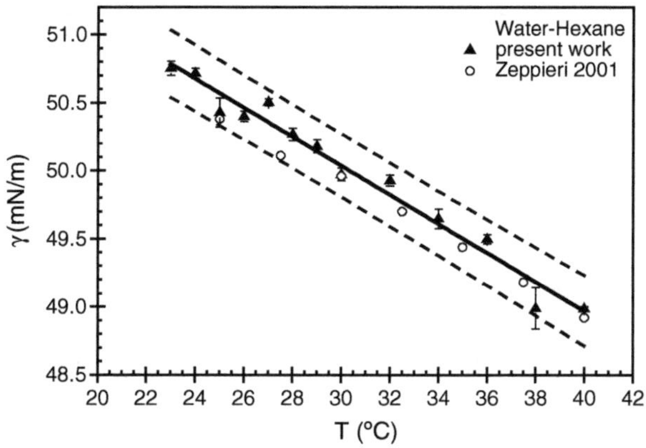

Fig. 4 Graph of interfacial tension (γ) versus temperature (**T**) for water–hexane interface

Giner et al. (2007) reported surface tension values only for the case of hexane, which are in good agreement with the results presented in this work. Values reported of surface tension with temperature for dodecane and hexadecane oils are limited.

Next, some results for the measurements of interfacial tension both in pure phases and also in the presence of hydrophilic and lipophilic surfactants are presented.

Figure 4 shows the graph of interfacial tension for variations in temperature, in this case for a water–hexane system. The range of temperature was from 22 to 40 °C. As already mentioned in the preceding paragraphs, as the temperature increases the system tends to reduce its interfacial tension.

According to the results obtained the interfacial tension for the system hexane/water in the region studied here, it shows a dependence with the temperature as follows: $\gamma = -0.10665\,(\mathrm{mN/m\,°C})\mathbf{T} + 53.2363\,(\mathrm{mN/m})$. Constructing the prediction interval (Walpole et al. 2012) that has a probability of 95 % for the interfacial tension measurements in a range from 22 to 40 °C, in this region the interfacial tension data reported by Zeppieri et al. (2001) are contained into the prediction limits (see Fig. 4). For example, estimating the 95 % prediction interval for the interfacial tension in the case of 30 °C is $49.8047 \leq \gamma(30\,°\mathrm{C}) \leq 50.2689$ in units of mN/m. For the same temperature, a value for interfacial tension of 49.96 mN/m ±0.04 was reported by Zeppieri.

Now comparing both results obtained for the surface tension and interfacial tension in the case of hexane, the rate of change of interfacial tension with temperature is greater than the one obtained for surface tension. This result shows a greater sensitivity of the interfacial tension to temperature changes.

Next, let us present the results obtained from introducing surfactant molecules as a third component in the system. Figures 5 and 6 show the results of interfacial tension with hexadecane as the organic phase with a temperature set to 25 °C.

Figure 5 shows the results when an hydrophilic surfactant was used, in this case the SDS concentration was dissolved in water. Interfacial tension measurements were

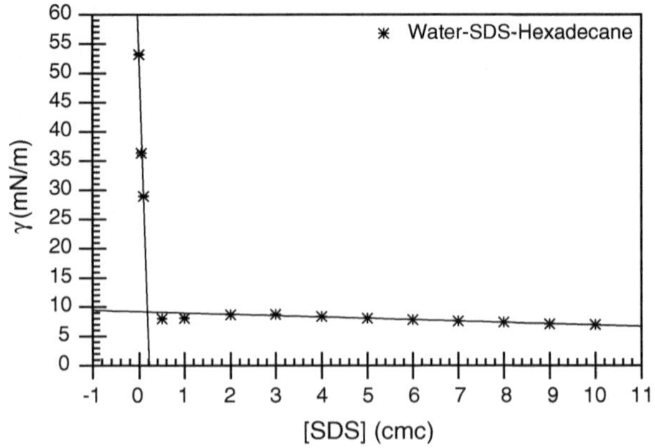

Fig. 5 Graph of the interfacial tension of water–hexadecane (γ) versus the surfactant concentration [SDS] dissolved in water

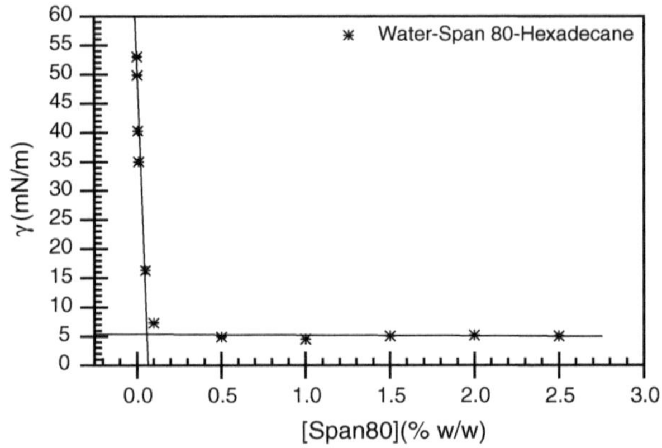

Fig. 6 Graph of the interfacial tension (γ) of water–hexadecane versus the surfactant concentration [Span 80] dissolved in oil

obtained for different SDS concentrations in a water–hexadecane system. The surfactant concentration of SDS was measured in terms of the critical micelle concentration (cmc) of SDS in water, which has a value of 8×10^{-3} mol/L. The SDS concentration values that were used in this investigation were: 0.05, 0.1, 0.5, 1, 2, 3, 4, 5, 6, 7, 8, 9 and 10 all in units of cmc. The first result in the graph corresponds to the interfacial tension value without surfactant.

It is known that the main function of surfactant molecules is to decrease the surface tension or the interfacial tension in a fluid system, this is clearly showed in the graph. Also it can be noted that the greater variations in the interfacial tension occur at the lower surfactant concentrations. Thereafter, the interfacial tension values remain unchanged when a certain surfactant concentration is reached. This can be explained from the fact that these molecules saturate the interface once a given surfactant concentration is reached, and the surfactant excess no longer modifies the properties of the interface. The intersection value can be determined by the trend lines in the two regions of the interfacial tension, this point shows a threshold-value of approximately 0.18 cmc.

Figure 6 shows the results for the interfacial tension of water–hexadecane system at different concentrations of a lipophilic surfactant, in this case different concentrations of Span 80 were dissolved in oil. Due to the high viscosity of Span 80, the corresponding surfactant concentrations are usually reported in weight fraction or weight percent in the organic or continuous phase (% w/w). The values used for the interfacial tension measurements were as follows: 0.0 01, 0.005, 0.01, 0.05, 0.1, 0.5, 1, 1.5, 2, 2.5 all in units of % w/w.

Again the effect of surfactant in the oil–water systems lowers the interfacial tension. The lower values for surfactant concentration produce greater variations in the interfacial tension measurements up to a given threshold value, and then the interfacial tension maintains a constant value. The intersection of the trend lines in the two regions of the interfacial tension, shows a threshold value of approximately 0.06 % w/w.

4 Conclusions

The technique of the pending drop was used to obtain surface and interfacial tension for water–oil systems at different temperatures in pure phase and in the present of surfactants. The threshold value that corresponds to the saturation was established, when an increase in surfactant concentration no longer causes a significant decrease in interfacial tension, both for the hydrophilic as well as the lipophilic surfactant.

References

Giner B, Villares A, Martín S, Artigas H, Lafuente C (2007) Study of the temperature dependence of surface tensions of some alkanol + hexane mixtures. J Chem Eng Data 52(5):1904–1907

Goebel A, Lunkenheimer K (1997) Interfacial tension of the water/n-alkane interface. Langmuir 13(2):369–372

Pardo JM, Tovar CA, González D, Carballo E, Romani L (2001) Thermophysical properties of the binary mixtures diethyl carbonate + (n-dodecane or n-tetradecane) at several temperature. J Chem Eng Data 46(2):212–216

Rolo LI, Caco AI, Queimada AJ, Marrucho IM, Coutinho JAP (2002) Surface tension of heptane, decane, hexadecane, eicosane, and some of their binary mixtures. J Chem Eng Data 47(6):1442–1445

Vargaftik NB, Volkov BN, Voljak LD (1983) International tables of surface tension of water. J Phys Chem Ref Data 12(3):817–820

Walpole RE, Myers RH, Myers SL, Ye K (2012) Chapter 11: Simple linear regression and correlation. In: Probability and statistics for engineers and scientists, 9th edn. Prentice Hall, New Jersey

Zeppieri S, Rodríguez J (2001) Interfacial tension of alkane + water systems. J Chem Eng Data 46(5):1086–1088

Zeppieri S, Ramírez Y, Molina J (2009) Estudio experimental de la tensión interfacial transitoria para sistemas alcanos-agua-tritón X-100 empleando la técnica de la gota emergente. Información Tecnológica 20(6):125–135

Experimental Measurement Process of a Volume Displacement of Oil Caught in a Fractured Rock by Gravity and Using Surfactant Foam

Arnulfo Ortíz Gómez

Abstract We present a series of experiments conducted in a cell with two parallel plates made of acrylic, the foundation plate used as a base is 0.5 in. thick. Inside it 4 mm deep and 6 in. wide channel was made, the wet perimeter was covered with sand and a cementing material, after the cap was set, it was filled with oil. We measured the filling time by means of a constant column, the time it took the cell to empty from different angles, where the only acting force displacing oil was gravity. We also measured the emptying time when using foam as displacing agent on different slopes, the foam was generated by injecting air into a mixture of water, glycerin and a surfactant, them it was introduced to the cell through a latex line of 10.0 mm diameter. The results found allowed us to establish meaningful conclusions in order to understand the rheology of foams, after being used as secondary methods of oil production in fractured reservoirs.

1 Introduction

Enhanced Oil Recovery (EOR) refers to all the processes used to recover more oil from a reservoir than the one being recovered by primary methods, it mostly involves the injection of gas or liquid chemicals and/or using thermal energy. Among the former ones, the most used are: gaseous hydrocarbons, CO2, nitrogen and flue gas. Among the liquid chemicals polymers, surfactants and hydrocarbon solvents can be found. Finally, thermal processes are related to the use of steam or hot water, or the in situ generation of thermal energy by internal combustion of oil in the rock of the reservoir (Donaldson and Chilingarian 1985; Baviere and Canselier 1997; Roehl and Choquette 1985).

A. O. Gómez (✉)
División de Ciencias Básicas, Facultad de Ingeniería, UNAM, Ciudad Universitaria,
CP 04510 Coyoacán, México City, D.F., México
e-mail: arnulfo64@gmail.com

J. Klapp and A. Medina (eds.), *Experimental and Computational Fluid Mechanics*, Environmental Science and Engineering, DOI: 10.1007/978-3-319-00116-6_17,

Such interactions can, for example, lead to a decrease in interfacial retention, oil swelling and reduction in viscosity, modifying the wettability or favorable behavior of phases. The latter ones are of special interest in this work, which deals with the process of oil displacement that is trapped in a crevice, by means of reducing the phenomena listed above.

The study was done by making an acrylic rectangular cell, where the wetted perimeter was covered with sand produced in the Cantarel area. Inside the cell a series of filling and emptying experiments were performed by means of gravity and by surfactant injection. We measured the filling and emptying times from different angles, using a surfactant, saturation and volume displacement percentages at emptying times. With the information gathered, capillary pressure variation, permeability of the material and percentages of saturation graphs were prepared. It was also possible to verify the formation of fingers due to the menisci advance.

2 Injection Foam (Surfactant)

The process demanding our attention is the one related to the injection of gases and mixed fluids in a surfactant, known as foam. In secondary recovery operations when water or gas is injected into the reservoir, also immiscible displacements occur, such kind of sweeping is due to the mixture of gas, liquid and additives which allow interfacial activity; *surfactants* is the term normally used to abbreviate the compounds with activity between phases (Morrow and Mason 2001; Shah and Schechter 1977). Chemically, the surfactants are characterized by having a molecular structure containing a group that has little attraction or antipathy for the solvent, known as liofóbico group, along with another group that has strong attraction, appetite for the solvent, called liofílico group. If the solvent is water, these groups are known as hydrophobic and hydrophilic portions of the surfactant. Usually, the hydrophobic group is a linear carbon chain or branched, while the hydrophilic portion is a group with some polar character. In the case of water injection the process is similar to the one mentioned above, this is due to the physicochemical characteristics of both fluids, a surfactant that allows the interaction between the present phases is required.

Assuming for simplicity that the porous medium containing a brine (water-crude, W-O), these two phases are distributed by the hydrostatic laws of capillarity. The basic Laplace Law capillary equation relates to the pressure difference between the sides of an interface (capillary pressure, Pc) with the curvature:

$$P_c = \Delta P = P_o - P_w = 2\gamma R \tag{1}$$

Where γ is the interfacial tension, R is the average curvature of the interface. A hemispherical surface to the average curvature radius is reversed. Pair any surface, R is half the sum of the reciprocal surfaces of the principal radius of curvature (high and low).

The interfacial tension is the Gibbs Energy per unit of area, this one depends on the substances absorbed in the interface. For water to be in equilibrium with a hydrocarbon phase, this energy is just of some sets of 10 dyne (din/cm or mN/m) (Johansen and Berg 1978; Shah 1981; Larry 1989). In the presence of a surfactant this energy is generally reduced to 1.0 or 0.1 din/cm.

Contact angle. The three-phase contact, water (W), oil (O) and solid (S) is identified as the contact angle between the surfaces.

If we assume that the solid is quite flat at the point of contact, you can set a vector equation of equilibrium between the forces, called Newman balance.

$$\gamma_{SW} + \gamma_{WO}\,(cos\ \theta_W) = \gamma_{SO} \tag{2}$$

$$\cos\theta_W = (\gamma_{SO} - \gamma_{SW})/\gamma_{WO} \tag{3}$$

The magnitude of the contact angle depends on the interfacial tension between O and W (γ_{WO}) and on the free interfacial energies of the solid with the two fluids as well, in other words, it depends on the natural fluids and the surface of the solid.

The fluid which has a contact angle, $\theta < 90°$ wets the solid surface. The reservoir rocks have a polar nature (silica carbonate) and thus the clean rock can be wet with water. Despite this, in many reservoirs $\theta > 90°$ something different is observed, because the solid surface is covered by a layer of absorbed substance which gives an oily character (Baviere 1991; Salager and Antón 1999; Pillai et al. 1999; Salager and Microemulsions 1999).

3 The Experiment

In order to understand the rheology of the surfactant interacting with the water and oil an acrylic cell with three sidewalls made of stone was designed, Fig. 1a. Subsequently it was filled with high density oil, so that it could not be displaced in the absence of gravity. Then we proceeded to inject a mixture of water, surfactant and glycerin, the latter one was to prevent *minimal surfaces* from dehydrating quickly and being broken down.

The idea is to simulate what happens in a fractured rock full of oil, when the lower part is connected with the reservoir and the upper part is in contact with the atmosphere, after an injection well was drilled (Fig. 1b).

In order to improve oil recovery, we developed a chemical formula that can alter the wettability of the oil and preferably water, lowering surface tension, so that the effect of capillary breaks and the oil can be conducted by effect of gravity, and can change the cross flow between layers permeability.

The experiment was performed as follows: firstly we connected a supply valve (Qe Fig. 1) to a conducting line (latex hose inner diameter 0.8 cm), secondly, we filled the container with a volume of oil equal to the volume of the cell ensure that the flow opened two keys; the supply and the output. The filling process was always the

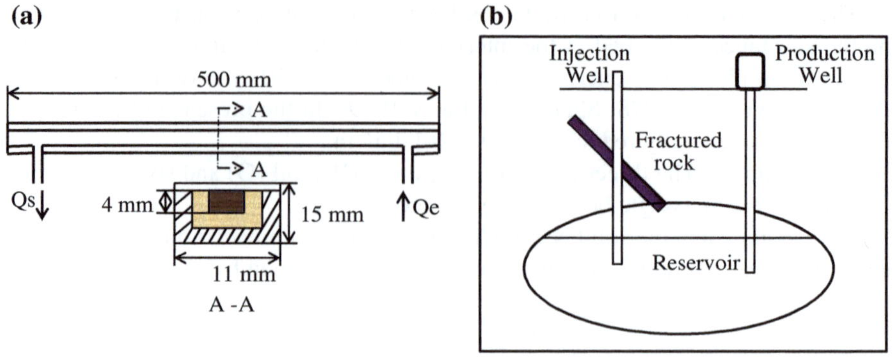

Fig. 1 **a** Experimental model; acrylic cell slot 50.0 mm by 4 m deep, the wetted perimeter is a limestone core Cantarel study. **b** Image of a fractured penetrated by an injection well

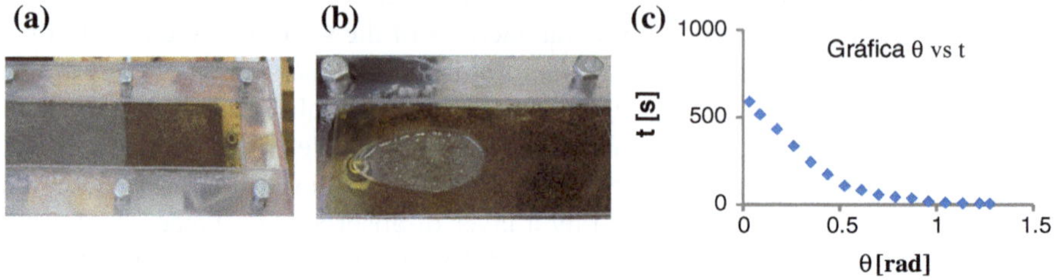

Fig. 2 **a** Advancing front during the filling process. **b** Advancing front during the casting process. **c** Graph of slope θ, versus time t, it is seen a shorter steeper emptying the cell

same, the total filling time associated with the cell was 15.0 min ($T_{LL} = 15$ min), maintaining a hydraulic head of 20.0 cm above the cell. This process is performed by simulating a source and a sink (*Qe* and *Qs* Fig. 1), since the supply valve and outlet are located at the base of the cell. Right in the source, oil gets radially into the cell but as soon as it finds walls, it advances or moves in a horizontal direction (*xi*), the face is convex, because the friction on the walls increases, the liquid is confined to four walls where the wetted perimeter of the landfill is sandy and the top cover is 4.0 mm acrylic to make it easy to view. In Fig. 2a an image of the forward advance in two different instants is shown. During this movement it was observed that the parabola formed by the advancing front is not parallel, sometimes it fills more in one side of the cell than in the other and vice versa, this is because the wall friction does not have a constant presence.

The times associated to the emptying of the cell to different slopes were measured too, now the walls stop the oil slick, emptying first in the center, so the front is concave Fig. 2b. The experiment was performed from different angles ($0° < \theta < 80°$) measuring the times for each slope, the results found are shown in the graph of Fig. 2c.

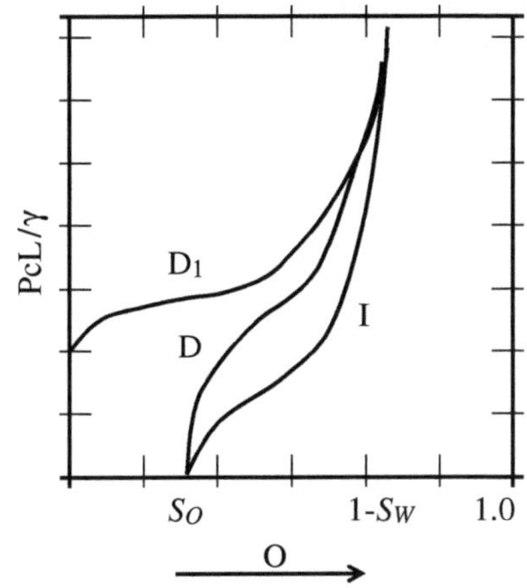

Fig. 3 Capillary pressure variation to successively injects of water and oil in a rock core, usually for a study of the perforations. In the experiment we do it in the cell

When two immiscible fluids coexist in equilibrium in a porous medium, they are distributed according to the laws of hydrostatics and capillarity. The distribution of the fluids depends on the dimension of the pores, the contact angle, the interfacial tension and the saturation S_W and S_O.

In practice, the variation of the capillary pressure was experimentally determined.

$$\Delta P_c = P_c L/\gamma \tag{4}$$

Where L is the characteristic length of the medium, for example the average pore diameter, in our case we use the channel height that was practiced in the cell of the cell (4 mm), Eq. 4 allows the drainage on determined imbibitions curves, Fig. 3.

Curves I and D comprise the hysteresis loop of capillary pressure. This means that for a certain state of saturation of the porous medium, the given capillary pressure in the real distribution of the fluids depends on the previous history of the system evolution.

The drain and saturation experiments show that it is not possible for one of the fluids to be displaced completely by the other. In both displacement cases there is a residual saturation of 20 %, which corresponds to a fluid in form of beads disconnected from one another and trapped by capillary force.

The motion of a single-phase fluid in a porous medium is dependent on one specific property of the medium called permeability, this can be found experimentally by determining the relationship between the velocity of a fluid movement and the pressure drop produced.

$$u = (k/\eta)(dP/dL) \tag{5}$$

Darcy's law (Eq. 5, is a linear relationship that satisfies the low speeds involved: where u is the specific speed or filtration speed, i.e. the volumetric flow per unit area of the medium it passes through, the viscosity η, fluid, dP/dL is the pressure gradient (also comprises hydrostatic pressure).

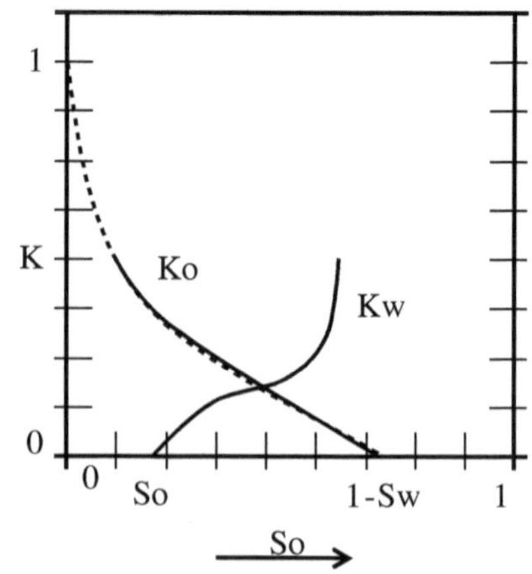

Fig. 4 Graphical permeability determined in a core of 20 cm height and 5.0 cm radius, extracted from Cantarel Reservoir

The Eq. 5 is equivalent to the Poiseuille Equation for a cylindrical capillary, and to involve all the pores, it incorporates a factor (0.125R2):

$$v = (R^2/8\eta)(dP/dL) \tag{6}$$

By defining Eq. 6 for each of the fluids involved in the experiment, it was found that the effective permeability was lower than the permeability of the porous medium.

The ratio of the effective permeability and the one of the medium is defined as the relative permeability *K*.

$$K_O = k_O/k, \qquad K_W = k_W/k \tag{7}$$

For a core of 20.0 cm high and 5.0 cm radius, graphs of relative permeability were constructed as a function of saturations SW and SW, Fig. 4.

4 Conclusions

In order for the displacement of fluids to take place it is necessary that the displacing fluid possess more energy than the displaced one. As the first fluid is injected, a separation front starts appearing and two zones in the cell are well distinguishable, one of them is called non-invaded, where an oil bank is being formed due to the oil displacing forward. Behind that bank is the invaded zone, which was formed by the injected fluid and the remaining oil.

In order to improve the recovery of oil it is required to develop a chemical formula that can alter the preference wettability of the oil and water, which reduces the low surface tension in such a way that the capillary effect of the oil is broken and can be driven by gravity effect, and the permeability between layers can be changed by the crossed flow.

Chemical products injected can be introduced in the zones of less permeability, not only by spontaneous imbibitions, but also by the foam that diverts the liquid in the lower permeability layer. The foam as mobility control can also improve the efficiency of sweeping.

As a result in each of the flow processes, the foam is usually injected after low surface tension. The disadvantage in lowering the surface tension of the surfactant is in the lack of mobility control, generating a poor swept. It is ideal that the formulation of surfactants can simultaneously reduce the surface tension to very low values and generate strong foams which generate microscopic displacement and an efficient swept of the reservoir.

The laboratory-scale dimensions are totally different from the reservoir, as a consequence, several aspects need to be considered when extrapolating results.

The effect of gravity cannot be neglected in a field of several meters thick, or in the case of a sloping site. The reservoir thickness produces gravitational segregation with higher saturation of oil in the upper part, therefore conditions along the vertical axis change.

References

Baviere M (ed) (1991) . Basic concepts in enhanced oil recovery. In: Kunieda H (ed), Springer, New York

Baviere M, Canselier JP (1997) In: Solanes C, Kunieda H (eds) Microemulsions in chemical enhanced oil recovery. Academic Press, New York

Donaldson E, Chilingarian GV (1985) Enhanced oil recovery, fundamentals and analysis. Elsevier, New York

Johansen RT, Berg RL (1978) Chemistry of oil recovery. American Chemical Society Series 91, Washington DC

Lake LW (1989) Enhanced oil recovery. Prentice Hall, New Jersey

Morrow N, Mason G (2001) Recovery of oil by spontaneous imbibitions. Curr Opin Colloid Interface Sci 6(4):321–337. doi:10.1016/SI1359-0294(01)00100-5

Pillai V, Kanick J, Shah DO (1999) Applications of microemulsions in enhanced oil recovery, (Chapter 24). In: Kumar P, Mittial KL (eds) Handbook of microemulsions science and technology. Dekker

Roehl P, Choquette P (eds) (1985) Carbonate petroleum reservoirs. Springer, New York

Salager JL, Antón RE (1999) Ionic microemulsions. In: Kumar P, Mittal LK (eds) Handbook of microemulsions, science and technology. Dekker, New York

Salager JL (1999) Microemulsions. In: Broze G (ed) Hand book of detergents-part A: properties. Marcel Dekker, New York

Shah DO, Schechter RS (eds) (1977) Improved oil recovery by surfactant and polymer flooding. Academic Press, New York

Shah DO (ed) (1981) Surface phenomena in enhanced oil. Plenum Press, New York

Modeling, Simulation and Scale-up of a Batch Reactor

René O. Vargas and Francisco López-Serrano

Abstract A simple model for the batch emulsion polymerization of styrene, considering different size reactors has been studied. The resulting differential equation-set was solved for the temperature (jacket and reactor) and conversion profiles as a function of time. The fundamental dimensionless numbers, appearing in the different terms of the reaction equation-set, involving the physics of the process, were compared for different size reactors allowing the basic description of scaling and its difficulties.

Nomenclature

A	Heat transfer area (m^2)
K_p	Propagation rate constant ($Lmol^{-1}s^{-1}$)
mo	Monomer charge (mol)
M_p	Monomer concentration inside particles($molL^{-1}$)
n	Average number of radicals per particle (dimensionless)
N_{av}	Avogadro's number (mol^{-1})
NT	Total particles concentration (L^{-1})
t	Time (s)
U	Overall heat transfer coefficient (W/m^2K)
V	Reactor volumen (L)

R. O. Vargas (✉)
SEPI ESIME Azcapotzalco, Instituto Politécnico Nacional, Av. de las Granjas 682, Col. Santa Catarina, CP 02250 Azcapotzalco, México D.F., México
e-mail: reneosvargas@yahoo.com.mx; rvargasa@ipn.mx

F. López-Serrano
Departamento de Ingeniería Química, Facultad de Química , Universidad Nacional Autónoma de México, CP 04510 México D.F., México
e-mail: lopezserrano@unam.mx

J. Klapp and A. Medina (eds.), *Experimental and Computational Fluid Mechanics*, Environmental Science and Engineering, DOI: 10.1007/978-3-319-00116-6_18,

1 Introduction

Emulsion polymerization has grown to become one of the major means for the production of synthetic polymers. There are substantial incentives for improved design and control of emulsion polymerization reactors (Fogler 1992; Schork et al. 1993). To achieve these aims the problem of laboratory data scale-up is crucial (Barton and Nolan 1987). Polymerizations are very complex reaction systems characterized by a strong change in the bulk viscosity and high exothermicity, often accompanied by autoacelerating kinetics that can lead the system toward uncontrollable situations (Soh and Sundberg 1982). Regarding the equation-set complexity, nonlinearity characterize the polymerization processes. To this date, many industrial polymerization reactions are carried out in batch reactors, especially at laboratory and pilot scales. In emulsion polymerization, monitoring and controlling, the monomer conversion trajectory is crucial both for proper process operation and for obtaining products with the desired properties (due to the influence of the conversion trajectory on the polymer molecular weight and particle size distributions) (Thickett and Gilbert 2007; Yoon and Kim 2004). Polymer reaction engineering is a discipline that deals with various problems concerning the fundamental nature of chemical and physical phenomena in polymerization processes. Mathematical modeling is a powerful tool for the development of process understanding and advanced reactor technology in the polymer industry (Asteasuain et al. 2006a,b; Nauman 2001). The application of process design of model-based reactors and their control with a dimensionless approach to scaling, is considered in this work.

2 Model Description

The model was developed for a batch reactor with a cooling/heating jacket, as shown in Fig. 1. For this reactor, geometrical similarity was keep for scaling-up considerations.

The reactor temperature control is achieved by means of a cooling/heating fluid through a jacket. Energy balances lead to the following set of ordinary differential equations, describing the batch reactor and jacket temperature dynamics:
Reactor

$$\frac{dT_r}{dt} = -\frac{UA}{\rho_r C_{pr} V}(T_r - T_j) + \frac{Q_{\text{loss}}}{\rho_r C_{pr} V} + \frac{\Delta H R_p V}{\rho_r C_{pr} V} + \frac{\mu V}{\rho_r C_{pr} V}\phi, T_r(0) = T_{r0} \quad (1)$$

Jacket

$$\frac{dT_j}{dt} = \frac{UA(T_r - T_j)}{\rho_j C_{pj} V_j} + \frac{Q_{lsj}}{\rho_j C_{pj} V_j} + \frac{F_h S C_{pj} T_h}{\rho_j C_{pj} V_j} + \frac{F_c(1-S) C_{pc} T_c}{\rho_j C_{pj} V_j} - \frac{F C_{pj} T_j}{\rho_j C_{pj} V_j};$$
$$T_j(0) = \mathrm{T_{j0}} \quad (2)$$

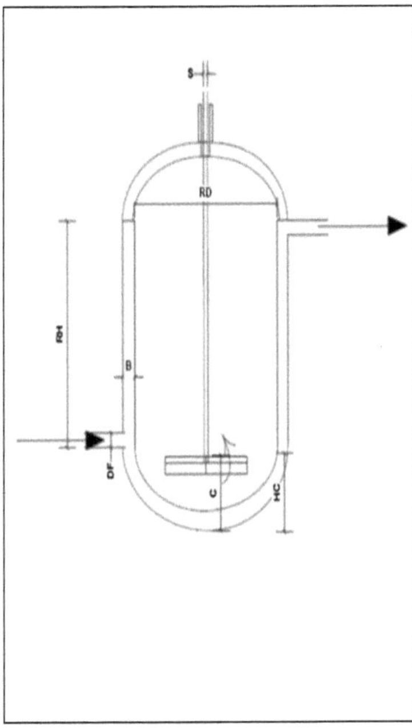

Parameter	Description
C_p	Specific heat capacity [J/kg K]
h	Partial heat transfer coefficient [W/m^{2}K]
k	Thermal conductivity [W/m K]
M_w	Molecular weight [kgmol/kg]
Q	Heat release rate [W]
T_r, T_j	Reactor and Jacket temperatures [K]
X	Conversion
ΔH	Enthalpy [J/mol]
η	Viscosity [Pas]
ρ	Density [kg/m^3]

Fig. 1 Reactor description and nomenclature

For simplicity, an on-off controller is used in this dynamical system, where the manipulated variable is either at maximum or at zero flow rate (*F*) (Asteasuain et al. 2006b; Nauman 2001; Thickett and Gilbert 2007; Yoon and Kim 2004).

$$\begin{aligned} \mathrm{T_{set}} &= \mathrm{K_{cm}}(\mathrm{T_{set1}} - \mathrm{T_{reactor}}) \\ \mathrm{set1} &= \mathrm{K_{cs}}(\mathrm{T_{set}} - \mathrm{T_{jacket}}) \end{aligned} \tag{3}$$

here, $\mathrm{T_{set1}}$ is the desired reactor temperature, for this process $\mathrm{T_{set1}}$ was 70 °C. The conversion evolution (x) is:

$$\frac{dx}{dt} = \frac{k_p M_p n N_T V}{m_0 N_{av}} \tag{4}$$

In stirred batch reactors heat transfer is governed by forced convection and conduction.

$$q = U(T_h - T_c) \tag{5}$$

The proportionality factor U is denoted as the overall heat transfer coefficient, several resistances in series determine the value of U.

$$\frac{1}{U} = \frac{1}{h_i} + \frac{1}{h_{di}} + \frac{D_i}{2k_w} * In\frac{D_0}{D_i} + \frac{1}{h_0}\frac{D_i}{D_0} + \frac{1}{h_{d0}}\frac{D_i}{D_0} \tag{6}$$

here, h_i and h_0 represent the partial heat transfer coefficients in the vessel and the jacket, respectively; k_w stands for the thermal conductivity coefficient of the wall; h_{di} and h_{d0} are the fouling factors for the inner and outer side of the wall, respectively; D_i and D_0 are the inner and outer diameter of the vessel, respectively. To derive the dimensionless equations, the following dimensionless variables and parameters are introduced:

$$T_r^* = \frac{T_r - T_0}{T_{SL} - T_c}; \quad T_j^* = \frac{T_j - T_{i0}}{T_{SL} - T_c}; \quad \phi_v = \phi_v^* t_B^2$$
$$T_0^* = \frac{(T_0 - T_{i0})}{(T_{SL} - T_c)}; \quad T_h^* = \frac{T_h - T_c}{T_{SL} - T_c};$$
$$T_c^* = \frac{T_c - T_{i0}}{T_{SL} - T_c}; \quad t^* = \frac{t}{t_B};$$

The dimensionless equations take the form:

$$\frac{dT_r^*}{dt^*} = Fo_r[-Nu_r(T_r^* + T_0^* - T_j^*) + Q_{\text{loss}}^* + DaNrDa_{III} + Br\phi_v^*]; \quad T_r^*(0) = T_{r0}^* \tag{7}$$

$$\frac{dT_j^*}{dt^*} = Fo_j[Nu_j\left(T_r^* + T_0^* - T_j^*\right) + Q_{lsj}^* + F(T_h^* S + T_c^* - T_j^*)]; \quad T_j^*(0) = T_{j0}^* \tag{8}$$

$$\frac{dx}{dt} = DaNr; \qquad x(0) = 0 \tag{9}$$

Where Fo, Nu, Q_{loss}, Da, Da_{III}, Br are Fourier, Nusselt, Damkohler I, Damkohler III and Brinkman dimensionless numbers, respectively.

3 Result and Discussions

The main objective was to obtain a product with the same characteristics. This can be warranted if the conversion profiles are kept similar in all scales. The operating conditions for the emulsion polymerization were: initial reactor temperature (T_{r0}) 25 °C, initial jacket temperature (T_{j0}) 25 °C and the desired reactor temperature (T_{set1}) 70 °C. It is a common practice to consider the overall heat transfer coefficient (U) constant for any size reactor (Asteasuain et al. 2006a). Figure 2a shows the effect of the impeller speed on U for different reactor sizes (1–10,000 L), in all cases U increases with the impeller speed and for small sizes this effect is more pronounced. Conversion as a function of the time, for different reactor sizes, is shown in Fig. 2b. This figure tells that increasing the size requires more time to reach the desired temperature for starting the reaction (heating time). In Fig. 2b all curves, except the one for 10,000 L, were shifted by the lag time, to a single one, depicting the same

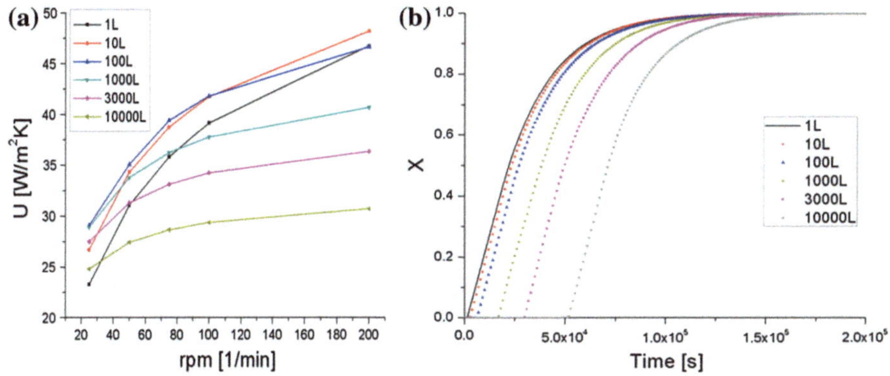

Fig. 2 **a** Overall heat transfer coefficient U as a function of the impeller speed. **b** Conversion of the reaction for different reactor sizes

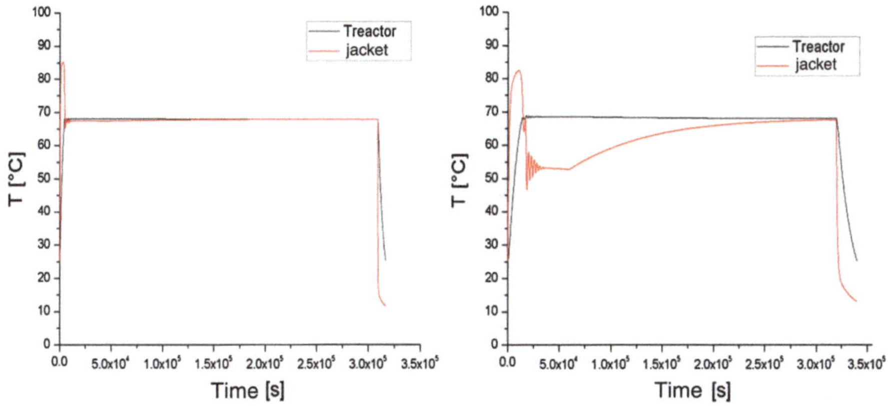

Fig. 3 Reactor and jacket temperatures as a function of time for rpm = 50. Reactor size 100 L (*left*) and 1,000 L (*right*)

chemical similarity. The polymerization process should be isothermal, to obtain a product with the same thermal history. This is one of the most difficult challenges in the scaling-up procedure for exothermic reactive systems due to the heat reaction generation can lead to a run away of the process.

The reactor and jacket temperatures, during the process for two different scales (100 and 1,000 L), are shown in Fig. 3. In both cases the reactors have been well controlled at T_{set1}, meaning that the thermal as well as the mechanical histories were kept constant. The latter because the stirrer operated at the same rpms. Figure 4a shows the required time to reach T_{set1} as a function of the jacket Reynolds number (*Re*). Increasing *Re* the time is shorter, as expected. In Fig. 4b the Damkohler I (Da_I) number is shown for different reactor sizes) from 1 up to 3,000 L), Da_I remains constant, this is a common industrial practice for scaling. This figure only verifies the initial assumption referred to as chemical similarity. Recalling that *Nu* is a strong

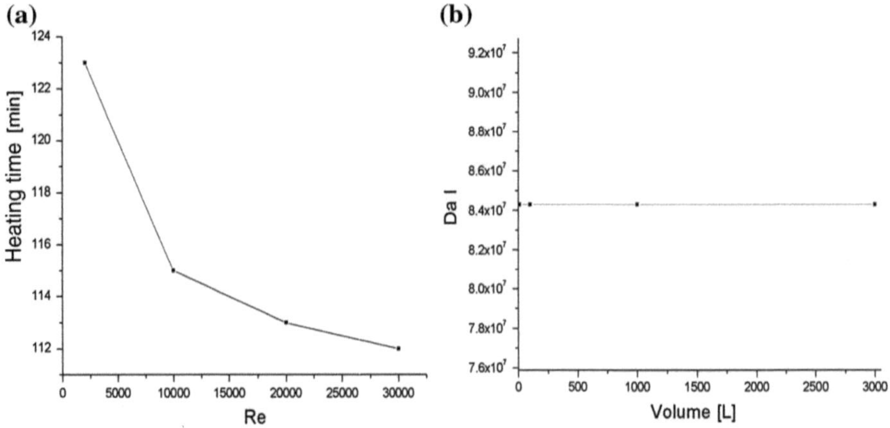

Fig. 4 **a** Heating time as a function of Reynolds number. **b** Damkohler I as a function of reactor size

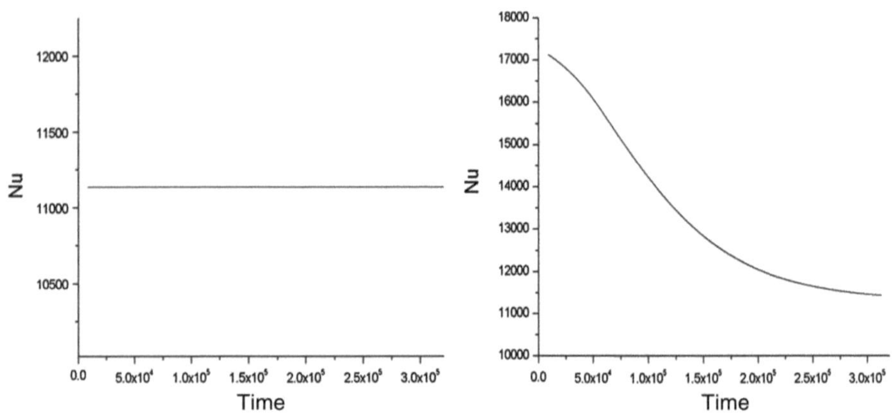

Fig. 5 Reactor Nusselt number (100 L) as a function of time for: constant viscosity (*left*) and conversion dependent viscosity (*right*)

function of *Re*, the effect of the viscosity on the Nusselt number, is shown in Fig. 5. This effect can be observed in the right graph. The viscosity increase, drastically affects the heat removal of the reactive system and could damage the final product.

4 Conclusions

A simple dimensionless model was developed for the scale-up batch polystyrene emulsion polymerization keeping geometrical, thermal, mechanical and chemical similitude. It was possible to scale-up the process from 1 up to 3,000 L. The largest reactor (10,000 L) could not be controlled keeping the above mentioned scale-up criteria.

Acknowledgments Funds provided by the research grants 20120934, SIP-IPN and PAPIIT IN114212, UNAM are gratefully acknowledged.

References

Asteasuain M, Bandoni A, Sarmonia C, Brandolin A (2006a) Simultaneous process and control system design for grade transition in styrene polymerization. Chem Eng Sci 61:3362–3378

Asteasuain M, Sarmoria C, Brandolin A (2006b) Integration of control aspects and uncertainty in the process design of polymerization reactors. Chem Eng Sci 61:3362–3378

Barton JA, Nolan PF (1987) J hazard mater 14:233

Fogler HS (1992) Elements of chemical reaction engineering. Prentice-Hall, Englewood Cliffs

Nauman Bruce E (2001) Chemical reactor design, optimization, and scale up. MacGraw Hill, NewYork

Schork FJ, Deshpande PB, Leffew KW (1993) Control of polymerisation reactors. Dekker, New York

Soh SK, Sundberg SC (1982) J Polym Sci Chem Ed 20:1299

Thickett SC, Gilbert RG (2007) Emulsion polimerization: state of the art in kinetics and mechanisms. Polymer 48:6965–6991

Yoon WJ, Kim YS et al (2004) Recent advances in polymer reaction engineering: modeling and control of polymer properties. Korean J Chem Eng 21(1):147–167

Fluid Flow Modelling Through Fractured Soils

Roberto González-Galán, Jaime Klapp-Escribano, Estela Mayoral-Villa, Eduardo de la Cruz-Sánchez and Leonardo Di G. Sigalotti

Abstract The main goal of this study is to simulate fluid flow across a fractured medium and to visualize its motion as a function of several parameters such as the tortuosity, the inlet pressure and the geometry of the fracture. Using the concept of double porosity, we have developed a hybrid model based on the Finite Element Method. The hydraulic characterization of the medium is realized with a 3D geometry, while the transport process of radionuclides (RNs) through the interface fracture—porous matrix is done in 2D. The results show that the sorption increases when the flow rate decreases. Moreover, an increment in the inlet pressure reduces the residence time of the RNs in the fracture.

1 Introduction

Deep geological repository (DGR) systems are needed to isolate residual long-lived radionuclides (RNs) produced by human activity. DGRs are based on the multiple barriers concept, in which the barriers work together to provide containment. The

R. González-Galán · J. Klapp-Escribano (✉) · E. Mayoral-Villa · E. de la Cruz-Sánchez
Instituto Nacional de Investigaciones Nucleares, Carretera México-Toluca S/N, La Marquesa, 52750 Ocoyoacac, Estado de México, México
e-mail: jaime.klapp@inin.gob.com

R. González-Galán
Facultad de Ciencias,Universidad Autónoma del Estado de México, El Cerrillo, Piedras Blancas C.P., 50200 Estado de México, México
e-mail: rgonzalez470@yahoo.com.mx

J. Klapp-Escribano
Departamento de Matemáticas, Cinvestav del I.P.N., 07360 México, D.F., México

L. Di G. Sigalotti
Centro de Física, Instituto Venezolano de Investigaciones Científicas, Apartado Postal 20632, Caracas 1020-A, Venezuela
e-mail: leonardo.sigalotti@gmail.com

J. Klapp and A. Medina (eds.), *Experimental and Computational Fluid Mechanics*, Environmental Science and Engineering, DOI: 10.1007/978-3-319-00116-6_19,

natural (or geological) barrier is the host rock formation itself. Since the retention of RNs within the natural barrier delays or prevents RN migration, it can be considered one of the most important safety functions of the deep geological repository (Astudillo 2001).

In crystalline rocks, fractures play an important role in the transport because the permeability of the fractured network is greater than the permeability of the rock and therefore fractures represent a highly effective pathway for transport. However, in most cases the flow occurs through a few preferential pathways which form channel clusters of flow or macro—channels that intercepts between them (Tsang and Neretnieks 1998). Therefore, the permeability of the fractured soil depends on the degree of interconnectedness of the fractures, and should also take into account a dependence on scale, which has been observed from laboratory to field scales (Brace 1980; Clauser 1992). The study of flow in a fractured medium is more difficult than the study of flow in a porous media because of the complex geometry characterizing fractured systems.

Early research work has mainly focused on the plane of the fracture and assumed that the flow through a fracture is similar to that between two parallel plates for which the Navier-Stokes equations can be applied. One of the most relevant results is the expression for calculating the velocity of the fluid in the fracture, which is known as the cubic law. Actually, several conceptual models have been developed for describing fluid flow in fractured porous media. Four concepts have dominated the research: (a) The explicit discrete fracture, (b) dual continuum, (c) discrete fracture network and (d) single equivalent continuum. Each method can be distinguished on the basis of the storage and flow capabilities of the porous medium and the fracture.

The aim of this study is to simulate the fluid flow across a fracture and visualize how its motion is affected when several parameters such as the tortuosity, the inlet pressure and the aperture are modified. The mathematical model is based on the substitution of the mass balance equation by the Darcy's law and the derivation of expressions for the pressure and the velocity in discrete blocks within the system. The pressure-velocity equations are related to the rock and fluid properties. In this model, one of the most important parameters is the "aperture" of the fracture. In order to understand fluid flow in a fractured medium, we need to analyze its motion and take into account the variation of its velocity at the fracture scale depending on the variable aperture and the inlet pressure.

The influence of the tortuosity of the fracture on the flow is studied by using a derivation of the Reynolds equation, also known as the "cubic law" equation. In this relation, it is very important the "aperture" term, which is assumed to be a non-constant function of the x- and y-coordinates (COMSOL 2008). We use the aperture data by defining an interpolation function, which is used as the aperture in the cubic-law equation.

Another related problem is to understand how the flow of the fluid affects the radionuclides' concentration in provisional or permanent sites for nuclear waste disposal. Since the retention of RNs within the natural barrier delays or prevents RN migration, the geophysical and chemical parameters can be considered as some of the most important safety functions in deep geological repositories.

In this work the conditions that govern the flow through the fracture and the relations governing the matrix-fracture exchanges along the fracture surface are studied. In our simulations, we use a pre-calculated result for the pressure-velocity fields, which is then used in the estimation of the RN transport rate along the fracture and across the porous matrix.

In addition, the water flow rate affects the residence time of the RNs in the fracture (Zimmerman et al. 2002) and their interactions with the rock, as well as the extent of the matrix diffusion. With respect to this problem, several simulations have been performed to establish a comparison between the different inlet pressure values of the fracture and the RNs' transfer rate at the fracture—porous matrix interface.

Using the concept of double porosity, we have developed a hybrid model based on the Finite Element Method and simulated it using the COMSOL software. The hydraulic characterization of the medium is realized with a 3D geometry, while the transport process of radionuclides (RNs) through the interface fracture—porous matrix is done in 2D.

The proposed model contains an interchange layer, where the contaminant travels by diffusion through the porous matrix, and this reduces the RNs' concentration in the aqueous stream of the fracture. The model has been validated through comparisons with analytical solutions (see Souley and Thoraval 2011).

The results obtained indicate that the sorption increases when the flow rate decreases. Moreover, an increment in the inlet pressure reduces the RNs residence time in the fracture.

2 The Problem

2.1 The Mathematical Model

Barenblatt et al. (1960) introduced the dual continuum approach which is based on an idealized flow medium which is constituted by a primary porosity or the "solid" matrix and a secondary porosity created by fracturing, jointing or dissolution. The porous medium and the fractures are envisioned as two separated but overlapping continua. This concept implies the definition of two coupled equations, the first one for the flow through the solid matrix and the second one for the flow in the fracture. Darcy's law governs the velocity in the matrix blocks, while the flow in the boundary of a fracture is established by taking into account the fracture's thickness. In this work we consider saturated conditions.

The time-dependent fluid flow in the matrix block is governed by Darcy's law

$$\left[\chi_f \theta_s + \chi_s \left(1 - \theta_s\right)\right] \frac{\partial p}{\partial t} - \nabla \cdot \left(\frac{k_m}{\eta} \nabla p\right) = 0; \quad \Omega \text{ matrix block}, \tag{1}$$

where the dependent variable p is the fluid pressure in the pore space [Pa] θ_s, is the void fraction, or porosity of the matrix blocks [m^3/m^3], χ_f and χ_s are the compressibilities [1/Pa] of the fluid and solid, respectively, k_m gives the permeability of the matrix blocks [m^2], and η is the fluid dynamic viscosity [Pa $\cdot$ s]. Here we use a predefined velocity variable that gives the Darcy velocity variable: $\boldsymbol{u}_{esdl} = -\left(\frac{k_m}{\eta}\nabla p\right)$, which is a volume flow per unit area.

In the fracture, we modify the coefficients of the Darcy's law to account for a relatively small flow resistance on the fracture and the fracture thickness:

$$\left(S_{frac}\, d_{frac}\right)\frac{\partial p}{\partial t} - \nabla \cdot \left(\frac{k_{frac}}{\eta} d_{frac} \nabla p\right) = 0; \qquad \Omega\, fracture \tag{2}$$

where S_{frac} is the fracture-storage coefficient [1/Pa], k_{frac} describes the fracture permeability [m^2], and d_{frac} is the fracture thickness or aperture [m]. Because the thickness appears in the fracture flow equation, the predefined variable $\mathbf{u}_{\mathrm{esdl}}$ gives the volume flow rate per unit fracture length on the fracture:

$$\mathbf{u}_{\mathrm{esdl}} = -\frac{k_{frac}}{\eta} d_{frac} \nabla_{\mathrm{T}}\mathrm{T}p; \qquad \Omega\, fracture \tag{3}$$

where $\nabla_{\mathrm{T}}\mathrm{T}p$ denotes the gradient operator restricted to the fracture's tangential plane.

Fluids in fractured porous media move quickly through the fractures but also migrate, albeit relatively slowly, through the tiny pores within the surrounding matrix blocks. In this work, the fluid mass transfer between the porous matrix and the fractures occurs at the interface layer. The transfer occurs according to the interface conditions described by dimensionless partition coefficients.

Based mainly on the formulation given in Sudicky and Frind (1982) and Gonzalez-Galan et al. (2013), we consider convection–diffusion within the fracture, here modified to take into account a finite length of fracture, diffusion in the interface layer and convection-diffusion within the porous matrix:

$$\nabla \cdot (-D\nabla C_1 + C_1\boldsymbol{u}) = 0; \qquad in\ \Omega_{fracture} \tag{4}$$

$$\nabla \cdot (-D_m\nabla C_2) = 0; \qquad in\ \Omega_{interface} \tag{5}$$

$$\nabla \cdot (-D\nabla C_3 + C_3\boldsymbol{u}) = 0; \qquad in\ \Omega_{matrix} \tag{6}$$

where C_i denotes the concentration of the contaminant (mol/m^3) in the respective phases, D denotes the diffusion coefficient (m^2/s) in the liquid phases, and D_m is the diffusion coefficient in the membrane, while $\boldsymbol{u}$ denotes the velocity (m/s) in the respective liquid phase.

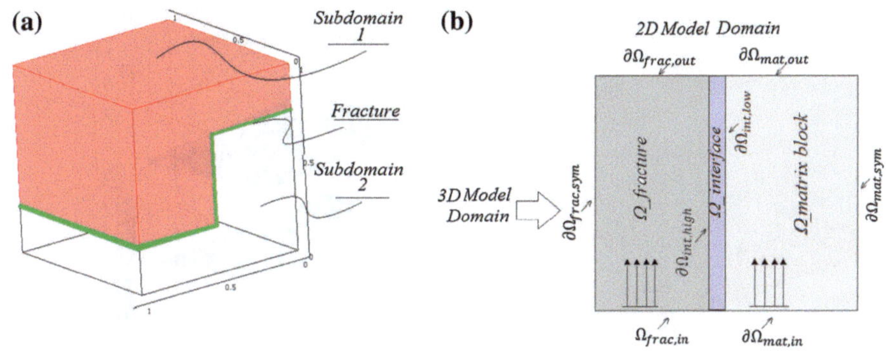

Fig. 1 Implementation of the suggested mix model in 3D (**a**) and in 2D (**b**)

2.2 Set up of the Model

A synthetic example is developed in order to show the resolution methodology for the RNs' transport problem in fractured porous media as is illustrated in Fig. 1.

In the 3D model we study the flow across the fracture and the matrix block. This model consists of a solid block that represents the porous matrix. The fracture in this model is represented by a sequence of interior boundaries. Because the thickness appears in the fracture flow equation, the predefined variable $\mathbf{u}_{esdl}$ gives the volume flow rate per unit fracture length on the fracture. In this model, we first calculate the pressure and velocity fields of both the fracture and the porous matrix. We then analyze the interchange mechanics between the porous matrix and the fracture with a 2D model.

2.3 Boundary Conditions

Along all faces of the block a zero flow boundary condition is applied. The boundaries of the fracture are edges that intersect the porous media block. Conditions in these edges are: at the inlet edge the pressure is constant, $p = p_0$, and at the outlet it decreases linearly with time: $p = p_0 - t \cdot 10\left[\frac{Pa}{s}\right]$. There is no flow through the other edges so that $-\frac{k_{frac}}{\eta} d_{frac} \nabla p = 0$.

For the RNs' transport, the contaminant must be dissolved in the interface layer in order to be transported through it. At the inlet of the model domain, we define the following concentration conditions: $c_1 = c_0$ at the boundary $\Omega_{frac,in}$ and $c_3 = 0$ at the boundary $\partial\Omega_{mat,in}$. At the outlet, we assume that the convective contribution to the mass transport is much larger than the diffusive contribution: $\nabla \cdot (-D\nabla C_i + C_i \boldsymbol{u}) \cdot \boldsymbol{n} = C_i \boldsymbol{u} \cdot \boldsymbol{n}$ at $\partial\Omega_{frac,out}$ and $\partial\Omega_{mat,out}$. Here $\mathbf{n}$ is the normal unit vector to the respective boundary. Furthermore, we assume that there is no transport on the symmetry boundaries: $\nabla \cdot (-D\nabla C_i + C_i \boldsymbol{u}) \cdot \boldsymbol{n} = 0$ at $\partial\Omega_{frac,sym}$ and $\partial\Omega_{mat,sym}$. We

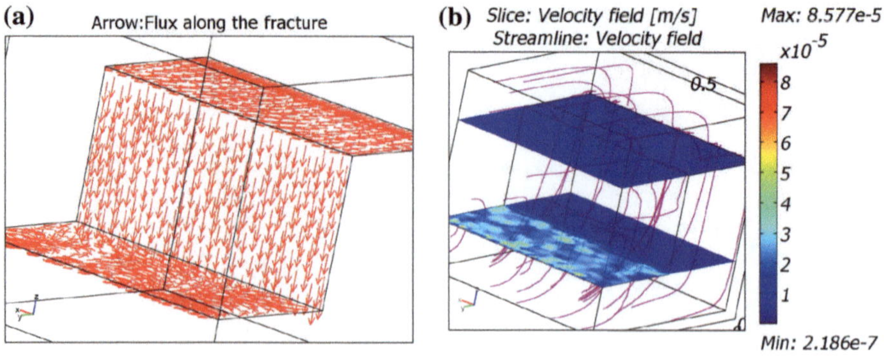

Fig. 2 **a** Flux along the fracture and **b** flow velocity field through the porous matrix

also assume symmetry at the horizontal boundaries of the interface: $(-D_m \nabla C_2)\cdot \boldsymbol{n} = 0$ at $\partial\Omega_{int,high}$ and $\partial\Omega_{int,low}$.

The interface conditions between the liquid and membrane phases for the concentration are described by the dimensionless partition coefficient:

$$K = \frac{c_2^{int}}{c_1^{frac}} - \frac{c_2^{int}}{c_3^{mat}}.$$

3 Results

In the simulation, the fluid moves from the left to the right through the block, entering at the upper fracture edge and exiting at the lower edge. Initially, the fluid does not move within the volume. The walls of the block are impermeable to the flow except at the fracture edges. The fracture is divided in three sections, the first is the upper one and is located in the z = 0.75 plane, the second section is a vertical plane at y = 0.5, and finally the third section is located in the z = 0.25 plane (see the geometry in Fig. 1). The fracture has a thickness of 0.1 mm in the first two sections. The third section of the fracture has a variable thickness defined by the aperture, which is an interpolation function over a sample data that corresponds to an aperture with a fractal dimension of 2.6.

In panel (a) of Fig. 2 the arrows show the flux along the fracture. In panel (b) we show the velocity streamlines of fluid through the porous matrix. The slice shows the velocity field between the planes z = 0.25 and z = 0.75.

The fracture is far more permeable to the fluid than the matrix block, and the influence of its variable thickness over the velocity field is clearly seen in Figs. 2 and 3. The first two sections have a constant aperture; its velocity field has a regular distribution, while in the third section the velocity field is not uniform because in this section the fracture has a variable thickness.

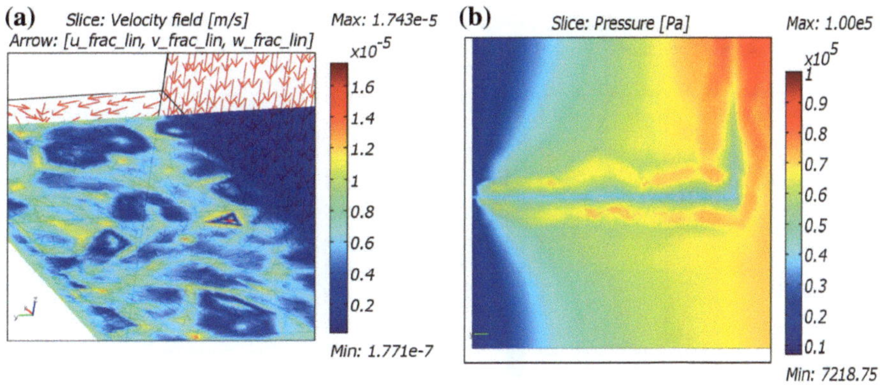

Fig. 3 Detail of the fracture. The velocity and pressure fields are strongly influenced by the variable thickness of the fracture in this region **a** Enlarged view of the velocity field in the near area to the exit of the fracture. In this zone, is shown the strong influence of the fracture's thickness. **b** Lateral view of the pressure field in the same area next to the out of the fracture

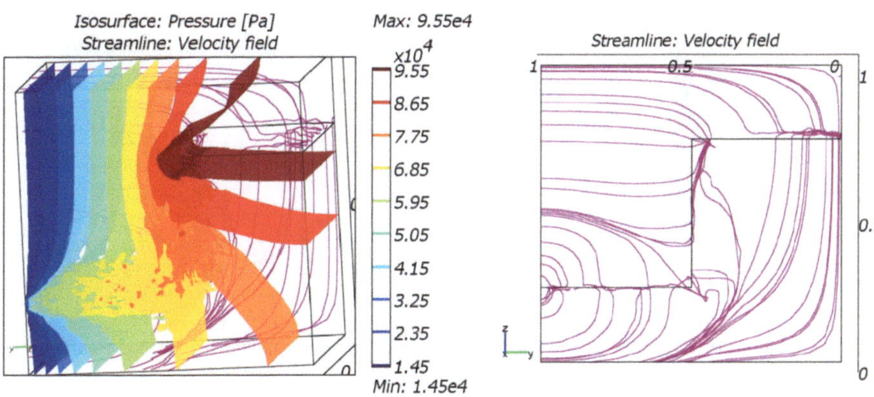

Fig. 4 In right panel, the isosurfaces show the pressure field in the porous matrix. In left panel, the streamlines are the velocity field through the domain of simulation, in a lateral view

In Fig. 3 we show in detail the region mentioned above. The left panel shows the velocity field while in the right panel we show the pressure from a lateral view of this section. We can see that the pressure field is highly affected by the variable thickness, and accordingly the velocity field (represented by arrows) follow an irregular trajectory according to the variable pressure field estimated.

The isosurfaces depicted in Fig. 4 are pressure contours throughout the block. The pressure is continuous across the fracture from block to block. Even so, the bends in the isosurfaces indicate different flow regimes in the fracture and the matrix blocks. This is observed in the region for a non-constant fracture aperture. The pressure and the velocity fields do not have a uniform structure as is indeed observed in other regions of the system.

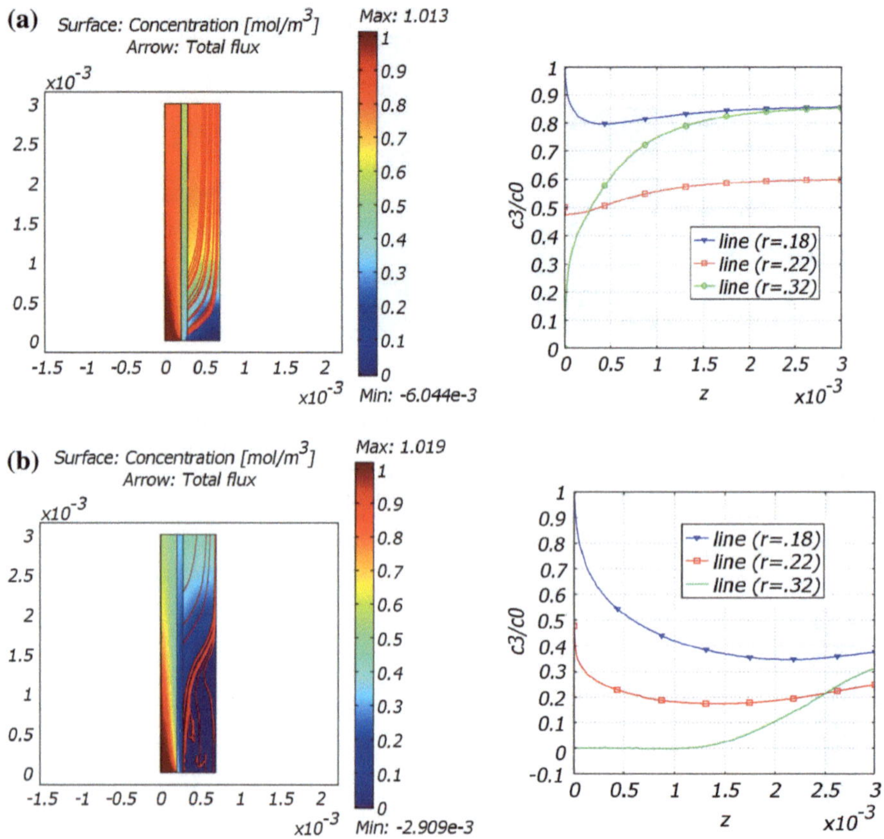

Fig. 5 Concentration profiles in the fracture interface (*left panels*) and the porous matrix (*right panels*) for **a** $p_0 = 10$ [Pa] and **b** $p_0 = 10^4$ [Pa]

The streamlines in the right panel indicate velocities in the porous matrix. The fluid moves from the inlet to the outlet along the fracture with a velocity field that is uniform across the block. The figure indicates that the linear velocity in the matrix is significantly smaller than the average linear velocity along the fracture.

On the other hand, the surface plot and the plot in Fig. 5 visualize the concentration distribution throughout the three model subdomains: the fracture region inside the model on the left side, the interface fracture-porous matrix in the middle, and the porous matrix to the right. In the left panels of Fig. 5, the concentration profiles in each region are shown (three lines in $r = 0.18$ [mm], $r = 0.22$ [mm] and $r = 0.32$ [mm] which correspond to the three regions mentioned above). The plots in the right panels show the distribution of concentration along each sub-domain. The figure also shows the concentration jump that arises at the boundary between the fracture and the interface membrane. Finally, we can also see that the concentration absorbed by the porous matrix is influenced by the filtration process and the velocity flux.

4 Conclusions

In this work, we described a mathematical model for simulating flow in a fractured porous media, using the dual media approach. According to the proposed model and the results obtained, one of the most important parameters is just the fracture aperture.

We have showed how to model the flow in a discrete fracture by considering the interactions between the fluid and the porous matrix. The fracture is defined as a 2D domain within the other 3D domains; this prevents us of using a large number of grid elements along the fracture that reduces the computational time significantly, thus making our method computationally very efficient.

The velocities in the porous matrix are of the order of 5×10^{-7} [m/s], while they are 2×10^{-6} [m/s] along the fracture. As expected, this result indicate that the linear velocity in the matrix is significantly smaller than the average linear velocity along the fracture. The variation of the aperture parameter in the fracture produced a large variation in the pressure, and therefore also in the velocity. There is enough evidence to conclude that on the fracture surface there is a large variation in the flow resistance due to the variability in the opening and contact areas, which creates tortuous flow lines over the plane of the fracture. As was shown, the concentration inside the fracture decreases markedly over the first sections of the interface from the inlet.

Moreover, the simulations were performed for a wide range of pressures at the inlet of the fracture and the obtained results clearly indicate that for high inlet pressures the radionuclides flow through the fracture with high velocities, which shortens the residence time and reduces the interaction with the porous matrix. Thus, the radionuclides' concentration in the porous matrix gets smaller as the inlet pressure increases.

Acknowledgments This work has been partially supported by ABACUS, CONACyT grant EDOMEX-2011-C01-165873.

References

Astudillo J (2001) El almacenamiento geológico profundo de residuos radiactivos: principios básicos y tecnología. ENRESA, Madrid

Barenblatt GI, Zheltov YP, Kochina IN (1960) Basic concepts in the theory of seepage of homogeneous liquids in fissured rocks. Sov Appl Math Mech Engl Transl 24:852–864

Brace WF (1980) Permeability of crystalline and argillaceous rocks. Int J Rocks Mech Min Sci 17:241–251

Clauser Ch (1992) Permeability of Crystalline Rocks. Eos Trans Am Geophys 73(21):232–237

COMSOL (2008) Model library. Earth science module & model library, multiphysics module

González-Galán R, De la Cruz-Sánchez E, Klapp-Escribano J, Mayoral-Villa E, Pérez-Quezadas N, Galindo Uribarri S (2013) Evaluation of a temporary repository of radioactive waste. Fluid dynamics in physics, engineering and environmental applications, environmental science and engineering. ISBN: 978-3-642-27722-1, Springer, Verlag, Berlin, Heidelberg, pp 439–445

Souley M, Thoraval A (2011) Nonlinear mechanical and poromechanical analyses: comparison with analytical solutions. COMSOL conference, Stuttgart, 26–28 Oct 2011
Sudicky EA, Frind EO (1982) Contaminant transport in fractured porous media: analytical solutions for a system of parallel fractures. Water Resour Res 18(6):1634–1642
Tsang CF, Neretnieks I (1998) Flow channeling in heterogeneous fractured rocks. Rev Geophys 36(2):275–298
Zimmerman MD, Bennett PC, Sharp JM Jr, Choi WJ (2002) Experimental determination of sorption in fractured flow systems. J Contam Hydrol 58:51–77

Part III
Convection and Diffusion

Numerical Analysis of the Conjugate Convection in an Open Cavity with and Without an Obstruction Inside

G. E. Ovando-Chacón, S. L. Ovando-Chacón, J. C. Prince-Avelino, A. Servín-Martínez and J. A. Hernández-Zarate

Abstract Conjugate mixed convection due to the steady state heat transfer and the incoming fluid dynamics inside a two-dimensional square cavity is studied numerically for a Reynolds number of 400. The boundary of the left side of the cavity is a hot thick wall. The other three sides are bounded by adiabatic walls. In order to investigate the effect of a solid body on the conjugate heat transfer, a circular obstruction is placed at the centre of the cavity. The fluid inlets are at the lateral lower left wall, and the exits are located at the lateral upper right wall. The analysis is carried out for different ratios between the solid and fluid thermal conductivities. The governing equations of continuity, momentum and energy for incompressible flow are solved by the finite element method combined with the splitting operator scheme. The temperature field, the streamlines, the velocity and the pressure field are studied, and the axial velocity profiles are analysed as a function of the transversal position. It is observed that an obstruction placed inside the cavity plays a major role on vortex formation and on the thermal behaviour of the flow inside the cavity.

1 Introduction

The numerical simulation of the conjugate convection in an open cavity with and without obstruction is an important issue in many technological processes. Manca et al. (2003) investigated the mixed convection in a U-shaped cavity, and evaluated its thermal performance in terms of the heated wall position. Mahmoudi et al. (2010)

G. E. Ovando-Chacón (✉) · J. C. Ovando-Chacón · A. Servín-Martínez · J. A. Hernández-Zarate
Departamento de Metal Mecánica y Mecatrónica, Instituto Tecnológico de Veracruz, Calzada Miguel A. de Quevedo 2779, Col. Formando Hogar, 91860 Veracruz, Veracruz, México
e-mail: geoc@itver.edu.mx

S. L. Ovando-Chacón
Departamento de Química y Bioquímica, Instituto Tecnológico de Tuxtla Gutiérrez, Carretera Panamericana Km. 1080, 29000 Tuxtla Gutiérrez, Chiapas, México

J. Klapp and A. Medina (eds.), *Experimental and Computational Fluid Mechanics*, Environmental Science and Engineering, DOI: 10.1007/978-3-319-00116-6_20,

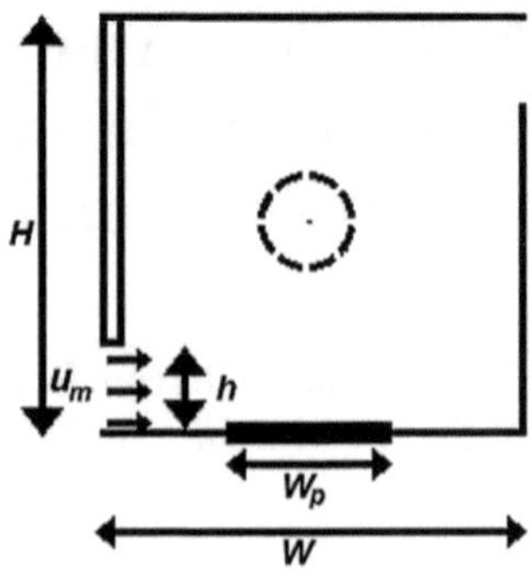

Fig. 1 Geometry of the cavity

numerically examined the effect of the inlet and the outlet locations on the mixed convection flow and on the temperature field in a vented square cavity. Rahman et al. (2011) analysed the mixed convection in a ventilated square enclosure with a heat generating circular block. Mariani and Coelho (2007) presented the simulation of natural convection due to the temperature difference between the left and the right walls, and an internal local heat source in open cavities. Mamun et al. (2010) studied the effect of a heated hollow cylinder on mixed convection in a ventilated cavity. Radhakrishnan et al. (2007) reported experimental and numerical investigation of mixed convection from a heat generating element in a ventilated cavity. Varol et al. (2008) investigated the conjugate natural convection in enclosures via entropy generation. The main aim of this numerical investigation is to study the effect of the thermal conductivity of a finite thickness wall on the thermal behaviour in an open cavity with and without an obstruction. The analysis is carried out for the laminar regime and for the case when the buoyancy effect is outweighed by forced convection.

2 Problem Formulation

This work presents 2D numerical simulations inside an open square cavity ($H/W = 1$) with and without an obstruction placed at the centre of the cavity, H is the height and W is the width of the cavity. The fluid, with a thermal conductivity k, enters at the left lower side wall and leaves the cavity at the right upper side wall, see Fig. 1. The thickness of the left wall, with thermal conductivity k_s, is 0.05 W. Different ratios k_s/k are investigated. The blockage ratio of the solid is fixed to $d/W = 0.2$, where d is the diameter of the obstruction. The Reynolds number ($Re = U_m W/\nu$), based on the velocity of the inlet flow U_m and the width W of the cavity, studied in this investigation was $Re = 400$ for a Richardson number of $Ri = 0.01$ and a Prandtl number of $Pr = \nu/\alpha = 10.0$, where ν is the kinematic viscosity and α is the thermal diffusivity. The entrance and exit of the cavity was fixed to 0.25 W.

The governing equations for a non-isothermal incompressible steady state flow are given as:

$$-\frac{1}{\mathrm{Re}}\Delta\vec{u}+\vec{u}\cdot\nabla\vec{u}+\nabla p=RiT\vec{J}\ \mathrm{in}\ \Omega, \tag{1}$$

$$\nabla\cdot\vec{u}=0\ \mathrm{in}\ \Omega, \tag{2}$$

$$\frac{1}{Pe}\Delta T+\vec{u}\cdot\nabla T=0\ in\ \Omega, \tag{3}$$

For the solid cylinder, the energy equation is

$$\Delta T_s=0\ \mathrm{in}\ \Omega, \tag{4}$$

In the above equations Δ is the laplacian operator, $\vec{u}=(u_1,u_2)$ is the velocity vector, being $u_1\,y\,u_2$ the horizontal and vertical velocity components, respectively; ν is the kinematic viscosity, p is the pressure, T is the temperature, T_s is the temperature of the solid and $\vec{J}$ is the vertical unitary vector. In the governing equations, the Richardson number, the Reynolds number and the Peclet number are defined as follows:

$$Ri=g\beta h\,(T_h-T_c)/U_m^2,\quad \mathrm{Re}=U_w h/\nu,\quad Pe=\mathrm{Re}\,\mathrm{Pr}, \tag{5}$$

where g is the gravity, β is the compressibility coefficient, T_h is the hot temperature, T_c is the cold temperature. No slip boundary conditions ($u_1=u_2=0$) were established in all the walls of the cavity, and adiabatic walls ($\partial T/\partial n=0$) were supposed except in the left wall where the heating takes place. The temperature of the inlet flow was fixed to $T=T_{c,}$ while the isothermal left wall was fixed to $T=T_h$. The non-dimensional values of this temperature were $T_c=0$ and $T_h=1$. The boundary conditions of the inlet flow were $u_1=U_m$, and $u_2=0$. On the outlet flow $\partial u/\partial n=0$ was imposed. The governing equations were solved with the finite element method combined with the operator splitting scheme, see Glowinski (2003). The convergence analysis was done for three different meshes with resolution of 4,500, 15,800 and 18,700 elements. An analysis of the temperature profiles on the middle horizontal and vertical lines indicates that the largest difference between the results of the meshes of 4,500 and 15,800 elements was 10 %, while the maximum difference of the results between the meshes of 15,800 and 18700 was 1.02 %. The analysis was also done for the velocity components. For all cases, the worst relative error between the meshes of 15800 and 18700 was less than 1.0 %. All the simulations presented in this paper were performed for a cavity with 18,700 elements.

3 Results

Figure 2 shows the velocity field and streamlines, for $Re=400$, $Ri=0.01$, $Pr=10$ and $k_s/k=1.0$. The left panel shows the behaviour of the fluid inside the cavity without an obstruction. For this case, three vortices can be observed along

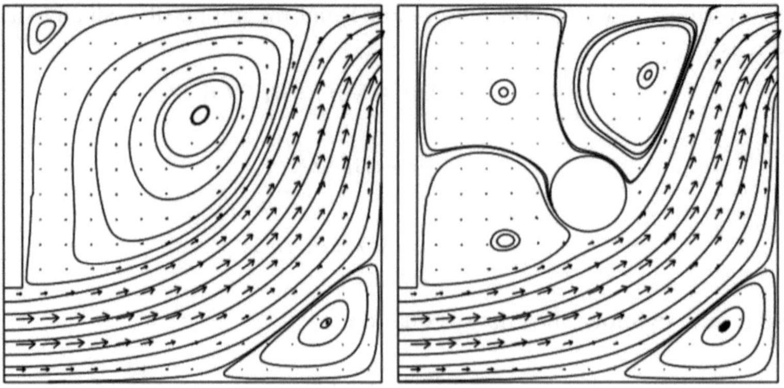

Fig. 2 Velocity fields and streamlines for $k_s/k = 1.0$. *Left*: without obstruction. *Right*: with obstruction

the diagonal perpendicular to the one that connects the inlet with the outlet. One of the vortices is anti-clockwise, big and strong, located at the central superior region of the cavity. At the upper left corner, a clockwise weak vortex appears, and at the inferior right corner another clockwise vortex emerges from the impingement of the fluid with the corner. The inlet jet emerges horizontally from the flow entrances of the cavity, but as it moves forwards, its horizontal component is decreased and the vertical component is increased. The right panel shows the behaviour of the fluid inside the cavity with a centred obstruction. For this case, the flow is characterized by four vortices, the clockwise vortex at the lower right corner remains, while three vortices appear around the obstruction as a consequence of the interaction of the solid with the flow that moves from the inlet to the outlet of the cavity. Two anti-clockwise vortices can be appreciated, one near the entrance and another near the exit of the cavity. Both are driven by the jet of fluid that crosses the cavity. The third vortex around the obstruction rotates clockwise and is driven by the motion of the other two vortices. The top panel of Fig. 3 shows the isotherms for the cases without obstruction. The contours tend to concentrate at the left wall of the cavity. The heat of the hot wall flows to the right upper part of the cavity due to the main stream of the fluid. As expected, high temperature occurs near the left wall. Farther, the contours are distorted and follow the main stream direction towards the outlet of the cavity. For low values of the thermal conductivity ratio ($k_s/k = 0.1$) the temperature contours remain undisturbed near of the left hot wall. As the thermal conductivity ratio ($k_s/k = 5$) is increased, the thermal boundary layer detaches from the lower part of the hot wall and a thermal plume moves toward the exit of the cavity due to the motion of inlet flow. This behaviour is intensified for $k_s/k = 100$. For this case, the flux of heat reaches the central part of the cavity. The bottom panel of Fig. 3 shows the isotherms for the case when the obstruction inside the cavity has the same thermal conductivity as the fluid. This explains the behaviour of the isotherms through the obstruction, which are not affected by the boundary of the circular solid. The interaction of the solid with the flow distorts the temperature contours towards

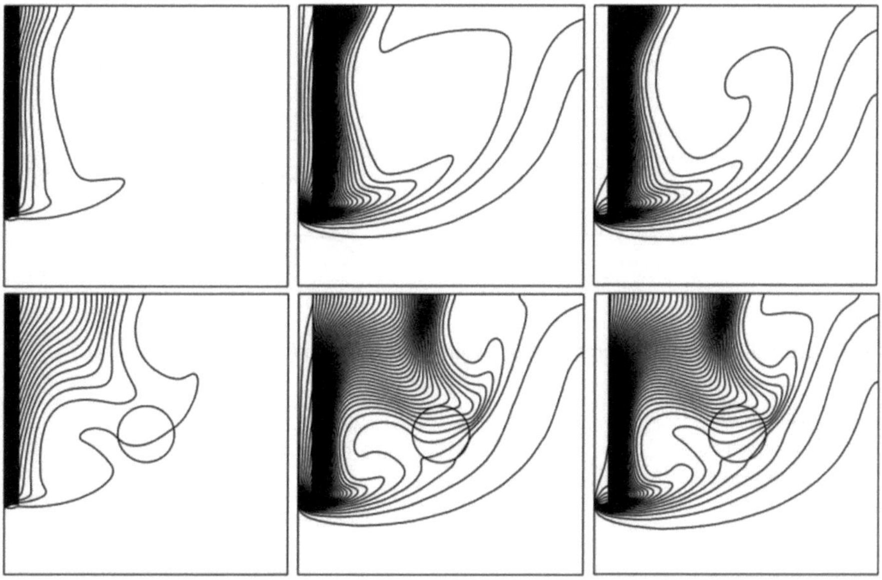

Fig. 3 Isotherms. *Top*: without obstruction. *Bottom*: with obstruction. *Left*: $k_s/k = 0.1$. *Middle*: $k_s/k = 5.0$. *Right*: $k_s/k = 100$

the central part of the cavity even for lower values of the thermal conductivity ratio ($k_s/k = 0.1$). As the ratio is increased to $k_s/k = 5$ and $k_s/k = 100$ the distortion of the isotherms towards the centre of the cavity are increased. When a solid is present inside the cavity, its interaction with the flow generates thermal boundary layer separation not only from the lower part of the hot wall, but also from the upper part of the hot wall, This phenomenon increases the flux of heat from the left hot wall towards the fluid.

Figure 4 shows the temperature profiles in the interior surface of the left hot wall for the cavity without obstruction (see left panel) and the cavity with obstruction (see right panel) for different values of the thermal conductivity ratio. For both cases the temperature is increased as the k_s/k ratio increases. The maximum temperature value, for a given k_s/k ratio, is reached at the upper part of the left interior wall. For $k_s/k \leq 1.0$ the temperature of the lower part of the interior left wall is $T \approx 0$. However, for $k_s/k \geq 10.0$ the temperature of the lower part of the interior left wall is $T > 0.5$. For low values of k_s/k the maximum temperatures are higher with, than without the solid blockage. For large values of k_s/k the temperatures tend to be uniform and the maximum temperature is $T \approx 1.0$.

Figure 5 shows the temperature profiles on the middle vertical line. The left panel shows the behaviour of the temperature without the effect of the obstruction. The temperature starts to increase, reaches a maximum value about $y = 0.43$ and then deceases. The larger the k_s/k ratio the greater the temperature values. The right panel shows the effect of the obstruction on the temperature behaviour, which is more

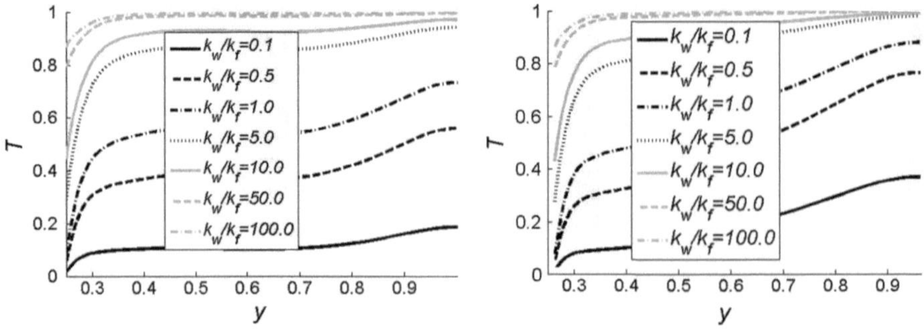

Fig. 4 Left wall interior temperature. *Left*: without obstruction. *Right*: with obstruction

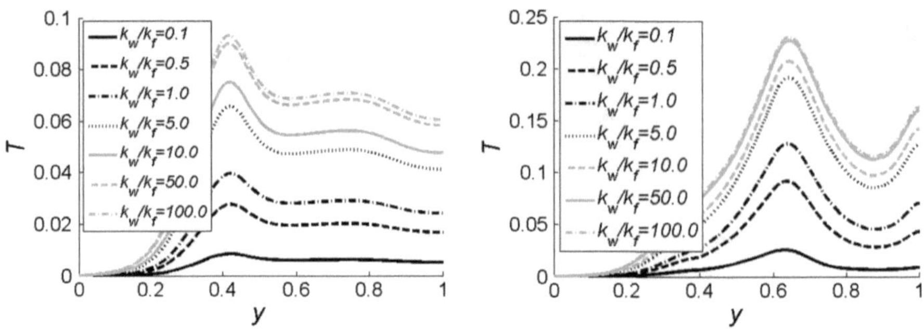

Fig. 5 Middle vertical line temperature. *Left*: without obstruction. *Right*: with obstruction

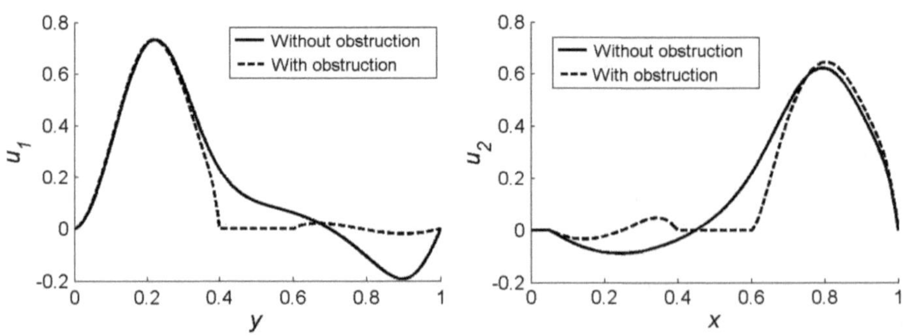

Fig. 6 Velocity profile for $k_s/k = 1.0$. *Left*: Horizontal component on $x = 0.5$. *Right*: Vertical component on $y = 0.5$

complex and is characterized by the increase of the temperature up to a maximum value of about $y = 0.63$. Beyond this value, the temperature starts to decrease to a minimum and then increases again. Given a k_s/k ratio for a particular vertical coordinate, the temperatures are higher when the obstruction is present. Figure 6 shows the horizontal velocity profiles on the middle vertical line (left panel) and the vertical velocity profile on the middle horizontal line (right panel) for the cavity with and without an obstruction. On the middle vertical line, it can be seen, for the flow without obstruction, that the fluid is accelerated as the vertical coordinate is increased, its horizontal velocity reaches a maximum at about $y = 0.22$, then the velocity decreases and changes its direction to generate the main vortical motion. The effect of the obstruction is insignificant near the bottom wall. On the obstruction the velocity becomes zero and above of the obstruction surface the horizontal velocity decays. On the middle horizontal line, it can be seen, that without obstruction, the flow goes downwards as the horizontal coordinate is increased, then starts to move upwards and reaches a maximum at about $y = 0.78$, then the velocity decreases toward the right wall. When the obstruction is present the vertical velocity near of the left wall becomes zero and after the obstruction the vertical velocity also reaches a maximum value.

4 Conclusions

We have presented in this work results of a finite element simulation of the flow inside an open cavity with and without a centre solid obstruction for different values of the ratio between the thermal conductivities of the solid and the fluid. The streamline patterns reveal that for a square cavity without obstruction three vortices are formed along the diagonal perpendicular to the line between the inlet and the outlet of the cavity. When the circular obstruction is placed at the centre of the cavity, the flow becomes more complex. Four vortices appear inside the cavity, three of them around the obstruction. The isotherms reveal that the presence of the obstruction intensifies the distortion of the thermal plume towards the centre and the exit of the cavity. So far, this study is limited to $Re = 400$, $Ri = 0.01$ and $Pr = 10$. The fluid dynamics for other values is the subject of on-going research, but it is useful to understand the thermal behaviour of the flow inside an open cavity with and without an obstruction in applications like the cooling of electronic and electrical devices, thermal design of building, greenhouse design and refrigeration.

References

Glowinski R (2003) Numerical methods for fluids, part 3. Garlet PG, Lions JL (eds) Handbook of numerical analysis, vol IX. North-Holland, Amsterdam

Mahmoudi AH, Shahi M, Talebi F (2010) Effect of inlet and outlet location on the mixed convective cooling inside the ventilated cavity subjected to an external nanofluid. Int Commun Heat Mass Transf 37:1158–1173

Mamun MAH, Rahman MM, Billah MM, Saidur R (2010) A numerical study on the effect of a heated hollow cylinder on mixed convection in a ventilated cavity. Int Commun Heat Mass Transf 37:1326–1334

Manca O, Nardini S, Khanafer K, Vafai K (2003) Effect of heated wall position on mixed convection in a channel with an open cavity. Numer Heat Transf Part A 43:259–282

Mariani VC, Coelho LS (2007) Natural convection heat transfer in partially open enclosure containing an internal local heat source. Braz J Chem Eng 24:375–388

Radhakrishnan TV, Verma AK, Balaji C, Venkateshan SP (2007) An experimental and numerical investigation of mixed convection from a heat generating element in a ventilated cavity. Exp Thermal Fluid Sci 32:502–520

Rahman MM, Parvin N, Rahim NA, Islam MR, Saidur R, Hasanuzzaman M (2011) Effect of reynolds and prandtl number on mixed convection in a ventilated cavity with a heat-generating solid circular block. Appl Math Model 36:2056–2066

Varol Y, Oztop HF, Koca A (2008) Entropy generation due to conjugate natural convection in enclosures bounded by vertical solid walls with different thicknesses. Int Commun Heat Mass Transf 35:648–656

Experimental Studies of a Steam Front in a Radial Porous Cell

A. Torres, S. Peralta, F. Aragón and A. Medina

Abstract In many Geothermal reservoirs water is injected through porous media in order to obtain steam. This process is basically used to generate geothermal energy. It could be expected that the water or steam injected in the homogeneous porous media will have a stable motion. It has been found experimentally that this is not always true. In porous media with high permeability, steam migrates in a finger-shape form reported in literature as the Saffman-Taylor Instability. In this work, a couple of experiments based on steam injected through a radial porous cell are shown. The main objective of this study is to present the observed instabilities in the steam front.

1 Introduction

In this paper the experimental study of a steam front in a porous radial cell is studied. This kind of phenomenon is observed in Geothermal Reservoirs which are used for electrical production, therefore the importance of this research. It is assumed that the flow of the interface vapour–liquid through the porous radial cell is planar. The experiments here developed show the existence of some instabilities. This kind of disturbance in the flow of the interface through the porous medium is known as Saffman-Taylor Instability (1958).

A. Torres (✉) · S. Peralta
Coordinación del Posgrado, Instituto Mexicano del Petróleo, Eje Central Lázaro Cárdenas No. 152, 07730 Col. Atepehuacan, México D.F., México
e-mail: higherintellect@hotmail.com

F. Aragón
ESIME Zacatenco, Instituto Politécnico Nacional, Av. Instituto Politécnico Nacional S/N, 07738 Col. Zacatenco, México, D.F., México

A. Medina
ESIME Azcapotzalco, Instituto Politécnico Nacional, Av. de las Granjas No. 682, 02550 Col. Sta. Catarina, México D.F., México

J. Klapp and A. Medina (eds.), *Experimental and Computational Fluid Mechanics*, Environmental Science and Engineering, DOI: 10.1007/978-3-319-00116-6_21,

The flow of a liquid–steam interface in low permeability Geothermal Reservoirs remains stable, it means that the steam front does not advance in a finger-shaped form (a characteristic form in presence of this kind of instability but in the case of Geothermal Reservoirs considered of high permeability the movement of the vapor front occurs in a finger-shaped form, due to that the vapour-liquid interface in the liquid phase experiments minor opposition to the movement and advances quickly, reaching more exposition to the super-heated rock and rapidly vaporizing (Islam and Azaies 2010).

2 Problem Description

Many geothermal systems are characterized by a low permeability and by the fact that their internal flows have low Reynolds numbers, hence the flow can be described with the law of Darcy

$$\dot{m} = -\frac{K}{\mu}\frac{\nabla P}{L}. \tag{1}$$

where $\dot{m}$ is the mass flow rate, K is the permeability, μ is the viscosity, ∇P is the gradient of the pressure and L is the total radial length of the radial porous cell.
The problem of the motion of the steam front through a radial porous cell is described by using the Conservation of Mass (Continuity) and Darcy's Law, Eqs. (1) and (2)

$$\varphi\frac{\partial \rho_v}{\partial t} + \nabla \cdot (\boldsymbol{v}_v \rho_v) = 0. \tag{2}$$

where φ is the porosity and ρ_{v} is the steam density.
Moreover, if we want to study the heat transfer in this system, we need to add the Conservation of Energy (First Law of Thermodynamics) and the Perfect Gas Law to this system in every moment during the phenomena.

$$\frac{\partial (C_a T)}{\partial t} + \nabla \cdot (\boldsymbol{v}_v C_v T) = \kappa_a C_a \nabla^2 \mathrm{T}. \tag{3}$$

$$P = \rho_v R_v T. \tag{4}$$

where C is the specific heat referred to the unit volume, κ is the thermal diffusivity, subindexes are a for the average and v for the steam, R is the Universal Gas Constant and T is the temperature.

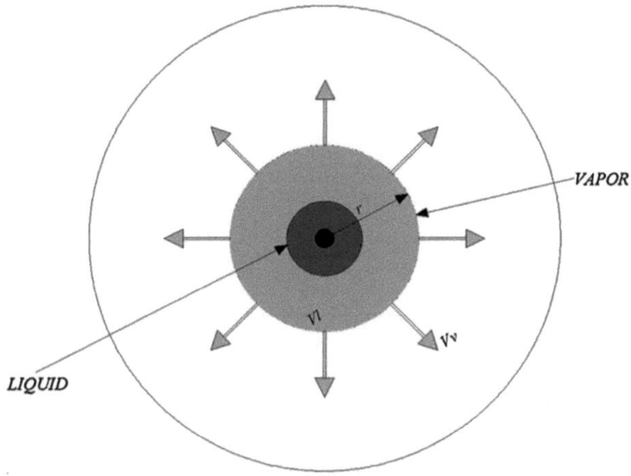

Fig. 1 Scheme showing the vaporization front phenomenon

If we observe Fig. 1, from the equality of the flow in the liquid–gas interface

$$\dot{m}_l - \dot{m}_v = (\rho_l - \rho_v) A \frac{dr}{dt} \tag{5}$$

where "r" is the radial coordinate, A is the area and is the massic flow.
Towards the steam front, part of the injected steam is condensed and we can see that (Fig. 1):

Hence, we can write the conservation of mass equation in this way:

$$\varphi(\rho_l - \rho_v) \frac{dr}{dt} = \rho_l v_l - \rho_v v_v \tag{6}$$

Darcy velocity for each phase can be expressed as:

$$v_v \sim \frac{k \Delta P_v}{\mu_v l_v} \tag{7}$$

and

$$v_l \sim \frac{k \Delta P_l}{\mu_l l_l} \tag{8}$$

If:

$$\rho_l \gg \rho_v$$

Table 1 Properties of different elements used in the experiment

Property	Value
Sand porosity	0.5
Atmospheric pressure	101 325 Pa
Water density	21 °C: 998 kg/m^3
Steam viscosity at 100 °C	1.37473×10^{-5} Pa*s
Sand permeability	1.8×10^{-6}
Water viscosity at 21 °C	2.30763×10^{-4} Pa*s
Steam pressure	392.26 kPa

Substituting (7) and (8) in (6) we have:

$$\frac{dr}{dt} = \frac{1}{\varphi}\frac{kP_T}{v_v L}\left(\frac{\Delta P_v L}{\Delta P_T l_v} - \frac{\Delta P_l L v_v}{\Delta P_T v_l}\right) \tag{9}$$

Finally, with this expression we can estimate the total advance of the liquid and the steam front:
The values used in the experiment are shown in Table 1. Hence, in order to estimate the velocity of the flow we use the next values (Table 1):
Also, an expression for the Conservation of Energy can be found as:

$$\varphi(h_l - h_v)\frac{dr}{dt} = \left[\kappa_a C_a \frac{\partial T}{\partial r}\right] + h_l v_l - h_v v_v \tag{10}$$

where $[r]$ denotes the change of the property r through the interface and denotes the specific enthalpy.
The problem is completely described with Eqs. (1)–(4), as it can be seen in (Fitzgerald and Woods 1995) and it can be solved numerically.

3 Experimental Procedure

A radial porous cell is manufactured with a circular plate of tempered glass with a 36 cm diameter. These dimensions allow us to see the complete behavior of the phenomenon. The glass plate is placed above a metallic container with the same diameter and a depth of 2 cm. One is filled with silica sand forming a nonconsolidated porous matrix. The injection port was drilled at the center of the glass plate in order to allow us to inject the steam (as can be seen in Fig. 2).

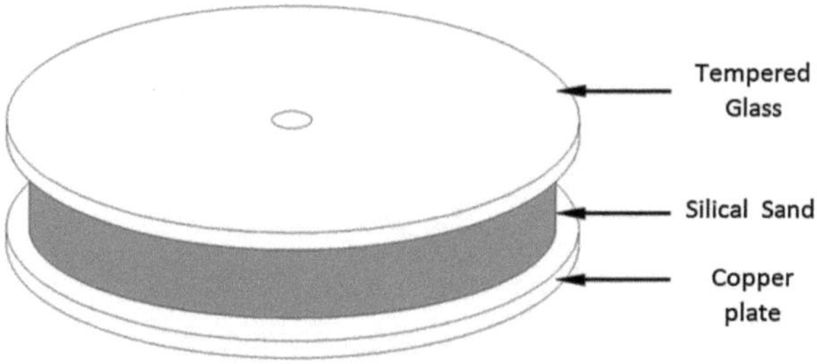

Fig. 2 Scheme showing the main conformation of the radial porous cell

Fig. 3 Plot showing the steam front advance as a function of time

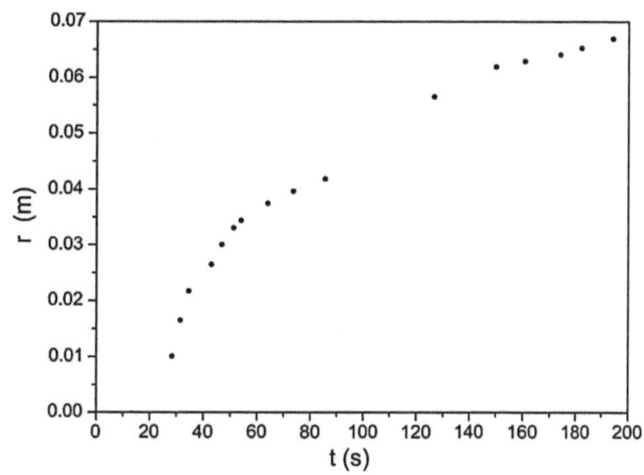

In the first experiment water is injected at ambient temperature (21 °C) through the center and it spreads through the matrix to a preheated porous matrix (about 100 °C). Part of the injected water boils almost instantly because of the heat exchanged with the porous matrix and the steam front moves through this system (Fig. 1). In Fig. 3, the advance of the steam front as a function of time can be observed.

In Fig. 4 water is injected and moving forward in a stable form, but due to the growth of the radial area, the rate area-injected water begins to decrease and the fluid begins to boil quickly, this steam front begins to move in a finger-shaped form, and we can say that the Saffman-Taylor instability is present.

In the second experiment, as in the previous one, water is injected constantly at an ambient temperature (21 °C) through the preheated porous matrix at about 120 °C over 20 °C more than in the first experiment. In this case, the steam front begins to destabilize in an asymmetric form at early stages of the phenomenon (Fig. 5). In Fig. 6 the advance of the vapor front as a function of time can be observed.

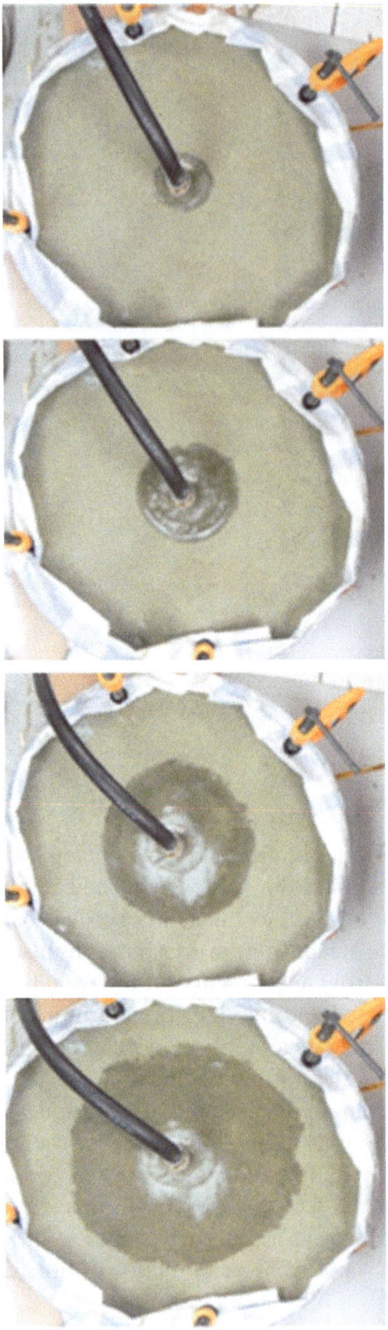

Fig. 4 Behavior of the vapor front moving through to the radial porous matrix in presence of a discrete Saffman-Taylor instability

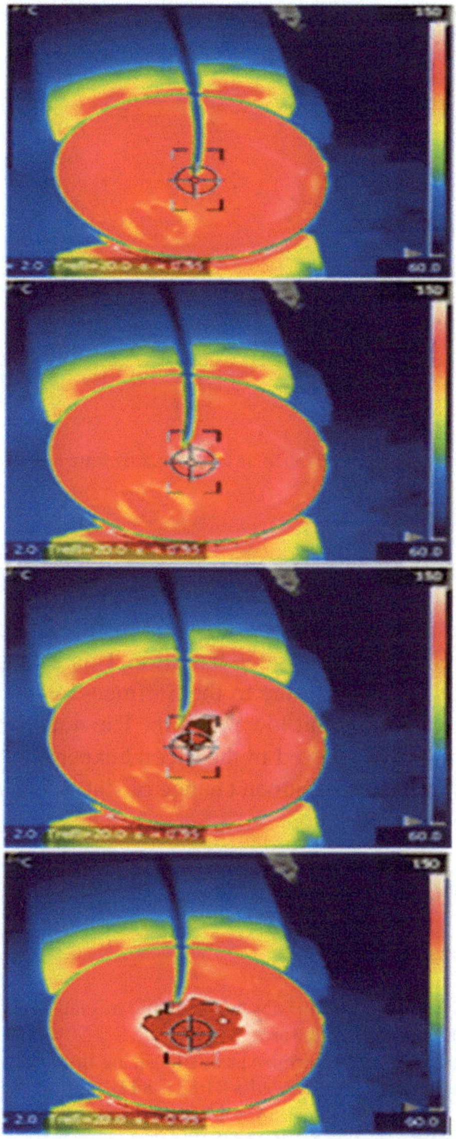

Fig. 5 Sequence showing a steam front moving through the radial porous matrix in presence of a strong "Saffman-Taylor" instability (Slide series was taken by using a thermal camera)

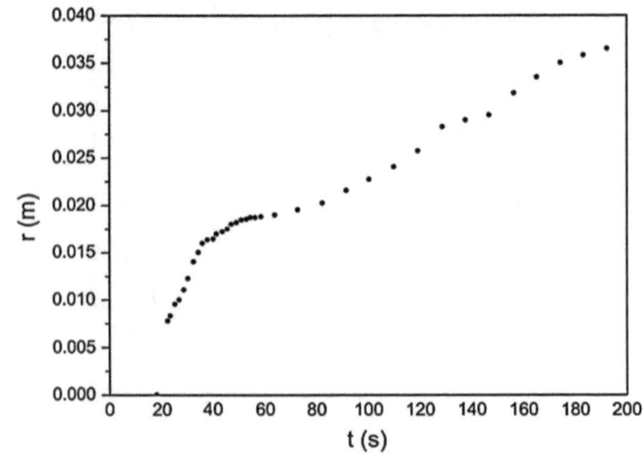

Fig. 6 Plot showing the steam front advance as a function of time

4 Conclusions

As it can be observed in this work, for both experiments the flow of the steam front through the porous cell showed the same behavior. This means that the steam front advanced in a finger-shaped form due to the Saffman-Taylor instability caused by the formation of zones of high and low pressure. The nonsymmetric finger-shaped behavior of the steam front observed in the second experiment (compared with the first one) can be attributed to the sudden change of state from water to steam due to the high temperature of the host core.

References

Fitzgerald SD, Woods AW (1995) Instabilities during liquid migration into superheated hydrothermal systems. Stanford University, Stanford

Islam MN, Azaies J (2010) Miscible thermo-viscous fingering instability in porous media, part 1: linear stability analysis. Transp Porous Media 84:821–844

Saffman PG, Taylor G (1958) The penetration of a fluid into a porous medium or Hele Shaw containing a more viscous liquid. R Soc Lond A 245:312–329

Transport of Particles in a Periodically Forced Flow

Erick Javier Lopez-Sanchez and Gerardo Ruiz-Chavarria

Abstract In oceanography particle transport is a constant: the ocean currents carry the plankton from one place to another. In shallow water trawling and sand deposition can affect positively or negatively certain human activities. For example, sandbars formed by the deposition in areas of low pressure may affect navigation near the coast, but at the same time they can reduce the intensity of a tsunami when approaching a populated coast. In this work we present a numerical solution of particle transport in a flow occurring in a system formed by the channel and an open domain and that is subject to a periodic forcing. For this purpose the equations of motion in the formulation vorticity- stream function are solved with a pseudo spectral method. After the velocity field is calculated, the trajectory of particles is obtained through the solution of a differential equation deduced from first principles. The goal is to model the transport of particles in a tide induced flow. The results we obtain are consistent with some experimental and observational data reported in previous works.

1 Introduction

An example of a system formed by a channel connected to an open domain (see Fig. 1) is a river flushing into a lake or the open sea. The tides induce a flow between both domains, in fact, during the stage of positive flow rate (positive is considered when flow is directed toward the open domain) a pair of vortices is formed. This is a coherent structure known as a dipole. The evolution of dipole depends on a dimensionless parameter, the Strouhal number S, defined as follow:

E. J. Lopez-Sanchez(✉) · G. Ruiz-Chavarria
Facultad de Ciencias, Universidad Nacional Autonoma de Mexico, Circuito Exterior s/n, C. P. 04510 Coyoacan, DF, Mexico
e-mail: lsej@ciencias.unam.mx

G. Ruiz-Chavarria
e-mail: gruiz@unam.mx

J. Klapp and A. Medina (eds.), *Experimental and Computational Fluid Mechanics*, Environmental Science and Engineering, DOI: 10.1007/978-3-319-00116-6_22,

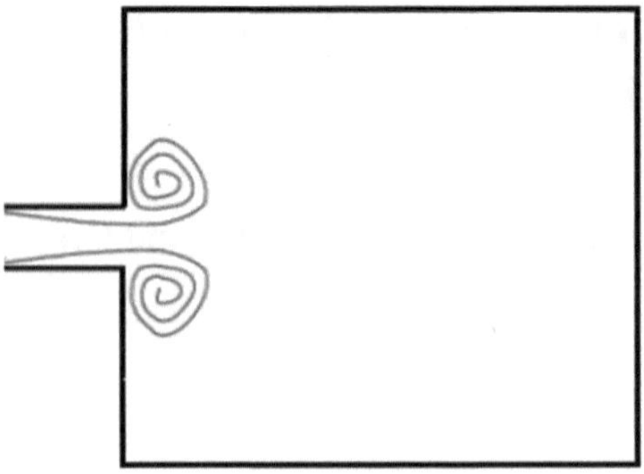

Fig. 1 System geometry under study, consisting of a channel connected to a basin

$$S = \frac{H_1}{UT} \qquad (1)$$

where H_1 is the channel width, U is the maximum speed at the channel and T is the forcing period. Considering a potential flow composed of a linear sink (or source, depending of the stage) and two counter rotating point vortices, Wells and Van Heijst (2003) shown that when $S < 0.13$ the dipole escapes, and when $S > 0.13$ the dipole returns to the channel. The numerical code described in this manuscript provides a detailed knowledge of the vortex formation or another features of the flow, like the coexistence of multiple dipoles, interaction among them or even the coalescence of vortices. All these facts are important because is well known the ability of dipoles to carry particles from one region to another. There are some previous works about both dipoles and particle transport. For instance Duran-Matute et al. (2010) made a numerical simulation of a dipole assuming as initial condition the velocity field of a Lamb-Chaplygin vortex in the horizontal plane and a Poiseuille velocity profile for the vertical coordinate. The relevant parameters are the Reynolds number Re and the aspect ratio $\delta = H/R_0$, where H is the fluid-layer depth and R_0 is the radius of the Lamb-Chaplygin vortex. They found that the three-dimensional nature of the flow depends on the single parameter $K = \delta^2 Re$. When $K < 6$ the flow remains bidimensional, and when $K > 15$ the flow becomes three dimensional with a spanwise vortex in front of dipole. From the experimental side Lacaze et al. (2010) and Albagnac (2010) produced a dipole by rotating two vertical plates in a rectangular basin. They also observed a spanwise vortex in front of the dipole and report that intensities of the spanwise vortex and the dipole are comparable.

Concerning the mass flow Angilella (2010) investigated the transport of dust in the vicinity of a pair of identical point vortices rotating about a common center and that remain in a vertical plane. To this end the equation of motion for trajectories is solved in a rotating reference frame, then Coriolis and centrifugal forces are included in addition to gravity and drag. The motivation of the research was to test the idea that the particle dispersion increases with the presence of a co-rotating vortex pair. When drag is the dominant force, the particle trajectories exhibit chaotic behavior, so mixing is enhanced.

Shaden et al. (2007) made experiments to determine the particle transport during the formation and growth of an annular vortex, which was produced with a piston-cylinder apparatus immersed in a water tank. They found that in the early stage, most of the fluid that enters the region of nonzero vorticity comes from the cylinder, and as the vortex ring grows and moves, fluid outside this cylinder is entrained.

The goal of this work is to calculate the trajectories of solid particles from an equation deduced from first principles (Maxey and Riley 1983) in which drag, added mass and history forces are included. For the integration of this equation the velocity field of the flow in the system shown in Fig. 1 is required. The solution obtained allows to determine the regions from which the particles are expeled and the regions where the particles are deposited .

The paper is organized as follows: In Sect. 2, we introduce the equation for particle motion equation, then it is written in dimensionless form. After we present the governing parameters and describe the methodology. In Sect. 3, we present data obtained for different values of S and Re and we compare them with some experimental results and, finally, we draw conclusions in Sect. 4.

2 Theoretical Background

In the flow the driving force is introduced through a periodic flow rate given by

$$Q = Q_0 \sin\left(\frac{2\pi}{T}t\right) \tag{2}$$

where T is the driving period. This choice allows periodic reversal of the flow. In dimensionless variables the flow rate is:

$$Q = \sin\left(2\pi St\right) \tag{3}$$

In order to describe all equations and results in dimensionless form we need to introduce a second parameter, namely, the Reynolds number, defined as: $Re = \frac{UH_1}{\nu}$. Here H_1 as the channel width, $U = Q_0/H_1$ is the maximum velocity in the channel and ν the kinematical viscosity. For the calculation of trajectories we solve a second order differential equation deduced by (Maxey and Riley 1983):

$$\begin{aligned} m_p\frac{d\mathbf{v}_p}{dt} &= (m_p - m_f)\mathbf{g} + m_f\frac{D\mathbf{u}}{Dt} \\ &+ 6\pi r\mu_f(\mathbf{u}-\mathbf{v}_p) + \frac{m_f}{2}\frac{D(\mathbf{u}-\mathbf{v}_p)}{Dt} \\ &+ 6r^2\sqrt{\pi\mu_f\rho_f}\int_0^t \frac{\frac{D(\mathbf{u}-\mathbf{v}_p)}{D\tau}}{\sqrt{t-\tau}}d\tau \end{aligned} \tag{4}$$

in which it is assumed that solid particles are spheres. Here, **u** is the fluid velocity at the particle position (if the particle was removed), $\mathbf{v}_p$ is particle velocity, m_p is its mass, m_f is the mass of fluid displaced by the solid sphere, μ_f is the dynamic viscosity, g is gravity, and r is the radius of the particle. The first term on the right-hand side is the sum of gravity and the buoyant force. The second term is the Stokes force, which is proportional to the difference between particle and fluid velocities. The third term is the added-mass term, and the last term is the history force (Mordant 2001). In the last equation the velocity field of the flow is required. This field has been obtained in a previous paper (Lopez and Ruiz 2013). Now we write this equation in non dimensional form:

$$\frac{d\mathbf{v}_p}{dt} = -\left(\frac{\rho_p}{\rho_f} - 1\right) Fr\hat{z} + \frac{3}{\left(\frac{2\rho_p}{\rho_m} + 1\right)} \frac{D\mathbf{u}}{Dt} \tag{5}$$
$$+ \frac{9(\mathbf{u} - \mathbf{v}_p)}{\left(\frac{2\rho_p}{\rho_m} + 1\right) Re} \left(\frac{H_1}{r}\right)^2 + \frac{9}{\left(\frac{2\rho_p}{\rho_m} + 1\right) \sqrt{\pi Re}} \left(\frac{H_1}{r}\right) \int_0^t \frac{\frac{D(\mathbf{u} - \mathbf{v}_p)}{D\tau}}{\sqrt{t - \tau}} d\tau$$

where ρ_p is the particle density, ρ_f is the fluid density and $\hat{z}$ is a unit vector in the vertical direction. $Fr = gH_1/U^2$ is the Froude number. For this calculation we make the approximation:

$$\frac{d(\mathbf{u} - \mathbf{v}_p)}{dt} \approx \frac{D\mathbf{u}}{Dt} - \frac{d\mathbf{v}}{dt} \tag{6}$$

Equation (5) is solved in two dimensions, so gravity and buoyancy are dropped and consequently the Froude number does not enter in the calculations.

For the integration we use $\rho_p = 2400\,\text{kg/m}^3$ (tipical sand density) and $r = 5 \times 10^{-4}$ m (the radius of the solid particles). We present results for particle transport in the flow using various values for S and Re. The calculation were made in some selected initial positions and we assume that the initial velocity is zero.

3 Particle Transport

Before to describe the motion of particles we present a result about the vorticity field obtained by solving equations of fluid dynamics in the vorticity - stream function formulation. Figure 2 shows the vorticity field for $S = 0.05$ and $Re = 667$. Multiples dipoles appear, each one is formed during a cycle. The solution was calculated using a pseudo-spectral method, based on Chebyshev polinomial for the space, and an Adams-Bashford semi-implicit schema for the time. In order to have a picture of the motion induced by the flow, some trajectories of solid particles and fluid elements are plotted in Fig. 3. The values of Reynolds and Strouhal numbers are respectively 67 and 0.0075. A comparison between motion of solid particles and fluid elements reveals that at the early stage both kind of trajectories resembles each to other.

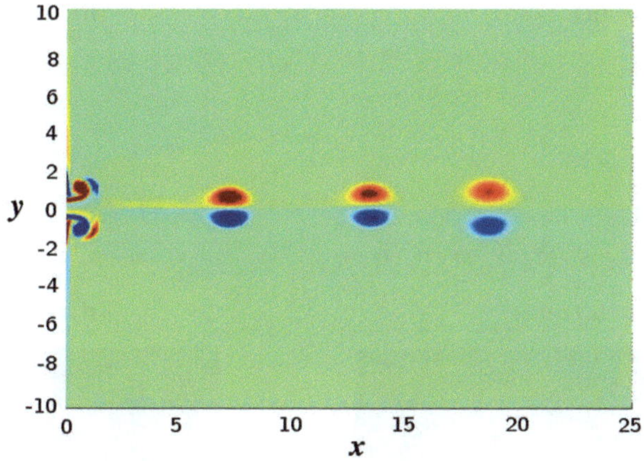

Fig. 2 Vorticity at $t = 3.24$ T. Multiple dipoles are present

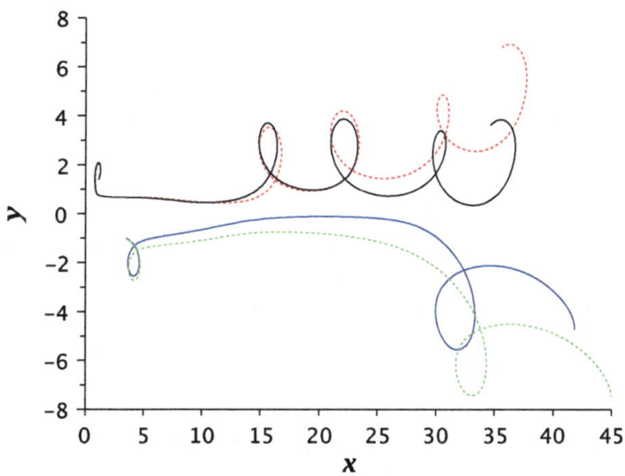

Fig. 3 Trajectories of both the solid particle (*continuous*) and the fluid elements (*dotted*) for two initial points $p_1 = (1.1, 1.4)$ and $p_2 = (3.5, -1)$, for the case $Re = 67$, $S = 0.0075$

However, after a while trajectories separate which is a signature of the influence of drag and other forces acting on the solid particles.

In Fig. 3, the particle initially located at the point $p_1 = (1.1, 1.4)$ moves away from the axis of symmetry. Then, due to the passage of the first dipole it is pulled back to the centerline and for a while it follows a nearly parallel path to the x-axis and after it make several curls. This kind of motion is induced by the presence of the dipole, which pulls and drives the particle. On the other side the particle initially located at the second point $p_2 = (3.5, -1)$ has in the early stage a similar behavior as first one, in the sense that it is moved away and pulled back to the axis of symmetry. Furthermore the path between $x = 6$ and $x = 28$ do not follow any curl. Finally, the curve made a curl, which is related to the presence of the second dipole.

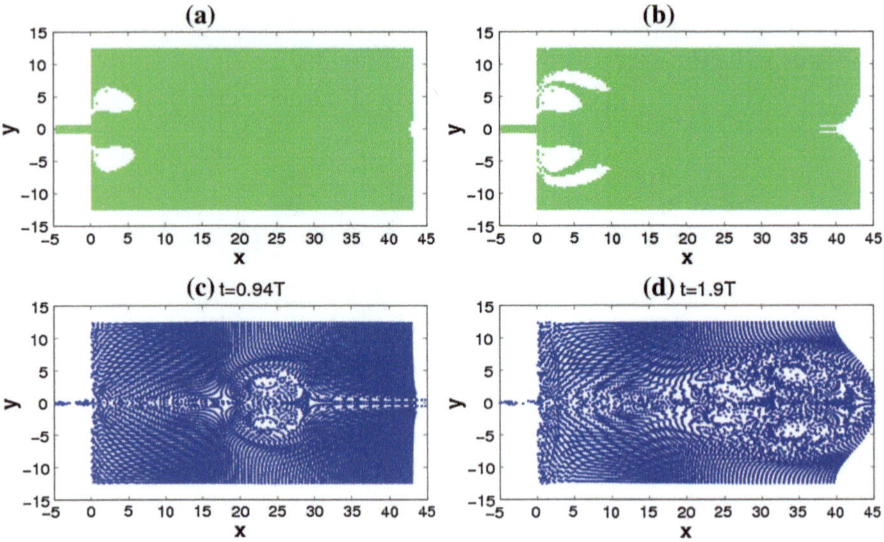

Fig. 4 Solid particles position at two times for $S = 0.0075$, and $Re = 67$. (**a**) and (**c**) Show the initial position of the particles and the white spaces indicate erosion zones. (**c**) and (**d**) Show the particles distribution at the time indicated in each panel

An overall picture of the transport of particles is obtained if the calculation of trajectories is performed for many particles. For this reason the entire domain is divided in a set of cells. In each one a fixed number of particles (9 in this work) is placed there at initial time. The size of a cell is 0.88×0.96. Then we integrate Eq. (5) for all these particles. With this procedure we can identify regions where particles accumulate or where particles are expelled. In Fig. 4 initial (upper graphs) and final (lower graphs) position are plotted. The assertion "initial position" includes only those particles that remain in the domain at the final time of integration. Blanks in the graphs indicate regions where particles are expelled outside the domain. These blanks correspond to erosion zones. The lower graphs correspond to final position of particles. Figure 4a shows the erosion zones until the time t=0.94T. Figure 4b shows areas of erosion until the time t=1.9T. The particle distributions at time t=0.94T and t=0.94T are shown in Figs. 4c and 4d respectively. In those graphs the passage of dipole is clearly observed. Particle transport can be outlined by the calculation of the histogram of the particle position. We proceed as follow: First, we chose a set of uniformly distributed particles, so that the probability density function (PDF) is the same elsewhere. After a while a new distribution of particles is obtained because the particles are constantly moving. The particles may remain inside the domain or may leave it. In the latter case the final particle position is unknown and consequently this particle position is not considered in the histogram.

A histogram of particle position for $t = 5.6$T, $S = 0.05$, and $Re = 333$ is presented in Fig. 5. For this histogram the size of cells is $\Delta x = 0.56$ and $\Delta y = 0.54$. In Fig. 5 left, we see that the particle concentration rises to 50 in a small region inside the channel.

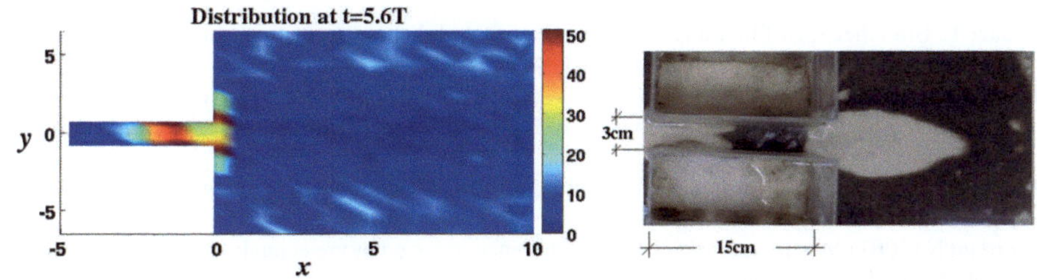

Fig. 5 *Left* Histogram of particle positions at $t = 5.6T$, $S = 0.05$, and $Re = 333$. *Right* Experimental setup

The right side of Fig. 5 shows a picture of a experiment in which the same parameters were used in the histogram. We place sea sand uniformly distributed and after various periods, we can observe an accumulation of particles within the channel near the outlet. In addition in front of the channel there is a region where the particles were removed by the flow.

A description of the sand barriers position in tidal inlet system is made by de Swart and Zimmerman (2009). The tidal currents induce the formation of a barrier in front of the inlet. The results presented above are in some agreement with these observational data.

4 Conclusions

In this paper we have made a study of the motion of solid particles in a flow with periodic forcing. In this flow one or more dipoles can coexist. The calculation of trajectories was made by the integration of a second order differential equation in which drag, added mas and history forces are included. The results show the existence of regions where particles are expelled outside the entire domain. Otherwise, there are also regions where particles concentrate. Some of these results are in agreement with observational and experimental data, even if trajectories were calculated in 2D and gravity was not taken into account.

Acknowledgments Authors acknowledge DGAPA-UNAM by support under project IN116312, "Vorticidad y ondas no lineales en fluidos".

References

Albagnac J (2010) Dynamique tridimensionnelle de dipoles tourbillonnaires en eau peu profonde. Thèse de doctorat, Université Paul Sabatier Toulouse III Institut de Mécanique des Fluides de Toulouse, France

Angilella J-R (2010) Dust trapping in vortex pairs. Physica D 239:1789–1797

Duran-Matute M, Albagnac J, Kamp LPJ, van Heijst GJF (2010) Dynamics and structure of decaying shallow dipolar vortices. Phys Fluids 22:116606

Lacaze L, Brancher P, Eiff O, Labat L (2010) Experimental characterization of the 3D dynamics of a laminar shallow vortex dipole. Exp Fluids 48:225–231

Lopez-Sanchez EJ, Ruiz-Chavarria G (2013) Vorticity and particle transport in periodic flow leaving a channel. Eur J Mech B: Fluids 42:92–103. Accessed (http://authors.elsevier.com/sd/article/S099775461300068X)

Maxey MR, Riley JJ (1983) Equation of motion for a small rigid sphere in a nonuniform flow. Phys Fluids 26:883–889

Mordant N (2001) Mesure lagrangienne en turbulence: mise en ouvre et analyse. Thése de doctorat, Ecole Normale Supérieure de Lyon, France

Shaden SC, Katija K, Rosenfeld M, Marsden JE, Dabiri JO (2007) Transport and stirring induced by vortex formation. J Fluid Mech 593:315–331

de Swart HE, Zimmerman JTF (2009) Morphodynamics of tidal inlet systems annu. Rev Fluid Mech 412:20329

Wells MG, Van Heijst G-JF (2003) A model of tidal flushing of an estuary by dipole formation. Dyn Atmos Oceans 37:223–244

Experimental Study of Heat and Mass Transfer During Steam Injection in Homogeneous Porous Media

S. Peralta, A. Torres, F. Aragón, G. Domínguez Zacarías, A. Medina and A. López-Villa

Abstract We carry out experiments in which we have already measured, in a short period of time, the physical parameters that appear during the steam injection in a homogenous porous media (Woods and Fitzgerald 1993). For this work, we measured the distribution of displacement and temperature that occurred when steam is injected at different pressures (0.25–1 kg/cm^2) in a homogeneous porous media by the control of the injection pressure and flow rate.

1 Introduction

The displacement of a fluid through a porous medium has been a topic of interest because of its relevance in the recovery of oil (Chung and Butler 1989b). Most thermal recovery methods have been applied to high viscosity oil reservoirs with the objective of increasing oil production by reducing oil viscosity. Heat can be injected into the reservoir as hot water or steam, or can be generated in-situ by burning part of

S. Peralta (✉) · A. Torres
Instituto Mexicano del Petróleo, Eje Central Lázaro Cárdenas 152, Col. Atepehuacán, Distrito Federal, 07730 México, DF, México
e-mail: peraltasalomon@hotmail.com

F. Aragón
SEPI-ESIME Zacatenco, Av. Instituto Politécnico Nacional s/n, Unidad Profesional "Adolfo López Mateos", Col. Lindavista, Del. Gustavo A. Madero, C.P. 07738 México, DF, México

G. D. Zacarías
Cordinación Tecnológica de Ingeniería de Yacimientos, Eje Central Lázaro Cárdenas Norte 152 Col. San Bartolo Atepehuacán, Del. Gustavo A. Madero, C.P. 07730 México, DF, México

A. Medina and A. López-Villa
SEPI-ESIME-Azcapotzalco. Instituto Politécnico Nacional, Av. de las Granjas 682, Col. Santa Catarina, Azcapotzalco, 02250 México, DF, México

J. Klapp and A. Medina (eds.), *Experimental and Computational Fluid Mechanics*, Environmental Science and Engineering, DOI: 10.1007/978-3-319-00116-6_23,

the reservoir crude oil. Of all these processes, steam injection is the most reliable, and has enjoyed by far the most commercial success. In 1988, 72 % of the total enhanced oil recovery in the U.S.A. was due to steam injection (Castanier and Gadelle 1991). Steam injection technology was the leading thermal method in the former U.S.S.R. and it represented the basement for modeling thermal recovery processes for heavy oil (Jabbour et al. 1996).

This chapter aims to describe some experiments on the method of steam flooding. Recovery by steam flooding is commonly used in heavy-oil reservoirs containing oil whose high viscosity is a limiting factor for achieving commercial oil-producing rates. It has also been considered, however, as a method for recovering additional light oil.

High-temperature steam is continuously injected into a reservoir. As the steam loses heat, it condenses into hot water which, coupled with the continuous supply of steam behind it, provides the drive to move the oil to production wells.

As the formation heats, oil recovery is increased by:

1. The heated oil which becomes less viscous, making it easier to move through the formation towards production wells.
2. Expansion or swelling of the oil aids in releasing it from the rock.
3. Lighter fractions of oil tend to vaporize, and as they move ahead of the steam they condense and form a solvent.
4. Finally, the condensed steam cools as it moves through the reservoir and results in an ordinary water flood ahead of the heated zone.

An added bonus from the use of steam in both steam flooding and cyclic steam stimulation is the flushing of liners and casing perforations, as well as the reduction of deposits that may build up in the wells. Possible flow restrictions to oil production through the wells are thus reduced (United Energy Group, Enhanced Oil Recovery 2008).

In order to understand this phenomenon, it is very important to have an experimental previous study. Therefore, we present some experiments in tubes filled with sand with different particle diameters, simulating the steam injection in a porous media.

2 Experiments

The experiments have been carried out by controlling the variables that are involved, such as the pressure, temperature and steam flow rate, all in cylindrical pipes filled with homogeneous sand.

The first problem was the control of the pressure, temperature and flow rate. We used a high resolution digital camera (Nikon, 16 megapixels) to measure the fluid velocity within the porous medium and to measure the temperature distribution a thermal camera FLIR SC660 was used.

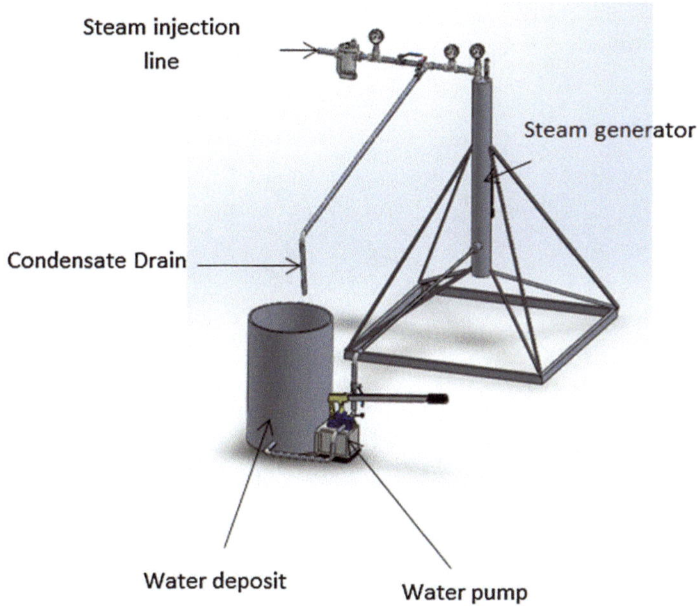

Fig. 1 Steam generator. The tube was made with an AISI 1040 steel pipe $d_t = 0.0508$ m and $L_t = 1.5$ m with a capacity of 4 lt

The steam generator was made with an AISI 1040 steel pipe with a diameter D = 0.0508 m and a length of L = 1.5 m with a capacity of 4 lt. The maximum pressure and temperature that we could reach were 8 kg/cm^2 and 120 °C respectively, see Fig. 1.

The test core is a cylindrical acrylic tube, with 0.006 m wall thickness, a length $L = 0.42$ m and a diameter $d_t = 0.039$ m, filled with a homogeneous mixture of two types of sand, see Fig. 2. The test sands are: Ottawa sand 850 μm mean diameter (d_O) and Veracruz sand with 315 μm mean diameter (d_V).

We made experiments with the material mentioned above, at different pressures $p = 0.25, 0.5$ and 1 kg/cm^2.

The first experiment was made under the following conditions: p = 0.25 kg/cm^2, $T_{initial} = 26$ °C, $T_{final} = 81.6$ °C. Volume of condensed fluid 50 ml, v = 0.03604 m/s and $Q = 0.00004305$ m^3/s, or 43.05 ml/s.

For this pressure, the gravitational force is very important in the displacement of the steam. We observe that in the bottom, the condensate forms a protuberance due to the force of gravity see Fig. 3.

For the second experiment we obtained a displacement within the matrix at a steam pressure of 0.5 kg/cm^2, an initial temperature of 26 °C and a final temperature of 83.3 °C, v =0.07157 m/s and Q = 0.00008549 m^3/s or 85.49 ml/s see Fig. 4.

In the third experiment steam injection pressure reached 1 kg/cm^2 in the porous media. Figure 5 shows the photograph of the displacement of the steam and a con-

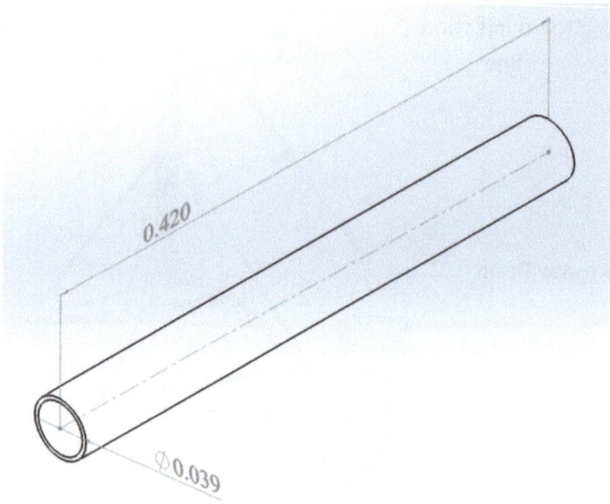

Fig. 2 Acrylic cylinder with a wall thickness of 0.06 m, $L = 0.42$ m and $\phi_{\text{test tube}} = 0.039$ m

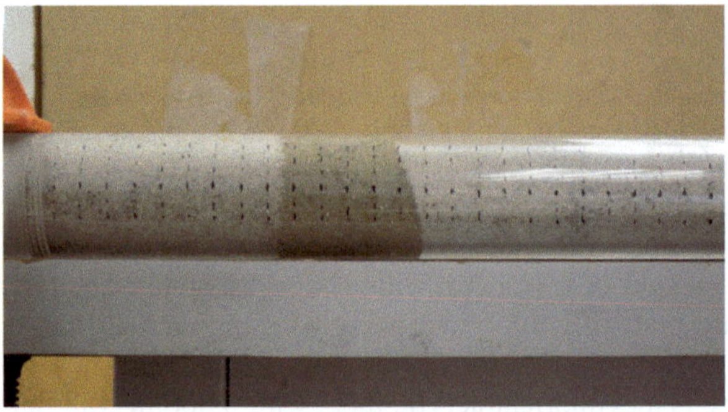

Fig. 3 Condensed profile in a homogeneous matrix formed in Ottawa sand, with an injection pressure of 0.25 kg/cm^2. As seen in the photograph, at the bottom of the specimen, bulges condensate due to the effect of the gravitational force on the experiment

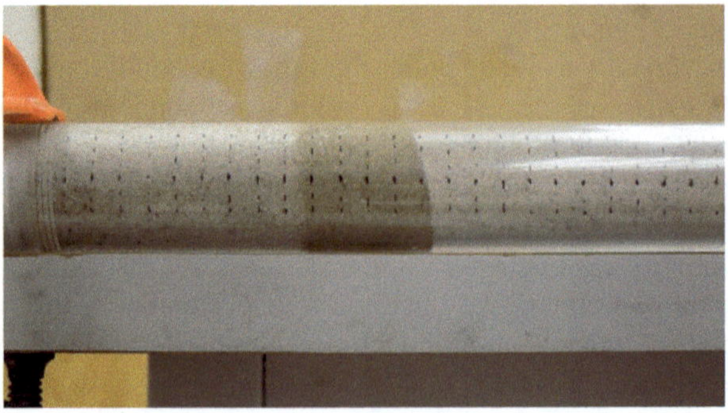

Fig. 4 Displacement with $p = 0.5$ kg/cm^2, the gravity force forms a protuberance

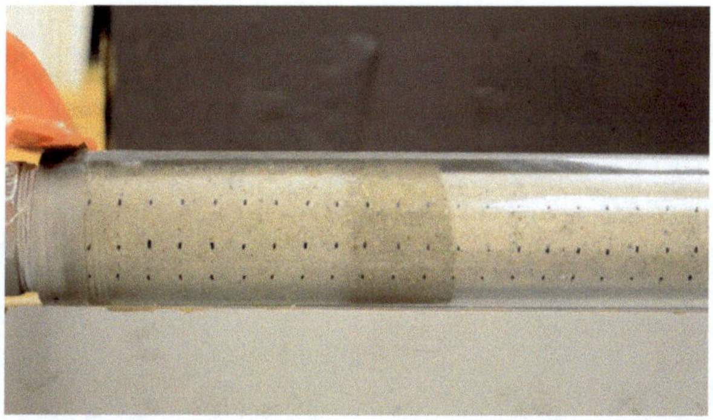

Fig. 5 Profile observed in the homogeneous matrix with steam injection at the pressure of 1 kg/cm^2

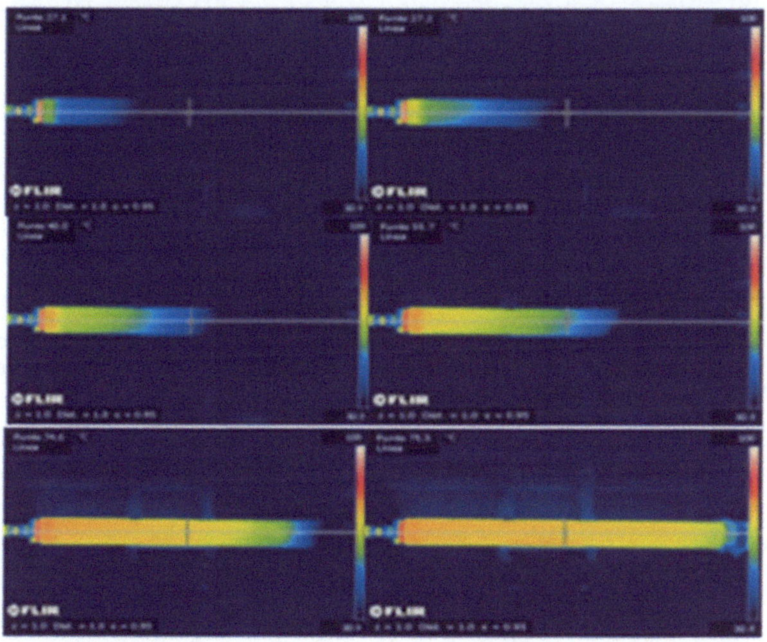

Fig. 6 Thermal sequence with steam injection at a pressure of 0.25 kg/cm^2, a homogeneous matrix consisting of sifted Veracruz sand with $d = 315$ m. The temperature is distributed evenly through the homogeneous matrix homogeneous, over time the temperature converges to 81.6 °C

densation front caused by the temperature difference in the porous media (the porous media temperature was 21 °C).

In Figs. 3, 4 and 5, we can see how the condensation front changes by the increase of the pressure and the steam flow rate. The profile shows how the flow is transformed to a type piston flow.

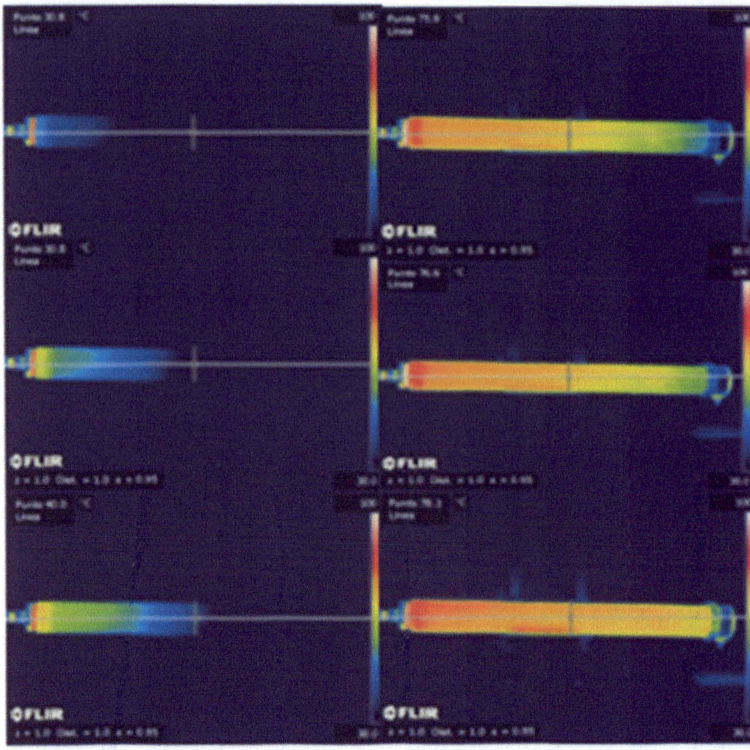

Fig. 7 Sequence of thermal photographs of the homogeneous matrix obtained with a FLIR camera, at a steam pressure of $0.5\,kg/cm^2$, with an initial temperature of 26 °C and a final temperature of 83.3 °C and volumetric flow of 85.49 ml/s

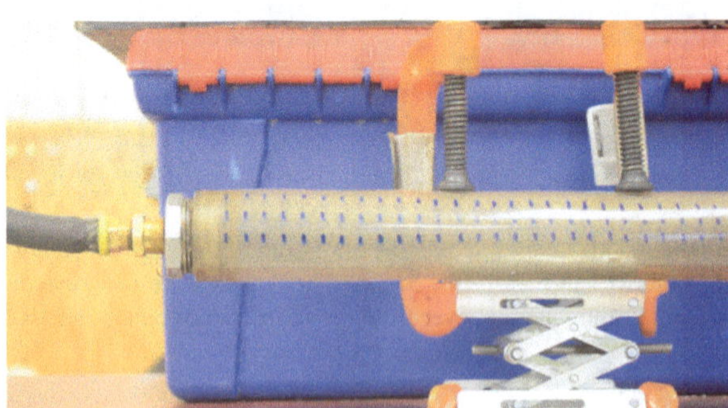

Fig. 8 Homogeneous matrix (Veracruz sand) where the thermal pictures were taken with a FLIR SC-660 camera each 10 s during all the experiment

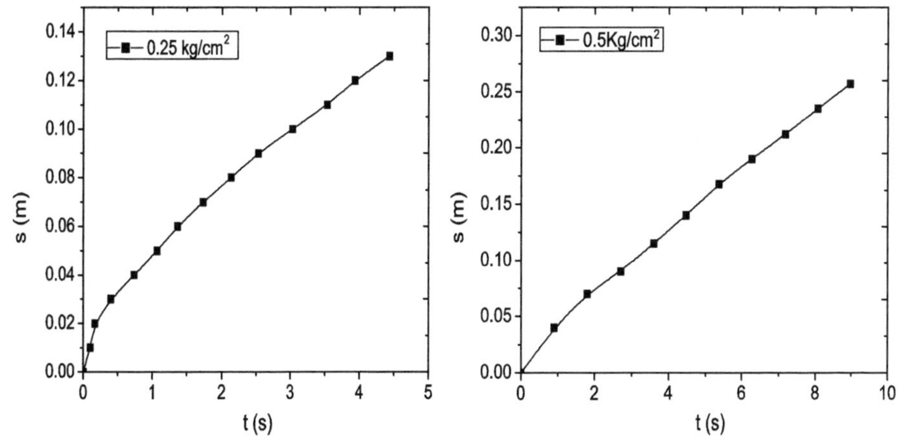

Fig. 9 Graphs of the steam displacement in the porous medium with Veracruz sand at $p = 0.25\,\text{kg/cm}^2$ and $p = 0.5\,\text{kg/cm}^2$

To observe the change of temperature, we have recorded the temperature gradient along the test tube in two different cases, with a thermal camera, as it can be seen in Fig. 6 for the case $0.25\,\text{kg/cm}^2$.

Figure 7 shows the sequence of the temperature distribution in the homogeneous porous matrix. The temperature distribution profiles are different because of the different pressures. In Fig. 6 $p = 0.25\,\text{kg/cm}^2$, and in Fig. 7 the pressure is $p = 0.5\,\text{kg/cm}^2$.

In the thermal pictures the homogeneous matrix is formed with Veracruz sand in both cases (see Fig. 8). This could be done because the low permeability of the sand from Veracruz allows a better reading of the thermal camera, and therefore shows profiles with better approximation.

3 Results

The graphs show the progress of the steam as a function of time. At the beginning, the steam shows a higher velocity. After a few seconds it slows down until it reaches the end of the test core. The final speed increases with increasing pressure (see Fig. 9).

For the case of the increase of temperature for $p = 0.25\,\text{kg/cm}^2$ and $p = 0.5\,\text{kg/cm}^2$, the maximum temperatures are 81.6 and 83.3 °C measured with a FLIR camera (Fig. 10), there is a difference of temperatures between the real measurement and the measured with the camera due to the acrylic cover, we know this difference is of 12 °C.

Figure 11 shows the displacement of the steam in the porous medium at a pressure of $1\,\text{kg/cm}^2$. Notice that the graph is a straight line.

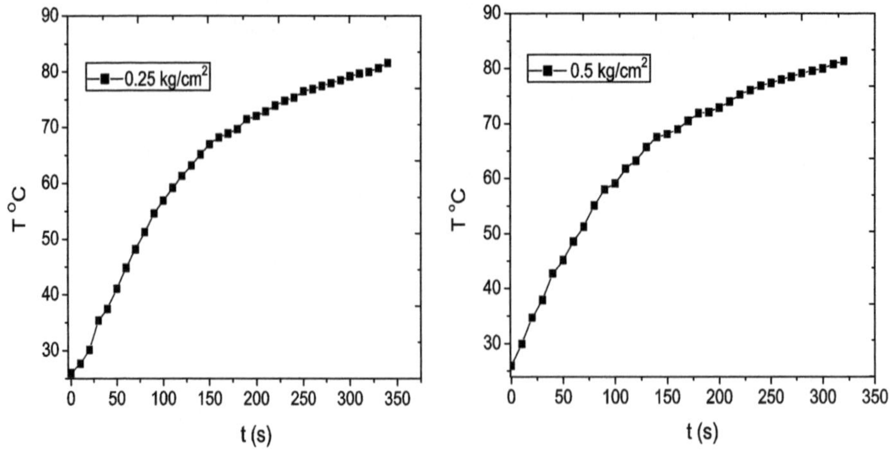

Fig. 10 Graphs of the increase of the temperature with time for Veracruz sand at $p = 0.25\,\text{kg/cm}^2$ and $p = 0.5\,\text{kg/cm}^2$

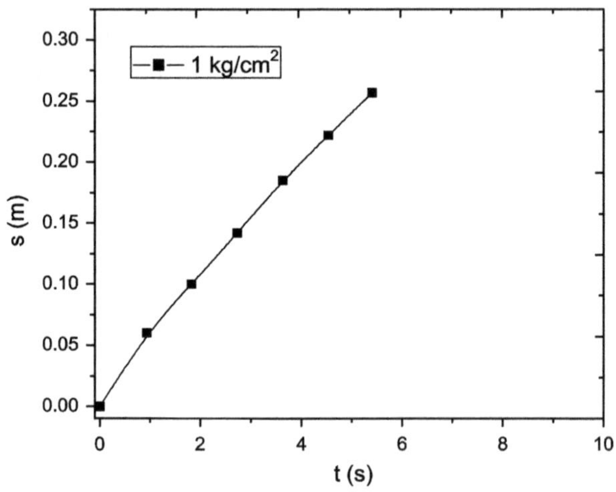

Fig. 11 Graph of the steam displacement in the porous medium with Ottawa sand for $p = 1\,\text{kg/cm}^2$

Comparing 0.5 and $1\,\text{kg/cm}^2$, see Fig. 12, the slope changes due to the pressure and the velocity inside the porous media. With the pressure of 0.25, see Fig. 10, we can see that the gravity force domain and the displacement is not proportional.

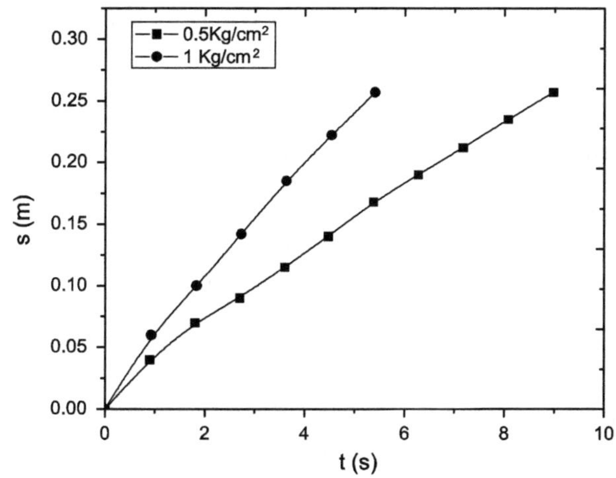

Fig. 12 comparison of the experiment with different steam pressures $p = 0.5\,\text{kg/cm}^2$ and $p = 1\,\text{kg/cm}^2$

4 Conclusions

In this work we have shown the basic experiments in which is involved the easiest arrangement (porous media and steam). This is the first step to understand their behavior into several conditions, for example at different pressures. The next step is to complicate the situation introducing a new fluid such as light and heavy oil for understanding how these fluids are going to behave during steam injection. The experiments show the real behavior of the steam in the porous media.

Acknowledgments SP and AT wish to thank the IMP and CONACYT for their fellowships. AM and AL wish to thank IMP-CONACYT-IPN for the support to the project "Análisis de mecanismos en procesos de recuperación mejorada por inyección de fluidos calientes y agentes espumantes con propiedades modificadoras de la mojabilidad". Finally, AM also thanks the IMP for his sabbatical.

Appendix

Properties of steam

Below we have the following characteristics of the steam.
Properties of steam at 0.25 kg/cm^2:

Pressure		0.25	kg/cm^2
Saturation temperature		106.188	°C
Specific entalphia of water	(h_f)	445.24	kJ/kg
Specific entalphia of evaporation	(h_{fg})	2240.21	kj/kg
Specific entalphia of steam	(h_g)	2685.45	kj/kg
Density of steam		0.731769	kg/m^3
Specific volume of steam	(V_g)	1.36655	kg/m^3
Specific entropy of water	(s_f)	1.37636	kj/kg K
Specific entrophy of evaporation	(s_{fg})	5.90557	kj/kg K
Specific entrophy of steam	(s_g)	7.28193	kj/kg K
Specific heat of steam	(c_v)	1.54032	kj/kg K
Specific heat of steam	(c_p)	2.06686	kj/kg K
Speed of sound		475.913	m/s
Dinamic viscosity of steam		1.25E$_{05}$	Pa s
Isentropic coefficient	(k)	1.31656	
Comppressibility factor of steam		0.982271	

Properties of steam at 0.5 kg/cm^2:

Pressure		0.5	kg/cm^2
Saturation temperature		111.45	°C
Specific entalphia of water	(h_f)	467.45	kj/kg
Specific entalphia of evaporation	(h_{fg})	2226	kj/kg
Specific entalphia of steam	(h_g)	2693.51	kj/kg
Density of steam		0.864345	kg/m^3
Specific volume of steam	(V_g)	1.15695	kg/m^3
Specific entropy of water	(s_f)	1.4346	kj/kg K
Specific entrophy of evaporation	(s_{fg})	5.78782	kj/kg K
Specific entrophy of steam	(s_g)	7.22243	kj/kg K
Specific heat of steam	(c_v)	1.55322	kj/kg K
Specific heat of steam	(c_p)	2.08789	kj/kg K
Speed of sound		478.49	m/s
Dinamic viscosity of steam		1.27E$_{05}$	Pa s

Properties of water at $1\,kg/cm^2$

Pressure		1	kg/cm^2
Saturation temperature		120.44	°C
Specific entalphia of water	(h_f)	504.426	kj/kg
Specific entalphia of evaporation	(h_{fg})	2202.01	kj/kg
Specific entalphia of steam	(h_g)	2706.44	kj/kg
Density of steam		1.12564	kg/m^3
Specific volume of steam	(V_g)	0.888386	kg/m^3
Specific entropy of water	(s_f)	1.52938	kj/kg K
Specific entrophy of evaporation	(s_{fg})	5.59889	kj/kg K
Specific entrophy of steam	(s_g)	7.12827	kj/kg K
Specific heat of steam	(c_v)	1.57627	kj/kg K
Specific heat of steam	(c_p)	2.12621	kj/kg K
Speed of sound		482.559	m/s
Dinamic viscosity of steam		$1.30E_{05}$	Pa s
Isentropic coefficient	(k)	1.3138	
Comppressibility factor of steam		0.975884	

References

Castanier LM, Gadelle C (1991) Hot fluid injection: basic concepts in enhanced oil recovery processes. In: Bavitre M (ed) Elsevier applied science. Crit Rep Appl Chem 33:271–312
Jabbour C, Quintard M, Bertin H, Robin M (1996) Oil recovery by steam injection: three-phase flow effects. J Petrol Sci Eng 16:109–130
United Energy Group, Enhanced Oil Recovery (2008) U.S. Department of Energy, National Energy Technology Laboratory
Chung KH, Butler RM (1989b) A theoretical and experimental study of steam-assisted gravity drainage process. In: Meyers RF, Wiggings EJ (eds) The fourth UNITAR/UNDP international conference on heavy crude and tar sands, vol 4. In-Situ recovery, AOSTRA, Edmonton, pp 191–210
Woods AW, Fitzgerald SD (1993) The vaporization of a liquid front moving through a hot porous rock. J Fluid Mech 251:563–579

Numerical Simulations of the Kelvin–Helmholtz Instability with the Gadget-2 SPH Code

Ruslan F. Gabbasov, Jaime Klapp-Escribano, Joel Suárez-Cansino and Leonardo Di G. Sigalotti

Abstract The method of Smoothed Particle Hydrodynamics (SPH) has been widely studied and implemented for a large variety of problems, ranging from astrophysics to fluid dynamics and elasticity problems in solids. However, the method is known to have several deficiencies and discrepancies in comparison with traditional mesh-based codes. In particular, there has been a discussion about its ability to reproduce the Kelvin–Helmholtz Instability in shearing flows. Several authors reported that they were able to reproduce correctly the instability by introducing some improvements to the algorithm. In this contribution, we compare the results of the Read et al. (2010) implementation of the SPH algorithm with the original Gadget-2 N-body/SPH code.

R. F. Gabbasov (✉) · J. Suárez-Cansino
Universidad Autónoma del Estado de Hidalgo, Instituto de Ciencias Básicas e Ingeniería, Ciudad Universitaria, Km 4.5 Carretera Pachuca—Tulancingo, 42184 Mineral de la Reforma, Hidalgo, México
e-mail: ruslan.gabb@gmail.com

J. Suárez-Cansino
e-mail: jsuarez@uaeh.edu.mx

J. Klapp-Escribano
Instituto Nacional de Investigaciones Nucleares, Carretera México-Toluca Km 36.5, 52750 La Marqueza, Estado de México, México
e-mail: jaime.klapp@inin.gob.mx

J. Klapp-Escribano
Departamento de Matemáticas, Cinvestav del IPN, 07360 D.F., México, México
e-mail: jaime.klapp@hotmail.com

Leonardo Di G. Sigalotti
Centro de Física, Instituto Venezolano de Investigaciones Cientficas, IVIC, Apartado Postal 20632 Caracas 1020-A, Venezuela
e-mail: leonardo.sigalotti@gmail.com

J. Klapp and A. Medina (eds.), *Experimental and Computational Fluid Mechanics*, Environmental Science and Engineering, DOI: 10.1007/978-3-319-00116-6_24,

1 Introduction

The Kelvin–Helmholtz Instability (KHI) is the instability that appears at the interface between two shearing fluid flows of different densities. Many experimental and numerical results have been published where such instability is reproduced. In particular, in astrophysics such instabilities may be responsible for many phenomena observed in regions of high gas density contrast (Murray et al. 1993). The necessity for subsonic velocities for the KHI survival was first explored in the context of astrophysical flows by Vietri et al. (1997). As discussed by these authors, the growth rate of the KHI is smaller than the gas speed of sound. The correct modeling of KHI is essential since the vorticity and shear flows appear in diverse hydrodynamic processes, such as for example, the onset of turbulence.

The Smoothed Particle Hydrodynamics method (SPH) is a Lagrangian meshless particle method used for simulation of fluid transport. Since its original formulation the SPH method has been constantly improved. In recent years a search for possible differences between grid-based and particle-based methods has been widely discussed. For instance, Agertz et al. (2007) compare the results for several hydrodynamic tests obtained with SPH and grid-based methods. They found a striking difference between the results. In particular, the inability to reproduce the vorticity rolls in the shearing flow was claimed to be a deficiency of the SPH method. Price (2008) showed that introducing an artificial thermal conductivity term into the standard SPH, in order to smooth the discontinuities in the thermal energy, allowed for similar results to those obtained by Agertz et al. (2007) using grid-based simulations.

On the other hand, several authors suggested that the artificial conductivity is not the main factor responsible for suppression of the instability. It is well known that the choice of the smoothing kernel also affects the results. For example, the standard cubic spline kernel tends to suffer from the clumping instability (known also as pairing instability) for large number of neighbors, introducing errors and lowering the resolution. Read et al. (2010) showed that kernels that produce a constant force term in the center prevent the clumping. Several alternative kernel shapes were proposed to remedy this problem [see for example, Dehnen and Aly (2012)]. Hubber et al. (2013) performed a comparison of KHI simulations obtained with SPH and with the AMR Eulerian code *Pencil*. They concluded that convergence between SPH and grid codes may be obtained if higher order kernels (i.e., quintic) that support larger numbers of neighbors are used. Cha et al. (2010) showed that inaccurate density gradients that are obtained with the standard SPH formulation are responsible for the suppression of instabilities and can be alleviated by using a Godunov formulation of SPH. These works concluded that the standard SPH formulation is unable to reproduce the KHI and that some improvements must be implemented.

Some of the causes that suppress the KHI are the following: when pressure discontinuities appear as a result of the lack of entropy mixing on the kernel scale, when large errors are introduced in the momentum equation due to a finite number of neighbor particles, due to the pairing instability, or due to contact discontinuities.

In this work, we compare the results for the KHI as obtained using the standard Gadget-2 SPH with the formulation proposed by Read et al. (2010).

2 Initial Conditions

As the initial conditions for the first part of the KHI tests (A), we use the example already included in the OSPH code (Read et al. 2010). This is done with the aim of providing direct comparison of our simulations with previously published results. In the second part (B), we use the so-called "well-posed" initial conditions described in McNally et al. (2012) [see also Robertson et al. (2010)]. The latter is regularized by smoothing the density and velocity on the interfaces in order to prevent abrupt pressure jumps. Such situations are frequently found for example in simulations of real astrophysical systems. In addition, the two sets of initial conditions differ by the type of density sampling, which is done by the spatial distribution of the particles (set A) or by varying the particle masses (set B). Set C differs from set B only in the magnitude of the densities and velocities.

For brevity, we describe only the initial conditions for the simulation sets B and C, which consist of a 3D slab of size $1.0 \times 1.0 \times 0.0325$, with the central part having a density $\rho_2 = 2\rho_1$ and moving with the velocity $\mathrm{v}_2 = -u$, while the upper and lower sections have a density ρ_1 and move with the velocity $\mathrm{v}_1 = u$ along the x-axis. A small velocity perturbation is added, $\mathrm{v}_y = \delta \mathrm{v}_y \sin(2\pi x/\lambda)$, with $\lambda = 0.5$. The gas has a constant pressure $P = 2.5$ everywhere and satisfies an ideal gas equation of state $P = (\gamma - 1)U\rho$, with $\gamma = 5/3$. The system of units is such that length, time and mass are equal to unity. For the simulation set B, we use $\rho_1 = 1.0$, $u = 0.5$ and $\delta \mathrm{v}_y = 0.01$, while for the set C we define $\rho_1 = 32.0$, $u = 0.1$, $\delta \mathrm{v}_y = 0.01$ and $\delta \mathrm{v}_y = 0.002$. The system is sampled with an equally spaced cubic grid of $256 \times 256 \times 8$ particles. The parameters are chosen with the purpose of staying in the subsonic regime, which guarantees the KHI formation.

The characteristic onset time of the KHI in the linear regime is given by:

$$\tau = \frac{(\rho_1 + \rho_2)\lambda}{\sqrt{\rho_1 \rho_2}\,|\mathbf{v}_2 - \mathbf{v}_1|}. \tag{1}$$

For the above initial parameters the characteristic times are $\tau_A \approx 3.4$, $\tau_B \approx 1.06$ and $\tau_C \approx 5.3$, and the total simulation times were 4, 2.1, and 8 respectively. Note that different units are used for set A.

We use the Gadget-2 code (Springel 2005) and its modification—OSPH—the Optimized SPH introduced by Read et al. (2010). In both codes the flags `-DPERIODIC`, `-DNOGRAVITY` and `-DLONG` were activated, and in OSPH the recommended flags were also switched on (`-DOSPHM`, `-DOSPHRT`, and `-DOSPHHOCT`). The artificial viscosity was set to $\alpha = 0.8$ in all simulations.

Table 1 Summary of the KHI simulations

Model	N_{ngb}	Code
AGADG33	33	Gadget-2
AGADG64	64[a]	Gadget-2
AOSPH96	96	OSPH
AOSPH442	442	OSPH
BGADG64	64[a]	Gadget-2
BOSPH64	64	OSPH
BOSPH96	96	OSPH
BOSPH442	442	OSPH
CGADG64L	64	Gadget-2
CGADG64S	64	Gadget-2
COSPH96L	96	OSPH
COSPH96S	96	OSPH

[a] The runs with a higher number of neighbors with the standard Gadget-2 were not completed because a maximum number of tree-nodes was reached due to the pairing instability

3 Results

We performed a number of tests using different combinations of the initial conditions and code parameters. In Table 1 we summarize some of them. The first column is the set of models coded by the first letter of the name, which are A, B or C, the second column gives the number of neighbors, and the third column shows the code employed for the simulation. Figure 1 (set A), Fig. 2 (set B) and Fig. 3 (set C) summarize the results of using different initial conditions and different numbers of neighbor particles.

From Fig. 1, we observe that model AOSPH442 is the only run that produces well distinguished rolls, while using either a smaller number of neighbors or the Gadget-2 code leads only to mild perturbations. This fact confirms the results of Read et al. (2010). A notable feature is that in model AOSPH96 with a smaller number of neighbors the perturbations are completely damped. On the other hand, run AGADG64, which uses a nearly optimum number of neighbors for the 3D cubic spline kernel, shows pronounced undulations, although without any rolls. Using their new SPHS code, which implements a dissipation switch, Read and Hayfield (2012) obtained KHI in both single mass and multimass particle models. Their initial conditions for single mass particles are identical to those used in set A. Note that in the set A a Mach number for the low density layer is $M \approx 0.11$.

The mass varying initial conditions in the simulations of set B lead to pronounced perturbations with well developed rolls being produced only in the runs with a maximum number of neighbors (Fig. 2). Small undulations observed at time $t = \tau_B$ (left column) are evolved into a KHI for $t = 2\tau_B$ (right column) in runs BGADG64 and BOSPH442. Surprisingly, standard SPH with multimass setup do develop KHI, contrary to previous claims. Comparing a BGADG64 density projection at $t = 2.1$ to a reference solution obtained with the *Pencil* code of

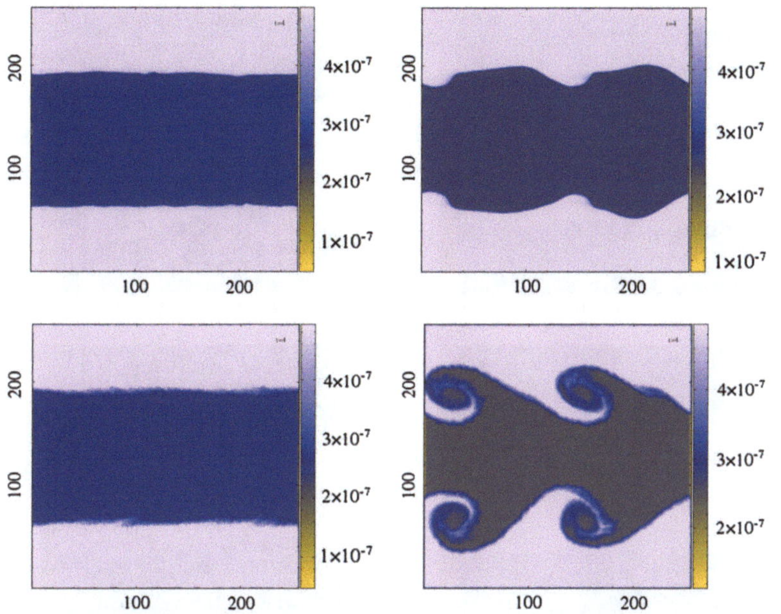

Fig. 1 Projected density plots for the simulations of set A showing the perturbations at $1.2\tau_A$. From *left* to *right* and *top* to *bottom*: AGADG33, AGADG64, AOSPH96, AOSPH442

McNally et al. (2012) (their Fig. 2) we observe a very similar shape and amplitude of density rolls. However, the reference solution shows the density projection at $t = 1.5$, indicating that in BGADG64 the KHI develops slowly. In the case of BOSPH442, the density projection which matches approximately the reference solution corresponds to $t = 1.7$. In order to check the effect of a low number of neighbors we repeated the BGADG64 with 33 neighbors (not shown), and obtained a similar situation to BOSPH64 with no KHI. The Mach number of the low density layer is $M_2 \approx 0.34$.

In order to explore the effect of different shear velocities and perturbation amplitudes, we performed some additional simulations with both codes reducing the initial velocities, and either reducing or keeping the same perturbation amplitude δv_y. These models are listed in Table 1 as set C, where the last letter in the name stands for large amplitude (L), $\delta v_y = 0.01$, and for small amplitude (S), $\delta v_y = 0.002$, respectively. In this case the gas is still subsonic with $c_1 = 0.36$ and $c_2 = 0.25$, giving the Mach number of low density layer $M_2 \approx 0.4$. The Fig. 3 compares the results obtained with both codes. While these appear very similar, the amplitude of the KHI looks smaller and lacks secondary ripples for the Gadget-2 code.

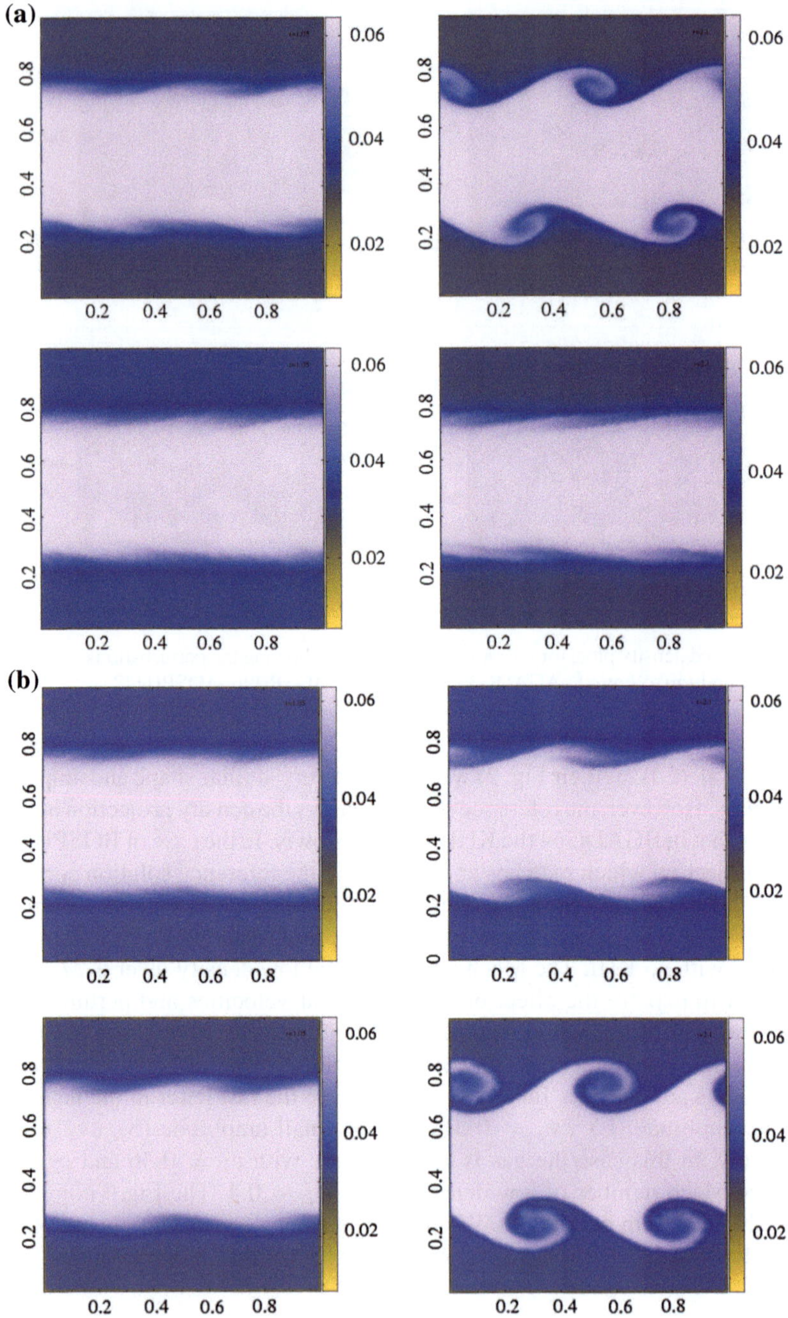

Fig. 2 Projected density plots for the simulations of set B. Each column shows the perturbations at τ_B (*left*) and $2\tau_B$ (*right*). (**a**) From *top* to *bottom*: BGADG64 and BOSPH64. (**b**) From *top* to *bottom*: BOSPH96 and BOSPH442

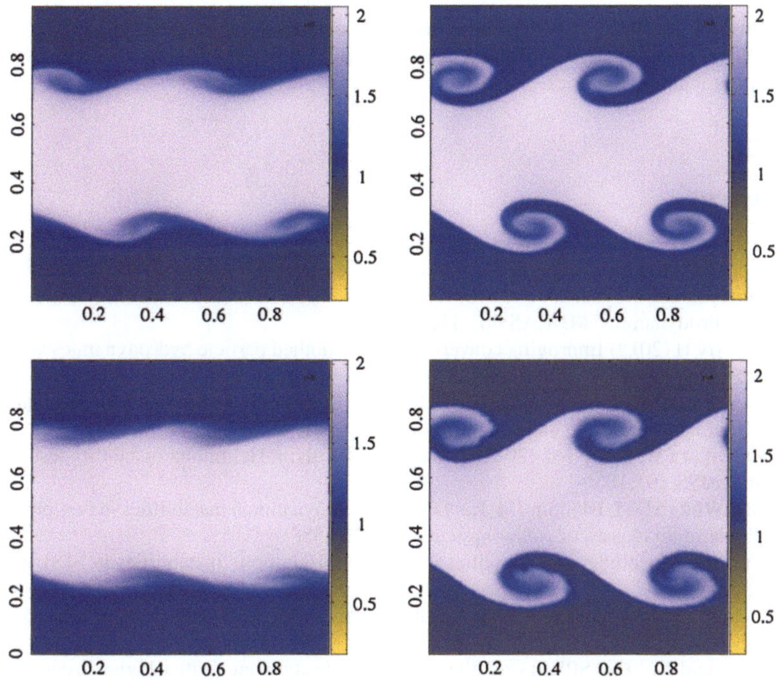

Fig. 3 Projected density plots for the simulations of set C showing the perturbations at $1.5\tau_C$. From *left* to *right* and *top* to *bottom*: CGADG64S, CGADG64L, COSPH96S, COSPH96L

4 Concluding Remarks

The aim of this work was to test several implementations of the SPH code and determine the necessary conditions for successful simulations of the mixing process in shearing flows. We have compared visually the projected densities for standard and optimized SPH codes for non-linear structure formation. Runs with single mass standard SPH formulations showed a similar behavior to that described in previous studies, [c.f. Fig. 13 of Agertz et al. (2007) and Fig. 7 of Price (2008)]. On the other hand, it seems that the multimass simulations do not depend on the SPH code implementation, but rather on the number of neighbors. The optimized SPH code is almost two times slower than the standard SPH for the same number of neighbors, – a disadvantage that was reported to be absent in the new SPHS code of the same authors (Read and Hayfield 2012). We have shown that for mixing problems using the SPH formalism, it is essential to take much care in setting the initial conditions and the number of neighbors. Further investigation of the problem is necessary in order to obtain robust results with SPH.

Acknowledgments This work has been partially supported by ABACUS, CONACyT grant EDOMEX-2011-C01-165873. We thank Acarus of the Universidad de Sonora, México, for the use of their computing facilities.

References

Agertz O et al. (2007) Fundamental differences between SPH and grid methods. MNRAS 380:963

Cha SH, Inutsuka SI, Nayakshin S (2010) Kelvin–Helmholtz instabilities with Godunov smoothed particle hydrodynamics. MNRAS 403:1165

Dehnen W, Aly H (2012) Improving convergence in smoothed particle hydrodynamics simulations without pairing instability. MNRAS 425:1068

Hubber DA, Falle SAEG, Goodwin SP (2013) Convergence of AMR and SPH simulations—I. Hydrodynamical resolution and convergence tests. MNRAS 432:711

McNally CP, Lyra W, Passy J-C (2012) A well-posed Kelvin–Helmholtz instability test and comparison. ApJSS 201:18

Murray SD, White SDM, Blondin JM, Lin DNC (1993) Dynamical instabilities in two-phase media and the minimum masses of stellar systems. ApJ 407:588

Price DJ (2008) Modelling discontinuities and Kelvin–Helmholts instabilities in SPH. J Comput Phys 227:10040

Read JI, Hayfield T, Agertz O (2010) Resolving mixing in smoothed particle hydrodynamics. MNRAS 405:1513

Read JI, Hayfield T (2012) SPHS: smoothed particle hydrodynamics with a higher order dissipation switch. MNRAS 422:3037

Robertson BE, Kravtsov AV, Gnedin NY, Abel T, Rudd DH (2010) Computational Eulerian hydrodynamics and Galilean invariance. MNRAS 401:2463

Springel V (2005) The cosmological simulation code GADGET-2. MNRAS 364:1105

Vietri M, Ferrara A, Miniati F (1997) The survival of interstellar clouds against Kelvin–Helmholtz instabilities. ApJ 483:262

Part IV
Vortex, Oceanography and Meteorology

Effect of the Inlet Flow Angle on the Vortex Induced Vibration of a Collinear Array of Flexible Cylinders

F. Oviedo-Tolentino, R. Romero-Méndez, F. G. Pérez-Gutiérrez, G. Gutiérrez-Urueta and H. Méndez-Azúa

Abstract In this work an experimental study of vortex-induced vibration (VIV) was carried out in an collinear array of ten identical cylinders. This investigation was conducted with a mass ratio ($m^*\xi = 0.13$) and a blockage ratio ($W/D < 1\%$). The inlet flow angle was fixed to 30° and the leading cylinder vibration amplitude was compared under the condition of 0° inlet flow angle. The free-end cylinders had two degrees of freedom with identical in-line and cross-flow natural frequencies in still fluid medium. The experimental essays were performed in a water tunnel in the lock-in region ($90 < Re < 450$). The results show that the cross-flow vibration amplitude is 68 % reduced when the inlet flow angle increases to 30°.

1 Introduction

In some industrial applications, such as marine risers, tall buildings, large suspension bridges, high voltage lines, tube and shell heat exchangers, smokestacks, etc., vortex induced vibration may occur. The particular importance of VIV on circular cylinders is partly due to industrial problems logged on this subject. The practical importance of VIV has led a number of fundamental studies, most of them on elastic cylinders with one degree of freedom. The physics involved in the VIV phenomenon is extensive, and some discussions on this topic are in the reviews of (Bearman 1984 and Gabbai and Benaroya 2005).

The fundamental studies on VIV involved the study of the mass-damping ratio, the cylinder natural frequency, vortex shedding modes and added mass on the lock-in region. The lock-in region is frequently represented by three distinguishable

F. Oviedo-Tolentino (✉) · R. Romero-Méndez · F. G. Pérez-Gutiérrez · G. Gutiérrez-Urueta · H. Méndez-Azúa
Facultad de Ingeniería, Universidad Autónoma de San Luis Potosí, Zona Universitaria Poniente, Av. Manuel Nava 8, 78290 S.L.P., San Luis Potosí, México
e-mail: francisco.oviedo@uaslp.mx

J. Klapp and A. Medina (eds.), *Experimental and Computational Fluid Mechanics*, Environmental Science and Engineering, DOI: 10.1007/978-3-319-00116-6_25,

responses: initial, upper and lower branches. High mass-damping cases are associated to lock-in regions with only two response branches, initial and lower. However, for low mass-damping cases the lock-in region consists of three branches: initial, upper and lower. This type of response is frequently referred as the Feng-type response, Feng (1968). On the other hand (Khalak and Williamson 1996, 1997a,b, 1999) conducted experimental studies for a very low mass damping ratio ($m^*\zeta = 0.013$). They concluded that the peak vibration amplitude is principally governed by the mass-damping ratio, whereas the mass ratio controls the lock-in range. Vikestad et al. (2000) found a mass ratio dependence on the Reynolds number. This dependence modifies the natural frequency of the cylinder in the lock-in region. The lock-in region has shown hysteretic behavior in the initial and lower branches. However for blockages of 1 % or less, the hysteretic behavior is completely eliminated in the initial branch.

There is evidence of the destructive nature of the vortex-induced vibration in cylinder arrays. Paidoussis (1980) cited various industrial problems registered in heat exchangers and nuclear reactors due to the VIV phenomenon. A good starting point in the study of cylinder arrays is a tandem configuration. The results on this configuration have shown a dependence of the vibration amplitude on the distance between cylinders. The important parameter in these studies was the ratio between the cylinder separation P to cylinder diameter D, P/D. Papaioannou et al. (2008) showed that for $P/D < 3.5$, the vibration amplitude is very similar to that of an isolated cylinder. However for $P/D > 3.5$ the vibration amplitude is significantly increased. This result confirmed the hypothesis made by Tanaka and Takahara (1981), who suggested that cylinder vibration is induced by the forces originated due to the fluid dynamics around both the neighbor cylinders and the cylinder itself. For the VIV in cylinder arrays, various excitation mechanisms are involved. Weaver and Fitzpatrick (1988) discussed the excitation mechanisms in typical tube array patterns in tube and shell heat exchangers. They pointed out the destructive nature of the fluid-elastic instability excitation mechanism in a heat exchanger. Recently, Zhao and Cheng (2012) investigated the effect of the inlet flow angle on the lock-in range in a square cylinder array configuration. They showed that as the flow inlet angle increases up to 30°, the lock-in range is also increased; after this flow inlet angle the lock-in range is reduced.

This investigation presents experimental results of VIV of ten identical flexible cylinders positioned in a collinear array at 30° angle with respect to the inlet flow direction. The experiments were conducted at a low mass-damping ratio $m^*\zeta = 0.13$ and blockage ratio $W/D < 1\%$. The natural cylinder frequency as well as the cross-flow and in-line vibrational response were determined in the lock-in region. The vibrational response of the leading cylinder in the array is compared with the case when the array is aligned with the free stream direction previously studied in Oviedo-Tolentino et al. (2013). The results show that the cross-flow vibrational response is reduced by 68 % at 30° inlet flow angle. This behavior is manly due to the low synchronization between cylinders in the lock-in region. On the other hand the lock-in range is slightly larger in the 30° inlet flow case.

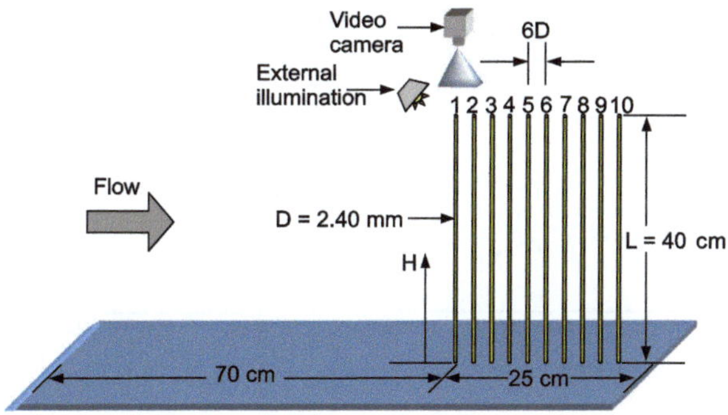

Fig. 1 Experimental model at 0° inlet flow angle

2 Experimental Procedure

2.1 Experimental Setup

Figure 1 shows a sketch of the experimental model. The model consisted of a 9 mm thick, 95 cm long and 37.5 cm wide acrylic flat plate. The ten identical circular cylinders were inserted into a drilled hole in a collinear arrangement. The cylinder array can be rotated with respect to the inlet flow angle. The cylinders were 2.40 mm in diameter and 40 cm in height with an elastic modulus of 10.5×10^{10} Pa. The experiments were carried out in a water tunnel with a test section 38.1 cm wide, 50.8 cm high and 1.5 m long. The water velocity can be varied from 0.01 to 0.3 m/s. A flow conditioning system maintains the turbulence levels at less than 1 % RMS at the inlet test section.

The vibration amplitude was obtained using a fast recording camera. This camera has an internal memory (2 GB) in which images can be stored digitally at 506 frames per second with full resolution, 1280×1024 px. For lower resolutions the frame rate can be increased up to 112,000 frames per second. The shutter time can be adjusted in the interval of 2 μs to 1 s. With these features, the free-end cylinder displacements could be measured up to an accuracy of the order of 11 μm.

2.2 Experimental Methodology

With the experimental setup described above, the cylinder free-end was recorded during 20 s at a rate of 240 frames per second using 100 μs as exposure time. Under this recording conditions, an external illumination was needed. The dynamic vibration amplitude was obtained using a particle tracking velocimetry technique on the recorded fotograme. Therefore, plots of the cylinder free-end position as a function of time were obtained. The cylinder frequency was determined through spectral analysis of the cylinder free-end time trace.

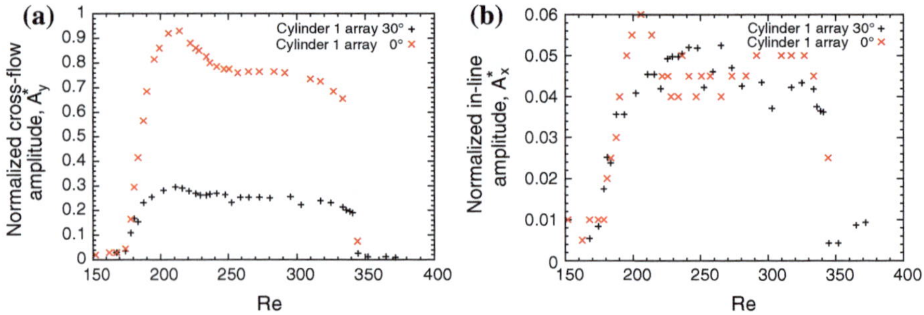

Fig. 2 Inlet flow angle effect on the vibrational response, **a** Cross-flow lock-in and **b** In-line lock-in

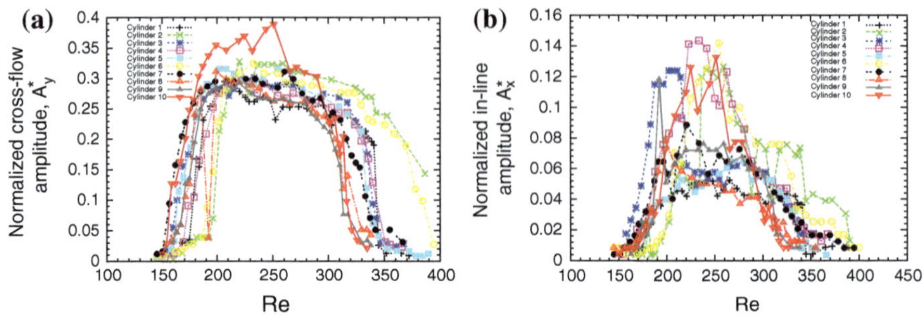

Fig. 3 Vibrational response in the array, **a** Cross-flow and **b** In-line

3 Results

Figure 2 shows the maximum cross-flow and the in-line vibrational responses for cylinder 1 as a function of the Reynolds number. In the cross-flow vibrational response (Fig. 3a), three branches are identified. This behavior confirms the mass-damping and the vibration amplitude branches relationship made by Feng (1968) who suggested that for low mass-damping ratio three response branches are associated. (Khalak and Williamson 1996, 1997a,b, 1999) related the peak amplitude with the mass-damping ratio in an isolated cylinder. The results in the collinear array reveal that the peak response is highly related to the inlet flow angle. On the other hand, the mass ratio has been associated with the lock-in range. The results in the collinear array show almost no influence of the inlet flow angle on the lock-in range.

The lock-in region for the ten cylinders in the collinear array is shown in Fig. 3. The peak vibrational response of cylinders 1 to 9 is, in general, of the same order of magnitude. However, cylinder 10 registers the maximum peak cross-flow vibrational response. This results suggest that the inlet flow angle prevents synchronization between cylinders. Figure 4 shows the non-dimensional frequency for the ten cylinders in the array. In order to show the preferred frequency at which the cylinders in

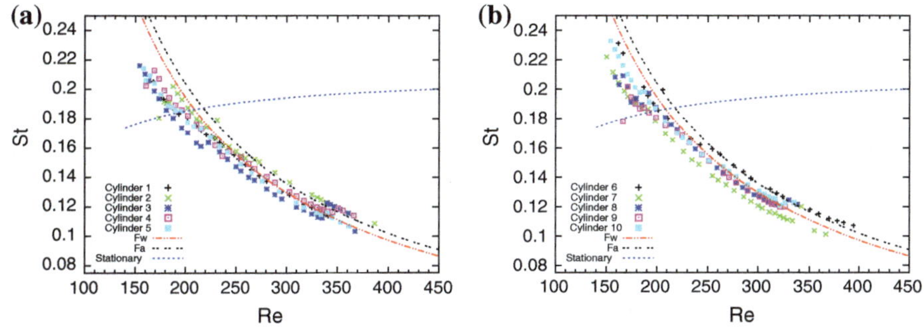

Fig. 4 Dimensionless frequency, **a** Cylinders from 1 to 5 **b** Cylinders from 6 to 10. The frequency was normalized using the Strouhal number

the array vibrate in the lock-in region, the cylinder frequency in still water (F_w), still air (F_a) and the vortex shedding frequency of an stationary cylinder were included in Fig. 4 as a reference. The non-dimensional cylinder frequency shows no synchronization between cylinders in the lock-in range. The low cylinder synchronization is partly due to the inlet flow angle. These experiments suggest that the inlet flow angle should be taken into account in the study of the synchronization between cylinders. The cylinder frequency in the lock-in region has, as a maximum value, the natural frequency in still air. In accordance with the results of (Vikestad et al. 2000), the added mass has a Reynolds numbers dependence in the lock-in region. The added mass changes leaded to natural cylinder frequency changes. The frequency and the added mass changes can be observed in Fig. 4. From this observation, it can be concluded that the results show low and high added mass dependence on the Reynolds number in the lock-in region.

4 Conclusions

An experimental study of vortex induced vibration in a collinear array was conducted. The inlet flow angle shows high influence on the peak cross-flow vibration. The cross-flow vibration amplitude is reduced by 68 % in the case of a 30° inlet flow angle. The reduction in the cross-flow vibration amplitude is due to a low cylinder synchronization in the lock-in. On the other hand, the inlet flow angle has almost no influence on the lock-in range. The cylinders response shows agreement with previous results, where three response branches were identified for the study mass-ratio case. The non-dimensional frequency results show low and high added mass dependence on the Reynolds number in the lock-in region.

Acknowledgments The authors thank CONACyT (Mexico) for support from project CB-2007/84618, and PROMEP (México) for a grant from the Extraordinary Funds of PROMEP- UASLP-12-CA04 and PROMEP/103.5/12/7964.

References

Bearman PW (1984) Vortex shedding from oscillating bluff bodies. Annu Rev Fluid Mech 16:195–222

Feng CC (1968) The measurements of vortex-induced effects in flow past stationary and oscillating circular and D-section cylinders. MS thesis. University of British Columbia, Vancouver

Gabbai RD, Benaroya H (2005) An overview of modeling and experiments of vortex induced vibration of circular cylinders. J Sound Vib 282:575–616

Khalak A (1996) Dynamics of a hydroelastic cylinder with very low mass and damping. J Fluids Struct 10:455–472

Khalak A, Williamson CHK (1997a) Fluid forces and dynamics of a hydroelastic structure with very low mass and damping. J Fluids Struct 11:973–982

Khalak A, Williamson CHK (1997b) Investigation of relative effects of mass and damping in vortex-induced vibration of a circular cylinder. J Wind Eng Indus Aerodyn 69–71:341–350

Khalak A, Williamson CHK (1999) Motions, forces and mode transitions in vortex-induced vibrations at low mass-damping. J Fluids Struct 13:813–851

Oviedo-Tolentino F, Romero-Méndez R, Hernández-Guerrero A, Pérez-Gutiérrez FG (2013) Vortex-induced vibration of a collinear array of bottom fixed flexible cylinders. J Fluids Struct 39:1–14

Paidoussis MP (1980) Flow induced vibration in nuclear reactors and heat exchangers. J Fluids Struct 1–80

Papaioannou GV, Yue DKP, Triantafyllou MS, Karniadakis GE (2008) On the effect of spacing on the vortex-induced vibrations of two tandem cylinders. J Fluids Struct 24:833–854

Vikestad K, Vandiver JK, Larsen CM (2000) Added mass and oscillation frequency for a circular cylinder subjected to vortex-induced vibrations and external disturbance. J Fluids Struct 14:1071–1088

Tanaka H, Takahara S (1981) Fluid elastic vibration of tube array in cross flow. J Sound Vib 77–1:19–37

Weaver DS, Fitzpatrick JA (1988) A review of cross-flow induced vibrations in heat exchanger tube arrays. J Fluids Struct 2:73–93

Zhao M, Cheng L (2012) Numerical simulation of vortex-induced vibration of four circular cylinders in a square configuration. J Fluids Struct 31:125–140

Forced Dynamics by Wind Normal to the Boundary

Alma Delia Ortíz Bañuelos and Federico Ángel Velázquez Muñoz

Abstract The ocean response is studied through the implementation of a numerical (POM, Princeton Ocean Model) with a constant Coriolis force, continuous vertical stratification, flat bottom and all boundaries closed. We considered four types of wind to force the model: Normal to the coastline; with Inertial path; in Fan shape and Realistic (sum of Inertial and Fan). The results of the numerical simulations show in each case the emergence of two counter rotating eddies in both sides of maximum wind and a cooling of sea superficial temperature. In the first case, the inertial wind adds asymmetry to the response, causing significant differences in the size and intensity of eddies. With the wind from a Fan-shape, the response remains similar to case one, but with more intense eddies and greater cooling. In the latter case with Realistic wind field, evident asymmetry is shown; it is used to explain in general, the dynamics produced by these winds.

1 Introduction

For a long time the effect of wind stress on the circulation in coastal regions has been studied. It produces upwelling and increased primary biological production and is responsible for the formation of anticyclonic and cyclonic eddies at the ocean surface offshore. This phenomenon has been studied in different parts of the world, such as in the Gulf of Tehuantepec, the Gulf of Papagayo, northwestern Baja California, the Canary Islands and Cape Verde, to name a few places where winds blow from land to sea controlling coastal dynamics significantly, (Lavín et al. 1992).

Observations with satellite imagery and in-situ measurements have shown that the ocean response to these wind jets is not the same on both sides of the wind, taking

A. D. O. Bañuelos (✉) · F. Á. V. Muñoz
Department of Physics, University of Guadalajara, Av. Juárez 976, Col. Centro, Guadalajara, Jalisco 44100 , México
e-mail: fisicalmaortiz@gmail.com

J. Klapp and A. Medina (eds.), *Experimental and Computational Fluid Mechanics*, Environmental Science and Engineering, DOI: 10.1007/978-3-319-00116-6_26,

a major component of asymmetry that can be noticed in the size and intensity of the eddies (Trasviña et al. 1995; Clark 1988; Lavín et al. 1992; Barton et al. 1993; Willett et al. 2006; Velázquez Muñoz et al. 2011). McCreary et al. (1989) mention that asymmetry between the eddies is due to the accumulation of cold water on the side of the cyclonic eddy caused by the difference in current directions (south) and countercurrent (opposite direction). Furthermore, Clark (1988) mentions that anticyclonic eddies can be strengthened by the introduction of negative vorticity by the inertial path, but only shows the antisymmetry in the displacement of the thermocline and symmetry in the Southern current of the surface layer of the ocean.

This chapter aims to provide new information and an explanation of the physical parameters involved in the asymmetric response of the ocean by the numerical model POM (Blumberg 1987) which is three-dimensional and solves the primitive equations of the ocean obtained from Navier-Stokes.

2 Methodology

2.1 Configuration and Implementation of the Numerical Model

The ocean is considered a flat-bottomed rectangular basin. The computational grid has a horizontal resolution of 5 km, with walls like closed boundaries, where there is free sliding, i.e. the sidewalls are simulating the earth. The initial condition is at rest, with temperature and salinity profiles varying only with depth. It is also assumed that the Coriolis parameter is constant.

This numerical model POM version 2k, is forced by four different types of wind, in the land-sea direction and a time period of 6 days. The first (Fig. 1a) is wind Normal to the boundary, similar to McCreary et al. (1989). In the second case (Fig. 1b); the wind opens in a Fan shape caused by an atmospheric pressure gradient. In the third case (Fig. 1c), the normal wind changes direction following an inertial path (curves to the right). Finally, the fourth type of wind (Fig. 1d), is generated by adding the two characteristics considered in the previous cases: the Fan shape caused by the atmospheric pressure gradient (case 2) and the inertial path curvature due to the Coriolis force (case 3), producing a wind with the same characteristics as shown by Steenburgh and Schultz (1998). Therefore, in this study the fourth type of wind was named Realistic wind. For the analysis of the ocean's response to wind stress forcing (Normal, Inertial, Fan shape and Realistic), we used the numerical model output data: surface temperature (Ts), current velocity components integrated in the vertical; (UA, VA) and the sea level (η) in 2D layer and along a zonal section at 150 km parallel to the Northern shore. The aim is to identify which of these physical parameters respond asymmetrically, and to investigate which kind of wind contributes to a greater asymmetry of the ocean's response.

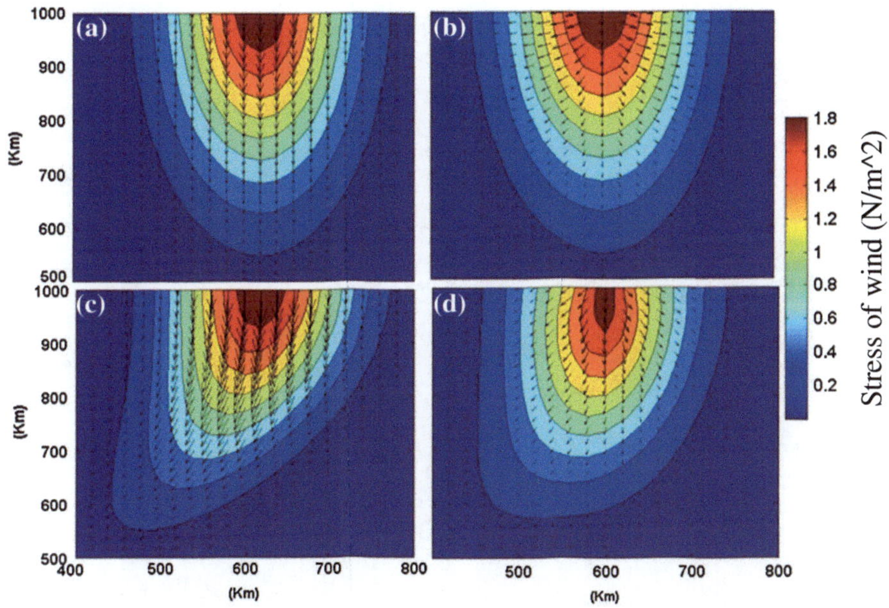

Fig. 1 Wind stress fields for **a** normal to the boundary, **b** fan shape, **c** inertial path and **d** realistic. Maximum magnitude for each case is 1.8 N m^{-2}, (domain not complete)

3 Results

3.1 Ocean's Response

Figures 2 and 3 show the response of the ocean's surface to the wind stress on the third and sixth day of simulation, respectively.

In every case there is an asymmetrical surface response when the wind reaches its maximum intensity on the third day of simulation (Fig. 2). This can be observed in a cooling along the northern boundary and in the emergence of two eddies: one anticyclonic to the left and one cyclonic to the right. Both eddies are markedly different in size and shape. It also shows that the sea level rises in the anticyclonic and falls into the cyclonic. Similarly, on day 6, Fig. 3 shows that the characteristics of the ocean response are maintained, but with increasing in intensity. In the normal wind case (Fig. 2a) there is cooling near to the coast, where the cold water mass is displaced by the water current in the same direction as the wind jet, i.e. offshore.

Figure 3a, shows the sea level for the case normal wind with an antisymmetric response because it has a positive displacement (anticyclonic region) and a negative displacement (cyclonic region) with the same magnitude, on day 6 (Fig. 3a) less cooling is observed, where cold water is displaced in the wind jet direction and slightly stretched by the anticyclonic eddy.

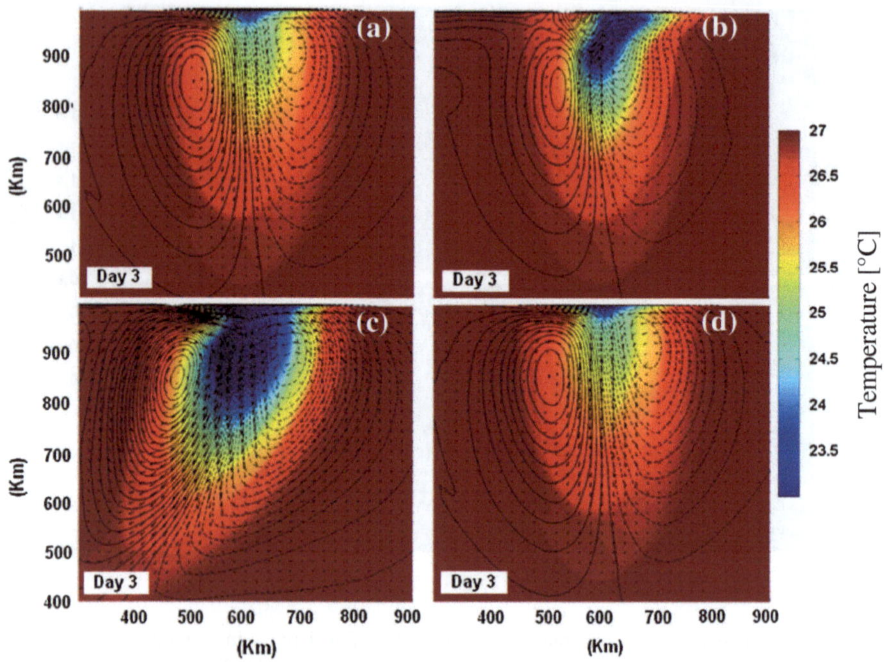

Fig. 2 Ocean's response to the wind stress **a** normal to the coast, **b** fan, **c** inertial and **d** realistic. The *color bar* indicates the temperature in °C, the contours correspond to the sea level (*solid line* represents elevation and *dotted line*, sinking) and the *black arrows* correspond to vertically integrated current. (The complete domain is not shown)

Figure 2b shows the ocean response to a fan shaped wind stress. One can observe the greater mass of cold water near the shore, in contrast with the case of normal wind. The sea level has a higher displacement in the cyclonic region than in the anticyclonic eddy.

For inertial wind (Figs. 2c, 3c), the ocean's response shows a greater difference from all the other cases with respect to cooling and to the sea level displacements produced (higher elevation than lowering). It also highlights the difference in the shape one diameter of the eddies. The anticyclonic eddy is more intense than the cyclonic, being the cyclonic eddy very weak and not well defined. In the case of realistic wind (Figs. 2d, 3d) an asymmetrical response is shown, very similar to that obtained in the cases of normal wind and fan shaped with respect to surface temperature, eddy size, sea level displacement and vertically integrated current, where there is a slight weakening of the cyclonic vortex.

The asymmetric response to the realistic wind can be attributed to its inertial component. This result is consistent with the mechanism proposed by Clark (1988), who remarks that anticyclonic vortex strengthening, is caused by the inertial trajectory of the wind. This allows us to suppose that either the inertiality or the fan shape

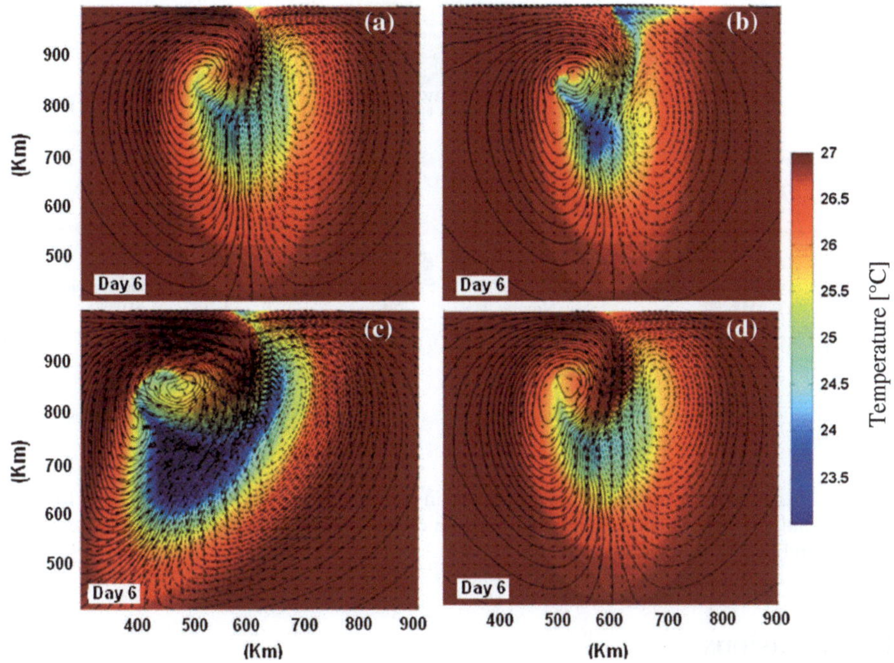

Fig. 3 Ocean response before wind stress **a** normal to the coast, **b** fan, **c** inertial **d** realistic. The *color bar* shows the temperature, the *outlines* represent sea level and the *black arrows* the vertically integrated current. (the domain of the complete model is not shown)

in the realistic wind could be important for the ocean's response. In this case, the realistic wind was built with equal contribution of these two features, but apparently, the shape of the realistic wind is affected more by contribution of the Coriolis force than the atmospheric pressure gradient.

Finally, with respect to the meridional southern current integrated in the vertical at 150 km of the coast, it is shown that the wind stress generates a current by dragging the superficial water mass offshore (Fig. 4). This current produces a vacuum near the north boundary, which is occupied by the masses of water from both sides of the wind jet, producing two opposite currents (countercurrent anticyclonic and cyclonic). It also cooperates with the elevation of the subsurface cold water mass, generating cooling near the coast. In all four cases studied, the current's velocity is higher than the velocity of the countercurrent, and the anticyclonic side countercurrent is greater than the current on the cyclonic side. The inertial wind contributes more to the southern velocity (the vertical), than the wind in all others cases.

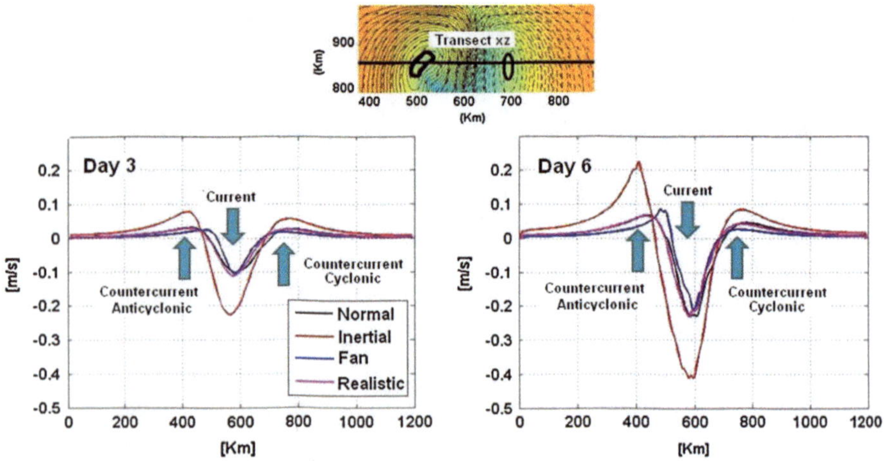

Fig. 4 Transect of the southern current integrated in the vertical (va) at 150 km to the coast for each kind of wind, (the image above shows the location of the transect, which crosses the center of the cyclonic and anticyclonic vortex)

4 Conclusions

This chapter provides an analysis of the ocean's response for four different kinds of offshore normal wind (1. normal, 2. inertial, 3. fan shape and 4. realistic), using a three-dimensional numerical model.

Model results show, in all cases, an asymmetrical ocean response with respect to the sides and shape of emergence eddies, with cooling in the sea surface, where the mass of cold water is stretched with greater intensity by the anticyclonic eddy. It is also noted that there is higher elevation than sinking at sea level. These results show that the ocean's response to normal wind (case 1) is similar to those shown by McCreary et al. (1989).

The inertial wind in contrast with the cases of normal wind, fan and realistic, provides greater asymmetry: higher cooling, strong meridional velocity and displacements of the sea level. These physical parameters show a greater change in the region of anticyclonic eddy than the cyclonic one.

References

Barton ED, Argote ML, Brown JP, Kosro M, Lavín MF, Robles JM, Smith RL, Trasviñia A, Vélez HS (1993) Supersquirt: dynamics of the of Tehuantepec Mexico Gulf. Oceanogr 6:23–30

Blumberg AF (1987) A description of a three-dimensional coastal ocean circulation model. American Geophysical Union, Washington DC

Clark AJ (1988) Inertial wind path and sea surface temperature patterns near the Gulf of Tehuantepec and Gulf of Papagayo. J Geophys Res 93:491–501

McCreary JP, Hyong SL, Enfield DB (1989) The response of the coastal ocean to strong offshore winds: with application to circulations in the Gulf of Tehuantepec. J Mar Res 47:81–109

Lavín MF, Robles ML, Argote ED, Barton R, Smith J, Brown M, Kosro A, Trasviña A, Vélez HS, García J (1992) Física del Golfo de Tehuantepec. Ciencia y Desarrollo. CONACyT. XVIII:Núm 103

Steenburgh WJ, Schultz DM (1998) The structure and evolution of gap outflow over the Gulf of Tehuantepec Mexico. Am Meteorol Soc 126:2673–2691

Trasviña A, Barton ED, Brown J, Vélez HS (1995) Offshore wind forcing in the Gulf of Tehuantepec, Mexico: the asymmetric circulation. J Geophys Res 100:20649–20663

Velázquez Muñoz FV, Martínez JA, Chavanne C, Durazo R, Flament P (2011) Wind-driven coastal circulation in the Gulf of Tehuantepec México. Ciencias Marinas 37:443–456

Willett, Leben RR, Lavín MF (2006) Eddies and tropical instability waves in the eastern tropical Pacific. Prog Oceanogr 69:218–238

Focusing of Surface Waves

Gerardo Ruiz-Chavarria, Michael Le Bars and Patrice Le Gal

Abstract In this chapter we present some original experimental results of the process of focusing of surface waves in a fluid. To this end, monochromatic waves of frequencies in the range 5–15 Hz are produced in a water layer of 10 cm depth using a parabolic wave maker. Experiments remain in the deep water approximation and both gravity and surface tension influence the evolution of waves. We find that, as in optics, the wave field exhibits phenomena such as diffraction, interference and the presence of two caustics intersecting at one point and forming a cusp. To investigate the properties of surface waves, the synthetic Schlieren method is used. Nonlinear behavior emerges during the process of focusing even for small amplitude waves. For example the peak amplitudes are more pronounced that the amplitude of the troughs. Some non expected results emerge from our experiments. The first is that the position of the maximum amplitude of the wave is dependent on the amplitude of the initial parabolic wave front, but in any case, is always in the vicinity of the origin of Huygens' cusp. Second, the predictions for linear waves are only in partial agreement with our experimental data. And finally, due to the fact that the ratio of the size of wave maker to the wavelength does not tend to infinity some finite size effects are observed.

G. Ruiz-Chavarria (✉)
Facultad de Ciencias, Universidad Nacional Autónoma de México,
Ciudad Universitaria,
04510 Mexico, México D.F., Mexico
e-mail: gruiz@unam.mx

M. L. Bars · P. L. Gal
Institut de Recherche sur les Phénomènes Hors Equilibre, UMR 7342,
CNRS - Aix-Marseille Université 49 rue F. Joliot Curie,
13384 Marseille, Cédex 13, France
e-mail: lebars@irphe.univ-mrs.fr

P. L. Gal
e-mail: legal@irphe.univ-mrs.fr

J. Klapp and A. Medina (eds.), *Experimental and Computational Fluid Mechanics*,
Environmental Science and Engineering, DOI: 10.1007/978-3-319-00116-6_27,

1 Introduction

In the theory of waves, the caustics are surfaces that separate illuminated regions from shaded regions. Geometrical optics establishes that in the illuminated region several rays reach each point, while in the shaded region no rays are present. The same theory predicts that the amplitude of the wave goes to infinity along the caustic. In reality this behavior does not hold, because the ray theory is only an approximation coming from geometrical optics forgetting about the wave properties. However, the behavior of a wave in the vicinity of a caustic has been investigated in the past (Paris and Kaminski 2001; Lewis et al. 1967; Stamnes and Spjelkavik 1983) for two rays superimposition. In this case, the wave field can be described in terms of the Airy function Ai(x). This function has an oscillating behavior for $x < 0$, while for $x > 0$ the amplitude has an exponential smooth behavior. In fact, the problem we present in this manuscript cannot be described only in terms of Airy function because inside the cusp the wave field is the result of the superposition of three rays. The method proposed by Pearcey (1946) must be used in this case instead of the classical theory. In surface waves, a field with a Huygens' cusp can be produced with a parabolic wave maker (Pomeau (1991), see Fig. 1). The initial wavefront is described by the equation

$$y_0 = ax_0^2 \tag{1}$$

A ray starting at the parabola of equation $y_0 = ax_0^2$ moves in a direction perpendicular to it, i. e. in a direction given by the normal vector $\hat{n}$ in the point: (x_0, y_0):

$$\hat{n} = \frac{(-2ax_0, 1)}{\sqrt{1 + 4a^2x_0^2}} \tag{2}$$

As the wave travels, its amplitude grows by focusing according to the following relation:

$$A = A_0\sqrt{\frac{\rho}{\rho - d}} \tag{3}$$

where A_0 is the initial amplitude, d is the distance traveled by the ray and $\rho = \frac{1}{K}$ is the inverse of the curvature of the parabola at the starting point, which in actual case is:

$$K = \frac{y_0''}{(1 + y_0'^2)^{3/2}} = \frac{2a}{(1 + 4a^2x_0^2)^{3/2}} \tag{4}$$

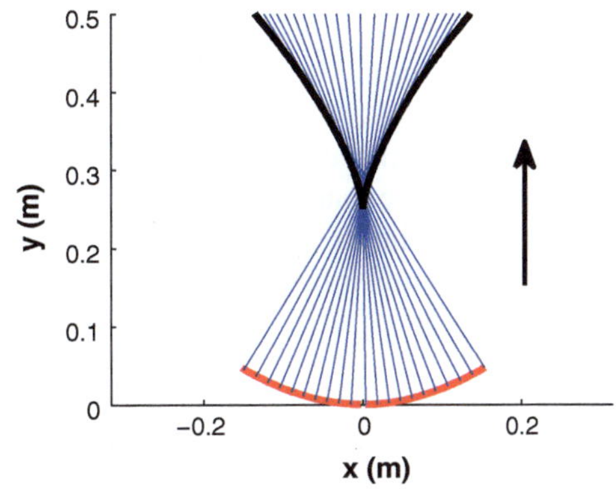

Fig. 1 Parabolic wave maker (*red line*). The equation of the curve is $y_0 = ax_0^2$. According to geometrical optics a ray starts at the parabola and moves in a direction normal to the curve (*blue lines*). In the figure, the Huygens' cusp is plotted as the *black line*. The *arrow* in the figure indicates the direction of propagation of waves

The space where the wave progresses is divided in two regions. In the first one, which is in contact with the parabolic wave maker, only one ray passes through each point. Far away a second region emerges; it is characterized by the fact that three rays reach each point. The curve separating both regions is the union of two caustics. One way to obtain the analytical expression of the caustics is to determine the limit of the region where only one real root exists for x_0 in the equation of a ray in parametric form $(x, y) = (x_0, y_0) + \hat{n}d$. An alternative method is to use the property that in the ray theory a caustic is the locus where the amplitude becomes infinite. This happens for $d = \rho$ (Eq. 3). That is:

$$(x, y) = (x_0, ax_0^2) + \hat{n}\rho = (-4a^2x_0^3, \frac{1}{2a} + 3ax_0^2) \tag{5}$$

The analytical expression for caustics can be obtained if we combine both components of this vector equation and eliminate x_0. This procedure leads to the following equation

$$x = \pm\frac{4\sqrt{a}}{3\sqrt{3}}(y - \frac{1}{2a})^{3/2} \tag{6}$$

The last equation has a cusp singularity at $(0, \frac{1}{2a})$ where both caustics intersect. Far from this point, the characteristics of the wave can be deduced using the method of the stationary phase. Near the cusp, the amplitude $h(x, y)$ of the surface wave can be calculated by the diffraction integral (Pearcey 1946; Paris and Kaminski 2001), which is an approximate solution of the wave equation:

$$h(x, y) = \int_{-\infty}^{+\infty} \frac{dx_0}{cos(\theta(x_0))} \frac{exp(ikd(x_0, x, y))}{\sqrt{d(x_0, x, y)}} \tag{7}$$

where $\theta(x_0)$ is the angle between the tangent to parabola at point (x_0, y_0) and the x axis. We assume that $cos(\theta(x_0)) \approx 1$. Interest is pointed out in the solution around the singular point $(0, \frac{1}{2a})$. For this purpose we made a Taylor expansion of $d(x_0, x, y)$ about this point. At first order we obtain:

$$d(x_0, x, y) \approx R + \frac{1}{2R}\left(a^2 x_0^4 + 2a\left(\frac{1}{2a} - y\right) x_0^2 - 2x x_0\right) \tag{8}$$

where $R = \frac{1}{2a}$. This expression is used only in the exponential term because $kd(x_0, x, y)$ is a rapidly varying variable and in the denominator we made the approximation that $d(x_0, x, y) \approx R$. The diffraction integral is then approximated by:

$$h(x, y) = \frac{exp(ikR)}{\sqrt{R}} \int_{-\infty}^{+\infty} exp\left(\frac{ik}{2R}\left[a^2 x_0^4 + 2a(\frac{1}{2a} - y) x_0^2 - 2x x_0\right]\right) dx_0 \tag{9}$$

Let us make the following change of variable $x_0 = \left(\frac{2R}{ka^2}\right)^{1/4} t$ and define two quantities $U = 2\left(\frac{k}{2R}\right)^{1/2}(\frac{1}{2a} - y)$ and $V = -\frac{2}{\sqrt{a}}\left(\frac{k}{2R}\right)^{3/4} x$.

Then we recover the Pearcey integral (Pearcey 1946; Berry 1992) for a linear wave in the vicinity of a cusp:

$$h(x, y) = \frac{k}{i2\pi} \frac{exp(ikR)}{\sqrt{R}} \left(\frac{2R}{ka^2}\right)^{1/4} \int_{-\infty}^{+\infty} exp\left(i\left[t^4 + Ut^2 + Vt\right]\right) dt \tag{10}$$

In the precedent equation the integral extends from $-\infty$ to $+\infty$. However, even if in a surface wave field λ is usually smaller than the length R , the ratio $\frac{R}{\lambda}$ does not go to infinity as in the case for light and the integral in Eq. 10 must be calculated for a finite interval. In Fig. 2 we show the envelope of the amplitude along the axis of symmetry x = 0 for a wave of frequency $\nu = 10$ Hz (black line). We assume that the surface is excited with a parabolic wave maker whose parameter a is 0.02 cm^{-1} and that extends in the interval $-15\,cm < x_0 < 15\,cm$. The curve is obtained from Eq. 10, but the integration is performed over a finite interval. For comparison we have also included the curve when the integration is made in an infinite domain $-\infty < t < \infty$ (blue line), the envelope calculated from a numerical solution of the wave equation (magenta line) (Ruiz-Chavarria et al. (2011)) and the asymptotic behavior resulting from the method of stationary phase (dashed red line). In all cases

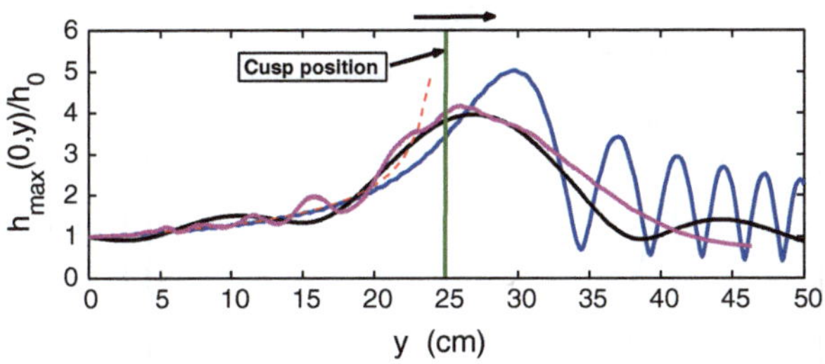

Fig. 2 Envelope of the wave amplitude along the axis of symmetry (*black line*) according to the diffraction integral calculated on a finite interval. h_0 is the amplitude at the wave maker. For comparison we also plot the curve when the integral extends from $-\infty$ to $+\infty$ (*blue line*), the curve obtained from the linear wave equation (*magenta line*) and the asymptotic behavior given by the ray theory (*red line*). In the last case, amplitude is proportional to $\sqrt{\frac{1}{(\frac{1}{2a}-y)}}$ and diverges at the cusp. The theoretical position of the cusp is indicated with the *vertical green line*

the initial amplitude A_0 is set to 1. As can be seen, the diffraction theory (both for finite and infinite integration domain) and the solution of the wave equation predict that the maximum amplitude happens further than the Huygens' cusp (on the right of the theoretical cusp position on the figure) . Moreover, the diffraction theory for a finite integration interval and the solution of the linear wave equation predict the same result for the position and the value of the maximum amplitude. On the other hand the Pearcey integral predicts the existence of oscillations of the envelope to the right of the cusp, whereas the solution of wave equation exhibits a monotonic decrease of the wave amplitude.

2 Experimental Setup

Experiments were carried out in a basin of size 120 cm × 50 cm × 15 cm made in plexiglass (see Fig. 3). The basin is filled with water up to a level of 10 cm. In order to produce the wave field the parabolic wave maker is connected to a mechanical vibrator. In all cases a monochromatic wave is produced, with a frequency between 5 and 15 Hz.

The shape of the free surface is determined with the method known as synthetic Schlieren (Moisy et al. 2009). This procedure is based on the fact that a change in the slope of the liquid-air interface causes a change in the direction of the light rays that cross this surface. Then, if a pattern of dots is placed at the bottom of the liquid layer, there is an apparent displacement of them when the free surface is deformed. The synthetic Schlieren method works well when the slope of the liquid-gas interface is small. In our case, the initial amplitude (at the edge of the parabolic wave maker) is of the order of tens of microns whereas the maximum amplitudes attained during focusing is about 150 μm. If we consider that the smallest wavelength measured in the

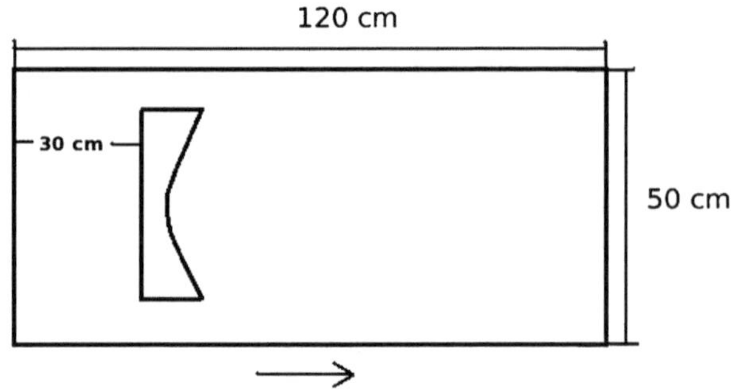

Fig. 3 The experiments of wave focusing were made in a plexiglass basin. The wave maker was colocated approximately to 30 cm away from the left border. Waves progresses from left to the right as indicated by the *arrow*. Frequencies in the mechanical vibrator moving the wave maker lies in the range 5–15 Hz. In most experiments camera is at a distance of 1 m from the free surface

experiments is 1.5 cm, the maximum value of the ratio of amplitude to wavelength is 0.15/15 = 0.01. Consequently the slopes remain small and the validity of the measurement method is guaranteed. Higher slopes could produce a failure of the method because the crossing of the light rays when travel from the bottom to the free surface.

The method uses a video camera to record the spatio-temporal evolution of the surface elevation. In order to have a good resolution we used a high definition camera, with an image size of 1,920 × 1,080 pixels. The area covered by a frame is 18.5 cm × 10.4 cm, so the conversion factor between pixels and length is 103.8 pixel / cm. To determine the wave features we printed in a paper sheet a pattern of dots randomly distributed. This sheet is placed at the bottom of the basin. A snapshot of the dot pattern is taken when the free surface is at rest (hereafter called the reference image). In a second step images of are taken when a surface wave passes. Apparent displacement is measured with a PIV software. As usually done in Particle Image Velocimetry the pictures are divided into a set of cells having a size of 32 × 32 pixels. The number of cells in each direction is 128. In order to reconstruct the form of the water-air interface we recall the relation between the apparent displacement δr and the gradient of the free surface h (Moisy et al. 2009):

$$\nabla h = -\frac{\delta \mathbf{r}}{h*} \tag{11}$$

where $\frac{1}{h*} = \frac{1}{\alpha h_p} - \frac{1}{H}$. H is the distance from the camera to the bottom of the fluid layer, h_p is the thickness of the fluid and $\alpha = 1 - n'/n$ (n' and n are the refraction indices of the gas and liquid respectively). The reconstruction of the topography of the free surface can be done by integration of Eq. 11. The system of equation is overdetermined and a least square routine is used to calculate h(x, y, t).

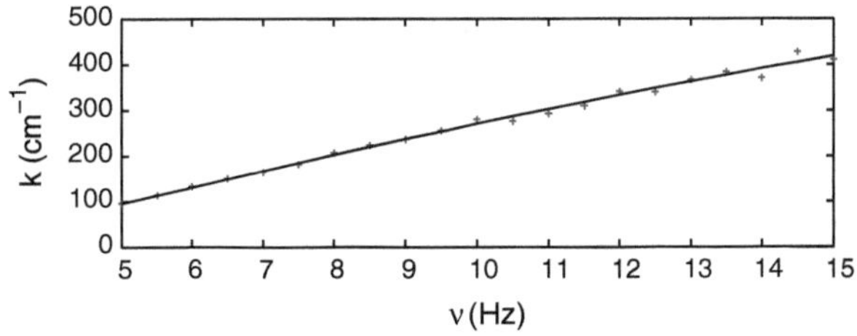

Fig. 4 Dispersion relation for a plane wave (wavenumber versus frequency). The determination of the wavelength was done with the method of periodograms. In the same figure the curve obtained from Eq. 12 has been also included. It can be observed a good agreement between the theory and experiments, which is a test of the validity of the synthetic Schlieren method

3 Results

The dispersion relation for a surface wave in a liquid is given by the following equation (Elmore and Heald 1969) :

$$\omega^2 = \left(gk + \frac{\sigma k^3}{\rho}\right) \tanh(\mathrm{kh}) \tag{12}$$

where $\omega = 2\pi\nu$ is the pulsation of waves, $k = \frac{2\pi}{\lambda}$ is the wavenumber, σ is the surface tension coefficient of the liquid and ρ is the density of the fluid. The first experiment we made was the measurement of λ for different frequencies. To this end, we have used a 30 cm long plane wave maker. In order to have a precise estimate of λ, we use the procedure based on periodograms. Figure 4 shows the graph of the wavenumber k versus frequency. For comparison we have included the prediction given by Eq. 12. The agreement is very good implying that the synthetic Schlieren method reproduces well the properties of waves in the system under study.

With regard to the process of focusing, measurements of the topography of the free surface were made. The liquid-gas interface was excited with the parabolic wave maker.. Measurements were carried out in a region between $15 < y < 45$ and $-8 < x < 8$. The diffraction theory predicts that the maximum amplitude is attained inside this region. Wave fronts-initially convergent- become divergent after passing the origin of Huygens' cusp. In Fig. 5 we present two graphs of the free surface shape versus (x, y) for a wave of frequency $f = 10\,\mathrm{Hz}$ corresponding to a wavelength $\lambda = 2.32\,\mathrm{cm}$. To the right of each figure there is a scale which gives the color to the value of the surface deformation h. As we can see in Fig. 5a (which covers the interval $15 < y < 25$) the focusing leads to an increase of the amplitude when wave moves from left to the right. In Fig. 5b the topography of the free surface versus the (x, y) coordinates, is plotted for the interval $25 < y < 35$. In the same figure, the Huygens' cusp is also plotted as a dashed line. The figure exhibits the change from a convergent (left side) to a divergent (right side) wave field. As expected, the

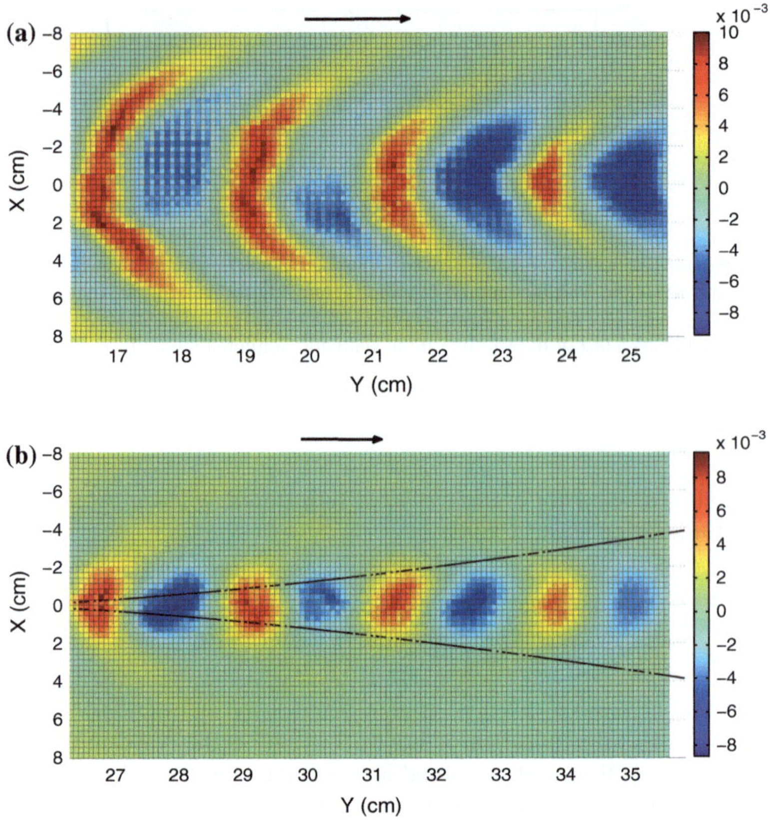

Fig. 5 Topography of free surface h in the plane (x, y) of a wave produced by a parabolic wave maker. Amplitude is proportional to the color intensity. *Red* stands for positive values and *blue* stands for negative values. The vibrator was driven at a frequency of 10 Hz, which corresponds to a wavelength $\lambda = 2.32\,\text{cm}$. **a** free surface in the range $15 < y < 25$. The focusing leads to an increase in amplitude of the wave. **b** free surface in the range $25 < y < 35$. The wave amplitude reaches a maximum and then there is a decrease in amplitude and an inversion in the wavefront is observed to the right

maximum amplitude is reached along the axis of symmetry after the wave traverses the cusp. In Fig. 6 we present a snapshot of the wave field for $35 < y < 45$. In this region, the wave is divergent and consequently the amplitude decreases as the wave travels. The amplitude of the wave does not exhibit oscillations in this region as predicted by Eq, 10, because away from the cusp interference does not happen.

A most appropriate way to exhibit the focusing is by means of the envelope of the wave along the axis of symmetry. The observed behavior is the combination of nonlinearities and a finite size effect. For a linear wave, the diffraction theory predicts that, when $\frac{R}{\lambda}$ approaches to infinity the maximum amplitude occurs at y = 28.6. Taking into account that $\frac{R}{\lambda}$ has a finite value, the maximum amplitude for a wave of frequency $\nu = 10\,\text{Hz}$ should occur at y = 25.5 according to the linear wave

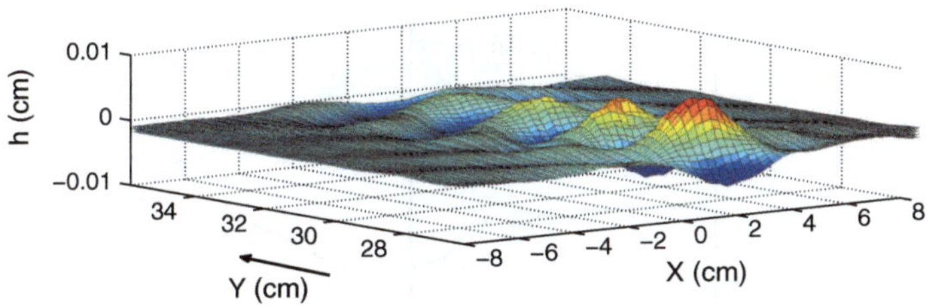

Fig. 6 Topography of the free surface h in the plane (x, y) of a wave produced by a parabolic wave maker ($a = 0.02\,cm^{-1}$). The vibrator was driven at a frequency of 10 Hz, which corresponds to a wavelength $\lambda = 2.32\,cm$. This figure clearly shows the decrease in amplitude as a function of y away from cusp. The amplitude becomes proportional to $\sqrt{\frac{1}{y-\frac{1}{2a}}}$ as predicted by the theory of linear waves

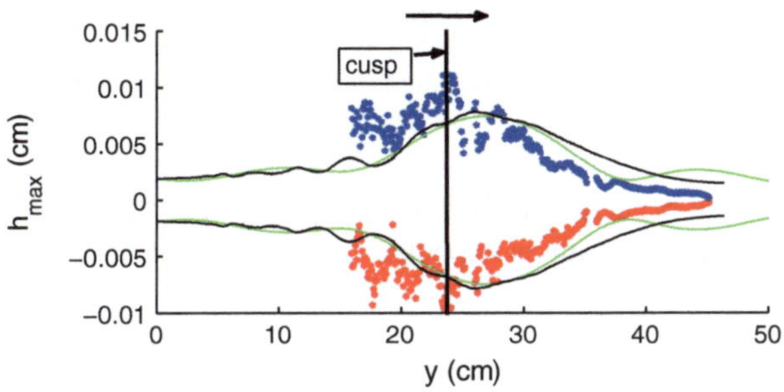

Fig. 7 Envelope of the wave along the axis of symmetry x = 0. The surface of the fluid is excited with a frequency of 10 Hz, which corresponds to a wavelength $\lambda = 2.32\,cm$. The positive and negative branches of the envelope have a small asymmetry. This is a signature of the appearance of nonlinearities. Otherwise, maximum amplitude is reached near the origin of cusp. For comparison, the prediction of diffraction theory is also included (*green line*) and the solution of the wave equation (*black line*). Amplitudes attained by the waves are greater than the prediction of the linear theory of waves

theory In experiments we have found that position of maximum is dependent of the initial amplitude A_0 of the wave front. Figure 7 shows the envelope of the wave on the axis of symmetry for an initial amplitude of $\approx 20\ \mu m$. In the same figure the curves of diffraction theory (green line) and the solution of wave equation (black line) are included. First, the positive and negative branches of the envelope have a small asymmetry and in this sense non linearities are weak. On the other hand the position of maximum is located to the left of black and red curves. Finally, the behavior of the envelope to the right of the figure shows a more pronounced decrease with respect to the predictions of the diffraction and the linear wave theories. For a greater value of the initial amplitude A_0 the experimental data show a shift of the position of the maximun to the right. This behavior can be observed in Fig. 8, in which the envelope is plotted for $A_0 \approx 25\ \mu m$. Under these circumstances the maximum amplitude is approximately that predicted by the linear theory. Concerning the behavior away from the cusp, experimental data and models show a similar trend. And finally the

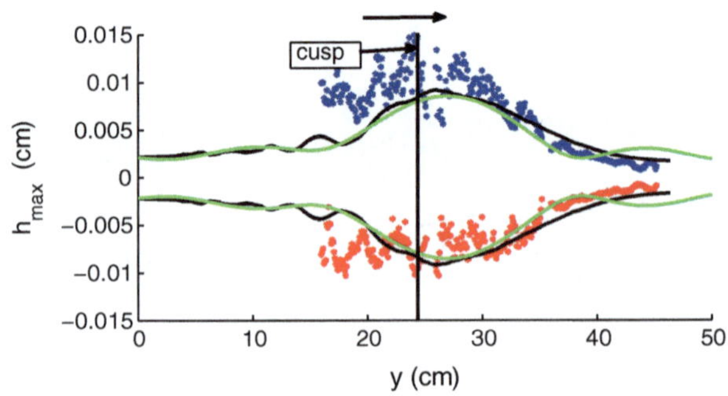

Fig. 8 Envelope of the wave along the axis of symmetry. The surface of the fluid is excited with a frequency of 10 Hz, which corresponds to a wavelength $\lambda = 2.32\,\text{cm}$. The position of maximum is shifted to the right with respect the previous figure, having a value close to the prediction by linear theory. Otherwise, positive and negative branches of the envelope become clearly asymmetric, which is a signature of the non linear effects. For comparison, the prediction of diffraction theory (*green line*) and the solution of the wave equation (*black line*) are also included

positive and negative branches of the envelope become clearly different, indicating that non linearities are relevant as expected for larger amplitude waves.

4 Conclusions

In this chapter we investigated the focusing of surface waves in water by the synthetic Schlieren method. For this purpose the liquid-gas interface was excited with a parabolic wave maker. Although the waves produced in experiments have small amplitude the non linear effects are important in the vicinity of the Huygens' cusp. In this respect, we have observed that the growth of peaks is greater than that predicted by the linear theory. In the same sense the positive and negative branches of the envelope become asymmetric as the initial amplitude grows. Another important result is that away from the Huygens cusp the wave field becomes divergent and non linearities stay weak in this region.

The experimental results presented here are the first step in the study of non linear waves near a cusped caustic, which is at present an open subject. Some new phenomena non present in linear waves will be investigated in the future. This is for instance the case of wave breaking that can be induced by the focussing process (Tejerina-Risso and Le Gal (2012)).

Acknowledgments G. Ruiz-Chavarria acknowledges DGAPA-UNAM by support for a sabbatical period at IRPHE between september 2010 and august 2011. Additionally the same author acknowledges support by DGAPA-UNAM under project 116312, "Vorticidad y ondas no lineales en fluidos"

References

Paris RB, Kaminski D (2001) Asymptotics and mellin-barnes integrals. Cambridge University Press. Cambridge, pp 442

Lewis RM, Bleistein N, Ludwig D (1967) Uniform asymptotic theory of creeping waves. Commun Pure Appl Math 20:295–328

Stamnes JJ, Spjelkavik B (1983) Evaluation of the field near a cusp of a caustic. Optica Acta: Int J Opt 30:1331–1358

Pearcey T (1946) The structure of an electromagnetic field in the neighbourhood of a cusp of a caustic. Phil Mag 37:311–317

Pomeau Y (1991) Caustics of nonlinear waves and related questions. Europhys Lett 11:713–718 and private communication

Berry M (1992) Rays, wavefronts and phase: a picture book of cusps in Huygens' Principle 1690–1990: theory and applications. Elsevier 299:97–111

Ruiz-Chavarria, G, Le Bars M, Le Gal P, Pomeau Y (2011) Focalisation d'ondes de surface et déferlement. In: Proceedings of the 20eme congres francais de mecanique. Besancon (2011) Art. p 552

Moisy F, Rabaud M, Salsac K (2009) A synthetic Schlieren method for the measurement of the topography of a liquid interface. Exp Fluids 46:1021–1036

Elmore WC, Heald MA (1969) Physics of waves. McGraw Hill, New York, p 477

Tejerina-Risso J, Le Gal P (2012) Around the cusp singularity and the breaking of waves. In: Proceedings of the conference "Water is in the Air", Marseille, 25–26 June 2012

Numerical Experiments of Wind Circulation off Baja California Coast

Torres-Navarrete Carlos, Larios-Castillo Sergio, Mejia-Trejo Adán, García-Toscano Jaime, Macias-Carballo Mariana and Gil-Silva Eduardo

Abstract In order to study the effect of orography on winds off Baja California, the momentum primitive equations describing an atmospheric flow over that region are solved numerically on a boundary-fitted grid. Numerical experiments are conducted for several flow conditions. Results show a remarkable agreement to available observations. Von Kármán vortex street shown in satellite images are also reproduced.

1 Introduction

Baja CaliforniaCoast (BCC) is characterized by the abundance of meso-scale eddies evidenced by cloud trails shown in several satellite images (True color composites images from MODIS-Aqua, http://earthobserva-tory.nasa.gov/) of the region. These features are related to the interaction of prevailing winds, local and seasonal, with the complex orography over they flow. In particular, strong cyclonic and anti-cyclonic eddies have been observed to emerge from major capes in Baja California such as Punta Baja (PB), Punta Eugenia (PE), Cabo San Lucas (CSL), Guadalupe Island (GI) and Cedros Island (CI) (Fig. 1). Winds over the coast of California and BCC have been modeled with ~10 km grid resolution by several mesoscale atmospheric models (e.g. Cerezo-Mota et al. 2006; Koracin and Dorman 2001; Koracin et al. 2004;

T.-N. Carlos (✉) · L.-C. Sergio · M.-C. Mariana
Centro Nacional de Datos Oceanograficos, Instituto de Investigaciones Oceanologicas/UABC, Km. 107 Carretera Tijuana-Ensenada, 22860 Ensenada, B. C. México
e-mail: ctorres@uabc.edu.mx

M.-T. Adán · G.-S. Eduardo
Depatamento de Oceanografia Fisica, Instituto de Investigaciones Oceanologicas/UABC, Km. 107 Carretera Tijuana-Ensenada, 22860 Ensenada, B. C. México

G.-T. Jaime
Facultad de Ingenieria, Instituto de Investigaciones Oceanologicas/UABC, Km. 107 Carretera Tijuana-Ensenada, 22860 Ensenada, B. C. México

J. Klapp and A. Medina (eds.), *Experimental and Computational Fluid Mechanics*, Environmental Science and Engineering, DOI: 10.1007/978-3-319-00116-6_28,

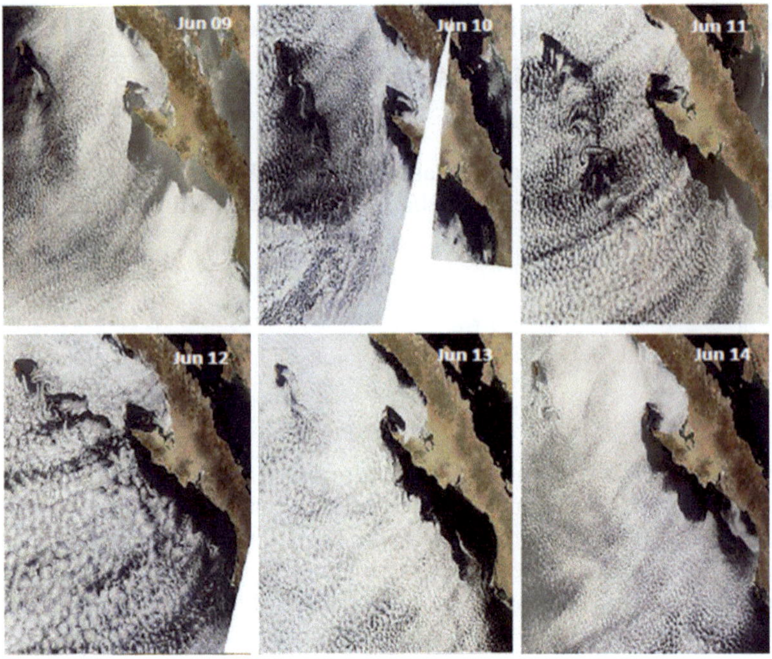

Fig. 1 Vortex shedding evidenced by trail clouds in Baja California coast (http://earthobservatory.nasa.gov/)

Pickett et al. 2003). However, the coarse spatial resolution of these simulations is not adequate to describe the locally driven processes evident in the Baja California region with detail.

In this study a high resolution numerical model, developed by first author, is used to study the effect of wind variability on wind circulation and eddy generation along the Baja California coast. Wind stress, wind stress curl and ocean surface currents are computed with the simulated winds. Coastal upwelling zones are also identified.

2 Study Area

The Baja California Coast, is at the Peninsula of Baja California in northwestern Mexico.The Peninsula extends 1,247 km (775 mi) from Mexicali, Baja California in the north to Cabo San Lucas, Baja California Sur in the south. It ranges for 40 km (25 mi) at its narrowest to 320 km (200 mi) at its widest point and has approximately 3,000 km (1,900 mi) of coastline and approximately 65 islands. The total area of the Baja California Peninsula is 143,390km2 (55,360 mi2). (http://en.wikipedia.org/wiki/Baja_California_peninsula).

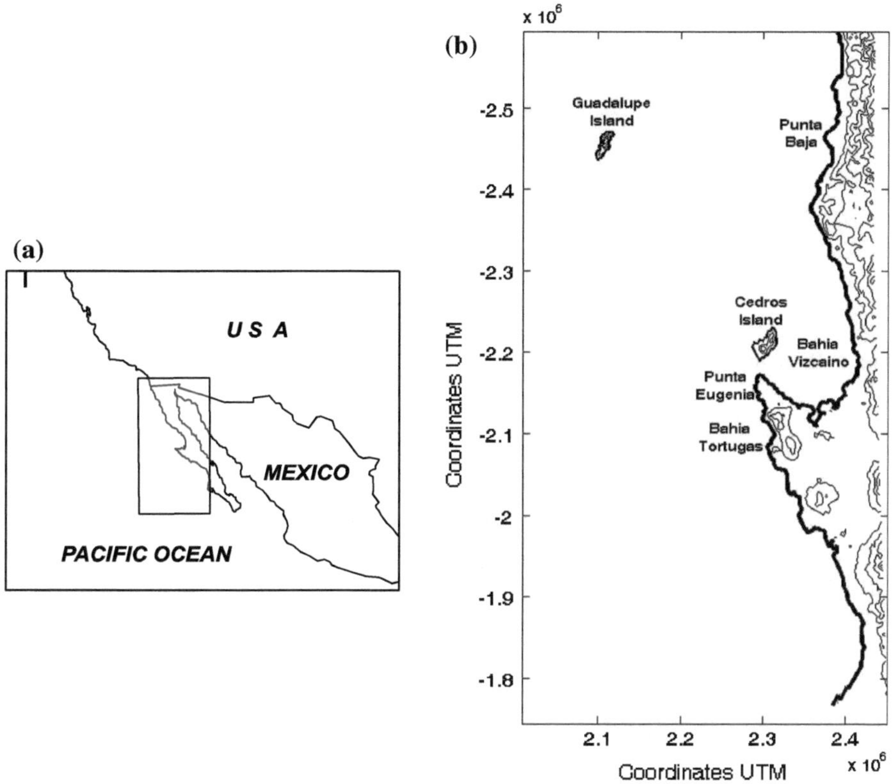

Fig. 2 (**a**) Location of study region, notice that in (**b**) it has been rotated for modelling purposes

Baja's mountains form a largely unbroken barrier running the entire length. The mountains are actually a series of ranges each with their own name. The mountains average between 2,000 and 3,000 ft. The tallest mountains on the peninsula are the twin peaks of Picacho del Diablo, or Devil's Peak, which rise to 10,154 ft (http://www.questconnect.org/baja_california.htm#Geography). Study region is shown in the box within Fig. 2.

The coastline bordering of the peninsula is characterized by bays, ports, keys, marshes and beaches. Around the Peninsula there are 35 islands, most Gulf side. Side of the Pacific Ocean, is the GI which is approximately 200 km seaward from coast, but the rest are close to the coast and share their geological and biological characteristics (Subsecretaría de Innovación y Modernización de la SPF 2011).

3 Wind Observations

Winds seasonal variability in BCC has been associated to changes in position and strength of the Pacific high pressure center, located off the west coast of California

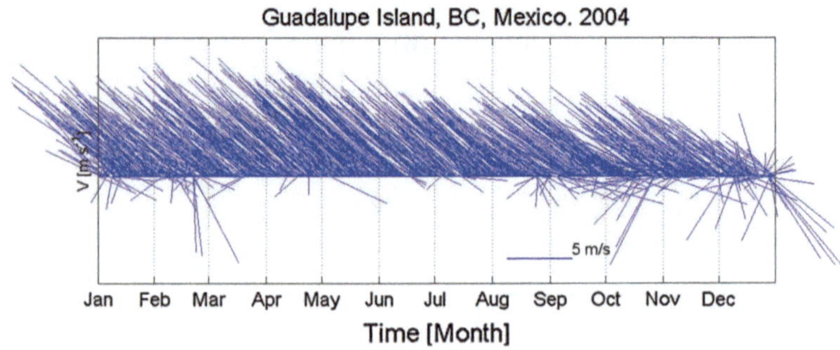

Fig. 3 Wind time series during 2004

(Huyer 1983). Most of such variability is in magnitude rather than in direction, which is predominantly from NW (Bakun 1975).

Wind observations in BCC are scarce. However, wind measurements at GI show prevailing northwesterly (southeastward) winds with mean velocities of 6 m/s in agreement with previous observations (Castro et al. 2005; Strub et al. 1987). A typical yearly time series (2004) at GI is shown in Fig. 3.

4 Numerical Methods

4.1 The Model

In this study, we consider a background of a linearly stratified fluid of uniform buoyancy, N, and constant Coriolis (i.e. f-plane dynamics) over irregular orography. The governing dimensionless set of primitive variables equations describing such situation is given as:

$$\mathrm{D}\mathbf{u}/\mathrm{Dt} = -\nabla p - (1/Ro)(v\mathbf{i} - u\mathbf{j}) - (1/F^2)\,\rho\,\mathbf{k} + (1/Re)\nabla^2\mathbf{u} \tag{1}$$

$$\mathrm{D}\,\rho\,/\mathrm{Dt} = 0 \tag{2}$$

$$\nabla^2 p = -(1/F^2)\nabla\bullet(\rho\boldsymbol{k}) - \nabla\bullet[(\mathbf{u}\bullet\nabla)\mathbf{u}] + (1/Re)\nabla^2 D - \partial D/\partial\mathrm{t} + (1/Ro)\nabla\bullet(v\mathbf{i} - u\mathbf{j}) \tag{3}$$

where bold letters represent vectors. This way, $\mathbf{u} = (u, v, w)$ is the velocity vector, and **i, j** and **k** are the unit vectors in the x, y, and z coordinates respectively, p is the pressure, ρ is the density of the air and $\mathrm{D}(= \nabla \bullet \mathrm{u})$ represents the divergence. The dimensionless numbers are the Reynolds number, $Re(= \mathrm{UL}/\nu)$, the Froude number, F(=U/NL) and the Rossby number *Ro*(=U/fL), where U and L are a typical velocity and length respectively, ν is the kinematic viscosity of air, f (=$2\Omega \sin\varphi$) the Coriolis parameter and $\mathrm{N}^2 = (-\mathrm{g}/\rho_0)\partial\rho/\partial\mathrm{z}$ is the Brünt-Väisälä frequency. As it is noted,

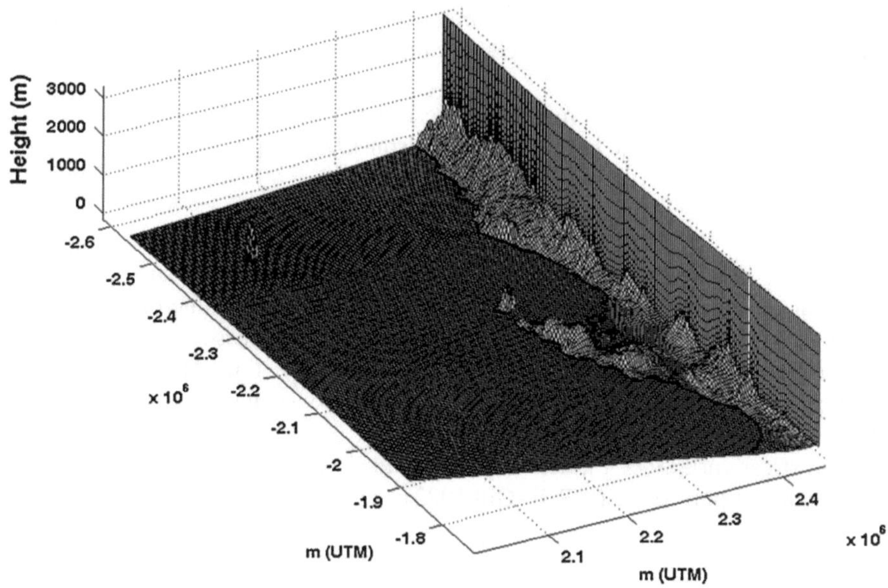

Fig. 4 Grid in physical space

in primitive variables, unknowns are (u,ρ, p) as indicated by Eqs. 1 and 2, while Eq. 3 for p substitutes the incompressibility $\nabla \bullet u = 0$ condition.

The coordinate transformation is defined by $x = x(\xi \times \eta \times \zeta)$, $y = y(\xi \times \eta \times \zeta)$, $z = z(\xi \times \eta \times \zeta)$, where (x, y, z) denotes the physical plane and $(\xi \times \eta \times \zeta)$ denotes the computational one. The boundary conditions are imposed on the computational plane $(\xi \times \eta \times \zeta)$ and are given as follows: on the upstream boundary u = (1, 0, 0) while on the downstream outer boundary $\partial u/\partial x = 0$, and $\partial \rho/\partial x = 0$. On solid boundaries **u** = 0, with no flux of density, $\nabla \rho \bullet n = 0$ (where n is the unit normal vector). For p (Eq. 4), the boundary condition is determined by substituting **u** = 0 into the transformed momentum equations to give (Torres et al. 2004):

$$p_\zeta = Re^{-1}(x_\zeta \nabla^2 u + y_\zeta \nabla^2 v + z_\zeta \nabla^2 w) - F^{-2} z_\zeta\, \rho, \tag{4}$$

where the subscript denotes partial differentiation. On the outer boundary of the computational domain the pressure is extrapolated assuming the zero normal derivative, $\partial p/\partial n = 0$ and the density perturbation is assumed to vanish, i.e. $\rho = 0$.

The computational domain consists of $193 \times 193 \times 11(\xi \times \eta \times \zeta)$ grid points in the x, y and z directions respectively with 3 km horizontal resolution and 11 vertical layers with 100 m minimum distance between points. The grid extends 523 km in the northeast direction (y-axis) and 297 km in the southeast direction (x-axis). Grid point distribution at $\zeta = 1$ is shown in Fig. 4 (note that coordinates are scaled). The dimensionless numbers used in this study were $F = 200$ for several Re numbers (to take into account wind intensities). Calculated Rossby number was $Ro > 1$, therefore rotation was neglected. In all the simulations the fluid started impulsively from rest, and several inertial periods were necessary until the fluid adjusted to this impulsive start.

For each wind velocity simulation, an estimate of the surface wind stress was calculated according to the bulk formula (Nelson 1977), Eq. 5:

$$(\tau_x, \tau_y) = \rho_a C_D\{|\boldsymbol{W}_{10}|U_{10}, |\boldsymbol{W}_{10}|V_{10}\} \quad (5)$$

where τ_x, τ_y denote the eastward and northward components of stress, ρa is the surface air density which was considered to have a constant value of 1.22 kg/m3, W_{10} is the observed wind speed, and U_{10} and V_{10} are the eastward and northward components of the wind velocity measured at a height of 10 m. The empirical drag coefficient C_D, referred to the 10 m level, was given a constant value of 0.0013 (Kraus 1972).

Wind stress curl, $(\nabla \times \tau)$, derived from wind simulations, was used to calculate the Ekman pumping velocity, WE (Gill 1982):

$$W_E = 1/f\ \rho_w(\nabla \times \boldsymbol{\tau}) \quad (6)$$

where ρ_w is seawater density (taken as 1,024 kgm^{-3}). This, way the effect of wind on ocean circulation and upwelling/downwelling generation can be assessed.

An estimation of ocean surface current velocity, u_s, generated by winds is given by the relation (Bowden 1983) $u_s = 0.875\ (C_D)1/2w10$; $C_D = 1.5 \times 10^{-3}$. For example, wind velocities of 10 m/s would generate a surface ocean velocity of $u_s = 0.34$ m/s.

5 Results and Discussion

Vorticity field for $Re = 20$ and $Re = 4 \times 10^6$ and slight stratification conditions ($F = 200$) is shown in Fig. 5a, b (top panel). Cyclonic and anti-cyclonic eddies are observed merging from GI and CI. New features unveiled by model simulations are the recirculating zones in coastal areas, particularly the jet generated at Punta Baja that deflects inland at Bahia Vizcaino (BV) and comes out to the coast around Bahia Tortugas. The effect of vortical structures in mixing and transporting of physical properties in the region could be of importance for local weather and transport of larvaes. Corresponding vertical distribution along the blue line shown in Fig. 5a is presented in the bottom panel (Fig. 5c, d). Another application of the 3D wind structure from this study could be in aeronautics by selecting the better routes for aircraft landing.

The velocity field corresponding to the above figure is shown in Fig. 6. Two jets are located at the flanks of CI, and the other one merging from Punta Baja (top panel). Calm zones (with 2 m/s winds) are identified at BV and between CI and Punta Eugenia. Velocity intensities are of the order of 5–7 m/s in the GI region and offshore. Measurements of a meteorological station in Punta Baja recorded wind speeds of 5

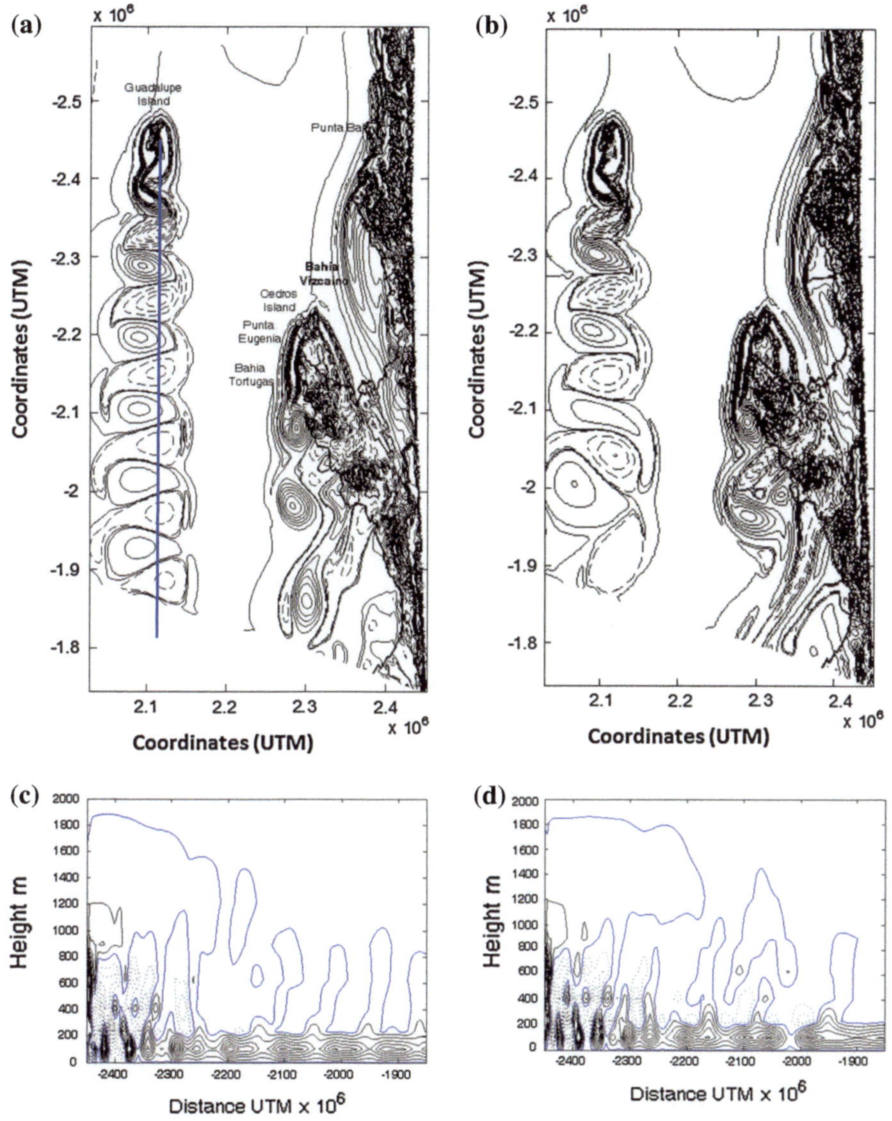

Fig. 5 Vorticity field for (**a**) $Re = 20$, and (**b**) $Re = 4 \times 10^6$. (**c**) and (**d**) is the vorticity field in the vertical along *blue line* shown in (**a**)

m/s, with hourly winds falling in the 4–8 m/s wind. Long-term mean monthly ship wind speeds in this region for June are 6–7 m/s (Nelson and Husby 1983). The annual average is close to 6.7 m/s during a period of 2000–2007 (Castro and Martinez 2010). Near the mountains, flow decelerates and winds are in the range 2–6 m/s (bottom panel). A novel result of these simulations is that we have an estimation of the wind

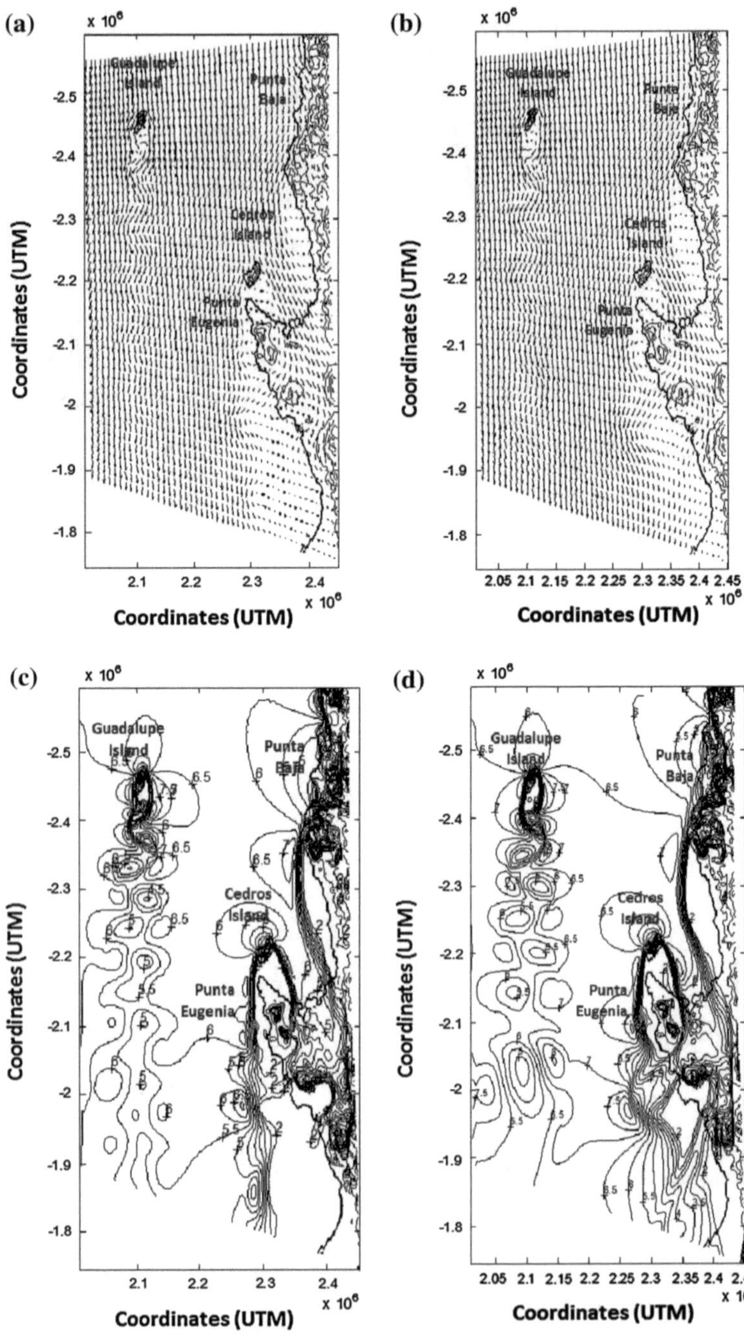

Fig. 6 Wind velocity vectors for (**a**) $Re = 20$, and (**b**) $Re = 4 \times 10^6$, (**c**) and (**d**) corresponding magnitudes (m/s)

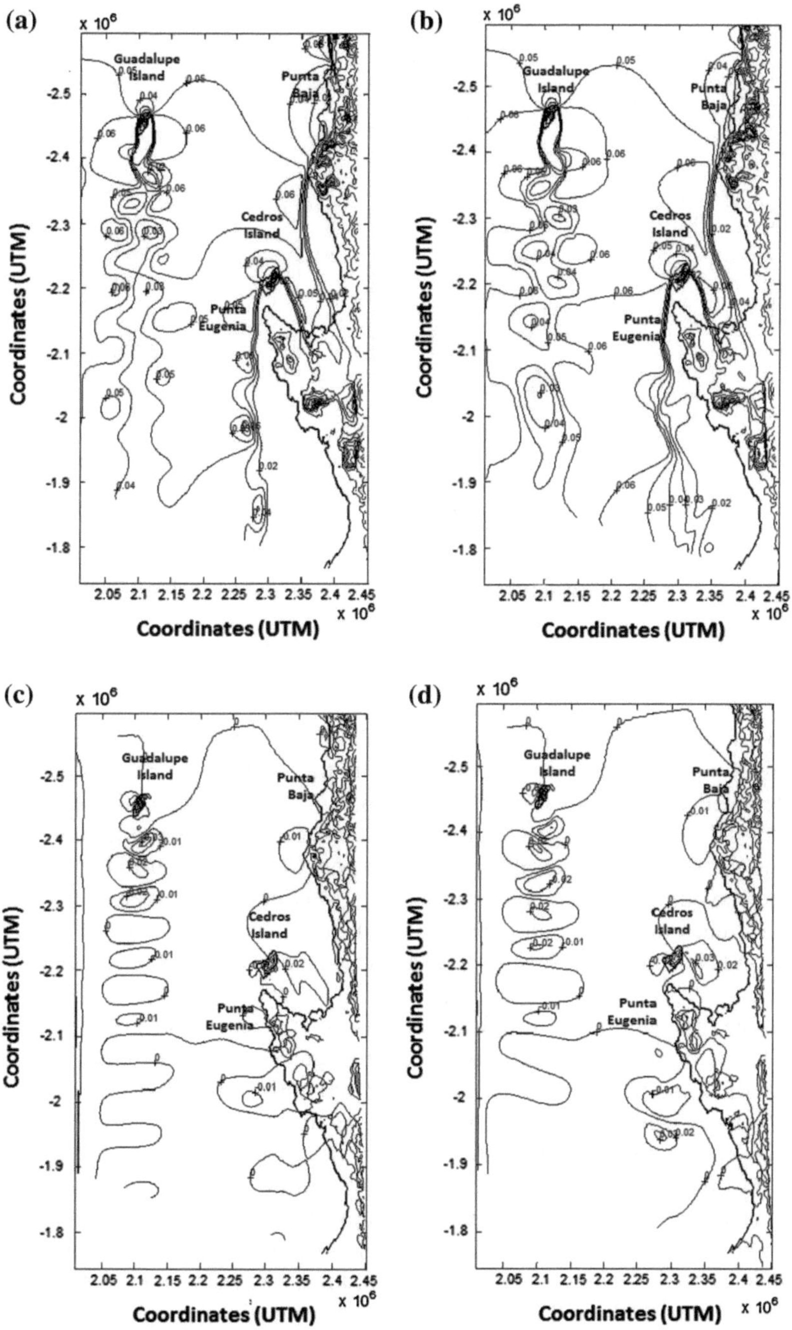

Fig. 7 Contours of Zonal (*top panel*) and Meridional (*bottom panel*) wind stress (Nm^{-2}) for $Re = 20$(**a**, **c**), and $Re = 4 \times 10^6$(**b**, **d**)

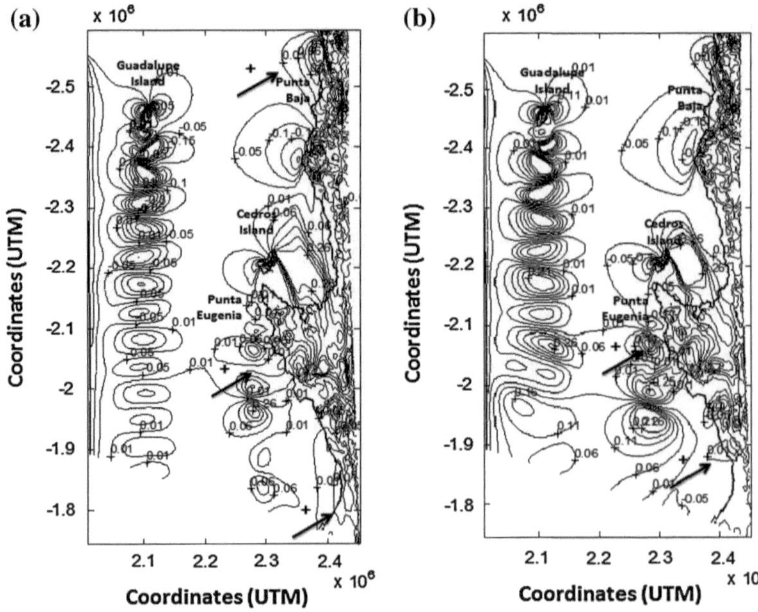

Fig. 8 W_E distribution for (**a**) $Re = 20$ and (**b**) $Re = 4 \times 10^6$

velocity of the vortices in the lee of GI and CI, this information could be of help for tracking of passive tracers or bodies floating in the vicinity of that regions.

Calculated wind stresses are presented in Fig. 7. A characteristic feature of the wind stress distribution is the occurrence of a line of zero wind stress at some distance from the coast. There is an alongshore component, equatorward (Fig. 7a, b), that promotes favourable conditions for upwelling. Maximum, values are found at Punta Baja, GI, and Punta Eugenia consistent with the values reported in the literature (Bakun and Nelson 1977; Nelson 1977) and those simulated by Koračin and Dorman (2001).

The role of upwelling in bringing nutrients into the surface layers where they are available for organic production is widely recognized (Pérez-Brunius et al. 2007; Rykaczewski and Checkley 2008). Favorable conditions for phytoplankton growth are maintained within surface photic layers by upwelling (positive W_E values) of nutrient-rich subsurface water. W_E fields for the cases simulated are presented in Fig. 8; positive values are marked with arrows within the figure. According to the preliminary model simulations, the major upwelling areas characterized with the positive curl of the wind stress are: Punta Baja, Punta Eugenia. These regions are in agreement with those upwelling signatures found by Zaytsev et al. (2003) using the SST temperature isotherm (-2°C). Also, dipoles of positive (upwelling) and negative (downwelling) curl appear in the lee of GI and CI, the strongest being in the lee of the GI where strong eddies have been observed to form. Comparison of 8a and 8b

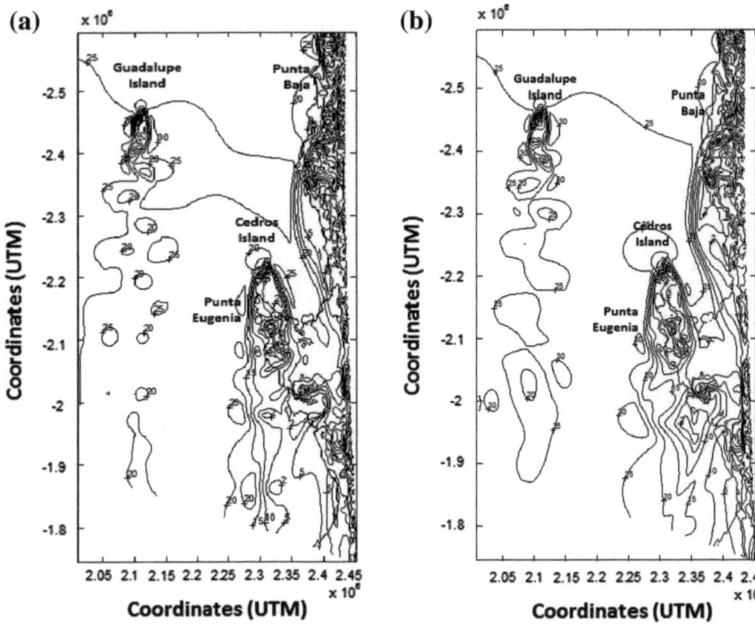

Fig. 9 Estimated ocean surface currents (cm/s) for (**a**) $Re = 20$ and (**b**) $Re = 4 \times 10^6$

seems to indicate that intensities of wind do not change the position of upwelling zones.

Figure 9 shows estimated ocean surface currents calculated with simulated wind fields. Numerical values are 5–20 cm/s in BV; 2 cm/s south CI; 20 cm/s Punta Baja; ~15 cm/s Bahia Tortugas; 25–30 cm/s GI; 20cm/s Punta Eugenia. These results show a remarkable agreement with the values reported by Palacios-Hernández et al. (1996) obtained with a 2D numerical circulation ocean model of the region.

6 Conclusions

The present study has demonstrated the utility of a 3D high resolution numerical simulation combined with boundary-fitted grids in describing details of wind generated surface structures over an area of the Pacific Coast of Mexico. We simulated eddies and coastal upwelling for different wind stresses and a complex orography. Positive values of Ekman pumping were consistent with regions of high productivity cited in the literature. It was found the existence of wind generated jets at Punta Baja, and at Cedros Island. Von Kármán vortex street generated at GI was also reproduced. Finally, an estimation of ocean surface currents due to calculated winds was in agreement with those obtained by an ocean model of the region.

Acknowledgments Authors want to thank Dr. Edmundo Lanz from Technologic Institute of Guaymas and to an anonymous reviewer for their valuable comments that that helped to improve this manuscript.

References

Bakun A (1975) Daily and weekly upwelling indices, West Coast of North America, 1967–73. U.S. Department of Commerce, NOAA Tech Rep NMFS SSRF-693, p 114

Bakun A, Nelson CS (1977) Climatology of upwelling related processes off Baja California. Calif Coop Oceanic Fish Invest Rep 19:107–127

Bowden KF (1983) Physical oceanoraphy of coastal waters. Camelot Press Ltd., Southampton, Great Britain, p 302

Castro R, Martinez JA (2010) Variabilidad especial y temporal del campo de viento. In: Gaxiola-Castro G, Durazo R (eds) Dinámica del Ecosistema pelágico frente a Baja California 1997–2007. SEMARNAT. Ensenada, Baja California, pp 291–304

Castro R, Mascarenhas A, Sanchez-Barba A, Durazo R, Gil-Silva E (2005) Condiciones meteorologicas en el sur de Isla Guadalupe. En: Santos-del Prado K, Peters E (comp.), Isla Guadalupe. Restauracion y conservacion. ISBN 968-817-725-3, pp 27–36

Cerezo-Mota RT, Cavazos, Farfán LM (2006) Numerical simulation of heavy precipitation in Northern Baja California and Southern California. J Hydrometeor 7:137–148. http://dx.doi.org/10.1175/JHM476.1

Gill A (1982) Atmosphere-ocean dynamics. Academic Press, New York, p 662

Huyer A (1983) Coastal upwelling in the California current system. Prog Oceanogr 12:259–284

Koracin D, Dorman CE, Dever EP (2004) Coastal perturbations of marine-layer winds, wind stress, and wind stress curl along California and Baja California, June 1999. J Phys Oceanogr 34:1152–1173

Koračin D, Dorman CE (2001) Marine atmospheric boundary layer divergence and clouds along California, June 1996. Mon Weather Rev 129:2040–2056

Kraus EB (1972) Atmosphere-ocean interaction. Oxford Univ Press, London, p 275

Nelson CS, Husby DM (1983) Climatology of surface heat fluxes over the California current region. NOAA Tech Rep NMFS SSRF, 763:155 (NTIS PB83-193631)

Nelson CS (1977) Wind stress and wind stress curl over the California current. NOAA Tech Rep NMFS SSRF, United States. National marine fisheries service. Department of Commerce, NOAA, 714:87

Palacios-Hernández E, Argote ML, Amador-Buenrostro A, Mancilla-Peraza M (1996) Simulación de la circulación barotrópica inducida por viento en la Bahía Sebastián Vizcaíno, Atmósfera BC, 9:171–188

Pérez-Brunius P, López M, Pares-Sierra A, Pineda J (2007) Comparison of upwelling indices off Baja California derived from three different wind data sources. Calif Coop Oceanic Fish Invest Rep 48:204–214

Pickett MH, Paduan JD (2003) Ekman transport and pumping in the California current based on the US Navy's high-resolution atmospheric model (COAMPS). J Geophys Res 108(C10):3327. doi:10.1029/2003JC001902

Rykaczewski RR, Checkley Jr DM (2008) Influence of ocean winds on the pelagic ecosystem in upwelling regions. PNAS, 15(6): 1965–1970. http://www.pnas.org/cgi. doi:10.1073_pnas.0711777105

Strub PT, Allen JS, Huyer A, Smith RL, Beardsley RC (1987) Seasonal cycles of currents, temperatures, winds, and sea level over the northeast Pacific continental shelf: 35oN–48oN. J Geophys Res 92:1507–1526

Subsecretaría de Innovación y Modernización de la SPF (2011). Archivos Históricos de Baja California. Gobierno del Estado de Baja California. http://www.bajacalifornia.gob.mx/portal/nuestro_estado/historia/baja_california.jsp Consultado el 20 de marzo 2013

Torres CR, Mascarenhas AS, Castillo JE (2004). Three-dimensional stratified flow over Alarcón Seamount, Gulf of California entrance. Deep-Sea Res II 51:647–657

Zaytsev O, Cervantes-Duarte R, Montante O, Gallegos-García A (2003) Coastal upwelling activity on the Pacific shelf of the Baja California Peninsula. J Oceanogr 59:489–502

Wake Patterns Behind a Flapping Foil

Alejandro Camilo Espinosa Ramírez and Anne Cros

Abstract We study in this work the different kind of patterns which develop in the wake of a flapping foil. The system which makes the foil flap is simple and consists of two pulleys and a motor. The Strouhal number *St* (associated with the foil frequency) and the Reynolds number *Re* (associated with the flow velocity) can be continuously varied; the oscillation amplitude A_D can take two values. We localize the regions of existence of the different patterns in the planes (St, A_D) and (*Re*, *St*). We conclude that for $140 < Re < 320$, the generated structures do not depend upon the Reynolds number.

1 Introduction

A typical system of propulsion used by insects, fish and birds, consists of flapping wings or fins. Fish and Lauder (2006), Triantafyllou et al. (2000), Wolfgang et al. (1999) among other describe how these animals use the vortices generated by their own flapping to propulse in rivers or in oceans. Basically, thrust is produced when the time-averaged velocity field behind the foil has the form of a jet. In order for the jet to occur, the inverted von Kármán street must have formed behind the obstacle. As was already observed in many works, the inverse Bénard-von Kármán street forms when the oscillation amplitude and the foil frequency take convenient values (see e.g. Anderson et al. 1998; Buchholz and Smits 2008; Dong et al. 2006; Jones et al. 1996; Schnipper et al. 2009).

Nevertheless Godoy-Diana et al. (2008) show that the inverse Bénard von Kármán wakes precede a positive thrust.

A. C. Espinosa Ramírez · A. Cros (✉)
Departamento de Física,Universidad de Guadalajara, Av. Revolución #1500,
Col. Olímpica, Guadalajara, Jalisco, México
e-mail: anne@astro.iam.udg.mx

J. Klapp and A. Medina (eds.), *Experimental and Computational Fluid Mechanics*, Environmental Science and Engineering, DOI: 10.1007/978-3-319-00116-6_29,

The experimental study performed by Schnipper et al. (2009) in a gravity-driven, vertical soap film, allowed these authors to draw a detailed transition diagram in the (St, A_D) space, where:

$$St = Df/u \tag{1}$$

is known as the Strouhal number, f being the foil oscillation frequency, D the width of the foil, u the free-stream flow speed, and:

$$A_D = A/D \tag{2}$$

is the non-dimensional amplitude where A is the peak-to-peak amplitude of the flap. Moreover, the Reynolds number is

$$Re = Df/\nu, \tag{3}$$

where ν is the fluid viscosity. In this work, the authors use the same terminology as used by Williamson and Roshko (1988) in the case of the patterns observed in the wake of a cylinder which oscillates transversally to the flow. They could observe wake structures as numerous as: various vortex pairs emitted by cycle (2P being two vortex pairs per cycle), the aligned Bénard-von Kármán vortices behind the obstacle denoted 2S and also asymmetric modes like P+S, where a pair is generated during the first half cycle and a single vortex during the other half cycle. Schnipper et al. (2009) observed in total nine kinds of patterns in the case of the pitching foil, one of them being up to eight pairs of vortices generated per oscillation cycle.

We propose in this work a simple experimental device to make the foil oscillate and visualize the different patterns which develop in the wake of the foil. We observe various patterns as complex as 16P when the three parameters *Re*, *St* and A_D are varied. We plot in two phase spaces (St, A_D) and (*Re*, *St*) the region of existence of the observed structures.

2 Experimental Device

Our experiments were performed in a water tunnel of squared section $10 \times 10\,\text{cm}^2$. A convergent and a honeycomb make the flow laminar before it enters the test section. Our foil is of aspect ratio four, as can be seen in Fig. 1, with a diameter $D = 5$ mm and a chord $c = 20$ mm. The axis of the foil is fixed to the centre of a pulley (2) which is itself connected to another pulley (1). A direct-current motor generates the continuous rotation of pulley (1). As a rigid bar is fixed between pulleys (1) and (2), when the pulley (1) makes a complete rotation, pulley (2) is only able to oscillate between two limit angles as shown in Fig. 1. The allowed peak-to-peak amplitudes for the foil are then $A = 1.2$ and 1.07 cm and the frequencies can be varied from 0 to 0.77 Hz.

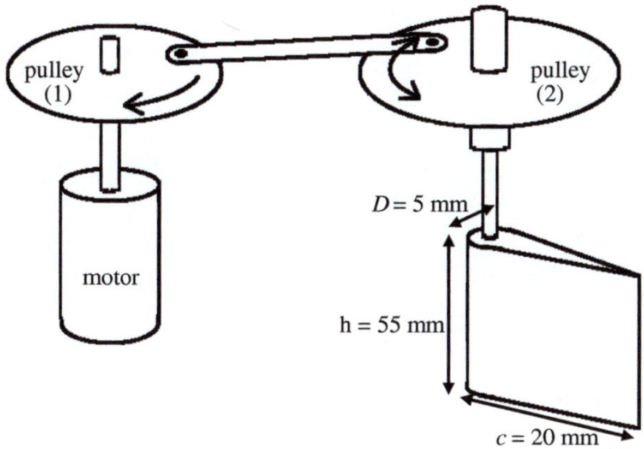

Fig. 1 Schema of the system which makes the foil oscillate. The *pulley* (1) rotates via a direct-current motor. It is connected to the *pulley* (2) via a rigid bar so that *pulley* (2) oscillates between two angles. The foil axis is fixed to the centre of the *pulley* (2) and oscillates with the peak-to-peak amplitudes $A = 1.2$ and 1.07 cm

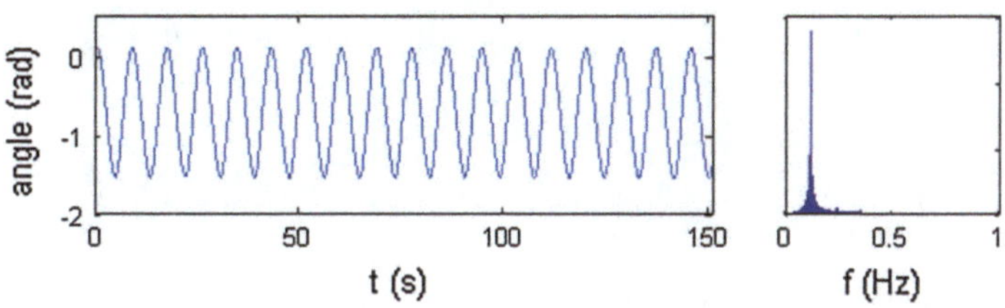

Fig. 2 Time evolution of the rotation angle of *pulley* (2) with the corresponding Fourier spectrum

In order to check that the foil oscillation is harmonic, we recorded the time evolution of the angle of pulley (2). An example is shown in the left-hand side of Fig. 2. The corresponding Fourier spectrum (Fig. 2, right-hand side) shows that a single frequency is indeed predominant.

Fluorescein or vegetable dye is used to visualize the structures which develop behind the foil. The needle of a syringe containing the dye is positioned to less than 2 mm upstream of the foil leading edge, in the center line. If fluorescein is injected, it is necessary to illuminate the flow with black light. The patterns are recorded by a video camera.

The experiments at *Re* constant are performed keeping the flow velocity constant and increasing the foil frequency. The experiments at A_D constant are done by keeping the foil frequency constant and increasing the flow velocity.

Fig. 3 Picture showing the Bénard-von Kármán wake

Fig. 4 Picture showing the inverse Bénard-von Kármán wake

3 Observations

Figures 3 and 4 show two typical patterns observed in the wake of the pitching foil. The pictures were taken with fluorescein to better visualize the patterns. As will be seen in Figs. 5 and 6, the von Kármán wake develops for intermediate values of the Strouhal numbers.

Nevertheless the transition diagrams were drawn visualizing the patterns with vegetable dye, as the water in the channel got dirty after a long time. As a comparison, the fluorescein allowed only a few minutes for visualizations.

4 Transition Diagrams

The regions of existence of the different patterns are firstly drawn in the (St, A_D) plane, as in the works of Godoy-Diana et al. (2008), Schnipper et al. (2009), Williamson and Roshko (1988). Fig. 5 was drawn for a Reynolds number

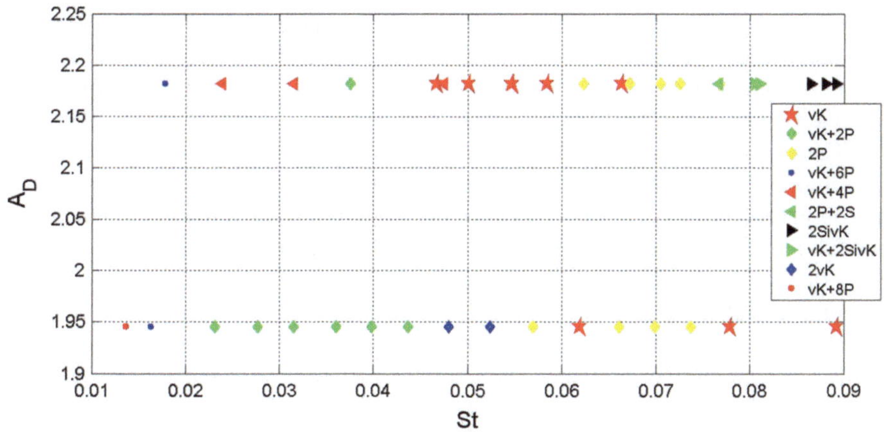

Fig. 5 Transition diagram for the two non dimensional amplitudes A_D of the foil as a function of Strouhal number *St*. This diagram is drawn for *Re* = 257

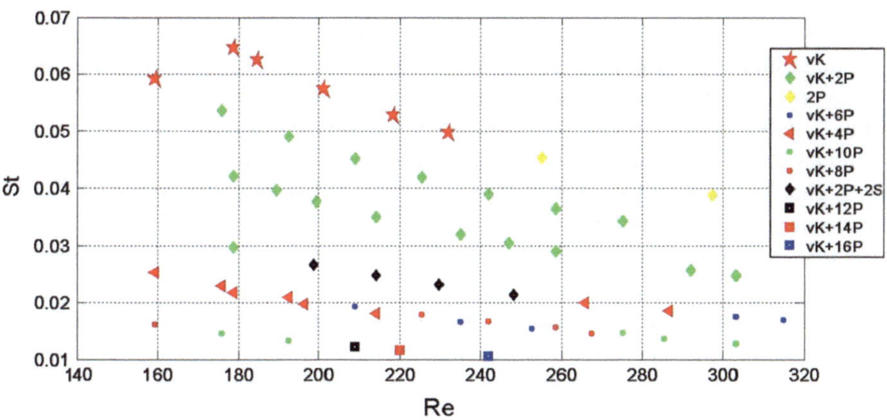

Fig. 6 Regions of existence of the patterns in the (*Re, St*) plane for $A_D = 2.18$

Re = 257. The oscillation amplitude of the foil is fixed, and the frequency is increased. In the legend, the symbol "P" represents a vortex pair generated by cycle, "vK" the classic von Kármán street as shown in Fig. 3, "2S" the von Kármán vortices aligned along the center line behind the foil, and "2SivK" the inverse von Kármán vortices aligned also along the center line behind the foil.

So Fig. 5 shows that a higher number of vortices are observed for small values of the Strouhal number and oscillation amplitude: for $A_D = 1.95$ and *St* =0.013, up to eight vortex pairs per cycle were observed. Simultaneously with these pairs, a classic von Kármán street is generated. When the Strouhal number is increased, the classical von Kármán street continues to be generated but the number of pairs decrease to six and then to two. The pattern denoted "2vK" which appears around $St = 0.050$ means that the vortices generated from the same side of the foil are co-rotative, and not counter-rotative as for "2P". Finally, for $0.055 < St < 0.075$, the pattern "2P" is advected without the lateral von Kármán vortices, and for $0.075 < St < 0.090$ the von Kármán street develops. These observations are very similar to the observations of Schnipper et al. (2009). We comment that the inverse von Kármán wake was not observed for this A_D value because our system did not allow us to reach high enough frequency values. Schnipper et al. (2009) observed the inverse vK street for $St > 0.10$ at $A_D = 2.0$.

For the higher value $A_D = 2.18$, a similar sequence is observed as the lower frequencies also allow the generation of numerous structures, whereas the increase of the Strouhal number provokes less complex patterns. As in the works of Godoy-Diana et al. (2008), Schnipper et al. (2009), the inverse von Kármán street first appears aligned as "2SivK" approximately for $St > 0.085$.

The diagram of Fig. 5 shows the observed patterns for a fixed value of the Reynolds number. In the previous mentioned studies, the authors also fix the Reynolds number: $Re = 220$ and 440 in Schnipper et al. (2009) and $Re = 255$ in Godoy-Diana et al. (2008), Williamson and Roshko (1988) mention that their transition diagram is unchanged for $300 < Re < 1000$. We were interested in studying the variation of the generated patterns when the Reynolds number is varied below this value. Fig.6 shows the region of existence of the different patterns for $A_D = 2.18$.

It can be seen that more structures are generated for small values of *St*, whereas for higher frequencies, a simpler pattern develops. Typically for $St > 0.03$, only the usual von Kármán wake, or two vortex pairs per cycle, or a combination of both structures, develop. For $St < 0.03$, it is difficult to delimit on first sight a region for each pattern.

Points that are roughly aligned along an inclined curve from low to high *Re* values and for slightly decreasing Strouhal number correspond to a given series of experiments. Along this type of curve, it can be observed that the complexity of the structures increases. For instance, for the "bottom" series from ($St = 0.016$, $Re = 160$) to ($St = 0.010$, $Re = 243$), the generated structures pass from (vK + 8P) to (vK + 16P). The same evolution in the number of vortex pairs per cycle is observed in the other series of experiments. For the run from ($St = 0.025$, $Re = 160$) to ($St = 0.012$, $Re = 305$), the wake form passes from (vK + 4P) to (vK + 10P). Nevertheless in the phase diagram (*Re*, *St*) the pattern (vK + 10P) is observed for very distinct *Re* values: $Re \in [170; 200]$ and $Re \in [270; 310]$, both intervals for *St* [0.013; 0.015]. It happens roughly the same for (vK + 6P): this pattern appears in distant *Re* intervals but for the same *St* values. We conclude that the kind of pattern which develops behind a flapping foil *does not* depend, or depends only very slightly, upon the Reynolds number, for $140 < Re < 320$.

5 Conclusions

We developed a simple experimental system to generate oscillations in a foil. When this flapping foil is immersed in a flow, a great variety of structures are generated depending upon three non dimensional parameters: the Strouhal number St, the Reynolds number Re and the non dimensional oscillation amplitude of the foil A_D. We visualized complex structures like 16 vortex pairs per cycle. We localized the region of existence of the different patterns in two phase diagrams (St, A_D) and (Re, St). The first diagram is very similar to the one drawn by Schnipper et al. (2009) and shows that the complexity of the structures increases when the Strouhal number and the non dimensional amplitude are small. The second permits us to conclude that the advected patterns do not depend upon the Reynolds number. We confirm with a systematic study the observations mentioned in other works.

References

Anderson JM, Streitlien K, Barrett DS, Triantafyllou MS (1998) Oscillating foils of high propulsive efficiency. J Fluid Mech 360:41–72

Buchholz JHJ, Smits AJ (2008) The wake structure and thrust performance of a rigid low-aspect-ratio pitching panel. J Fluid Mech 603:331–365

Dong H, Mittal R, Najjar FM (2006) Wake topology and hydrodynamic performance of low-aspect-ratio flapping foils. J Fluid Mech 566:309–343

Fish FE, Lauder GV (2006) Passive and active flow control by swimming fishes and mammals. Annu Rev Fluid Mech 38:193–224

Godoy-Diana R, Aider JL, Wesfreid JE (2008) Transitions in the wake of a flapping foil. Phys Rev E 77:016308

Jones KD, Dohring CM, Platzer MF (1996) Wake structures behind plunging airfoils: a comparison of numerical and experimental results. AIAA J 36:1240–1248

Schnipper T, Andersen A, Bohr T (2009) Vortex wakes of a flapping foil. J Fluid Mech 633:411–423

Triantafyllou MS, Triantafyllou GS, Yue DK (2000) Hydrodynamics of fishlike swimming. Annu Rev Fluid Mech 32:33–53

Williamson CHK, Roshko A (1988) Vortex formation in the wake of an oscillating cylinder. J Fluids Struct 2:355–381

Wolfgang MJ, Anderson JM, Grosenbaugh MA, Yue DKP, Triantafyllou MS (1999) Near-body flow dynamics in swimming fish. J Exp Biol 202:2303–2327

Environmental Fluid Dynamics

Elizabeth Teresita Romero-Guzmán, Lázaro Raymundo Reyes-Gutiérrez and Jaime Lázaro Klapp-Escribano

Abstract The environmental fluid dynamics provides a comprehensive overview of the basic principles, it emphasizes the relevance of environmental fluid dynamics research in society, public policy, infrastructure, quality of life, security, and the law. It then discusses established and emerging focus areas. As communities face existential challenges posed by climate change, rapid urbanization, and scarcity of water and energy, the study of environmental fluid dynamics becomes increasingly relevant. This document is a review of the application of environmental fluid dynamics in atmosphere, lithosphere and hydrosphere.

1 Introduction

1.1 Definition

A fluid is a substance that has no fixed shape and yields easily to external pressure; http://oxforddictionaries.com/definition/english/fluid. A continuous, amorphous

E. T. Romero-Guzmán (✉) · J. L. Klapp-Escribano
Carretera México-Toluca S/N, Instituto Nacional de Investigaciones Nucleares, La Marquesa, 52750 Ocoyoacac, Estado de México México,
e-mail: elizabeth.romero@inin.gob.mx

L. R. Reyes-Gutiérrez · J. L. Klapp-Escribano
Departamento de Matemáticas, Cinvestav del I.P.N., 07360 México D. F. México
e-mail: raregu@gmail.com

J. L. Klapp-Escribano
e-mail: jaime.klapp@inin.gob.mx

L. R. Reyes-Gutiérrez
Universidad Autónoma Metropolitana-Lerma, Av. Hidalgo Poniente 46. Col. La Estación, 52006 Lerma de Villada, Estado de México, México
e-mail: raregu@gmail.com

J. Klapp and A. Medina (eds.), *Experimental and Computational Fluid Mechanics*, Environmental Science and Engineering, DOI: 10.1007/978-3-319-00116-6_30,

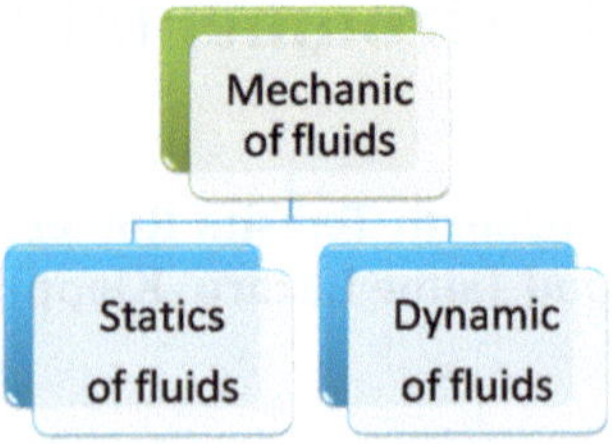

Fig. 1 Classification of mechanic of fluids

substance whose molecules move freely past one another and that has the tendency to assume the shape of its container.

Many things behave like a fluid: air, gas, liquid, electricity and traffic. Fluids are hard to contain since they take many shapes based on their environment (White 2003).

1.2 Classification of the Mechanic of Fluids

Fluid mechanics is one of the four branches of mechanics: elastic-body mechanics, fluid mechanics, relativistic mechanics and quantum mechanics. The study of fluid mechanics subdivides into statics and dynamics (Fig. 1) which in turn divide into incompressible and compressible flows. Incompressible and compressible flow divides into real and ideal. Real divides into laminar and turbulent. And so on.

Fluid mechanics is based upon five great principles of physics:

1. Conservation of mass
2. Conservation of linear momentum
3. Conservation of angular momentum
4. Conservation of energy
5. Conservation of thermodynamics (Granger 1995).

1.3 Kinds of Fluids

Some differentiate fluid from solid by the reaction to shear stress. It is a known fact said that the fluid continuously and permanently deformed under shear stress while solid exhibits a finite deformation which does not change with time. It is also said that fluid cannot return to their original state after the deformation. This differentiation leads to three groups of materials: solids and liquids. This test creates a new material group that shows dual behaviors; under certain limits; it behaves like solid and under others it behaves like fluid. The study of this kind of material called rheology and it will (almost) not be discussed in this chapter. It is evident from this discussion that when a fluid is at rest, no shear stress is applied.

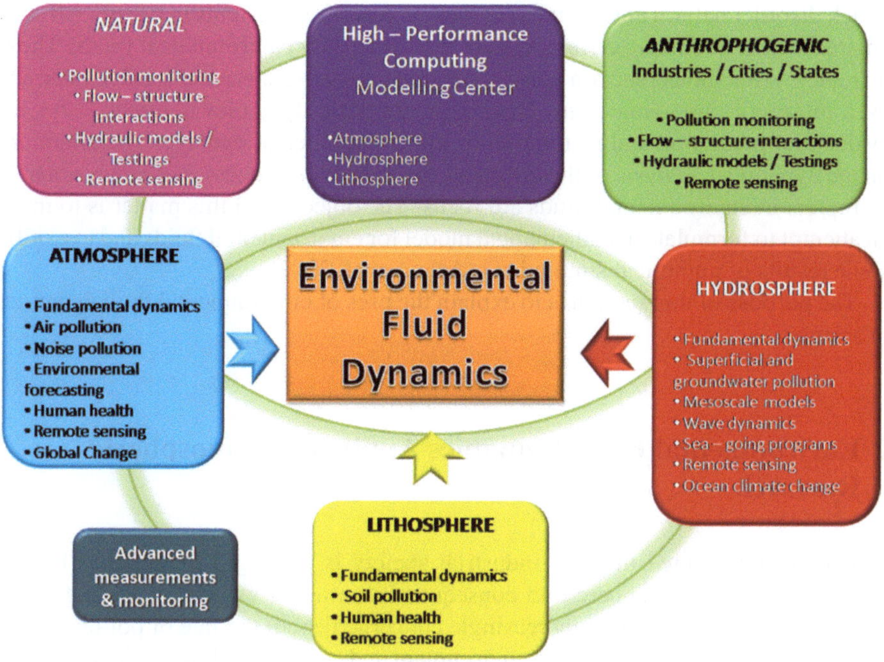

Fig. 2 Applications of Environmental Fluid Dynamics (EFD)

The fluid is mainly divided into two categories: liquids and gases. The main difference between the liquids and gases state is that gas will occupy the whole volume while liquids has an almost fix volume. This difference can be, for most practical purposes considered, sharp even though in reality this difference isn't sharp. The difference between a gas phase to a liquid phase above the critical point are practically minor. But below the critical point, the changes of water pressure by 1,000 % only change the volume by less than 1 %. For example, a change in the volume by more 5 % will required tens of thousands percent change of the pressure. So, if the change of pressure is significantly less than that, then the change of volume is at best 5 %. Hence, the pressures will not affect the volume. In gaseous phase, any change in pressure directly affects the volume. The gas fills the volume and liquid cannot. Gas has no free interface/surface (since it does fill the entire volume). The fluids have many properties which are similar to solid (Bar-Meir 2013).

Figure 2 shows a diagram which includes several applications of the Environmental Fluid Dynamics (EFD) in the three main areas of researches: atmosphere, lithosphere and hydrosphere.

Importance and applicability of numerical flow analysis to environmental science are outlined. Fluid phenomena in the ocean, rivers, atmosphere and the ground are investigated by means of numerical methods and in turn proposals for the control, restoration and counter plans against the so-called environmental disrupters which disorder natural environment as well as ecological systems in nature. All such environmental disrupters diffuse and are transported by environmental fluids. Those disrupters sometimes react on some other chemicals to generate offensive odor and even more poisonous materials. Environmental fluid dynamics is effective for the evaluation, prediction and restoration of the environmental damage. The attentions on the dynamical analysis of the diffusion and advection processes of environmental disrupters in environmental fluids are focus. The objective in this matter is to make an attempt to formulate a mathematical model for environmental fluids and to exhibit some results of numerical simulations (Arima et al. 2007).

The purpose of this section is to explain the uses of environmental fluid dynamics in science.

2 Environmental Fluid Dynamics Applied to Atmospheric Research

Having acknowledged that our industrial society is placing a tremendous burden on the planetary atmosphere and consequently on all of us, scientists, engineers, and the public are becoming increasingle concerned about the fate of pollutants and greenhouse gases dispersed in the environment and especially about their cumulative effect.

Atmospheric general circulation models used for climate simulation and weather forecasting require the fluxes of radiation, heat, water vapor, and momentum across the land-atmosphere interface to be specified. These fluxes are calculated by submodels called land surface parameterizations. Over the last 20 years, these parameterizations have evolved from simple, unrealistic schemes into credible representations of the global soil-vegetation-atmosphere transfer system as advances in plant physiological and hydrological research, advances in satellite data interpretation, and the results of large-scale field experiments have been exploited (Guohui 1995).

The developments of the Computational Fluid Dynamics (CFD) method as a powerful tool for prediction of wind environmental conditions around buildings are presented in this study. CFD techniques have been applied in predicting wind flow conditions: around a single building, between two parallel buildings and around a multiple building configuration. Also presented is a limited model validation for those simulated configurations. Finally, the application of CFD techniques for a case study in simulating an existing site together with proposed buildings and the local landscape (Baskaran and Kashef 1996; Zhang et al. 2011).

Some modern schemes incorporate biogeochemical and ecological knowledge and, when coupled with advanced climate and ocean models, will be capable of modeling the biological and physical responses of the Earth system to global change, for example, increasing atmospheric carbon dioxide.

Will the accumulation of greenhouse gases in the atmosphere lead to global climatic changes that, in turn, will affect our lives and societies? What are the various roles played by the oceans in maintaining our present climate? Is it possible to reverse the trend toward depletion of ozone in the upper atmosphere? Is it safe to deposit hazardous wastes on the ocean floor? Such pressing questions cannot find answers without, first, and in –depth understanding of atmospheric and oceanic dynamics and, second, the development of predictive models. In this twin endeavor, geophysical fluid dynamics assumes an essential role, and the numerical aspects should not be underestimated in view of the required predictive tools (Cushman-Roisin and Beckers 2011).

Research activities on various types of jet flow, with particular reference to their application in the field of environmental fluid dynamics are presented (Wang and Tan 2010).

Some projects which are actually developing are: chemistry of climate change, fluid dynamics and the atmospheric sciences, air flow and dispersion over an urban downtown area, data assimilation of a dual doppler lidar, coherent doppler lidar deployment and data analysis for terrain-induced rotor experiment (T-REX), tracking of PM plumes using a doppler lidar, modeling the behavior of oil dispersion (Fig. 3).

3 Environmental Fluid Dynamics Applied to Lithosphere Research

The object of geophysical fluid dynamics is the study of naturally occurring, large-scale flows on Earth and elsewhere, but mostly on Earth. Although the discipline encompasses the motions of both fluid phases—liquids (waters in the ocean, molten rock in the outer core) and gases (air in our atmosphere, atmospheres of other planets, ionized gases in stars)—a restriction is placed on the scale of these motions. Only the large-scale motions fall within the scope of geophysical fluid dynamics. For example, problems related to river flow, microturbulence in the upper ocean, and convection in clouds are traditionally viewed as topics specific to hydrology, oceanography, and meteorology, respectively.

Geophysical fluid dynamics deals exclusively with those motions observed in various systems and under different guises but nonetheless governed by similar dynamics (Henry 1986).

Typical problems in geophysical fluid dynamics concern the variability of the atmosphere (weather and climate dynamics), oceans (waves, vortices and currents), and, to a lesser extend, the motions in the earth's interior responsible for the dynamo effect, vortices on the planets and convection in stars (the sun, in particular) (Cushman-Roisin and Beckers 2011). The modeling of this type of system is very important but complex (Blocken and Gualtieri 2012).

Importance of geophysical fluid dynamics: Without its atmosphere and oceans it is certain that our planet would not sustain life. The natural fluid motions occurring in these systems are therefore of vital importance to us, and their understanding extends beyond intellectual curiosity—it is a necessity. Historically, weather vagaries have baffled scientists and laypersons alike since times immemorial. Likewise, conditions

Time: 65 to 23 million years ago: Ocean temperatures are about 10 to 15 degrees Celsius higher than today. Palm trees are growing in Greenland and Patagonia. The Indian tectonic plate collides with the Asian plate, forming the Himalayas. This sparks a new period of glaciation and Earth starts to cool down again. (Source: Shutterstock)

A jet airliner leaves condensation trails. The trails are formed by soot and water vapor from the plane's engines. The Intergovernmental Panel on Climate Change (IPCC) estimates that aircraft emissions of water vapor, nitrous oxides, aerosols and CO_2 could be 2 to 4 times stronger than emissions of CO_2 alone. Aviation is responsible for 3.5 percent of man-made global warming, says the IPCC, and it is the fastest growing source of greenhouse gases. (Source: Reuters)

Carbon capture artificial trees suck CO_2-from-air

On April 20, 2010 the Deepwater Horizon off shore oil rig in the Gulf of Mexico exploded, killing 11 workers and leading to the worst oil spill and environmental catastrophe in U.S. history. A ruptured underwater pipe spewed almost 5 million barrels of oil into the Gulf over three months, threatening hundreds of miles of beaches, wetlands, and estuaries. Thousands of animals, including turtles, crabs, fish, and birds fell victim, and the local fishing and tourism industries suffered badly. (Source: Reuters)

Fig. 3 Application of environmental fluid dynamics to atmospheric research

at sea have long influenced a wide range of human activities, from exploration to commerce, tourism, fisheries and even wars.

Thanks in large part to advances in geophysical fluid dynamics, the ability to predict with some confidence the paths of hurricanes has led to the establishment of

Flood History Effects on Sediment Entrainment and Transport in Gravel-bed Rivers

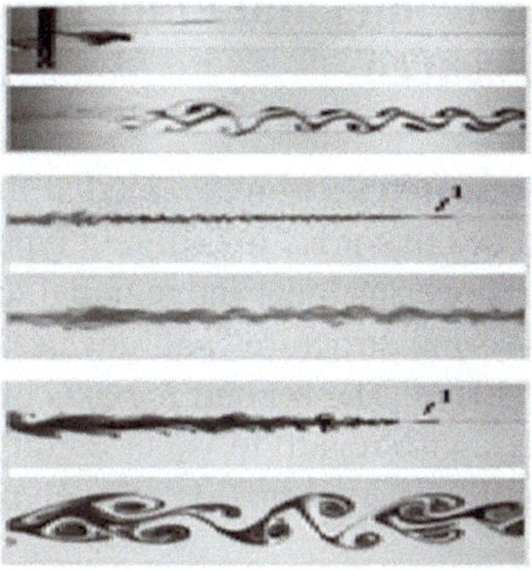

wakes of maneuvering bodies in stratified fluids

The beds of most alluvial river channels are not flat, but comprise a series of undulating sedimentary accumulations termed 'bedforms' that include ripples and dunes.

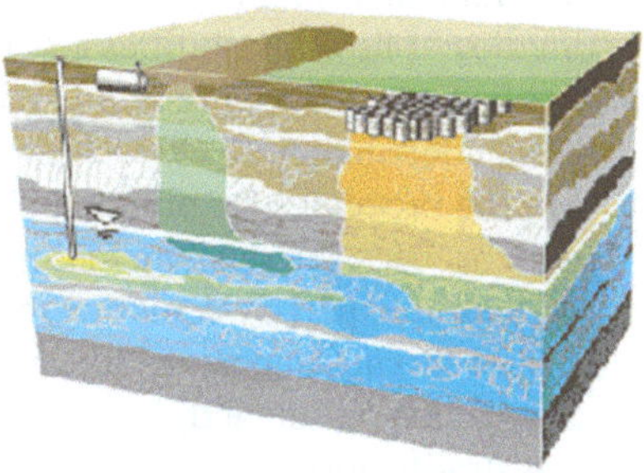

Migration of radionuclides through the unsaturated zone

Fig. 4 Application of environmental fluid dynamics to lithosphere research

a warning system that, no doubt, has saved numerous lives at sea and in coastal areas (Abbott 2004).

Some current projects in this area are: applied fluid mechanics, geophysical fluid dynamics, sediment transport, physics-based modeling support for the generation of synthetic ship stern wake, wakes of maneuvering bodies in stratified fluids, ripple dynamics and benthic transformations under variable wave forcing, migration of radionuclides through unsaturated zone (Fig. 4), geophysical and environmental fluid dynamics, numerical simulation of climate analysis and prediction, nonlinear partial differential equations, dynamical systems, scientific computing, geophysical fluid dynamics and turbulence.

4 Environmental Fluid Dynamics Applied to Hydrosphere Research

This research section is aimed at fluid flow and associated heat and mass transport processes that arise in different types of water bodies (river, lakes, resevoirs, coastal water) covering local and regional scales. The interaction of these flow processes with engineering structures and devices is the first major focus of these studies (Daraio et al. 2010). The second focus is on their response to anthropogenic heat and mass sources with regard to water quality aspects. Examples of current research projects that utilize a combination of analytical, numerical and experimental methods are:

a. Pollutant dispersion in rivers and river systems with pronounced heterogeneities, such as groynes or run-of-the-river reservoirs, including development of existing pollutant alarm modules used by river authorities.
b. Studies of flow instabilities and large-scale coherent structures in shallow flows (wide rivers, coastal regions), including transport and mixing of dissolved and suspended matter.
c. Mixing processes for pollutant discharges into water bodies in form of momentum or buoyant jets, including multiport diffuser devices. Development of PC-based expert systems for the qualitative description and quantitative forecast of mixing intensities for different water bodies.
d. Suspended sediment dynamics in river and estuarial systems, including effect of salt stratification and associated contaminant adsorption processes.
e. Gas exchange at the air-water interface under the influence of different turbulence generating mechanisms.
f. Topograpical effects for stratified flows in lakes or reservoirs.

On December 26, 2004, the tsunami generated by the earthquake was not detected, its consequences not assessed and authorities not alerted within the 2 h needed for the wave to reach beaches in the region. On a large scale, the passage every 3–5 years of an anomalously warm water mass along the tropical Pacific Ocean and the western coast os South America, Known as the El Niño event, has long been blamed for serious ecological damage and disastrous economical consequences in some countries. Now, thanks to increased understanding of long oceanic waves, atmospheric convection, and natural oscillations in air-sea interactions scientists have successfully removed the veil of mystery on this complex event, an numerical models offer reliable predictions with at least one year of lead time, that is, there is a year between the moment the prediction is made and the time to which it applies (Cushman-Roisin and Beckers 2011), Fig. 5.

With major implications for applied physics, engineering, and the natural and social sciences, the rapidly growing area of environmental fluid dynamics focuses on the interactions of human activities, environment, and fluid motion.

We have to that emphasized the relevance of environmental fluid dynamics research in society, public policy, infrastructure, quality of life, security, and the law. It then discusses established and emerging focus areas.

Tsunami in Japan, March 2011

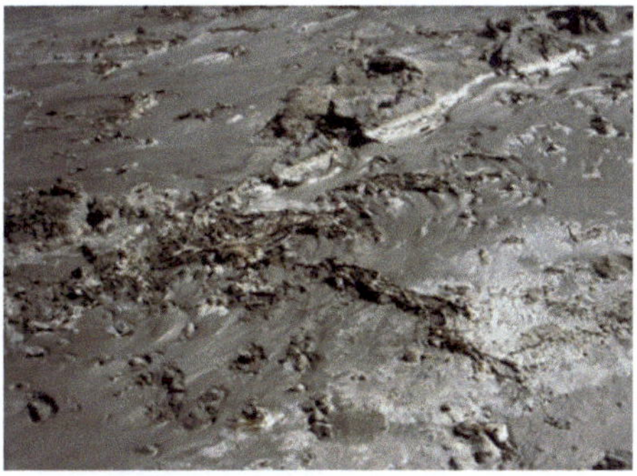

Fergana valley oil spill is the biggest oil spill to be occurred on land in the planet. It happened on march 2, 1992 when a blowout took place at well 5 in Mingbulak oil field, Fergana valley, Uzbekistan. It took place at populated industrial and agricultural zone in Uzbekistan. It spilled 88 million gallons of oil into the land and also the fire due to spill burnt for two months. The losses were estimated at over $250 million.

Yangtze river, China. Red river contaminated

Oklahoma disaster, May 20th, 2013

Fig. 5 Application of environmental fluid dynamics to hydrosphere research

Environmental Fluid Dynamics explores the interactions between engineered structures and natural flows, the environmental pollution, with a focus on numerical methods, predictive modeling, and computer infrastructure developments. The EFD looks at practical aspects of laboratory experiments and field observations that validate quantitative predictions and help identify new phenomena and processes. As communities face existential challenges posed by climate change, rapid urbanization, and scarcity of water and energy, the study of environmental fluid dynamics becomes increasingly relevant (Harindra 2012; Imberger 2012).

5 Conclusions

It is necessary to emphasize the relevance of environmental fluid dynamics research in society, public policy, infrastructure, quality of life, security, and the law, to discuss established and emerging focus areas. The application of environmental fluid dynamics in atmosphere, lithosphere and hydrosphere is necessary to know the behavior of the fluids and give some solutions to the society problems.

Acknowledgments The authors wish to thank the financial support provided by the Consejo Nacional de Ciencia y Tecnología (CONACyT) - EDOMEX-2011-C01-165873.

References

Abbott PL (2004) Natural disasters, 4th edn. McGraw-Hill, New York, pp 460

Arima T, Matsuura Y, Oharu S (2007) Computation of air flows and motion of environmental pollutants over complex geographical topographies. J Comput Appl Math 204(1):187–196

Bar-Meir G (2013) Basics of fluid mechanics Version 0.3.3.0. Chicago, p 557

Baskaran A, Kashef A (1996) Investigation of air flow around buildings using computational fluid dynamics techniques. Eng Struct 18(11):861–873, 875

Blocken B, Gualtieri C (2012) Ten iterative steps for model development and evaluation applied to computational fluid dynamics for environmental fluid mechanics. Environ Model Softw 33:1–22

Cushman-Roisin B, Beckers JM (2011) Introduction to geophysical fluid dynamics: physical and numerical aspects, 2nd edn, vol 101. Elsevier Inc., USA, p 815

Daraio JA, Weber LJ, Newton TJ, Nestler JM (2010) A methodological framework for integrating computational fluid dynamics and ecological models applied to juvenile freshwater mussel dispersal in the Upper Mississippi River. Ecol Model 221(2):201–214

Granger RA (1995) Fluid mechanics, 2a edn. Dover Publications Inc., New York

Guohui G (1995) Evaluation of room air distribution systems using computational fluid dynamics. Energ Build 23(2):83–93

Harindra JF (ed) (2012) Handbook of environmental fluid dynamics, Vol. 1. Edited by CRC Press, p 635

Henry S (1986) Climatological atlas of the world ocean: by Sydney Levitus. NOAA profesional paper 13, environmental research laboratories, geophysical fluid dynamics laboratory, Princeton, NJ and Rockville, MD. Deep Sea Res Part A Oceanogr Res Pap 33(7):998

Imberger J (2012) Environmental fluid dynamics: overview and fundamentals. Flow processes, scaling, equations of motion, and solutions to environmental flows. 1st Edn. Academic Press, p 460

Wang X, Tan SK (2010) Environmental fluid dynamics-jet flow. J Hydrodyn Ser B 22(5):1009–1014

White FM (2003) Mecánica de fluidos. 5a edn. MacGrawhill, España, p 821

Zhang Z, Wieland R, Reiche M, Funk R, Hoffmann C, Li Y, Sommer M (2011) Wind modelling for wind erosion research by open source computational fluid dynamics. Ecol Inform 6(5):316–324

Part V
General Fluid Dynamics and Applications

Inductionless Magnetohydrodynamic Laminar Flow in a Narrow-Gap Annular Duct

Raúl A. Ávalos-Zúñiga, Martín J. Nieto-Pérez and Gonzalo A. Ramos-López

Abstract We consider laminar flow of an electrically conducting fluid in an annulus between two infinitely long, perfectly conducting cylinders subject to an azimuthal magnetic field. The conducting fluid is driven by a pressure gradient, either constant or harmonic. The magnetohydrodynamic equations governing the flow have been solved under the inductionless approximation assuming a small annular gap compared to the inner radius. As the magnetic force increases, the flow slows down as in the Hartmann problem, notwithstanding the absence of Hartmann boundary layers.

1 Introduction

The analytical solutions to magnetohydrodynamic (MHD) laminar flow in an annular duct under a purely radial magnetic field are well known [see Globe (1959); Elco et al. (1962); Edward (1966), more recently Zhao et al. (2011), and for pulsating driven flow Narasimhan (1964); Rudraiah (1966)]. However, solutions for an applied azimuthal magnetic field have been less investigated, see for instance Edward (1966) for a solution of steady MHD flow in a sector of an annular duct. In fact, solutions for a harmonic driven flow in an azimuthal magnetic field have not been found. In this paper, we describe a study of the problem when the flow is driven through an annular duct by a pressure gradient, either constant or harmonic, subject to an azimuthal magnetic field. In contrast to previous models, we have obtained solutions considering the so called inductionless approximation (Priede 2009; Davidson 2001; Müller and Bühler 2001) together with the narrow-gap approximation (Rudraiah 1966).

R. A. Ávalos-Zúñiga (✉) · M. J. Nieto-Pérez · G. A. Ramos-López
CICATA-Querétaro, Instituto Politécnico Nacional, Cerro Blanco No. 141,
Colinas del Cimatario,76090 Santiago de Querétaro, QRO, México
e-mail: ravalosz@ipn.mx

J. Klapp and A. Medina (eds.), *Experimental and Computational Fluid Mechanics*,
Environmental Science and Engineering, DOI: 10.1007/978-3-319-00116-6_31,

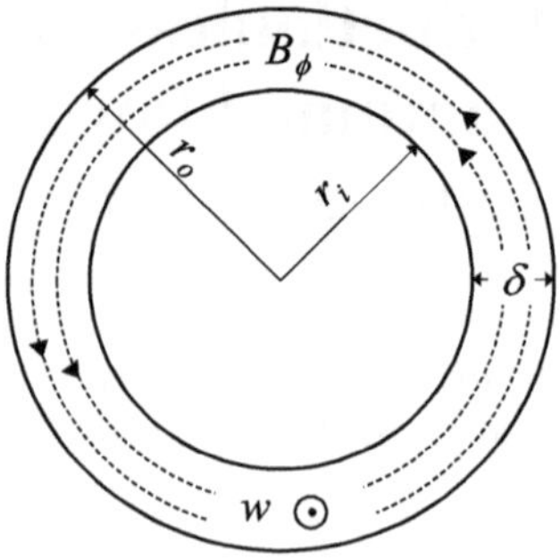

Fig. 1 Sketch of two concentric cylinders subject to an azimuthal magnetic field

2 General Formulation

Consider the fully developed flow of an incompressible, electrically conducting fluid within a gap between two infinitely long, perfectly conducting concentric cylinders subject to an imposed azimuthal magnetic field, $\mathbf{B_0}$. The field is defined by $\mathbf{B_0} = \dfrac{S}{r}\mathbf{e}_\phi$, in which S is a measure of the field strength and $\mathbf{e}_\phi$ the azimuthal unit vector in cylindrical coordinates (r, ϕ, z). If the fluid motion slightly perturbs $\mathbf{B_0}$, then the induced magnetic field, $\mathbf{b}$, associated with induced currents ($\sim\sigma\mathbf{u}\times\mathbf{B_0}$) is negligible with respect to $\mathbf{B_0}$. That is, the total magnetic field $\mathbf{B} \approx \mathbf{B_0}$. This leads to the inductionless approximation which is characteristic of very small magnetic Reynolds number $Rm = \mu_0\sigma u_0 l \ll 1$, where μ_0 is the permeability of vacuum, σ the electrical conductivity, u_0 and l are the characteristic velocity and length scales. Then, the fluid velocity $\mathbf{u}$ is governed by the Navier-Stokes equation with electromagnetic force:

$$\frac{\partial \mathbf{u}}{\partial t} + (\mathbf{u}\cdot\nabla)\mathbf{u} = -\frac{\mathbf{1}}{\rho}\nabla\mathbf{p} + \nu\nabla^{\mathbf{2}}\mathbf{u} + \frac{\mathbf{1}}{\rho}\mathbf{j}\times\mathbf{B_0} \tag{1}$$

where the induced current $\mathbf{j}$ is described by Ohm's law for a moving media:

$$\mathbf{j} = \sigma\left(-\nabla\varphi + \mathbf{u}\times\mathbf{B_0}\right) \tag{2}$$

where φ is the electrostatic potential. In Eq. (2), the electric field $\mathbf{E}$ has been replaced by $-\nabla\varphi$ because $\nabla\times\mathbf{E} \approx \mathbf{0}$ due to the associated inductionless approximation. Additionally, mass and charge conservation require $\nabla\cdot\mathbf{u} = \nabla\cdot\mathbf{j} = 0$.

The model is confined to an ideal axially unbounded system shown in Fig. 1 which allows a pure axially velocity distribution, $\mathbf{u} = w(r,t)\mathbf{e_z}$, where $\mathbf{e_z}$ is the axial unit vector. The simplified model is then described by Eq. (1) reduced to:

$$\frac{\partial w}{\partial t} = -\frac{1}{\rho}\frac{dp}{dz} + \nu\frac{1}{r}\frac{\partial}{\partial r}\left(r\frac{\partial w}{\partial r}\right) + \frac{j_r}{\rho}\frac{S}{r} \tag{3}$$

where w is the axial velocity, p the pressure, ν the kinematic viscosity, and ρ the density. The last term of Eq. (3) is the Lorentz force and j_r is the induced radial current, which is given by Ohm's law (2) as:

$$j_r = -\sigma \left(\frac{\partial \varphi}{\partial r} + \frac{S}{r} w \right) \tag{4}$$

Moreover, the charge conservation condition requires $\frac{\partial r j_r}{\partial r} = 0$, implying that $j_r = I/r$, where I is some constant defined for convenience as $I = -\sigma c$. Here, c is a new constant to be determine later. Then, inserting j_r in (3), we obtain the final governing equation for the velocity:

$$\frac{\partial w}{\partial t} = -\frac{1}{\rho}\frac{dp}{dz} + \frac{\nu}{r}\frac{\partial}{\partial r}\left(r \frac{\partial w}{\partial r} \right) - \frac{\sigma c}{\rho}\frac{S}{r^2} \tag{5}$$

2.1 The Small Gap Approximation

As developed by Rudraiah (1966), we perform the dimensionless variable change $\chi = \frac{r - r_i}{\delta}$ ($\chi \in [0, 1]$) to Eq. (5) leading to:

$$\frac{\partial w}{\partial t} = -\frac{1}{\rho}\frac{dp}{dz} + \frac{\nu}{\delta^2}\left(\frac{\partial^2 w}{\partial \chi^2} + \frac{1}{(\chi + r_i/\delta)}\frac{\partial w}{\partial \chi} \right) - \frac{\sigma c}{\rho \delta^2}\frac{S}{(\chi + r_i/\delta)^2} \tag{6}$$

where $\delta = r_o - r_i$ is the annular space, in which r_o and r_i are the outer and inner radii respectively. Assuming a very small annular space defined by $r_i/\delta \gg 1$, the coefficient $(\chi + r_i/\delta)^{-1} \approx \delta/r_i$ and $(\chi + r_i/\delta)^{-2} \approx (\delta/r_i)^2$. Then Eq. (6) simplifies to:

$$\frac{\partial w}{\partial t} = -\frac{1}{\rho}\frac{dp}{dz} + \frac{\nu}{\delta^2}\frac{\partial^2 w}{\partial \chi^2} - \frac{\sigma c}{\rho}\frac{S}{r_i^2} \tag{7}$$

Within this approach, the Lorentz force becomes constant and the effect of curvature in viscous terms is just negligible.

3 The Steady Flow Case

We assume a constant pressure gradient, $-\frac{dp}{dz} = G$, which transforms (7) to:

$$\frac{G}{\rho} + \frac{\nu}{\delta^2}\frac{\partial^2 w}{\partial \chi^2} - \frac{\sigma c}{\rho}\frac{S}{r_i^2} = 0 \tag{8}$$

or in dimensionless form to:

$$\frac{\partial^2 \tilde{w}}{\partial \chi^2} - \frac{\tilde{c}}{\tilde{r}_i^2} Ha^2 + \tilde{G} = 0 \tag{9}$$

The variables in tilde indicate dimensionless quantities measured in terms of δ, u_0 and S, for length, velocity and field scales. The Hartmann number defined as $Ha = S\sqrt{\sigma/\nu\rho}$ is the ratio between magnetic and viscous forces. Equation (9) has a standard analytical solution that combined with non-slip boundary conditions at the inner and outer radii is:

$$w(\chi) = \left(\frac{G}{2} - \frac{c\,Ha^2}{2\,r_i^2}\right)(1-\chi)\chi \tag{10}$$

From here until the end of the section we shall not remove the tilde symbol for simplicity.

Moreover, Ohm's law (4) is equal to $j_r = -\sigma \frac{c}{r}$, and then integrated from inner to outer radius, the resulting equation in dimensionless form is:

$$\frac{c}{r_i} = \Delta\varphi + \frac{1}{r_i}\int_0^1 w(\chi)\,d\chi \tag{11}$$

Equation (11) is solved for c, by considering a short circuit, such that, $\Delta\varphi = 0$. We get then:

$$c = \frac{G\,r_i^2}{Ha^2 + 12r_i^2}$$

Inserting this last result into (10), we obtain a final expression for the velocity:

$$w(\chi) = \frac{6G\,r_i^2}{Ha^2 + 12r_i^2}(1-\chi)\chi$$

Velocity profiles at different values of the Hartmann parameter (Ha) for a given annular space are plotted in Fig. 2a. In this case, as Ha increases the flow slows down until it completely stops. This behavior is typical of Hartmann flow between parallel plates. However, in our model the magnetic field is parallel to the walls: consequently, the Hartman layers are not formed as in the Hartmann's problem. In Fig. 2b, the effect of varying the ratio between the inner to outer radius, $k = r_i/r_o$, on velocity profile for a given Hartmann value is shown. In this case, as k tends to unity, the hydrodynamic regime dominates due to a vanishing effect of Lorentz forces. In other words, as the annular space is reduced the effect of magnetic field on the flow tends to vanish.

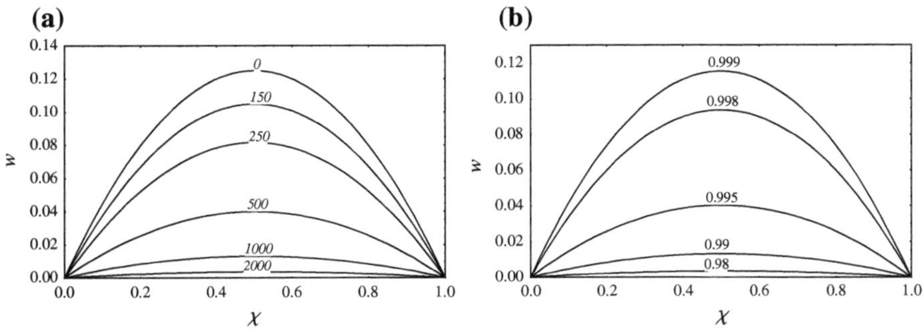

Fig. 2 Velocity profiles for several **a** Ha values and **b** radius ratio $k = r_i/r_o$

4 The Oscillatory Flow Case

In this case, the flow is driven by a harmonic gradient pressure which results in a harmonic fluid motion with oscillating frequency ω, that is:

$$-\frac{dP}{dz} = \Re\left[G \exp\left(\imath\omega t\right)\right]$$

The symbol $\Re$ refers to the real part of the expression. It is usual to express quantities as:

$$\left\{\begin{array}{c} w(\chi, t) \\ c \end{array}\right\} = \Re\left[\left\{\begin{array}{c} f(\chi) \\ c_o \end{array}\right\} \exp\left(\imath\omega t\right)\right] \tag{12}$$

where $f(\chi)$ and c_o are complex quantities. This reduces Eq. (7) to:

$$\frac{\partial^2 f}{\partial \chi^2} - \frac{\imath\omega\delta^2}{\nu} f + \frac{G\delta^2}{\rho\nu} - \frac{\sigma c_o \delta^2}{\rho\nu}\frac{S}{r_i^2} = 0$$

or in dimensionless form to:

$$\frac{\partial^2 \tilde{w}}{\partial \chi^2} - \imath R\omega \tilde{w} - \frac{\tilde{c_o}}{\tilde{r}_i^2} Ha^2 + \tilde{G} = 0$$

where $R\omega$ is the oscillatory Reynolds number defined by $R\omega = \dfrac{\omega\delta^2}{\nu}$ and variables in tilde are dimensionless quantities as defined in Sect. 3. The solution of the last equation considering non-slip boundary conditions at the wall is:

$$w(\chi) = \frac{2\imath\left(G\, r_i^2 - c_o\, Ha^2\right)}{r_i^2\, R\omega \cosh\left(\frac{1}{2}\sqrt{\imath R\omega}\right)} \sinh\left(\frac{1}{2}\sqrt{\imath R\omega}(x-1)\right) \sinh\left(\frac{1}{2}\sqrt{\imath R\omega x}\right) \tag{13}$$

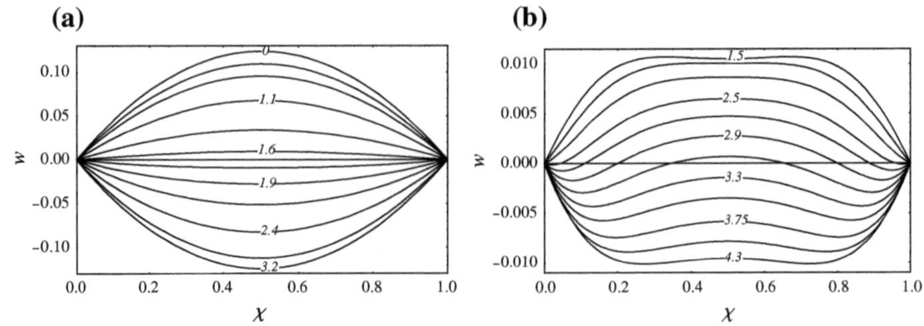

Fig. 3 Hydrodynamic velocity profiles for half of a period **a** $R\omega = 1$, **b** $R\omega = 100$. Labels indicate dimensionless temporal evolution, t.

As before, the tilde symbol has been removed for simplicity.

We insert Eq. (12) into Ohm's law (4) and into $j_r = -\sigma \frac{c}{r}$, the resulting equations are equal and integrated from the inner to the outer radius, which results in dimensionless form to:

$$\frac{c_o}{r_i} = \Delta\varphi + \frac{1}{r_i}\int_0^1 w(\chi)\, d\chi \tag{14}$$

This equation can be solved for c_o in the case of short circuit, that is, for $\Delta\varphi = 0$. Explicitly we obtain:

$$c_o = \frac{r_i^2\left(-\imath G\sqrt{R\omega} + 2\sqrt{\imath}\, G \tanh\left(\frac{1}{2}\sqrt{\imath R\omega}\right)\right)}{\sqrt{R\omega}\left(r_i^2 R\omega - \imath M^2\right) + 2\sqrt{\imath}\, M^2 \tanh\left(\frac{1}{2}\sqrt{\imath R\omega}\right)}$$

Inserting this result into (13) and simplifying, we obtain a final expression for the velocity:

$$w(\chi) = \frac{2\imath r_i^2 G\sqrt{R\omega}\operatorname{sech}\left(\frac{1}{2}\sqrt{\imath R\omega}\right)\sinh\left(\frac{1}{2}\sqrt{\imath R\omega}(x-1)\right)\sinh\left(\frac{1}{2}\sqrt{\imath R\omega}\, x\right)}{\sqrt{R\omega}\left(r_i^2 R\omega - \imath M^2\right) + 2\imath M^2 \tanh\left(\frac{1}{2}\sqrt{\imath R\omega}\right)}$$

In order to understand the effect of Ha on the oscillatory MHD flow, we first analyze the hydrodynamic problem ($Ha = 0$) for two extreme values of $R\omega$ for a given annular space as shown in Fig. 3. In both figures, the oscillating motion is shown for half of a period starting from the maximum velocity in one direction to the maximum velocity in the opposite direction. This corresponds for example to $t = 0$ until $t = 3.2$ in Fig. 3a, or to $t = 1.5$ until $t = 4.3$ in Fig. 3b. The effect of $R\omega$ on the velocity profile is observed in Fig. 3b. In fact, as $R\omega$ increases two distinct regions of the flow appear: a boundary layer region where viscous effects dominate

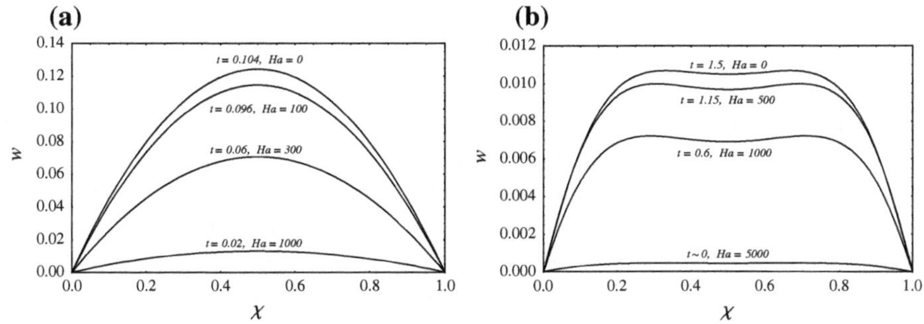

Fig. 4 Maximum velocity profiles at several Ha numbers for **a** $R\omega = 1$ and **b** $R\omega = 100$.

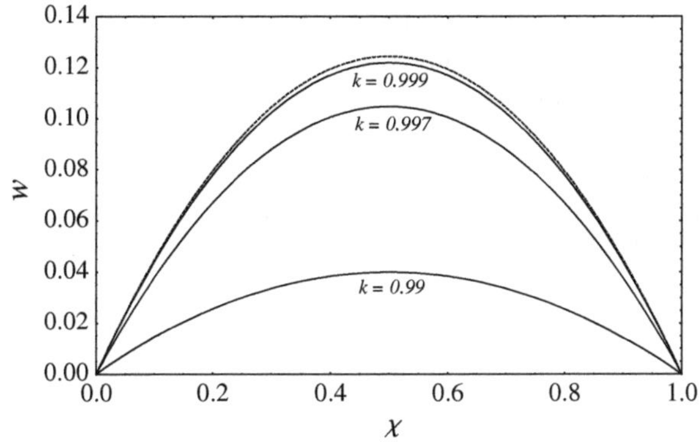

Fig. 5 Maximum velocity profiles at various ratio $k = r_i/r_o$ for $Ha = 500$ and $R\omega = 1$. *Dashed curve* corresponds to hydrodynamic profile ($Ha = 0$).

and a non-viscous core region. Another interesting effect is the phase shift between the regions which appear as $R\omega$ increases. For example, in Fig. 3b for $t = 2.9$ the fluid in the boundary layer flows in the opposite direction with respect to the flow in the core.

Once the main features of hydrodynamic oscillatory flow have been described, we proceed to analyze the effect of Ha on the MHD oscillatory profile for a given $R\omega$, as depicted in Fig. 4.

For $R\omega = 1$, shown in Fig. 4a, only maximum velocity profiles for various Ha numbers are compared. As seen in the plot, the increase of magnetic field (or Ha) slows down the flow as in the steady case. This fact is confirmed in Fig.4b which also corresponds to maximum velocity profiles at $R\omega = 100$.

The effect of varying the ratio between the inner to the outer radius, $k = r_i/r_o$, on the flow for given Ha and $R\omega$ numbers is shown in Fig. 5. As in the steady case, the hydrodynamic regime dominates as k tends to unity due to the vanishing effect of Lorentz forces.

5 Conclusion

We have considered the laminar MHD flow between two infinitely long, perfectly conducting cylinders subject to a purely azimuthal magnetic field. The fluid motion is driven by a pressure gradient, either constant or harmonic. The governing MHD equations have been solved by considering the inductionless approximation and assuming a small annular gap compared to the inner radius. For the case of steady motion, the effect of magnetic field is to slow down the flow as the Hartmann value increases. This effect is well known in the MHD flow between parallel plates known as the Hartmann problem. However, a distinct feature in our problem is the absence of Harmann layers which remain unchanged the velocity profile but just slow down the flow by Lorentz force. For the case of harmonically driven flow, the magnetic field reduces the amplitude of oscillation, and due to the absence of Hartmman layers, the velocity profile also remains unchanged as in the steady case. This can be understood since the Lorentz force is homogenous in the annular space for the approximation with a narrow gap.

Acknowledgments We thank to Paul Wiley for his comments to this manuscript. This project was supported by SIP-IPN grant 20121171.

References

Globe S (1959) Laminar steady-state magnetohydrodynamic flow in an annular channel. Phys Fluid 2(4):404–408

Elco RA, Hughes WF, Young FJ (1962) Theoretical analysis of the radial field vortex magneto-gas dynamic generator. Zeitschrift für angewandte Mathematik Physik 13(1):1–13

Edward R.D'Sa. (1966) Magnetohydrodynamic duct flow under circular and radial magnetic fields. Zeitschrift für angewandte Mathematik Physik 17(1):140–156

Zhao Y, Zikanov O, Krasnov D (2011) Instability of magnetohydrodynamic flow in an annular channel at high Hartmann number. Phys Fluids 23(084103):357–388

Narasimhan MNL (1964) Pulsating magnetohydrodynamic flow superposed on the steady laminar motion of an incompressible viscous fluid in an annular channel. J Phys Soc Jpn 19:117–120

Rudraiah N (1966) Pulsating magnetohydrodynamic flow in an annular channel. Proc Indian Acad Sci Sect A 63(5):293–304

Priede J (2009) Pseudo-magnetorotational instability in a Taylor-Dean flow between electrically connected cylinders. Phys Rev E 79:066314–1-066314-9

Davidson PA (2001) An introduction to magnetohydrodynamics. Cambrige University Press, United Kingdom

Müller U, Bühler L (2001) Magnetofluiddynamics in channels and containers. Springer-Verlag, Berlin

Surface Waves Generated on Electrolytes by a Traveling Electromagnetic Force

Gerardo Alcalá and Sergio Cuevas

Abstract This paper presents an experimental study of gravity-capillary waves generated at the free surface of a thin-film of electrolyte ($NaHCO_3$) due to the presence of an electromagnetic force created by the interaction of a direct electric current and a traveling magnetic field.The field is generated by a permanent magnet moving in straight line, localized externally to the bottom wall of the fluid container. The dominant component of the magnetic field is perpendicular to the plane of the fluid surface in equilibrium. The current is applied transversely to the motion of the magnet through a pair of parallel electrodes, in such a way that the force points either in favour or against the motion of the magnet, depending on the polarity of the electrodes and the magnet orientation. A vertical force component is also generated near the edges of the magnet. It is shown that the electromagnetic force acts as an obstacle for the flow (a *magnetic obstacle*) and, similarly to a moving solid object, it is able to generate a stationary wave pattern. This patten is reconstructed by optical methods for several magnet velocities. Differences produced by the force acting in favour or against the magnet motion are discussed.

1 Introduction

Surface waves are a common phenomenon in nature. In many situations they are generated by obstacles interacting with the free surface as occurs, for instance, when a stone is dropped in a pond, a duck swims at the surface of a lake, or a river flow passes through a fishing line, among many other situations. Such waves establish a competition between gravity and surface tension, and the equilibrium shape taken by the surface of the fluid is a consequence of this balance (Landau and Lifshitz 1959).

G. Alcalá (✉) · S. Cuevas
Instituto de Energias Renovables, Universidad Nacional Autónoma de México,
A.P. 34, 62580 Temixco, Morelos, México
e-mail: gealp@ier.unam.mx

J. Klapp and A. Medina (eds.), *Experimental and Computational Fluid Mechanics*,
Environmental Science and Engineering, DOI: 10.1007/978-3-319-00116-6_32,

In the present work, we are interested in the experimental study of capillary-gravity waves generated by localized electromagnetic forces that act as *obstacles* for the flow. In a biological context capillary-gravity waves are involved in the mechanisms of locomotion, hunting or courtship for insects or arachnids (Bush and Hu 2006; Bleckmann and Bender 1987). On the other hand, magnetic fields can be used for the control of surface waves in different metallurgical applications (Sreenivasan et al. 2005). Before addressing the analysis of capillary-gravity waves created by electromagnetic forces, we recall some important features of surface waves generated by solid obstacles.

When an object partially immersed in a fluid is moving uniformly, it will experience drag forces acting on it. Different contributions to the drag or resistance forces can be identified (Burghelea and Steinberg 2002). In fact, in addition to the common viscous resistance, R_f, an eddy resistance, R_e, caused by either laminar or turbulent wakes will be present. But the generation of capillary-gravity waves can also contribute to another form of resistance called *wave resistance*, R_w, which represents the momentum removed by the waves from the object and taken away to infinity (Havelock 1919). Due to the viscosity of water and electrolytes, usually R_f and R_e can be considered negligible compared with the capillary-gravity wave resistance (Shliomis and Steinberg 1997).

In the frame of an object moving in the fluid with velocity $\mathbf{U}$, we will observe a stationary wave pattern where the wave speed, c, satisfies the following condition (Lighthill 1979)

$$c(\mathbf{k}) = U \cos\theta(\mathbf{k}), \tag{1}$$

where $\mathbf{k}$ is the wave vector and θ is the angle between $\mathbf{U}$ and $\mathbf{k}$, so that the component $U\cos\theta$ of the object velocity, at right angles to the crest, can cancel the crest's motion at the wave speed c.

In general, due to the dispersive nature of the capillary-gravity waves, complex surface structures will be observed. By neglecting viscous effects, the phase velocity, c, can be expressed by the following equation

$$c = \left[\tanh kh\,(g/k + \gamma k/\rho)\right]^{1/2}, \tag{2}$$

where ρ is the fluid density, γ is the liquid-air surface tension, g is the gravitational acceleration, h is the fluid depth, and k is the wave number (Acheson 1990). As we can see, the velocity c depends on the wavelength, so that among all wavelengths, Eq. (1) fixes some of them which can be stationary.

According to Eq. (2), for deep waters ($kh \gg 1$) phase velocity has a minimum $c_c = (4g\gamma/\rho)^{1/4}$ at the capillary wave number $\kappa = \sqrt{\rho g/\gamma}$. For water at room temperature, we have $c_c = 23$ cm/s and $\lambda_c = 1.7$ cm, where λ_c is the wavelength. This means that Eq. (1) cannot be satisfied for $U < c_c$, and no such waves can be present for a slow moving object. In fact, some waves will be found upstream and some others downstream the obstacle, which is related with the way the energy is propagated.

Now, let us consider a frame of reference moving with the stream and an obstacle traveling with velocity U in the upstream direction. The longer waves generated by the motion of the object, with wave speed $c = U$, greater than the energy propagation (group) velocity c_g, are then left behind the obstacle and travel in the downstream direction. In other words, after energy is generated, the long waves lag behind the object. Shorter waves, for which $c = U$ is smaller than c_g, will be found ahead of the obstacle (Lighthill 1979). The case of surface waves generated by objects of small size, relative to the capillary length $\kappa^{-1} = (\gamma/(\rho g))1/2$, has also been considered by many authors Raphaël and deGennes (1996); Richard and Raphael (1999); Sun and Keller (2001); Chepelianskii et al. (2008); Closa et al (2010); Le Merrer et al. (2011). Let us now consider surface waves generated by a localized electromagnetic force.

2 Waves Generated by Electromagnetic Forces

In this paper, we show that localized electromagnetic forces are an alternative to the use of solid obstacles for generating surface waves in electrically conducting fluids. In fact, it has been shown that when a uniform flow of a highly conducting fluid (*v.e.* a liquid metal) passes through a localized non-homogenous magnetic field, for instance produced by a dipolar permanent magnet, the field acts in a similar way as a solid obstacle by deflecting the incident flow and generating a wake behind the magnetic field spot (Cuevas et al. 2006; Votyakov et al. 2007; Votyakov and Kassinos 2009). In this case, the currents induced by the relative motion of the liquid and the magnetic field interact with the applied field producing an opposing Lorentz force $\mathbf{j} \times \mathbf{B}_0$, where $\mathbf{j}$ is the induced current density and $\mathbf{B}_0$ is the applied field. Therefore, it is possible to conceive the localized magnetic field as an obstacle for the flow; incidentally, the term *magnetic obstacle* was coined for that purpose (Cuevas et al. 2006). It has to be mentioned, however, that flows past solid and magnetic obstacles present very important differences.

If the conducting fluid is an electrolyte, induced currents are negligibly small and in order to observe the effect of a localized magnetic field in an electrolytic flow, an external current, $\mathbf{j}_0$, has to be applied. In fact, experiments of flows of electrolytes past localized magnetic fields have been performed by dragging an external permanent magnet at constant velocity close to a quiescent thin layer of electrolyte, while a steady current was imposed on the fluid layer transversely to the motion of the magnet (Honji 1991; Honji and Haraguchi 1995; Afanasyev and Korabel 2006). In these studies, the attention was focused on the velocity patterns created in flow planes parallel to the bottom wall, while the effects of the localized Lorentz force on the free surface of the electrolyte were completely disregarded. Here, we are interested in studying flow regimes where three-dimensional effects produced by localized Lorentz forces are able to perturb the free surface and generate surface wave patterns.

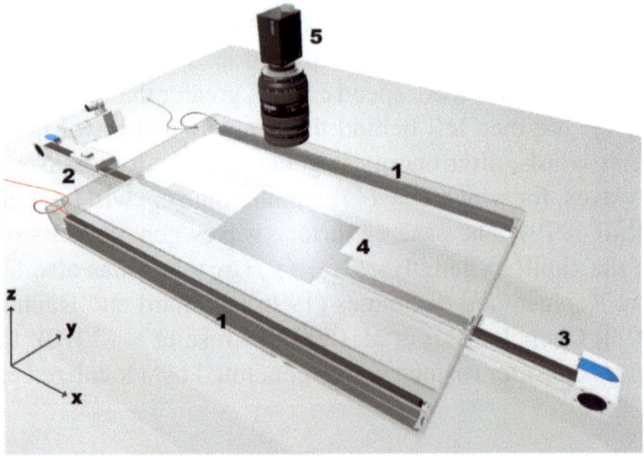

Fig. 1 Experimental setup. A localized Lorentz force is created by the interaction of a D.C. current applied through parallel electrodes (1) and the field of a permanent magnet (2) dragged axially outside the bottom wall with a linear actuator (3). The perturbed free surface is reconstructed at the observation zone (4) with the aid of a camera (5), using the synthetic Schlieren method (Moisy et al. 2009)

3 Experimental Setup and Procedure

A sketch of the experimental setup is shown in Fig. 1. The experiments were conducted in a rectangular crystal tank (35 cm $\times$ 60 cm $\times$ 5 cm), set horizontally and filled with a 8.6% aqueous electrolytic solution of $NaHCO_3$ up to the depth of 3 mm. Two electrodes (square graphite rods of 1 cm $\times$ 1 cm of cross-section and 50 cm in length) were placed along the two long sides of the container and connected to a power supply that provided a constant voltage of 60 V, so that a direct electric current (D.C.) of 2 A circulated in the fluid layer in the transversal direction. Electrodes were inside chambers (*bubble traps*) that allow the electric current to flow at the bottom while preventing the bubbles to invade the main flow region. A square permanent magnet of side length $L = 2.54$ cm, located below the bottom wall of the tank, was moved along the symmetry axis of the tank at a constant speed, U, using a linear actuator (FESTO EGC-70-500-TB-KF-0H-GK) driven by a servo motor (FESTO EMMS-AS-70-M-RS), in which U can be carefully controlled. In the experiment, starting at rest, the manget was suddenly set in motion reaching a constant velocity before it enters below the tank. The interaction of the D.C. current and the localized magnetic field of the traveling magnet, produced a Lorentz force that perturbed the free surface and generated a wave pattern. The Lorentz force also affected the bulk flow but this is not analyzed here.

Since the main component of the magnetic field is normal to the bottom wall and the applied current is transversal to the motion of the magnet, the Lorentz force points either in the direction of the motion of the magnet or against it, depending on the polarity of the electrodes and the orientation of the permanent magnet. If the polarity of the electrodes (*i.e.* the direction of the current) is fixed, the force direction

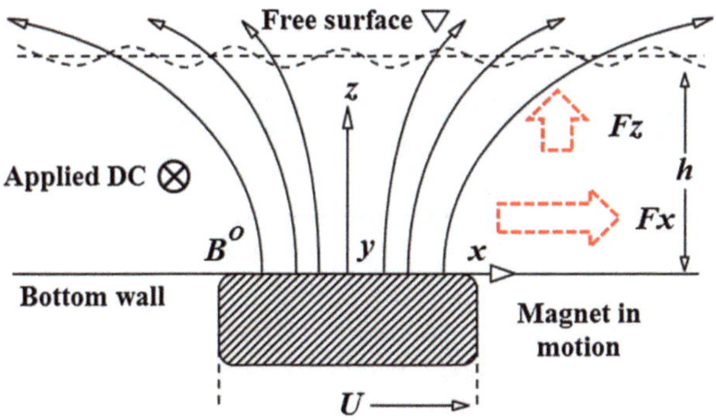

Fig. 2 Sketch of the physical situation when the north pole of the traveling magnet (moving to the *right* with speed U) is oriented upwards and the transversal current enters the plane of the figure. In this case, one component of the localized Lorentz force points in the direction of the motion of the magnet and another points vertically

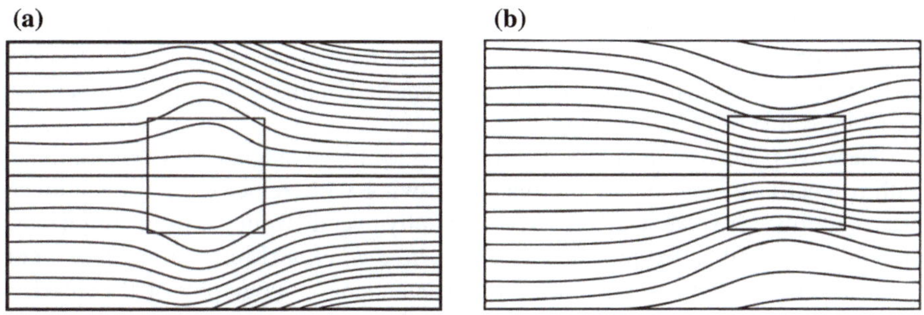

Fig. 3 *Top* view of streamlines in the flow of an electrolyte past a localized magnetic field (represented by the *square*). Streamlines expand for a positive force and contract for a negative one. **a** Positive force. **b** Negative force

depends only on the magnet orientation (see Fig. 2). Let us call *positive force* the case when the horizontal component of the Lorentz force points in the direction of the motion of the magnet, and *negative force* when it points against the motion of the magnet. For a positive force the streamlines expand around the region affected by the localized magnetic field, while for the other case the streamlines contract in this region, as shown in the Fig. 3a, b, respectively. Notice that the effects of a positive force are similar to those of a solid object while the effects of a negative force have no mechanical analogy. We have to consider, however, that near the magnet borders the component of the field in the axial direction gives rise to a Lorentz force component in the vertical direction that is able to perturb the free surface. In fact, the strength of the perturbation depends on the velocity of the magnet.

Two relevant dimensionless numbers can be identified in this problem. First, the Reynolds number, $Re = UL/\nu$, based on the speed of the magnet U, the side length of the magnet L, and the kinematic viscosity of the fluid ν. Secondly, the Lorentz force

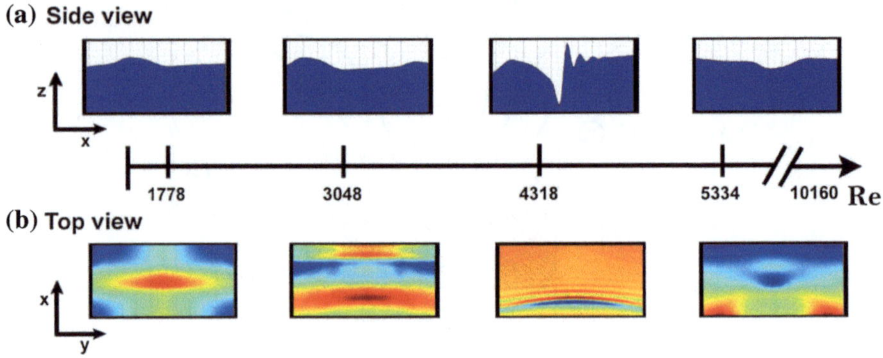

Fig. 4 Wave patterns for a positive force for different Reynolds numbers. **a** *Vertical* wave patterns profiles at the mid symmetry plane. **b** *Horizontal* proyection of free surface topography

parameter defined as $Q = j_0 B_{max} L^3/\rho \nu^2$ where j_0 is the applied current density, B_{max} is the maximum strenght of the magnetic field at the surface of the magnet, and ρ is the mass density. The experiments reported in this work were conducted by fixing the value of Q and varying Re.

The wave pattern generated by the traveling localized magnetic force is stationary with respect to the system of the magnet moving with velocity U. The stationary wave patterns were characterized experimentally over a 10 cm $\times$ 10 cm region using an optical method called *free surface synthetic Schlieren* (Moisy et al. 2009). This method allows the measurement of the instantaneous topography of the interface between two transparent fluids. In order to implement this technique, a camera located above the tank with the fluid at rest (see Fig. 1) records an aleatory dot pattern placed below the transparent bottom wall. When the free surface is perturbed, an apparent motion of the dots is observed due to refraction. By using a digital image correlation (DIC) algorithm to measure the apparent displacement field between the refracted image and the reference image obtained when the surface is flat, it is possible to calculate the height, h, of the free surface by solving the following equation

$$\nabla h = -\frac{\mathbf{r}}{h^*}, \tag{3}$$

where $\mathbf{r}$ is the displacement and the effective distance h^* is a constant that depends on the refraction indexes and thicknesses of different media (air, glass, and the electrolyte) through which light travels from the dots to the camera.

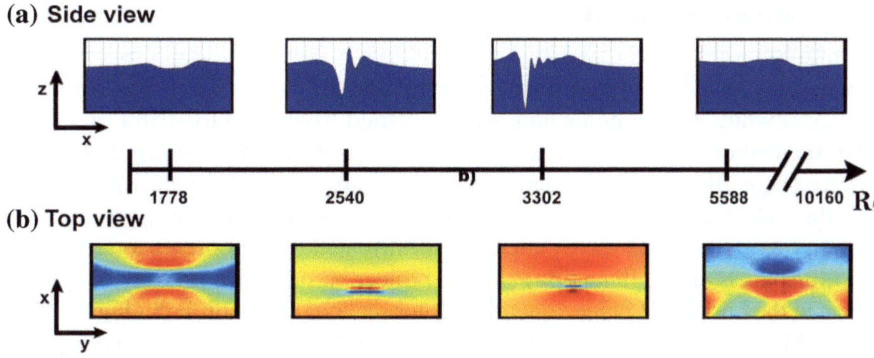

Fig. 5 Wave patterns for a negative force for different Reynolds numbers. **a** *Vertical* wave patterns profiles at the mid symmetry plane. **b** *Horizontal* proyection of free surface topography

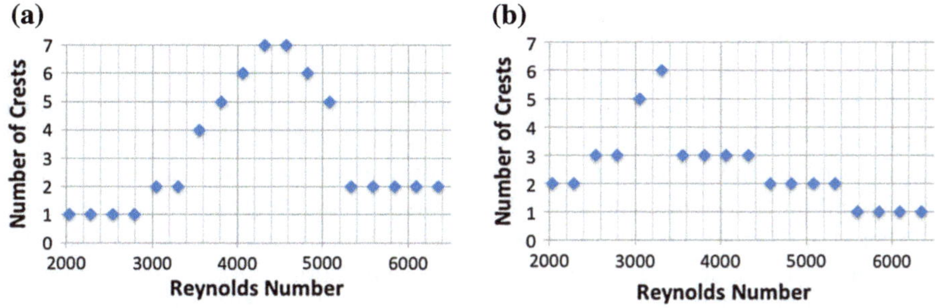

Fig. 6 Variation of the number of crests with *Re* due to emission of capillary-gravity waves. **a** Positive force. **b** Negative force

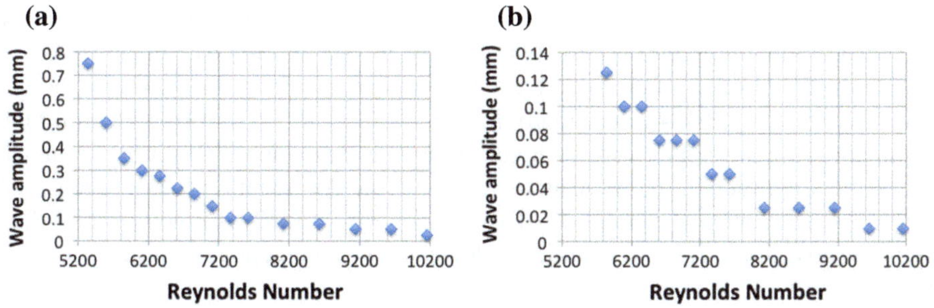

Fig. 7 Amplitude of the persisting wave in the large Reynolds number range as a function of *Re*. **a** Positive force. **b** Negative force

4 Experimental Results

Experiments were performed for a constant value of the Lorentz force parameter ($Q = 5.2$) while the Reynolds number, Re, varied from 1778 to 10160. In Figs. 4 and 5, the evolution of wave patterns for the cases of positive and negative forces, respectively, is illustrated schematically for growing Reynolds. In the upper panels (Figs. 4a and 5a), a side view of the vertical central plane that crosses the axial symmetry axis shows the wave pattern profiles created by the Lorentz force. In turn, the lower panels (Figs. 4b and 5b) show a projection of the topography of the free surface in the horizontal plane, where the color scale indicates different hights of the surface. Maximum wave amplitudes were 1.5 and 0.8 mm for positive and negative forces, respectively.

For a positive force, a dominant crest is observed in the low Re regime ($1778 \leq Re \leq 2794$). For higher Re, short waves appear in the front due to emission of capillary-gravity waves. Wave emission is observed within the range $3048 \leq Re \leq 5080$. In fact, the number of crests increases as Re increases up to reaching a maximum number of seven for an intermediate value ($Re \approx 4318$). This is shown in Fig. 6a, where the number of crests is plotted versus the Reynolds number for the positive force case. When Re is increased further, the number of crests decreases. Finally, in the high Re regime ($5334 \leq Re \leq 10160$), only a pair of crests can be identified, corresponding to a dominant valley that persists in this range of Re (see Fig. 6a). The depth of this valley decreases as Re grows. In Fig. 7a, the amplitude of the wave associated with the dominant valley is shown as a function of Re. It can be observed a sharp decrease of the amplitude from 0.8 mm for $Re = 5334$ up to practically disappearing for $Re = 10160$.

In the case of a negative force, a dominant valley is found for low Re ($2032 \leq Re \leq 2286$). As the Reynolds number increases, wave emission appears and the number of crests increases. Wave emission was observed in the range $2540 < Re < 5334$ and a maximum number of six crests was found for $Re \approx 3302$ (see Fig. 6b). As Re decreases, the number of crests reduces first to three ($3556 \leq Re \leq 4318$), then to two ($4572 \leq Re \leq 5334$), and finally to one ($5588 \leq Re \leq 10160$), indicating the dominant crest that appears in the higher range of Re explored (Fig. 6b). Figure 7b shows the amplitude of the wave associated with the dominant crest as a function of Re. In fact, the original amplitude (0.13 mm for $Re = 5842$) is much smaller than in the positive force case and similarly, decreases up to practically disappearing for $Re = 10160$.

5 Concluding Remarks

The generation of surface waves through localized traveling electromagnetic (Lorentz) forces in a thin layer of electrolyte was analyzed experimentally. The Lorentz force was created by the interaction of a direct electric current applied transversally to the motion of a permanent magnet external to the fluid container. In this work, it was shown that the localized Lorentz force acts as a magnetic obstacle that is able to generate not only vortical flows, as has been demonstrated by previous investigations (Honji 1991; Honji and Haraguchi 1995; Afanasyev and Korabel 2006), but also to perturb the free surface and produce free surface wave patterns. The instantaneous topography of the surface was reconstructed using the free surface synthetic Schlieren method (Moisy et al. 2009). The evolution of the wave patterns was characterized as a function of the Reynolds number (based on the velocity of the traveling magnet), while the Lorentz force parameter, Q, remained fixed. The Reynolds number varied from 1778 to 10160. In contrast with surface waves generated by moving solid obstacles, the Lorentz force can point either in favour (positive force) or against (negative force) the motion of the traveling magnetic field, depending on the orientation of the magnet and the direction of the applied current. In fact, the observed wave patterns differ according to the direction of the Lorentz force. It was found that, as the Reynolds number increases, there is a transition regime where short capillary-gravity waves are created by the localized electromagnetic force. For a positive force the transition goes from a dominant crest in the low Reynolds number regime to a dominant valley in the high Reynolds number regime. For a negative force, the transition occurs inversely from a dominant valley to a dominant crest. Emission of capillary-gravity waves starts at smaller Reynolds numbers in the case of a negative force. For both positive and negative forces, the amplitude of the persistent wave in the high Reynolds number regime was found to decrease and almost disappear as Re reached the largest explored value.

Acknowledgments This work has been supported by CONACyT, Mexico, under project 131399. G. Alcalá also acknowledges a grant from CONACyT.

References

Acheson DJ (1990) Elementary fluid dynamics. Clarendon Press, Oxford
Afanasyev YD, Korabel VN (2006) J Fluid Mech 553:119
Bleckmann H, Bender M (1987) J Arachnol 15:363–369
Burghelea T and Steinberg V, (2002) Phys Rev E **66**(5)
Bush JWM, Hu DL (2006) Annu Rev Fluid Mech 38:339–369
Chepelianskii A, Chevy F and Raphaël E, (2008) Phys. Rev. Lett. **100**(7)
Closa F, Chepelianskii AD, Raphaël E (2010) Phys Fluids **22**(5)
Cuevas S, Smolentsev S, Abdou MA (2006) J Fluid Mech 553:227–252
Havelock T H, (1919) Proc R Soc London A Math Phys Sci **95**(670):354–365
Honji H (1991) J Phys Soc Japan 60(4):1161–1164
Honji H, Haraguchi Y (1995) J Phys Soc Jpn 64(7):2274–2277
Landau LD, Lifshitz EM (1959) Fluid mechanics. Pergamon Press, Oxford
Le Merrer M, Clanet C, Quere D, Raphaël E, Chevy F (2011) Proc Nat Acad Sci **108**(37):15064–15068
Lighthill J (1979) Waves in fluids, 6th edn. Cambridge University Press, Cambridge
Moisy F, Rabaud M, Salsac K (2009) Exp Fluids 46(6):1021–1036

Raphaël E, deGennes PG (1996) Phys Rev E 53(4):3448–3455
Richard D, Raphael E (1999) Europhys Lett 48(1):49–52
Shliomis MI, Steinberg V (1997) Phys Rev Lett 79(21):4178
Sreenivasan B, Davidson PA, Etay J (2005) Phys Fluids 17:117101
Sun SM, Keller JB (2001) Phys Fluids 13(8):2146
Votyakov EV, Zienecke E, Kolesnikov YB (2008) J Fluid Mech 610:131–156
Votyakov EV, Kassinos SC (2009) Phys Fluids 21:097102

Study of a Surface of a Liquid Layer in Evaporation

Julio César Rubén Romo Cruz, Sergio Hernández Zapata and Gerardo Ruiz Chavarría

Abstract In a liquid layer of alcohol or another volatile substance a pattern of hydrodynamic cells are formed when relative humidity decreases under a certain threshold. We describe the thermodynamical conditions under which the cells appear. We also study the temperature as a function of the vertical coordinate in the layer and we find that the temperature gradient can not be used to explain the origin of the cells. The deformation of the free surface is determined by the Schlieren technique. For that a sheet of paper with a random pattern of dots is placed in the bottom of the container. Using PIV software the displacement of the dots in the pattern is determined. The statistical study of the displacement allows us to study the slope of the surface and eventually its shape. We calculate the correlation lenght and we find that it is of the order of the size of the cells.

1 Introduction

In a liquid-gas interface the expression for the relative humidity (RH) is:

$$RH = \frac{\text{Partial pressure}}{\text{Saturation pressure}}. \tag{1}$$

On the other hand, the chemical potential between the liquid and gaseous phases can be estimated by the following equation (http://en.wikipedia.org/wiki/Ideal_solution):

$$\mu_v = \mu_l + RTln(RH). \tag{2}$$

J. C. R. Romo Cruz (✉) · S. Hernández Zapata · G. Ruiz Chavarría
Facultad de Ciencias, Universidad Nacional Autónoma de México, Av. Universidad 3000, Circuito Exterior S/N Delegación Coyoacán, 04510 Ciudad Universitaria, DF, México
e-mail: juliocesar@fisica.unam.mx

J. Klapp and A. Medina (eds.), *Experimental and Computational Fluid Mechanics*, Environmental Science and Engineering, DOI: 10.1007/978-3-319-00116-6_33,

Here R is the universal constant of the gases, μ_l is the chemical potential for the liquid, μ_v is the chemical potential for the vapor, T is the temperature in Kelvin degrees. In a system in equilibrium the vapor is saturated. This implies $RH = 1$. Then $\mu_v = \mu_l$, i.e. the evaporation rate is equal to the condensation rate. On the other hand if $RH < 1$ then $\mu_l > \mu_v$, and the vapor is subsaturated and there is evaporation. Finally if $RH > 1$ then $\mu_v > \mu_l$, the vapor is supersaturated and there is condensation.

In the present paper it is described a system that consists of a liquid-gas interface (alcohol in the present case) where the liquid is in an evaporation process (Ward and Fang 1999; Bohren 2001; Vázquez et al. 1995; Aspe and Depassier 1990; Mancini and Maza 2004; Bestehorn 2007; Zhang 2006). The system is in equilibrium when there is not evaporation in the liquid. On the contrary, if there is evaporation the system becomes unstable and a pattern of *hydrodynamic cells* appears. These cells can be observed with the aid of a visualizer (Kalliroscope). If the relative humidity is less than 100 % but close enough to this value there are no patterns in the fluid. If the humidity decreases under a threshold value then the fluid becomes unstable. The hydrodynamic cells appear in the body of the liquid in a convective process. In the center of each one the fluid rises and in the borders it descends.

The liquid is poured in a crystallizer. The system is in equilibrium when the crystallizer is covered. The instability is reached when the crystallizer is uncovered. Then it is observed the gradual formation of the cells. Eventually they fill all the body of the liquid without overlapping. When the crystallizer is covered the cells disappear gradually until the alcohol layer becomes uniform. When this happens the liquid is in equilibrium with its vapor. That is, the relative humidity is 100 %.

It is possible to make an analogy between the performance of the hydrodynamic cells and the famous physical toy known as *"the drinking bird"*. This device can be considered like a thermodynamical machine that works when there is not equilibrium between a liquid and its vapor (Romo Cruz Julio 2012). If there is thermodynamical equilibrium the toy stops working. The analogy is shown in the Fig. 1.

In general, the convection processes produced in a liquid by evaporation have importance on many industrial applications. This includes, for example, many drying techniques, distillation (water separation processes in a salty mixture), heat exchangers, and specially the cooling of microelectronics equipment (Liu and Kabov 2012). In the latter case, it is usual to deposit thin films (of a certain liquid) on hot surfaces in microelectronics devices, where is very important avoid that evaporation instabilities generate fractures in the liquid thin film. This kind of fractures lead to disturbances in the layer behavior during the cooling process. From the point of view of fluid mechanics, the symmetry breaking due to evaporation involves a new mechanism that can be contrasted with the well-studied Marangoni and Rayleigh-Bénard mechanisms for instability generation. It should be stressed that this is really a particular case of the general topic of hydrodynamic instabilities that is central in the understanding of a huge variety of natural phenomena associated to fluid dynamics.

The paper is organized as follows: in Sect. 2 we study how the cell size depends on the liquid layer height. Then the relaxation time of the system when the relative humidity tends to 100 % is analyzed. This characteristic time depends on the

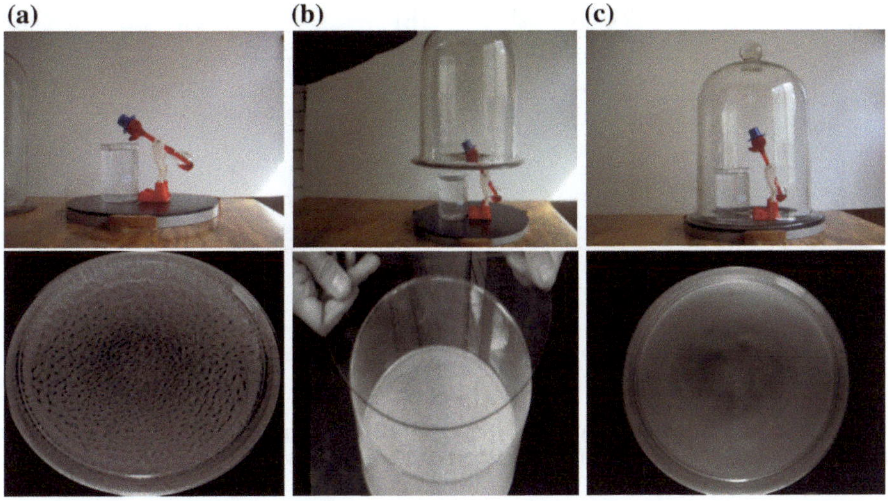

Fig. 1 **a** The bird working normally and at the same time dynamics in the liquid. **b** The bird is covered with a bell and the crystallizer is covered with a flat glass. **c** When both systems are saturated, they reach a stationary state of equilibrium

initial condition on which the system is prepared. In Sect. 3 we present experimental measurements on the dependence of the temperature with the vertical coordinate in the liquid layer. The goal is to study if the temperature gradient leads to the cell emergence. In order to establish if the cell dynamics is revealed on the behavior of the layer surface, in Sect. 4 we study the surface while the process of evaporation is taking place. With this purpose we use the Schlieren technique. Finally, in Sect. 5 we present our conclusions.

2 Characteristic Size of the Cells and Relaxation Time

In order to study the general characteristics of the hydrodynamic cells a set of experiments was performed. In a first stage it was determined the characteristic size of the cells as a function of the height of the liquid layer. The system of cells was recorded with a videocamera for different heights of the liquid layer. The digital images were treated with a software of image processing to measure the size of the cells. It is done in all the surface of the layer.

As it is observed in the Fig. 2a, the average size of the cells is an increasing function of the height of the liquid layer. A polynomial of the second degree was adjusted to the experimental data. For layers of height around 2.5 cm the cells begin to be imperceptible in the system.

Next, the time that is needed for the cells to completely disappear, from the moment in which the crystallizer is covered, is measured. This time is called the *relaxation time of the system*. In order to realize the measurement the *wet-bulb temperature* (WBT) and the *dry-bulb temperature* (DBT) are measured (Dossat 1995).

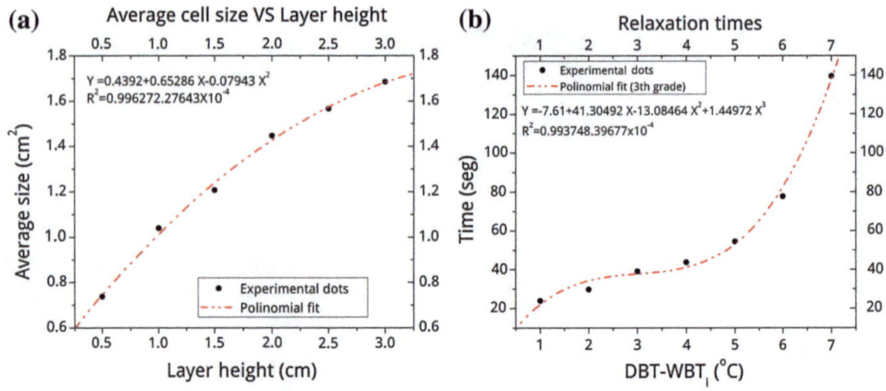

Fig. 2 Experimental results. **a** Average size of the cells as a function of the height of the layer. **b** Relaxation time as a function of the difference $DBT—WBT_i$ in Celsius degrees

To this end a pair of thermocouples is used. One is covered by a cloth dampened with alcohol. The two thermocouples are introduced in the crystallizer and the time elapsed for the WBT to become equal to the DBT is measured. At the beginning the WBT is less than the DBT because the alcohol in the dampened cloth becomes colder by evaporation. This is an indicator of the degree in which the system is far from equilibrium. It can be observed that the cells disappear when the WBT becomes equal to the DBT. The experiment is realized many times. In each one the initial WBT is different. In the Fig. 2b a graphics of the relaxation time as a function of the difference between the dry-bulb temperature and the initial wet-bulb temperature ($DBT—WBT_i$) is presented. A cubical adjustment to the experimental data is shown.

3 Measurement of the Temperature Gradients in the Liquid

We want to explore if the emergence of the cells is due to a temperature gradient. With the use of the thermo-physical properties of the alcohol (Romo Cruz Julio 2012), the threshold difference of temperature (ΔT_c) in the Rayleigh-Bénard instability is calculated. It must satisfy the condition (Guyon et al. 2001):

$$Ra = \frac{\alpha \Delta T_c g a^3}{\kappa \nu} = 1708 = Ra_c \quad (3)$$

where α is the thermal expansion coefficient, g is the gravity, a is the liquid layer height [for this case (Guyon et al. 2001)], κ is the thermal diffusivity coefficient, ν is the kinematic viscosity and Ra_c is the Rayleigh's critical number. From this equation the value ΔT_c can be obtained (Romo Cruz Julio 2012):

$$\Delta T_c \approx 0.01478\,\text{K}. \tag{4}$$

Analogously for Marangoni's instability we have the expression for ΔT_c (Guyon et al. 2001):

$$Ma = \frac{\frac{d\gamma}{dT}\Delta T_c a}{\eta\kappa} = 80 = Ma_c. \tag{5}$$

Here, $\frac{d\gamma}{dT}$ is the rate of change of surface tension with the temperature, η is the dynamic viscosity and Ma_c is the Marangoni's critical number. The value ΔT_c for the alcohol in this instability is calculated [with help of tables (Vázquez et al. 1995; Romo Cruz Julio 2012)], resulting:

$$\Delta T_c \approx 0.007335\,\text{K}. \tag{6}$$

Note that the differences for the thresholds critical values in both instabilities (see values in 4 and 6) are very small in alcohol. A very complex and expensive system of measurement is needed in order to measure temperature differences of this magnitude. It was proposed to work with a fluid where ΔT_c be greater.

A second calculation for the critical difference of temperature is realized. In this case for glycerol. ΔT_c is calculated for the Rayleigh-Bénard instability Vázquez et al. (1995):

$$\Delta T_c \approx 37.367\,\text{K} \tag{7}$$

In an analogous way, in the Marangoni instability, it is obtained (Romo Cruz Julio 2012):

$$\Delta T_c \approx 16.09\,\text{K} \tag{8}$$

In both cases it is obtained a much greater ΔT_c than the one obtained for pure alcohol. For this reason glycerol and alcohol mixtures are prepared. That is, liquids with measurable threshold difference of temperature. Also, a high precision thermometer based in a thermistor was constructed. A circuit for amplifying (around 50 times) the potential difference and with the possibility to adjust the offset was constructed. The calibration of the thermometer was done with water in a temperature interval from 15 to 35 °C.

The temperature difference through the liquid layer was measured. The results of the experiment are shown in the Fig. 3. An inverse temperature gradient was obtained, that is, the bottom of the liquid layer is colder than the surface. This implies that the difference of temperature can not be used to explain the origin of the cells, not in a situation like Marangoni (in this case the more natural since the surface is free) nor in a situation like Rayleigh-Bénard.

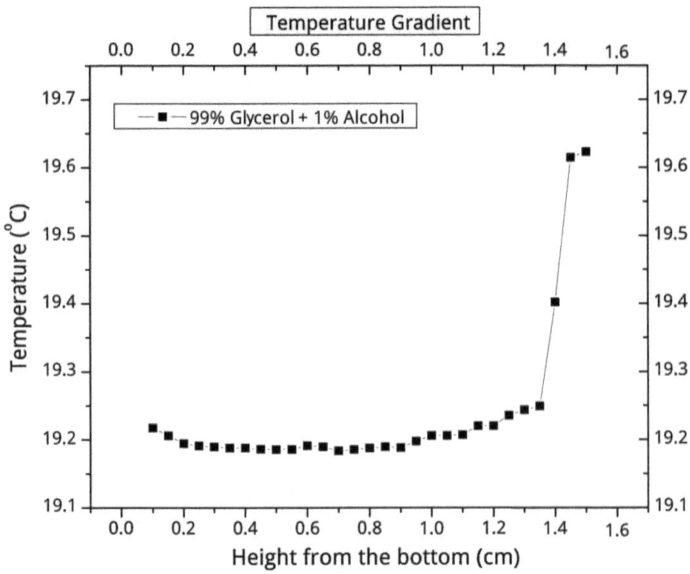

Fig. 3 Experimental data of the temperature as a function of the height taken from the bottom for: 99 % of glycerol and 1 % of alcohol

4 The Surface Deformation

The next experiment consisted in the verification of the existence of some deformation in the surface of the liquid layer when the cells were present. For this purpose the Schlieren technique was used (Moisy et al. 2009). The alcohol in the crystallizer was taken but this time without Kalliroscope. At the bottom of the crystallizer a pattern of random dots printed on a paper was placed. A high definition video camera was placed on top of the system. A section of the pattern of dots was focused (Fig. 4). The experiment consists on recording from the moment in which the crystallizer is covered. While recording the crystallizer is uncovered, the evaporation begins and the hydrodynamic cells appear.

The video is then transfered into a computer and fractionated in fifty images per second. A PIV software is then used to do the analysis. A reference image is taken (when the crystallizer is covered and the system is in equilibrium) and also images with the cells present (once the crystallizer is uncovered).Then the displacement of the dots between the latter images and the reference image is studied.

Using geometric optics in a paraxial approximation the following relation (Moisy et al. 2009) can be proved for the displacement field of the dots on the deformed surface:

$$\nabla h = -\left(\frac{1}{\alpha h_p} - \frac{1}{H}\right) \delta \mathbf{r}. \tag{9}$$

Fig. 4 Experimental assembly that consists of a pattern of *dots* printed on a piece of paper, a styrofoam container with alcohol and the video camera on top

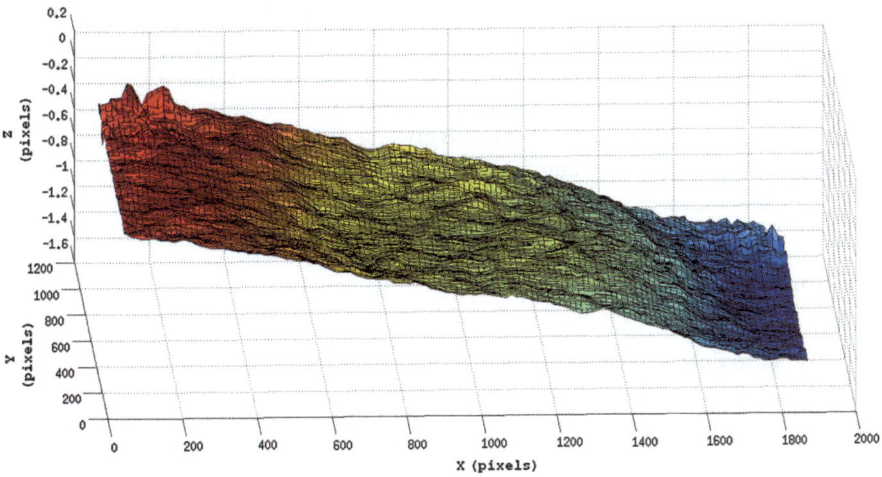

Fig. 5 Topography of the displacement field

where $\delta\mathbf{r}$ are the apparent displacement, H is the distance from the videocamera to the bottom, $\alpha = 1 - n/n'$ is determined by the index of refraction between the liquid and the external medium and h_p is the height of the liquid layer.

Finally, a numerical integration of the obtained displacement field is done and the surface deformed by the dynamics of the cells is rebuilt. We take advantage of this to calculate the lenght correlation.

A set of data representing a section of the surface is taken (Fig. 5) and the correlation lenght is obtained. For the case of a liquid layer with 0.5 cm of depth, the

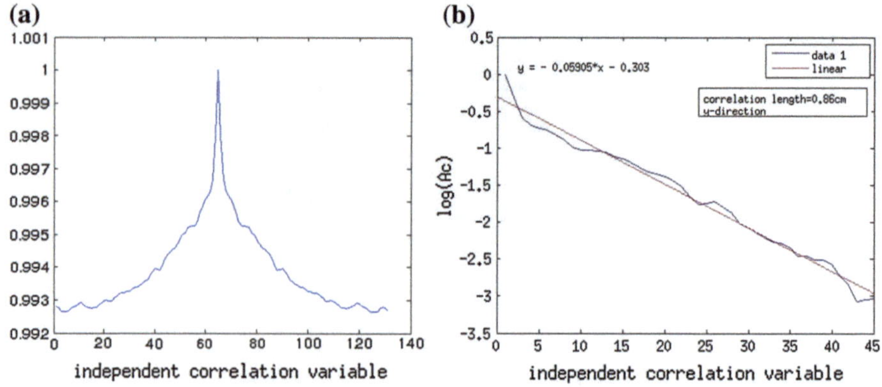

Fig. 6 Experimental results. **a** Curves (symmetrical) for the correlation length, was used the decreasing curve. **b** Linear fit in semi-log scale

correlation lenght is $0.35 \leq \delta \leq 0.86\,\text{cm}$, this is an acceptable value which means the method is reliable (Fig. 6).

A last experiment was measuring the density gradients in the Rayleigh-Bénard experiment using the Schlieren technique. The method is entirely analogous to the former. In this case, for a 1 cm distance between the bottom and the surface of the liquid layer, the correlation length is $0.8568 \leq \delta \leq 1.0043\,\text{cm}$.

5 Conclusions

An inverse temperature gradient was found in the body of the liquid, which rules out that dynamics of the cells is due to a direct temperature gradient in the liquid layer. We are studying the possibility that the instability is due to a pressure gradient. With the Schlieren technique it was possible to rebuilt the deformation of the surface and the study of the correlation length allows us to make the hypothesis that this deformation is due to the dynamics of the cells. The size of the cells is approximately of the order of the height of the layer in which they are studied, instabilities like Taylor, Marangoni and Rayleigh-Bénard present properties alike. We want to study the adimensional numbers that determine the threshold in relative humidity in order that the instability appears. The behavior of the cells is quite similar to the functioning of the drinking bird; we want to find more thermodynamic machines that function under these principles.

Acknowledgments Authors acknowledge support by DGAPA-UNAM under project IN-116312, "Vorticidad y Ondas no Lineales en Fluidos."

References

Aspe H, Depassier MC (1990) Evolution equation of surface waves in a convecting fluid. Phys Rev A 41(6):3125–3126

Bestehorn M (2007) Convection in thick and in thin fluid layers with a free surface: the influence of evaporation. Eur Phys J Special Topics 146:391–405

Bohren CF (2001) Clouds in a glass of beer: simple experiments in atmospheric physics. Dover science books, Mineola

Dossat RJ (1995) Principios de refrigeración. Compañía Editorial Continental, Mexico

Guyon E, Hulin J-P, Luc P, Mistecu CD (2001) Physical hydrodynamic. Oxford University Press, Oxford

Liu R, Kabov OA (2012) Instabilities in a horizontal liquid layer in cocurrent gas flow with an evaporating interface. Phys Rev E 85:066305

Mancini H, Maza D (2004) Pattern formation without heating in an evaporative convection experiment. Europhys Lett 66(6):812–818

Moisy F, Rabaud M, Salsac K (2009) A synthetic Schlieren method for the measurement of the topography of a liquid interface. Exp Fluids 46:1021–1036

Romo Cruz Julio CR (2012) Inestabilidades hidrodinámicas provocadas por evaporación. Tesis, Facultad de Ciencias, UNAM. http://132.248.9.195/ptd2012/agosto/302129094/Index.html

Vázquez G, Alvarez E, Navaza JM (1995) Surface tension of alcohol + water from 20 to 50. C J Chem Eng Data 40:611–6144

Ward CA, Fang G (1999) Expression for predicting liquid evaporation flux: statistical rate theory approach. Phys Rev E 59(1):429–440

Zhang N (2006) Surface tension-driven convection flow in evaporating liquid layers. Research Signpost 37/661 (2), Trivandrum

Analysis of the Performance of a Seven-Stage Axial Compressor Carried Out with a Numerical Simulation

I. González, M. Toledo, L. A. Moreno and G. Tolentino

Abstract This chapter presents the numerical analysis of a seven-stage axial compressor which was carried out with commercial computational fluid dynamic software. The results showed that the considerations made in the design method were not satisfied. On the other hand, the numerical analysis allows having an estimate of compressor performance in its operating range, with the aim to determinate the feasibility of the compressor without a physical model. This considerably reduced the cost in the design cycle of the compressor.

Nomenclature

C	Absolute velocity, velocity
h	Enthalpy
N	Rotational speed
P	Pressure
r	Radius
s	Entropy
T	Temperature
U	Blade velocity
α	Flow angle
β	Blade angle
π	Pressure ratio
ρ	Density
$\dot{m}$	Mass flow
η	Efficiency

I. González (✉) · M. Toledo · L. A. Moreno · G. Tolentino
SEPI, ESIME U. P. Adolfo López Mateos, Av. Instituto Politécnico Nacional S/N,
Instituto Politécnico Nacional, Edif. 5, 3er Piso, Col. Lindavista,
C.P. 07738, México D.F, México
e-mail: igonzalezn0900@alumno.ipn.mx

J. Klapp and A. Medina (eds.), *Experimental and Computational Fluid Mechanics*,
Environmental Science and Engineering, DOI: 10.1007/978-3-319-00116-6_34,

Subscripts

a	Axial velocity
s	Streamline component
w	Tangential component
0	Stagnation property

1 Introduction

The axial compressor is one of the principal components of the gas turbine in which the fluid is compressed as it passes through each stage. In the early days the design of the compressor was carried out by methods that were based in empirical knowledge and then they were evaluated in a rig test. After that, the geometry of the compressor was changed, with the aim to increase the performance (Horlock and Denton 2005). This is a very expensive procedure and it is one of the reasons that led the designers to look for a new design tool.

Nowadays with constant development and the more usual use of the CFD software the costs are considerably reduced. This tool makes feasible the analysis of a considerable amount of configurations, the investigation of new concepts and the analysis of the compressor performance. This is the foremost reason of why it is used in this study, even knowing that the results tend to be more qualitative than quantitative.

2 Compressor's Design

In the present study the free vortex design methodology was used to obtain the geometry of the blades and vanes of the compressor. Being its main consideration is that the fluid is in equilibrium in the radial direction. This means that pressure forces and inertia forces acting on a fluid element are in balance. A simplified expression of this is given by

$$\frac{\mathrm{dp}}{\rho\,\mathrm{dr}} = \frac{\mathrm{C_w^2}}{\mathrm{r}} + \frac{\mathrm{C_s^2}}{\mathrm{r_s}}\sin\alpha_\mathrm{s} + \frac{\mathrm{dC_s}}{\mathrm{dt}}\sin\alpha_\mathrm{s}, \quad (1)$$

Assuming that $\mathrm{r_s}$ is so large, and α_s is so small, the Eq. (1) is therefore

$$\frac{dp}{\rho dr} = \frac{C_w^2}{r}, \quad (2)$$

This is known as the radial equilibrium equation. Now taking the variation of the stagnation enthalpy with the radius as

Table 1 Design point

T_0	288.15 K
P_0	101325.00 Pa
π	4.15
$\dot{m}$	20.00 kg/s
Ca	150.00 m/s
U_{tmax}	354.72 m/s
N	15000 rpm

$$\frac{dh_0}{dr} = T\frac{ds}{dr} + \frac{dp}{\rho dr} + C_a\frac{dC_a}{dr} + C_w\frac{dC_w}{dr}, \tag{3}$$

and substituting the Eq. (2) in Eq. (3), it gives the basic equation for the analysis of flow in the compressor annulus as

$$\frac{dh_0}{dr} = T\frac{ds}{dr} + \frac{C_w^2}{r} + C_a\frac{dC_a}{dr} + C_w\frac{dC_w}{dr}, \tag{4}$$

In this study the term on the left-hand side of Eq. (4) was neglected, because the stagnation enthalpy was assumed to be uniform far from the wall of the annulus at the entry of the compressor. The first term on the right-hand side was neglected also, because the entropy variation related with the shock losses is significant with supersonic Mach numbers. However in the present study the relative Mach number was barely above the unity. On the other hand, the axial velocity was assumed to be constant across the annulus, then the Eq. (4) reduces to

$$\frac{\mathrm{dC_w}}{\mathrm{C_w}} = -\frac{\mathrm{dr}}{\mathrm{r}}, \tag{5}$$

Integrating the Eq. (5) gives

$$C_w * r = constant, \tag{6}$$

where the tangential component of the absolute velocity varies inversely with the radius, which is known as the free vortex condition. It is the foremost characteristic of the design method used in the present study.

3 Characteristics of the Compressor

The parameters of the design point are summarized in Table 1. These were taken from the Saravanamuttoo's book (Saravanamuttoo et al. 2009).

The main characteristics of the blades are summarized in the Table 2. They show that the first two stages are transonic and all the blades have a prime number. Another interesting characteristic is that beyond the third stage the flow and the blade angles

Table 2 Characteristics of the blades

Stage	No. of blades	Span at leading edge (mm)	α at mid span (°)	β at mid span (°)	Maximum relative Mach number
1	23	112.91	0.00	60.58	1.15
2	35	97.38	10.96	57.67	1.02
3	43	84.10	28.52	50.89	0.86
4	55	71.65	27.60	51.36	0.80
5	65	61.83	27.60	51.36	0.76
6	75	53.86	27.60	51.36	0.72
7	85	47.31	27.60	51.36	0.68

Table 3 Characteristics of the vanes

Stage	No. of blades	Span at leading edge (mm)	α at mid span (°)	β at mid span (°)	Maximum Absolute Mach number
1	14	105.14	27.20	51.55	0.54
2	20	90.74	41.04	42.08	0.66
3	38	77.88	50.89	28.52	0.78
4	60	66.74	51.36	27.60	0.75
5	68	57.84	51.36	27.60	0.71
6	78	50.58	51.36	27.60	0.67
7	90	44.23	51.36	27.60	0.64

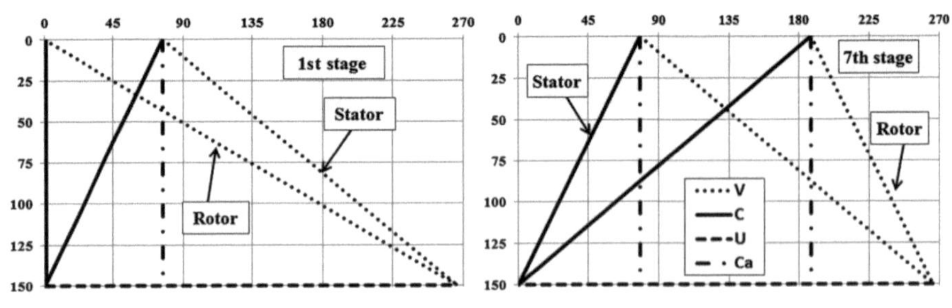

Fig. 1 Velocity triangles

are repeated. This is because the value of the degree reaction is 0.5 at mid span. It is worth saying that in the present study the mean radius was taken as constant. So the compressor has a tapered shape.

The main characteristics of the vanes are summarized in Table 3. In this case the numbers of the vanes are even. This helps to prevent the common multiples numbers in successive rows, leading to reduce the resonant forcing frequencies (Saravanamuttoo et al. 2009). As in the case of the blades, we can see that beyond the third stage, the flow and the blade angles are repeated.

The Fig. 1 shows the velocity triangle for the first and the last stages at mid span. Here we can see that the fluid enters in the axial direction at the entry of the compressor, and that the axial velocity is keeping constant through it. The triangles for

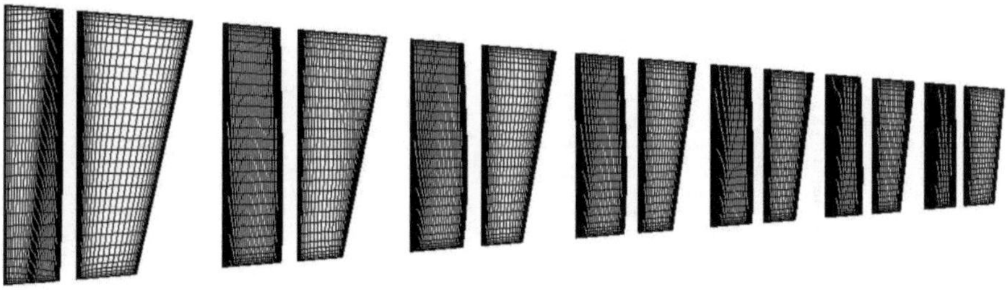

Fig. 2 Meridional view of the compressor mesh

the seventh stage show that the diffusion of the relative velocity is the same for the absolute velocity. This means that both the rotor and stator have an equal increase in the pressure ratio of the stage. This effect is present when the value taken for the degree reaction is 0.5.

4 Numerical Simulation

The simulation was carried out with the finite volume based Navier-Stokes solver Fine^{TM} Turbo by NUMECA International. The turbulence model used is the one-equation Spalart-Allmaras. The discretization was carried out with H-grids for the inlet and outlet ducts, meanwhile O-grids were used for the blades and vanes.

The mesh of the compressor has a total of 1,790,448 points. The Fig. 2 shows the mesh of the seven-stage compressor in a meridional view. The boundary conditions imposed at the inlet were the total pressure, the total temperature and the axial flow. At the outlet only the static temperature or the mass flow was set. These conditions were imposed based in the computations realized by Braun (Braun and Seume 2006) and Benini (Benini and Biollo 2007), which were similar to the one carried out in the present investigation.

5 Results

Figure 3 compares the axial velocity profile considered in the design method and the calculated one in the simulation, at the entry of the compressor. It Shows that the profile obtained in the simulation is not constant along the radial direction. Instead, it varies from root to tip. In around 20 % of the span it got the maximum value (161 m/s). On the other hand, near the tip and the root it shows a decrease below the design value (150 m/s).

Figure 4 compares the blade angle profile obtained in the methodology and that obtained in the simulation. As can be seen around the tip, the calculated values showed an increase of the angle. However, from the 60 % to the 0 % of the span there

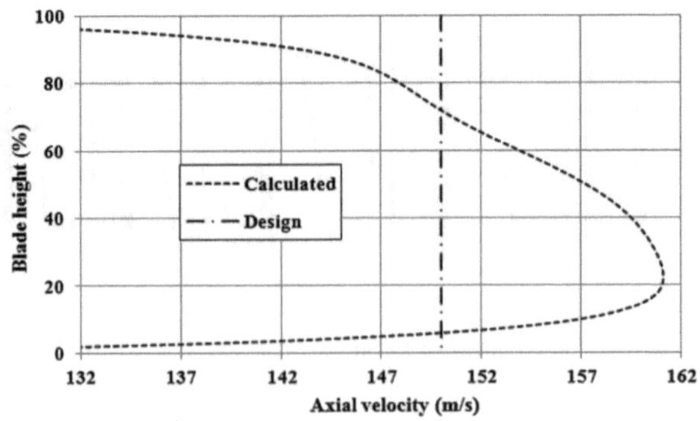

Fig. 3 Axial velocity profile at the entry of the compressor

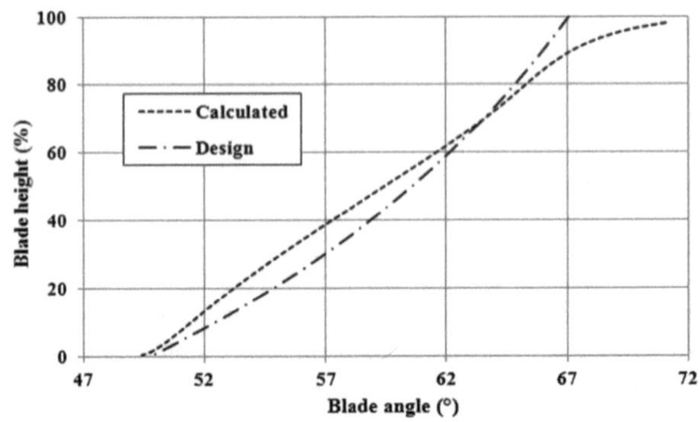

Fig. 4 Blade angle profile at the entry of the compressor

is a decrease. This happen because the axial velocity variation, directly influences the incidence on the blade. When the axial velocity is reduced the blade angle is increased, the opposite happens when the velocity is increased.

Figure 5 shows the variation on the axial velocity near the surfaces of the blades and vanes in the seven-stage compressor. In the blades of the first two stages there is an abrupt change in the velocity all along the span. After the second stage this abrupt change is concentrated at hub of the blade. Meanwhile, on the vanes it becomes considerable only after the second stage at the hub also. These abrupt variations are generated because of the separation of the flow in the suction surface of the blades and vanes of the compressor.

Figure 6 shows the flow separation on the suction surface of the first stage blade. This occurs because decreasing the axial velocity leads to the blade angle to have an excessive value. On the other hand, as the consideration of the equilibrium in the radial direction is not satisfied, the fluid in the hub tends to migrate to the tip of the blade. This causes a decrease in the compressor performance.

Figure 7 shows the flow separation in the blade row of the first stage. This can be taken as a representation of the rotating stall. In this phenomenon, the separated

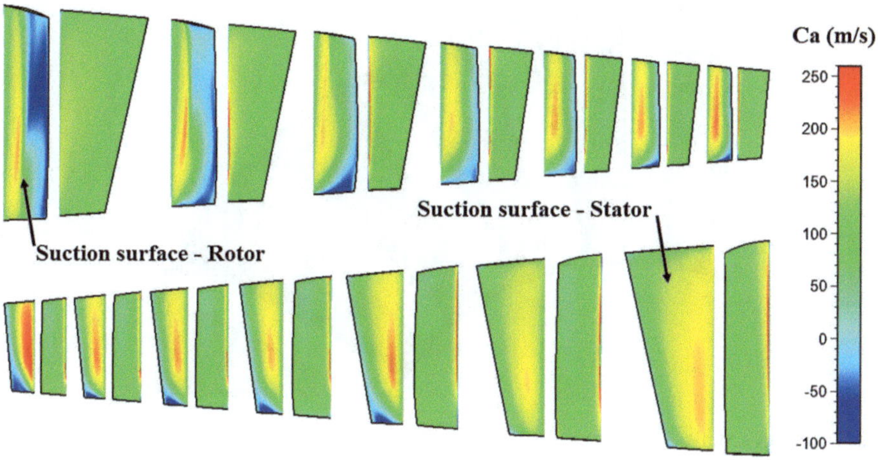

Fig. 5 Contours of Axial Velocity

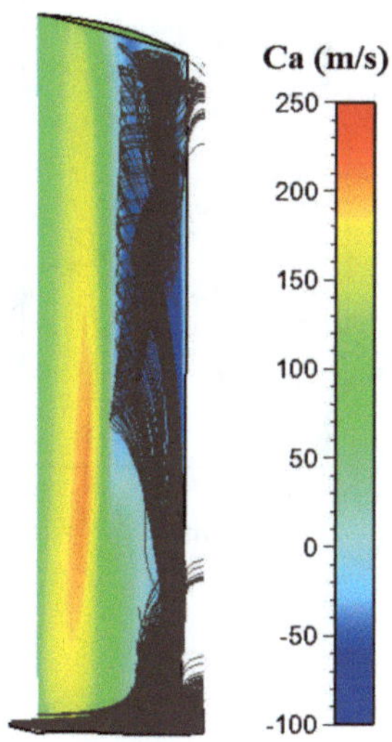

Fig. 6 Separated flow on the blade of the first stage

flow moves progressively in the direction of the rotation at about the half rotational speed.

Figure 8 shows the performance map of the compressor. Here it can be seen that the pressure radio of the design point is not obtained, showing a reduction of 23 %, and that the operating range for the rotational speed design is very narrowed. This is because the separated flow limits the useful operating range of the compressor.

Fig. 7 Rotating stall

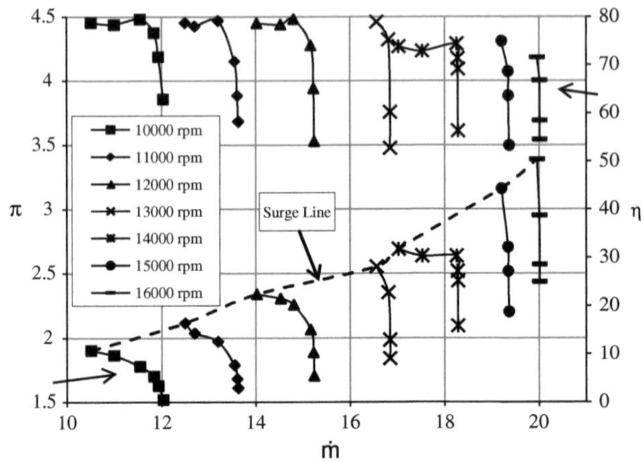

Fig. 8 Performance map

6 Conclusions

The numerical simulation shows that the assumptions taken in the design method are not satisfied. This leads to have a separated flow and a migration from the hub to the tip of the blades and vanes. All this produces a reduction of 23 % in the pressure radio, a poor performance and a narrowed operating range of the compressor, because the losses are increased. In order to obtain a better performance this compressor must be redesigned. Future researches will pursue this aim with the sweep of the blades and vanes of the compressor.

Acknowledgments The authors would like to thank IPN, CONACYT and COMECYT for the support of this investigation.

References

Horlock J.H, Denton JD (2005) A review of some early design practice using computational fluid dynamics and a current perspective. J Turbomach 127:5–13
Saravanamuttoo H.I.H, Rogers GFC, Cohen H (2009) Gas turbine theory. Prentice Hall, 6th edn
Braun M, Seume JR (2006) Forward sweep in a four-stage high-speed axial compressor. ASME turbo expo 2006: Power for land, sea and air, Barcelona, Spain, PP 141–152
Benini E, Biollo R (2007) Aerodynamics of swept and leaned transonic compressor-rotors. J Appl Energy 84:1012–1027

The Drag Coefficient and the Navier-Stokes Fractional Equation

J. R. Mercado-Escalante, P. Guido-Aldana, W. Ojeda-Bustamante and J. Sánchez-Sesma

Abstract The purpose of this work is to relate the Navier-Stokes fractional equation with the formulae for the drag coefficient, as are those of the Kármán-Schönherr, Prandtl-Kármán, and Nikuradse. The thickness of the boundary layer induces a multifractal description and a generalization of Blasius experimental result for the friction factor; whereas the dimensions are obtained by the approximations of Blasius and Falkner-Skan of the pressure gradient. The number associated with the multifractal characteristics are adjusted, and formulae, under study, are inferred. They are represented as a bi-multifractal, which provides an analytical way to find a critical Reynold's number, which draws the Kármán-Schönherr formula as appropriate to limit the right of the viscous sublayer. The frictional force is generalized to represent the fractional derivative as a multifractal whose resolution is the reciprocal of the Reynolds number and of appropriate dimension. Thus, the transformation of the moldable wall shapes is described as the result of the action of the frictional force on them.

1 Introduction

The purpose of this paper is to find the relationship between the formulae of Kármán-Schönherr, Prandlt-Kármán and Nikuradse for the drag coefficient and Navier-Stokes fractional equation.

There are ample experimental evidences supporting those formulae, but they are not analytically connected with the partial differential equation that models the motion of fluids. Our claim is that the classical form of the Navier-Stokes equation can not provide that connection, but instead its fractional form is able to do it.

J. R. Mercado-Escalante (✉) · P. Guido-Aldana · W. Ojeda-Bustamante · J. Sánchez-Sesma
Mexican Institute of Water Technology, Paseo Cuauhnáhuac 8532, Progreso, Jiutepec, C.P. 62550 Morelos, México
e-mail: rmercado@tlaloc.imta.mx

J. Klapp and A. Medina (eds.), *Experimental and Computational Fluid Mechanics*, Environmental Science and Engineering, DOI: 10.1007/978-3-319-00116-6_35,

The Navier-Stokes fractional equation was presented in references (Mercado et al. 2009, 2010, 2012a, b). The essential approach is that the viscous stresses produce a dispersive momentum flow that is described by a Darcy fractional law, while the dispersive flux divergence coincides with the change of momentum, according to Newton's law.

Similar to the classic case, the substantial simplification of the Navier-Stokes fractional equation, which leads to the boundary layer equations, is its relatively thin thickness, which means that the main velocity set in the downstream direction, with a great vertically velocity gradient compared with the longitudinal, which leads the velocity to perform the non-slip condition at the channel bottom, and, conversely, with the slight pressure gradients in the vertical-transverse direction compared with the strong in the longitudinal (Landau and Lifshitz 1987).

In the Falkner-Skan approximation, equilibrium is established through a dynamic triangle formed by the following forces: viscous, inertial and the longitudinal pressure gradient, the last one is given by the free or outer velocity dependent of a power of the longitudinal coordinate. In contrast, in the approximation of Blasius, the pressure gradient is absent, due to the elimination of the power in the outer velocity.

We obtained the friction force and in consequence that the drag coefficient is represented by a power of the indexed Reynolds number. The power is determined by both the spatial occupancy ratio and the power of the external velocity in a coupled way.

Alternatively and more generally, it is noted that the frictional force can be described as a fractional derivative of a multifractal, where the order of the derivative depends on the spatial occupancy index, and the dimension of the multifractal depends, in a coupled manner, of the power of the external velocity and the space occupancy index.

The classical equation of Navier-Stokes lacks from a parameter that reflects the fractal nature of the fluid motion. While fractional version expresses that character through the order of the derivative that gives the viscous friction force. If you want to see what is the value for that order, you can study the frictional force on a large flat bottom, through the boundary layer equations. The results encompass the experimental result of Blasius, which is interpreted as a multifractal. If this multifractal interpretation is maintained for the other formulae, we obtained that the fractal dimensions do not exceed the value 1, and it is smaller as the motion is more turbulent.Therefore, our claim is that the fractional Navier-Stokes equation applied to the boundary layer, with a scaling that reflects the thinness of this layer, which is interpreted as a multifractal, produces the formulas of Blasius, Kármán-Schönherr, Prandtl-Kármán and Nikuradse, as ways to establish ties of dependence, with varying degrees, between Euler and Reynolds numbers.

2 The Navier-Stokes Fractional Equation

The fluid motion is described from the Eurelian viewpoint, assuming a volume of fluid limited by a boundary surface with its momentum per unit volume given by $\rho\mathbf{v}$. For the interaction by internal friction, the fractional gradient is expressed by $\nabla_M^{\beta}\rho\mathbf{v}$, wherein ρ is the mass density, $\mathbf{v}$ the velocity, β the spatial occupancy rate and M the measure of mixture in the different spatial directions. The diffusivity of momentum is the kinematic $\alpha-$viscosity, so the Darcy's momentum flow is $\mathbf{q}_D$. The time rate of change of momentum is the negative divergence, or convergence, of the Darcy flow, and M is chosen such that the flow be proportional to the negative of the fractional Laplacian (Mercado et al. 2012a, b); we obtain (1) considering $\alpha = 1 + \beta$,

$$\mathbf{q}_D = -\nu_\alpha \nabla_M^\beta \rho\mathbf{v},\ \tfrac{d}{dt}\rho\mathbf{v} = -\nabla \cdot \left(-\nu_\alpha \nabla_M^\beta \rho\mathbf{v}\right) = -\nu_\alpha \left(-\Delta\right)^{\alpha/2} \rho\mathbf{v} \tag{1}$$

Pressure variations are incorporated to momentum change through the force provided by the pressure gradient; composition of the viscous friction stress arises and so the hydrostatic pressure which shapes the tensor $\mathbf{T} = \nu_\alpha \nabla_M^\beta \rho\mathbf{v} - p\mathbf{I}$ resulting the deformation law. Then an external potential force per unit volume of the type $-\nabla\rho\phi$ is included and the assumption of fluid incompressibility is incorporated. It makes explicit the material derivative which composes the local variation with the advective one. The requirement of objectivity includes invariance under coordinate changes, so modifying advective contribution and vorticity arises. Finally, the contribution of vorticity is conceived as an external force that energizes the evolution of the velocity field through its vorticity and that conflicts with the viscous force, while the third term in (2), is interpreted as a restriction that along the current lines contains the Bernoulli equation. The ν_α coefficient is compared with the turbulent viscosity of Boussinesq,

$$\frac{\partial}{\partial t}\mathbf{v} = -\nu_\alpha \left(-\Delta\right)^{\alpha/2} \mathbf{v} + \mathbf{v} \times rot\mathbf{v} - \nabla \left(\frac{1}{2}\left(\mathbf{v}\cdot\mathbf{v}\right) + \frac{p}{\rho} + \phi\right) \tag{2}$$

The boundary layer equations are obtained from the Navier-Stokes fractional equation by simplifications that are induced from the premise of a relatively thin thickness. Now the equation of the two-dimensional boundary layer is considered in its permanent or stationary form, together with the conservation of mass in the form of non-divergence, as shown in equation (3), where the pressure gradient is given by $-\partial_x\left(p/\rho\right) = U\frac{d}{dx}U,\ U = u\left(x, y\right)|_{y\to\infty}$, and U is the free velocity,

$$u\partial_x u + v\partial_y u = \nu_\alpha \partial_y^\alpha u - \partial_x\left(p/\rho\right),\ \partial_x u + \partial_y v = 0 \tag{3}$$

The main velocity, downstream, is presented by way of the potential through $\psi\left(u, v\right)$ function as $u = \partial_y\psi,\ v = -\partial_x\psi$; $u,\ v$ are longitudinal and vertical components of velocity respectively. In Blasius approach a balance between two forces is established: the viscous force and the inertial force. In Falkner-Skan approximation is present the longitudinal pressure gradient as the third input to equilibrium, allowing that the free velocity depends on the longitudinal coordinate. Then the underpotential function $g\left(\xi\right)$ arises as a solution of the Blasius fractional or Falkner-Skan fractional equation, ξ is auto-similar variable (Landau and Lifshitz 1987; White 2006).

Stress is calculated by $\tau_{xy} = \mu_\alpha \partial_y^\beta \partial_y \psi\,(u, v)$, being $\mu_\alpha = \rho \nu_\alpha$ and $R_{l\beta} = ul^\beta/\nu_\alpha$ the indexed Reynolds number. Subsequently we determine the friction force per binormal unit length as $F_f = 2\int_0^l \tau_{xy} dx$. In the Falkner-Skan approximation it is considered $U = Kx^m$, with $m \neq 0$. The friction force becomes dimensionless and the drag coefficient arises, $C_f = C_{g,\beta,m}\left((U|_l)^{-1}\right)^{\frac{\beta}{\alpha}(1+1/m\beta)}$.

The exponent of the drag coefficient will be called the Blasius exponent $\theta\,(\beta, m) = (\beta/\alpha)\,(1 + 1/m\beta)$, but also in the Blasius approximation it is considered $m = 0$; we integrate and the exponent obtained is $\theta\,(\beta, m = 0) = \beta/\alpha$, and the two can come together in a single expression.The Blasius exponent, as well as by dimension or decay must be $\theta \geq 0$, so that: either $m > 0$, that means a favourable pressure gradient, or $m \leq -1/\beta$, resulting in an adverse pressure gradient. In particular, the condition given by $\beta \to 0$, or with $m \leq -1/\beta$, will be called turbulent. Since the vertical axis is noted that $m = 1$ horizontal line separating two intervals: $m > 1$ the range of the spectrum, decreasing and convex, and $m < 1$ the range increasing and concave.

It is noted that the following representation is possible, $\theta = \beta\,(\sigma + 1/\,(1 + \beta))$, because for $\sigma > 0$, there is a $m > 0$, given by $m = 1/\sigma\beta\,(1 + \beta)$, which allows such representation, it is considered a generalization of Blasius experimental result. For $m = 0$, there is a representation as $\beta/\,(1 + \beta)$, which also contains the experimental result of Blasius for $\beta = 1/3$, and also the result for the laminar regime $\theta = 1/2$ for $\beta = 1$ (Mercado et al. 2012a, b). It is seen that the value of $\sigma > 0$ can become quite large under the condition that $m > 0$ be sufficiently close to zero.

In Falkner-Skan approach, to define the similarity variable the next dilemma is present: either $m = 1$ or $\beta = 1$, because they are the two options under which the coefficients of the equation of the underpotential function are independent of x. In the first option, the pressure variation is linear with the velocity as in the Hagen-Poiseuille model. For the second, the exponent of Blasius is $\theta\,(1, m) = (m + 1)\,/2m$, and m is an arbitrary value and $m \neq 0$, persevering the restriction: $m > 0$ or $m \leq -1/\beta$, which reduces to $m \leq -1$. The first, is part of the flow against a wedge angle $(m/(m + 1))\,\pi \geq 0$, with velocity lines converging and pressure decrease; the second is associated with flows through a depression angle $(m/(m + 1))\,\pi \leq 0$, clockwise, with divergent velocity lines or growth of pressure (White 2006; Rouse 1946). Furthermore, this exponent contains the experimental result of Blasius with $m = -2$. Now the law of pressure variation with the velocity is turbulent, and Chézy formula is exactly reproduced with $m = -1$. Moreover, in (4), it is possible to describe the friction factor by fractional derivative of order $\gamma = \beta^2/\,(1 + \beta)$, as a multifractal transformation of dimension $\sigma\beta = 1/m\,(1 + \beta)$ in other with dimensions θ,

$$f_\beta = 8\tilde{C}\,(\beta)\,D_x^\gamma\,\left(1/R_{x\beta}\right)^\sigma,\ \theta = \beta\,(1/m\beta\,(1 + \beta) + 1/\,(1 + \beta)) \tag{4}$$

The experimental results of the drag coefficient on a flat surface, have a first phase with exponent 1/2, for low Reynolds numbers, thick or high resolution, then a transition region with slight growth of the drag coefficient, a plateau and then a descent that approaches by an exponent of the type 1/5, for medium resolution, and then the representation of Kármán-Schönherr (Rouse 1946).

The experimental result of Blasius is generalized and interpreted as a multifractal taking the inverse of the Reynolds number $1/R_e$ as resolution, under the condition $0 < 1/R_e < 1$, and the exponent of Blasius θ (β), as dimension or the spectrum of singularities. This law can explain the phases of exponents 1/2 and 1/5, as described in the previous paragraph and also the plateau even at the limit where the exponent $\theta \to 0$ tends to zero, which could be interpreted as fully developed turbulence, where the participation of viscosity is removed through the annulment of the exponent (Mercado et al. 2012a, b). In the four formulae, the characteristic is that the number of features grows with friction force reduction.

Blasius. In the first formula, the friction factor is described as a multifractal, with the dimension θ, varying with β, and being $\theta = \theta_B = \frac{\ln(8B_\beta/f_\beta)}{\ln(R_e)}$, where $8B_\beta/f_\beta$ is the number of features.

Kármán-Schönherr. In the second one, the features are weakened from B/C_f to proportional to $K\exp\left(1/A\sqrt{C_f}\right)$, with two constants A and K, and dimension $1-\theta_B$. Being $A = 4.13\log e$, and $KB = 1$, then, rebuilt Kármán-Schönherr formula or KS as $1/\sqrt{C_f} = 4.13\log_{10}\left(R_e C_f\right)$; or $1/\sqrt{C_f} = 2AW\left(\sqrt{R_e}/2A\right)$, being Lambert W function.

Alternatively, the Kármán-Shönherr formula corresponds to an approximation establishing a weak dependence between Euler and Reynolds numbers given by $1/\sqrt{C_f} \approx A\left(1 - 1/B\,(R_e)^{1-\theta_B}\right)$, with the particularity that between the stronger the turbulence, the greater value of the exponent of the Reynolds number $1-\theta_B$. This exponent depends on both β and m, and $m \geq 1$ restriction is required to ensure the range increasing and concave.

Prandtl-Kármán. In the third one, it is assumed that we have a multifractal with weakened features proportional to $\left(P\sqrt{f_\beta}\right)\exp\left((1/A)\left(\left(1/\sqrt{f_\beta}\right) + D\right)\right)$, and $1-\theta_B$ dimension. Then, with $A = 2\log e$, $D = 0.8$ and $PB = 1$, the Prandtl-Kármán formula is recovered $1/\sqrt{f} = 2\log_{10}\left(\sqrt{f}R_e\right) - 0.8$, or $1/\sqrt{f} = AW\left(R_e/Ae^{D/A}\right)$.

Nikuradse. In the fourth one, the features of the Blasius law are weakened in the form proportional to $\left(NC_f\right)\exp\left(1/A\sqrt{C_f}\right)$, with dimension $1-\theta_B$ and $NB = 1$, so you get $1/\sqrt{C_f} = A\ln\left(R_e\right)$.

Therefore, the different formulae can be seen as consequences of the Navier-Stokes fractional equation, with its reduction to the boundary layer, multifractal interpretation, and the readjustment of its features or its related decays, hence growing its number of features with the growth of the Reynolds number. You can then highlight one bimultifractal (5), such that for relatively high Reynolds numbers, features are proportional to $(1/B)\exp\left(1/A\sqrt{C_f}\right)$, with resolution $1/R_e$ and $1-\theta_B$ dimensions, while for low Reynolds numbers features are inversely proportional to the friction factor: $8B_\beta/f_\beta$, the resolution also as $1/R_e$ and dimensions as θ_B,

$$N_{R_e} = \begin{cases} 8B_\beta/f_\beta & R_e \downarrow \\ (1/B)\exp\left(1/A\sqrt{C_f}\right) & R_e \uparrow \end{cases} \tag{5}$$

3 Applications

3.1 Critical Reynolds

The change in the representation of the number of features of multifractals offers a way to find an analytical critical Reynolds number that indicates the change from laminar to turbulent regime or from the viscous sublayer to the mixture or transition sublayer; which is defined by the law of change in pressure with the variation of the velocity, and which results in a decrease of the dimension from $\theta = 1$ to a lower value $\theta < 1$ (Sommerfeld 1950). For relatively low Reynolds numbers the behaviour is given by Blasius law, while for relatively high, the Kármán-Schönherr formula could be select, and the Reynolds number that makes the transition can be considered as the critical Reynolds. Thus, $\theta = 1$ is taken in Blasius and KS formulas, so that the critical Reynolds number can be estimated by $R_{ec} = B\,(A \ln B)^2$. Alternatively, using the Lambert W function we can obtain $R_{ec} = 4A^2BW^2\left(\sqrt{R_{ec}}/2A\right)$. In particular, for $B = 64$ and $A = 4.13 \log e$, it is obtain $R_{ec} = 3561.1$ that is of the same order of magnitude as the experimental value estimated in 2000, where the value of both A and B are experimentally determined (Rouse 1946).

Moreover, in the case of Nikuradse's formula we obtain $\left(1/\sqrt{R_e}\right) \ln \left(1/\sqrt{R_e}\right) = -1/2A\sqrt{B}$, which solution is explicitly expressed through Lambert W function as $R_{ec} = 4A^2BW^2\left(-1/2A\sqrt{B}\right)$; numerical evaluation with $B = 64$ and $A = 4.13/\ln 10$, produces $R_{ec} = 1.0749$, whose order is outside of magnitude in comparison with 2000. Similarly for the KP formula an estimate of 0.86709 is also obtained by means of the Lambert function. Therefore, it is concluded that the KS formula, within the three is the best suitable to define the lower limit of the blending or transition sublayer.

3.2 Bottom and Vortices

The vortices or coherent structures that form near the wall are the most significant flow patterns of the boundary layer. In general, a coherent structure can be defined as a flow pattern with a life time and spatial extent greater than the average of the turbulence scales in a flow. In their research, noted the occurrence of this phenomenon, using optical measurement techniques such as non-intrusive laser Doppler anemometry and particle image velocimetry in channel flow with and without sediment transport. The bed's form causes a rupture of the developed boundary layer and a recirculation downstream, whose main characteristic is the low velocity flow and vortex formation that are released and finally entering to the high velocity flow, above the shear layer. Kaftory states that three types of structures have been observed: low velocity swept very close to the wall, horseshoe-shaped or hairpins vortices that appears at higher heights and fluid ejections from the bottom to the mean flow (known as bursting), complemented with flow sweeps from the middle zone toward the wall region.

In the bottom shapes shows a characteristic spatial frequency, so that for Froude numbers greater than 1 exhibits wave-like forms with symmetric slopes, while for slopes less than 1 are displayed asymmetric and are called dunes. If you still have Froude numbers much smaller than unity, but with less height, curls are called (Niño 1996). For example, the action of the wind on the dunes change their ways and carry the sand.

Fragments resulting from the analysis of the Falkner-Skan classical equation, have a wedge-shaped with positive angle or counterclockwise, which produces a convergence of velocity lines and hence a decrease in pressure. Or else on the contrary, there is a depressed positive or clockwise angle, which produces a divergence of the velocity lines and therefore an increase in pressure. For the bed forms, a periodic extension of the fragments is performed and is approximated by a Fourier series.

The rectangular wave of spatial period l and height h is represented by $r(x) = \left(4/\pi^2\right) hl \sum_{n=1}^{\infty} s_n(x)$, $s_n(x) = (1/2n-1)\sin(2n-1)(\pi/l)\,x$. To do the graphical representations Lanczos filter are used; this type of filter regularizes the approximation in the corners, due to the failure of convergence in them. So the filter is introduced as $sc(n(\pi/\kappa)) = (1/(2n-1)(\pi/2\kappa))\sin((2n-1)\pi/2\kappa)$ function, in particular $\kappa = 50$ is chosen. Regularized fragments are described by $r_f(x) = \left(4/\pi^2\right) hl \sum_{n=1}^{\infty} sc(n(\pi/50))\, s_n(x)$, and its derivative fractional Weyl type and order $\gamma(m)$, is $D_x^{\gamma} r_f(x) = \left(4/\pi^2\right) hl \sum_{n=1}^{50} sc(n(\pi/50))\, D_x^{\gamma} s_n(x)$, $D_x^{\gamma} s_n(x) = (1/2n-1)((2n-1)(\pi/l))^{\gamma} \sin((2n-1)(\pi/l)\,x + (\pi/2)\,\gamma)$. Such fragments are represented by standardized period $hl = \pi$ and 3π, are plotted by $f_{\gamma}(x) = (1/2)\left(1 + D_x^{\gamma} r_f(x)\right)$, and illustrated for three values of γ. Thus, for $\gamma = 0$ are rectangles of unit height, for $\gamma = -0.5$ we get a shaped like shark fins, and for $\gamma = -1$ a triangular wave train is obtained. The three figures (Fig. 1a–c) show changes produced by the action of the fractional integral from rectangles, through fin shapes, until the triangles, while the fractional derivative transformation performed in reverse order.

With roughness representation by Fourier series, there are two lengths: the period and the height of roughness and the ratio of the two lengths could be considered, l/h. As result, l is relatively high with respect to h, or $l/h >> 1$, and is observed that vortices are knotted and circling each fragment, extending to the next. The aspect ratio is the ratio of the vortex size knot r_0 with his thick a, and can be matched with l/h, so $l/h \approx r_0/a$. These vortices are caused by roughness or detachment, such as collar or collars types, and the horseshoes vortices are deformed and raised by the current; they can reproduce and subsequently form knotted vortices that are integrated into the external medium flow stream and feeding the layer cutting free from the flow separation zone. To the other end, l is relatively low with respect to h, i.e.: $l/h << 1$ and the above events have little space to develop so that the flow behaves ground.

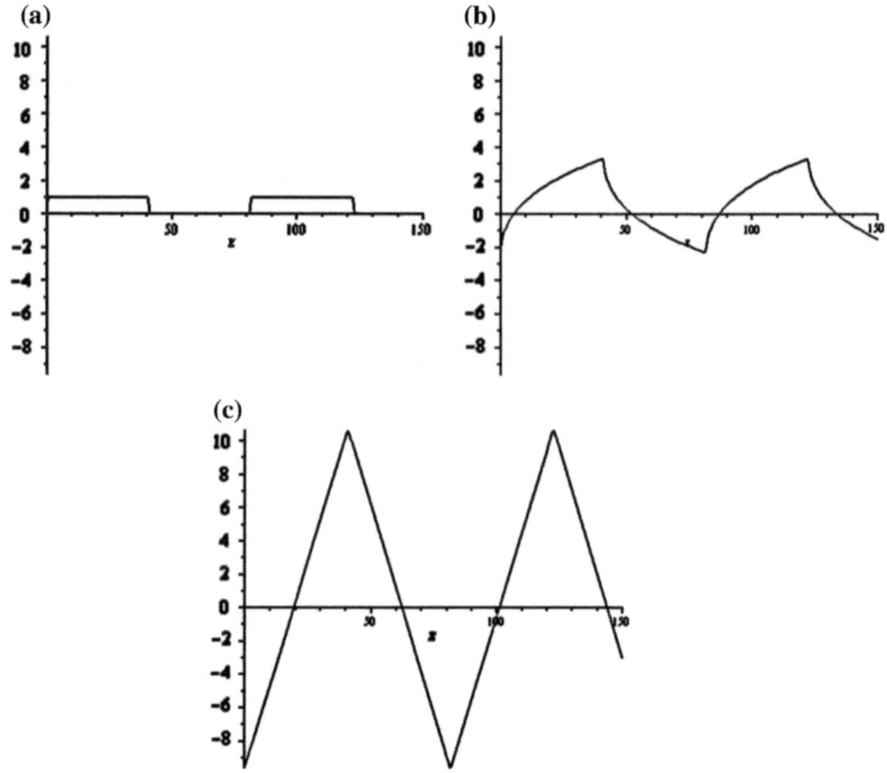

Fig. 1 Action of the friction on bottom shape. **a** $\gamma = 0$, **b** $\gamma = -051020$, **c** $\gamma = -10000$

4 Conclusions

When we studied the interaction of a fluid with a flat surface it could be established that the frictional force, and therefore the drag coefficient, is characterized by a power of the inverse of the indexed Reynolds number, power that we have called exponent of Blasius. It presents a coupled dependence of two parameters, one as a manifestation of the viscosity, and the other, as the pressure gradient.

Result is generalized by presenting the friction force as a fractional derivative. This allows describing the change in the bed forms by the action of a fluid as a fractional derivative on them.

Features are increased and multifractal structure is maintained so that it is possible to obtain the formulas of Kármán-Schönherr, Prandtl-Kármán and Nikuradse for the friction factor, for uniform surfaces and pipes, both smooth and rough.

Bi-multifractal description provides an analytical way to find the critical Reynolds number, and it is observed that only the Kármán-Schönherr formula produces an estimated according to the order of magnitude of experimentally known results in the change from the viscous sublayer to the mixed.

References

Landau LD, Lifshitz EM (1987) Fluid mechanics. Pergamon Press, Oxford

Mercado JR, Guido P, Ojeda W, Sánchez-Sesma J, Olvera E (2012b) Saint-Venant fractional equation and hydraulic gradient. J Math Syst Sci 2(8):494–503

Mercado JR, Guido P, Sánchez J, Íñiguez M, González C (2012a) Analysis of the Blasius' formula and the Navier-Stokes fractional equation. In: Klapp J et al. (eds) Fluid dynamics in physics, engineering, and environmental applications, environmental science and engineering, Springer, Berlin, pp 475–480

Mercado JR, Olvera E, Perea H, Íñiguez M (2010) La ecuación Saint-Venant fraccional. Int J Math: Theor Appl. Submitted: feb. 16, 2010; RMTA-030-2010

Mercado JR, Ramírez J, Perea H, Íñiguez M (2009) La ecuación Navier-Stokes fraccional en canales de riego. Int J Math: Theor Appl. Submitted: june 14, 2009; RMTA-082-2009

Niño Y (1996) Inestabilidades en un Lecho Granular Móvil: Análisis Matemático de Formas de Fondo. Ingeniería del Agua, vol 3, no 4, dic. 1996, 25–36. U. de Chile

Rouse H (1946) Elementary mechanics of fluids. Dover Publications, New York, p 376

Sommerfeld A (1950) Mechanics of deformable bodies. Academic Press, New York, p 396

White FM (2006) Viscous fluid flow. McGraw-Hill, New York, p 629

Numerical Solution of the Swift–Hohenberg Equation

S. Sánchez Pérez-Moreno, S. Ruiz Chavarría and G. Ruiz Chavarría

Abstract The Swift–Hohenberg equation accurately models the formation and evolution of patterns in a wide range of systems. However, in the field of fluid dynamics, two particular patterns arise during the Rayleigh-Bénard convection, rolls and hexagons, and the formation of both has been simulated in this work. The Swift–Hohenberg (S–H) equation is a nonlinear partial differential equation of fourth order, and through an implicit finite differences method it has been numerically solved. A set of snapshots of the evolution of these patterns is shown.

1 Introduction

Nonequilibrium physics typically studies the dynamics of systems driven far away from thermal equilibrium, which inherently require nonlinear equations to be modelled. Such systems may eventually reach a steady state as long as a net energy flux through it is constant. Meanwhile, in this state of maximum entropy, spatio-temporal patterns may arise as a result of self-organization, called *dissipative structures* (Nicolis and Prigogine 1977).

One example of such a system is the well known Rayleigh-Bénard convection, in which a thin layer of fluid is heated from below, such that above a critical temperature of the lower plate convection rolls and hexagonal convection cells appear. These patterns have been widely studied experimentally, theoretically and numeri-

S. Sánchez Pérez-Moreno (✉) · S. Ruiz Chavarría · G. Ruiz Chavarría
Facultad de Ciencias, Universidad Nacional Autónoma de México, Ciudad Universitaria, C.P.04510 Distrito Federal, México
e-mail: sebasanper@gmail.com

S. Ruiz Chavarría
e-mail: sabina@ciencias.unam.mx

G. Ruiz Chavarría
e-mail: gruiz@ciencias.unam.mx

J. Klapp and A. Medina (eds.), *Experimental and Computational Fluid Mechanics*, Environmental Science and Engineering, DOI: 10.1007/978-3-319-00116-6_36,

cally, making it one of the best understood processes in fluid dynamics (Cross and Hohenberg 1993). The Swift–Hohenberg equation models, up to a very good approximation, the formation and evolution of these patterns and will be further detailed in the next section.

2 The Swift–Hohenberg Equation

The S–H equation was introduced phenomenologically by studying phase transitions in the behaviour of the Rayleigh-Bénard convection, as an analogy to thermodynamic systems (Swift and Hohenberg 1977), but soon after it revealed itself as a very approximate model for nonlinear pattern formation. The importance of the S–H equation relies on its simplicity to yield very much the same results as the Navier–Stokes equations, which can be hard to solve both analytically and numerically. It is worth mentioning it has been implemented in studies of a very different nature, including but not limited to economy, biology, sociology, optics and of course fluid dynamics (Cross and Greenside 2009).

The Swift–Hohenberg equation reads:

$$\frac{\partial \psi}{\partial t} = \varepsilon\psi - (\nabla^2 + 1)^2\psi + g\psi^2 - \psi^3, \tag{1}$$

It is a nonlinear partial differential equation of fourth order, in which the order parameter ψ is a real scalar field proportional to the vertical velocity of the fluid at the plane $z = \frac{L}{2}$, but it may also represent the temperature deviation from the linear temperature profile between the lower and upper plates (Greenside and Cross 1985). The parameter ε is the reduced Rayleigh number and is expressed as:

$$\varepsilon = \frac{Ra - Ra_c}{Ra_c} \tag{2}$$

where Ra_c is the critical Rayleigh number at which instability sets in and convection begins. At last, the parameter g breaks the reflection symmetry so that the equation is no longer invariant under a change of the sign of the field.

3 Numerical Method

Domain

The Swift–Hohenberg equation has been numerically solved in the square domain:

$$\begin{cases} 0 \le x \le L \\ 0 \le y \le L, \end{cases} \tag{3}$$

where L is the length of both the height and width of the domain. This square has been discretized by the two dimensional space grid:

$$\begin{cases} x_i = \frac{iL}{N}, \ i = 0, 1, 2 \ldots N - 1 \\ \\ y_j = \frac{jL}{N}, \ j = 0, 1, 2 \ldots N - 1. \end{cases} \tag{4}$$

In this case, the number of grid points was set to 100 per axis ($N = 100$), in order to approximate the solution at 10,000 points within the domain.

On the other hand, the time parameter has been discretized by a temporal grid:

$$t_n = \Delta t n, \quad n = 0, 1, 2 \ldots, \tag{5}$$

where Δt is the size of the time step.

Along this section and the following, the notation $\psi_{i,j}^n$ will be used as a short for the function ψ evaluated at grid point $x = i\Delta x$, $y = j\Delta y$, where Δx and Δy are the respective length intervals of the spatial grid, at time step $t = n$.

Discrete S–H Equation

In order to solve (1), the easiest approach was to reduce the fourth order differential equation into the following system of two coupled second order differential equations:

$$\begin{aligned} \phi &= \nabla^2 \psi \\ \frac{\partial \psi}{\partial t} &= -2\phi - \nabla^2 \phi + (\varepsilon - 1)\psi + g\psi^2 - \psi^3. \end{aligned} \tag{6}$$

The Finite Difference Method (FDM) was chosen to solve (6) to a second order of accuracy. To begin with, the time derivative at time step $t = n$ and at grid point (i, j) was approximated by the following backward finite difference obtained through a Taylor series expansion:

$$\frac{\partial \psi}{\partial t} \approx \frac{3\psi_{i,j}^n - 2\psi_{i,j}^{n-1} + \psi_{i,j}^{n-2}}{2\Delta t}, \tag{7}$$

from which it can be seen that the solution of $\psi_{i,j}$ at time $t = n$ depends of $\psi_{i,j}$ at the two previous time steps.

Secondly, the Laplacian operator found in (6) was approximated by the central finite difference:

$$\nabla^2 \psi = \frac{\partial^2 \psi}{\partial x^2} + \frac{\partial^2 \psi}{\partial y^2} \approx \frac{\psi_{i-1,j}^n - 2\psi_{i,j}^n + \psi_{i+1,j}^n}{(\Delta x)^2} + \frac{\psi_{i,j-1}^n - 2\psi_{i,j}^n + \psi_{i,j+1}^n}{(\Delta y)^2}, \tag{8}$$

where it can be seen that to find the solution at every grid point, only four other adjacent points are needed. If ϕ is substituted into (8) instead of ψ, one will obtain the discretization of the second equation of (6). And finally, the non linear term was approximated by:

$$g\psi^2 - \psi^3 \approx 2g(\psi_{i,j}^{n-1})^2 - 2(\psi_{i,j}^{n-1})^3 - g(\psi_{i,j}^{n-2})^2 + (\psi_{i,j}^{n-2})^3 \qquad (9)$$

which is known at time step $t = n$.

Boundary and Initial Conditions

For this simulation, an infinitely large domain was considered, in other words, periodic boundaries were enforced. This can be mathematically expressed as:

$$\psi(x, y) = \psi(x \pm mL, y \pm mL), \quad m = 0, 1, 2 \ldots \qquad (10)$$

Initial conditions were set randomly so that $-5 \times 10^{-5} \leq \psi_{i,j}^{t=0} \leq 5 \times 10^{-5}$ for all i and j. In order to trigger the instability, a small random perturbation was implemented at all points at $t = 1$ in the form of:

$$\psi_{i,j}^{t=1} = \psi_{i,j}^{t=0} + \delta\psi, \quad -5 \times 10^{-5} \leq \delta\psi \leq 5 \times 10^{-5} \qquad (11)$$

Iterative Method

Since the Swift–Hohenberg equation is a fourth order partial differential equation, an implicit method was necessary to attain numerical stability regardless of the spatial and temporal step sizes. A linear system of 10,000 equations and unknowns is so obtained, and thus a successive over-relaxation (SOR) method is appropriate to find its solutions.

Let us write $\psi_{i,j}^n = F(\psi_{i,j}^{n-1}, \psi_{i,j}^{n-2}, \psi_{i+1,j}^n, \psi_{i-1,j}^n, \psi_{i,j+1}^n, \psi_{i,j-1}^n)$, so the implementation of the latter results in:

$$(\psi_{i,j}^n)^k = (1 - w)(\psi_{i,j}^n)^{k-1} + wF(\psi_{i,j}^{n-1}, \psi_{i,j}^{n-2}, \psi_{i+1,j}^n, \psi_{i-1,j}^n, \psi_{i,j+1}^n, \psi_{i,j-1}^n), \qquad (12)$$

where k is the iteration step and $1 < w < 2$ the relaxation factor.

The program was written in C, and run in an Intel Core 2 Duo processor at 2.40 GHz.

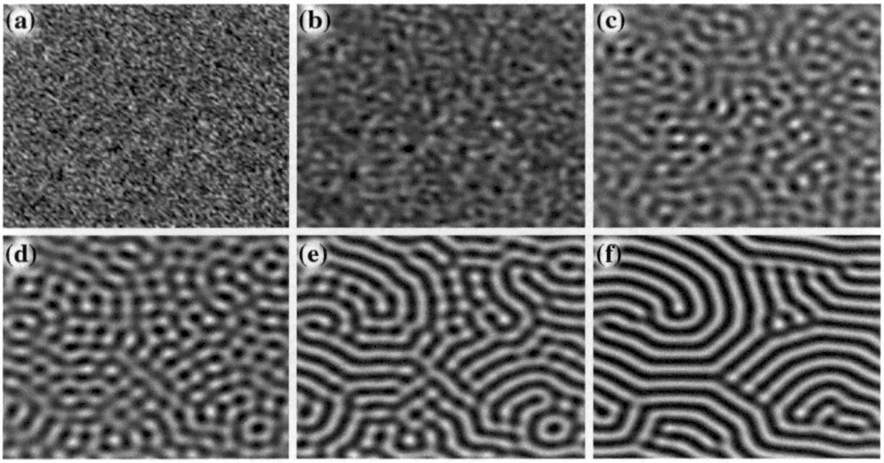

Fig. 1 Evolution of a roll pattern with the following parameters: $\Delta x = \Delta y = 0.9$, $\Delta t = 0.01$, $g = 0.0$, $\varepsilon = 0.3$, $w = 1.3$, $k = 15$. The value of parameter ψ is plotted at each grid point. *Brighter zones* represent positive values of ψ, while *darker zones* represent negative values. **a** Initial conditions at $t = 0.01$. **b** At $t = 1.2$ the instability sets in and self-organization begins. **c** $t = 6$. **d** $t = 20$. **e** $t = 60$. **f** At $t = 198$ the system barely evolves any more, for a steady-state has been reached

4 Results

Rolls

Figures 1, 2 and 3 show the formation and evolution of patterns that arise in the Rayleigh-Bénard convection, when simulated with the Swift–Hohenberg equation.

If the S–H equation is solved with the parameter $g = 0$, then no second order non-linearities appear and the system achieves an equilibrium state while forming rolls (see Fig. 1).

Furthermore, it can be seen in Fig. 2 that the system indeed complies with the periodic boundary conditions.

Since the algorithm takes a different seed to output random numbers at every execution, different position of the rolls appear. Three more examples of these roll patterns are shown in Fig. 3.

Hexagons

By setting $g = 1.0$ and $\varepsilon = 0.1$ the formation of hexagonal patterns can be obtained. In Fig. 4 one can witness the dynamics of this particular pattern at different time intervals.

Fig. 2 Periodicity of the roll pattern from Fig. 1

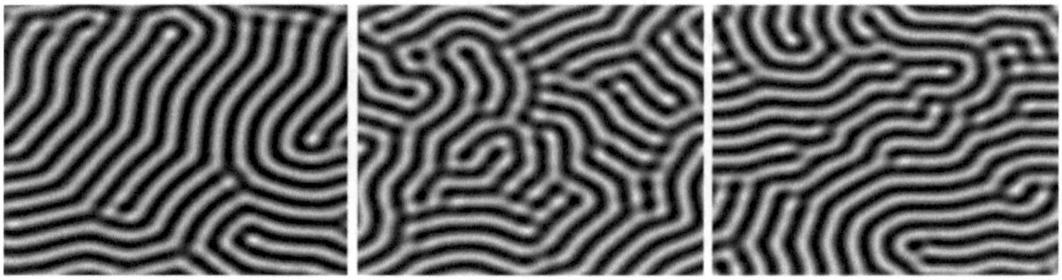

Fig. 3 Different patterns formed following different initial random conditions. The following parameters were set: $\Delta x = \Delta y = 0.9$, $\Delta t = 0.01$, $g = 0.0$, $\varepsilon = 0.3$, $w = 1.3$, $k = 15$

5 Discussion

In Figs. 1 and 3 one can see locally periodic rolls, whose wavelength is always approximate to $L/13$, where L is the length of each side of the domain. That results from using the same parameters at every execution. The spatial resolution allows for 7.7 length intervals per wavelength unit, which is enough to have a clear view of the pattern formation. The time resolution is also small enough to see the onset of the instability, considering also that this time step was the maximum possible to achieve numerical stability. Grain boundaries appear as the consequence of patterns following vectors with different wave number, which make the pattern to be non uniform globally.

In Fig. 4 it was easy to follow the dynamics of the hexagonal instability. One can see that at the beginning, the pattern evolves quite similarly to that of rolls, though at a certain point lines break up giving way to single droplets that take hexagonal symmetry.

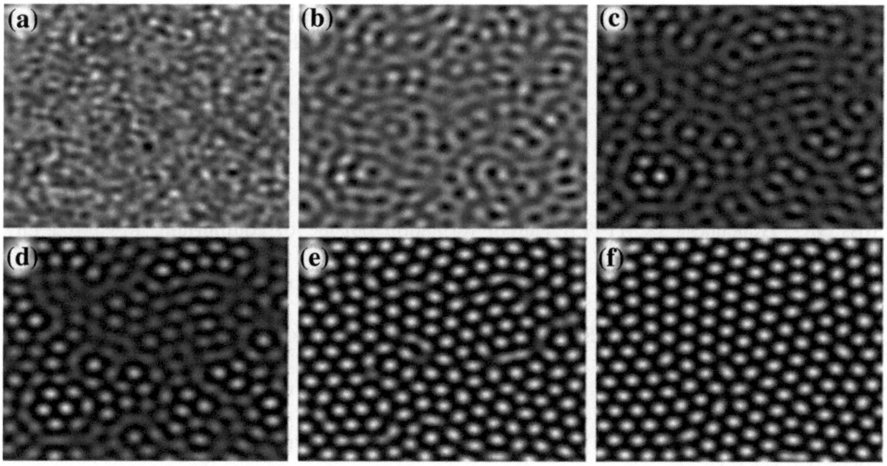

Fig. 4 Evolution of a hexagonal pattern with the following parameters set: $\Delta x = \Delta y = 0.9$, $\Delta t = 0.01$, $g = 1.0$, $\varepsilon = 0.1$, $w = 1.3$, $k = 15$. The value of parameter ψ is plotted at each grid point. *Brighter zones* represent positive values of ψ, while *darker zones* represent negative values. Initial conditions have been omitted since Fig. 1a is very similar. **a** Instability sets in at $t = 1.2$ and self-organization begins. **b** Certain structures may be seen at $t = 6$. **c** At $t = 100$ rolls and isolated dots coexist. **d** Rolls are broken into single droplets at $t = 108$. **e** By $t = 120$ a clear hexagonal pattern can already be seen. **f** Equilibrium is reached and hexagons move just slightly from $t = 198$ on

6 Conclusion

It has been proved that an implicit finite difference method is a precise computational algorithm to solve the Swift-Hohenberg Equation. Converting the equation into a system of two coupled equations was a crucial step towards finding a stable and steady-state solution; a discretization of the fourth order partial differential equation made it harder to code and numerically unstable. It is also highly recommended to use an implicit integration scheme, so as to not bound the time step by the fourth power of the spatial step.

Regarding these results, it is possible to confirm the qualitative similarity of the roll patterns obtained with those found in nature, such as the convection of Rayleigh-Bénard or the skin of zebras. On the other hand, the hexagonal pattern clearly resembles the convection cells found in dry lakes or the skin of certain animals like the jaguar. Pattern formation is the result of self-organization systems and all these are good examples of this phenomenon, in spite of the different mechanisms that trigger and amplify the instability.

In the field of fluid dynamics it is convenient to use this equation when speaking of pattern formation in Rayleigh-Bénard convection, instead of solving the more difficult Navier–Stokes equations. This work can be modified to study the dynamics of pattern formation in different parametric domains of thermal convection (e.g. annulus, Sensoy and Greenside 2001) and with non random initial conditions.

Acknowledgments We want to express our gratitude to Prof. Pablo de la Mora for his invaluable help and DGAPA-UNAM for their financial support under grant PAPIIT-IN110210.

References

Cross MC, Hohenberg PC (1993) Pattern formation outside of equilibrium. Rev Mod Phys 65(3):851–1112

Cross M, Greenside H (2009) Pattern formation and dynamics in nonequilibrium systems. Cambridge University Press, Cambridge

Greenside HS, Cross MC (1985) Stability analysis of two-dimensional models of three-dimensional convection. Phys Rev A 31(4):2492–2501

Nicolis G, Prigogine I (1977) Self-organization in nonequilibrium systems: from dissipative structures to order through fluctuations. Wiley, New York

Sensoy B, Greenside H (2001) Pattern formation near onset of a convecting fluid in an annulus. Phys Rev E 64(4)

Swift J, Hohenberg PC (1977) Hydrodynamic fluctuations at the convective instability. Phys Rev A 15(1):319–328

Experiments of Mercury Jets Impinging on Water

F. Wong, A. Medina, A. López Villa and G. J. Gutiérrez

Abstract In this work we have analyzed experimentally the impact of mercury jets on water surfaces, i.e., between liquids having a high contrast in their densities and their surface tensions. Mercury jets can have different lengths but maintained the condition of equal diameter. Measurements of the length of penetration of the jets under normal and oblique impact on water at room and low temperatures reveal interesting physical facts.

1 Introduction

Here we analyze experimentally the problem of the impact of mercury jets impinging on water surfaces which are sufficiently away from the bottom so that its presence is not important for the dynamics of penetration.

The motivation for these studies arises because in the theoretical analysis of the dynamics of high speed impact between a long projectile (solid bar) and a thick solid target is assumed that, because the nature of the impact, inviscids liquids of different densities are involved (Birkhoff et al. 1948; Allen and Rogers 1961; Zukas et al. 1982; Zukas 1990). The latter means that in such formulation it is possible to use the Bernoulli equation for describing approximately the penetration phenomena (or rebound) for normal or oblique impact.

Far as we know, in literature there are few studies in which analyzes in detail the impact between mercury and water, between two liquids of low viscosity and high contrast between their densities and surface tensions. The fact that liquids are inviscids allows the use of Bernoulli equation, the high contrast between densities is necessary because usually the bullets are made with a high density so that they can

F. Wong(✉) · A. Medina · A. L. Villa · G. J. Gutiérrez
ESIME-Azcapotzalco, Instituto Politécnico Nacional, Av. de las Granjas 682,
C.P. 02250 Azcapotzalco, Col. Santa Catarina, México D.F.
e-mail: franciscowong@hotmail.com

J. Klapp and A. Medina (eds.), *Experimental and Computational Fluid Mechanics*, Environmental Science and Engineering, DOI: 10.1007/978-3-319-00116-6_37,

pierce the target, and the condition of high surface tension is to maintain the integrity of the jet in the air (where it was fired), as well as it enters in water.
In experiments the length of penetration of a jet of initial length L is P, the velocity of impact is V and the velocity in water is U. The density of the projectile is ρ_p and it impinges on a semi-infinite target of density ρ_T. To describe the main experimental facts we calculate, for each test, the Froude, Weber and Reynolds number as Kersten et al. (2003) did to see how their variations determinate the different profiles and penetration lengths.

$$Fr = \frac{V^2}{gd};\ We = \frac{\rho d V^2}{\sigma};\ Re = \frac{\rho d V}{\mu}$$

Here d is the diameter of the capillary tube from the jet emerges, g is the acceleration of gravity, ρ is the mercury density, σ is the interfacial surface tension and μ is the dynamic viscosity.

2 Experimental Set-Up

In experiments we used a capillary glass-tube 2.5 mm inner diameter, a reservoir of $6 \times 16 \times 30$ cm with water to a level of 13 cm, mercury ($\rho = 13{,}600\,\mathrm{kg/m^3}$), a thermal bath PHYSICA, a fast camera "100K REDLAKE HG MotionXtra" and a 500 W lamp for illumination. In addition to the experimental set-up, it was necessary an image management program to measure the progress of the mercury jet.

3 Experiments

The experiments presented here are intended to demonstrate the hydrodynamic hypothesis used to describe the high-speed impact between solids, describe the physical facts developed between mercury jets impinging water in a container.
The variances between each one of them will be explained at each point (Fig. 1).

Normal Impact at Constant Target Temperature, the Role of L

Here we demonstrated that penetration length of a jet of high density depends mainly of the ratio between target and projectile.

The series of photographs presented below shows the importance of the length of the projectile (in this case the jet of mercury on water has a fixed densities ratio) to achieve a penetration length depending on the length, L, of the mercury jet (Fig. 2).

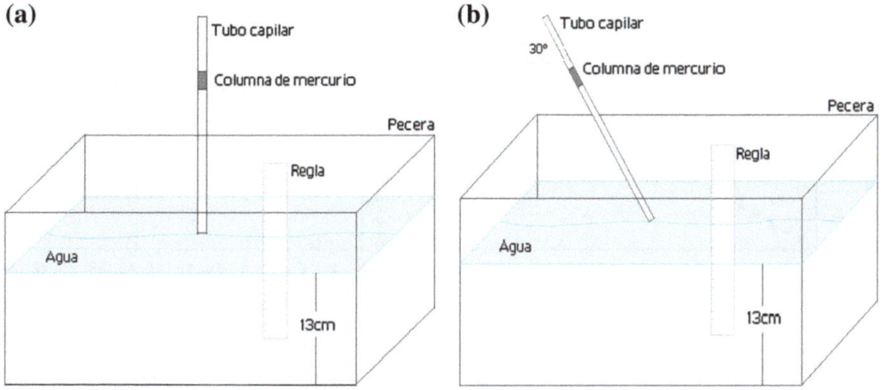

Fig. 1 Schematic of the jet: **a** jet impinging in a normal direction and **b** oblique jet

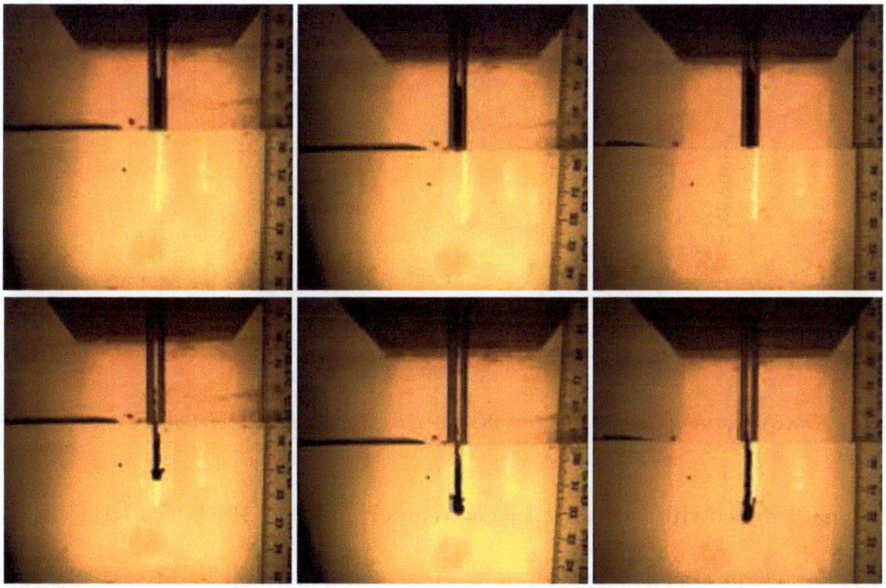

Fig. 2 Images of the initial (*upper*) and final shapes (*lower*) of different mercury jets impinging on water. The details of each jet are the next ones: from *left* to *right*, 2 cm (Fr: 73.72, We: 150.84 and Re: 29,324), 2.5 cm (Fr: 91.85, We: 188 and Re: 32,734) y 3 cm (Fr: 80, We: 162.6 and Re: 30,450)

The following plots describe the penetration length, velocity and acceleration of different lengths of mercury jets once they impact water. We observed the profile they draws on the water and we established a criteria to stop the experiments (the same to each experiment) (Figs. 3 and 4).

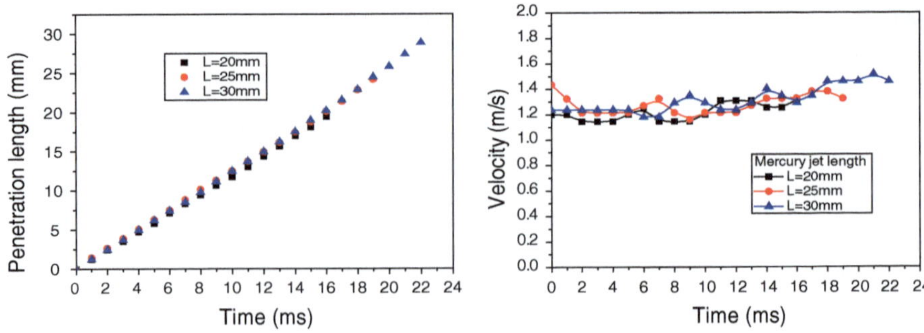

Fig. 3 Plot at the *left hand side*(lhs) shows the importance of L (jet length) to ensure a deeper penetration in the same target, plot at the *right hand side* (rhs) shows the velocity variations once the mercury jet is in water

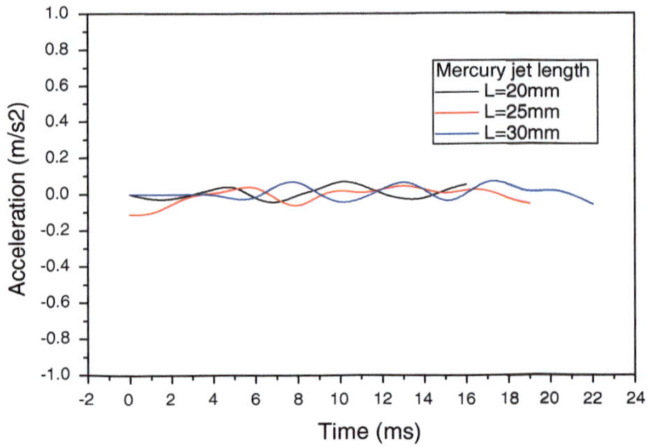

Fig. 4 Once the mercury jet penetrates in the water, its changes of acceleration are oscillatories

Normal Impact with Different Target Temperature, the Role of ρ_p/ρ_t.

Here it is demonstrated that penetration length of a jet of a high density fluid depends mainly of the ratio of densities of the target and projectile (Figs. 5, 6 and 7).

Oblique Mercury Jet Impact, the Role of θ, the Angle of Incidence

In cases of oblique impact of mercury jets we study the effect of the entry angle. We found that it directly affects the projectile length penetration; In Fig. 8 we analyze the input cases 30°, 45° and 60°.

Fig. 5 In these pictures are shown the initial (*upper*) and final shapes (*lower*) of different mercury jets impinging on water at 10 °C temperature. The details of each jet are the next: from *left* to *right*, 2 cm (Fr: 94.19, We: 192.73 and Re: 33,148), 2.5 cm (Fr: 111.75, We: 228.67, and Re: 36 106) and 3 cm (Fr: 106, We: 216.7 and Re: 35,150)

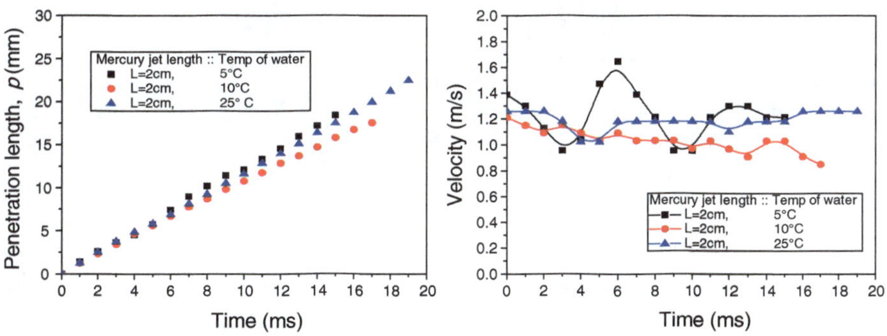

Fig. 6 Plot of the *LHS* shows that if we *decrease* the temperature of the water, it tends to increase its density, so that the penetration of the jet will *decrease*, and plot on the *RHS* shows the velocity variations once the mercury jet is in water

4 Conclusions

Normal Impinging

Summary of the experiments made:

1. Under normal impact, and with an similar impact velocity (V), the mercury jet length is the main variable to ensure a deeper pierce if we got a bigger L.

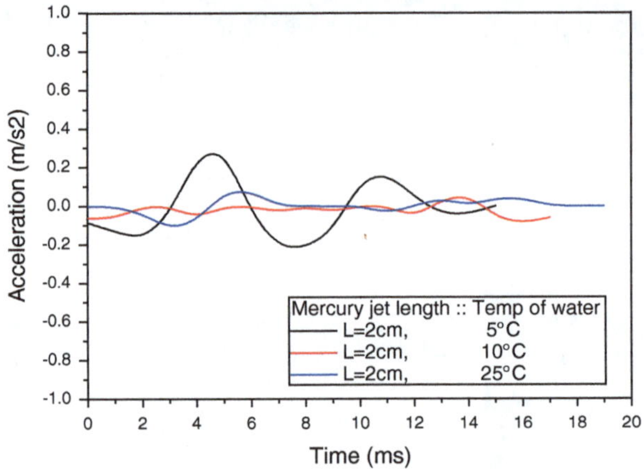

Fig. 7 Evolution of the jet acceleration different temperatures of water

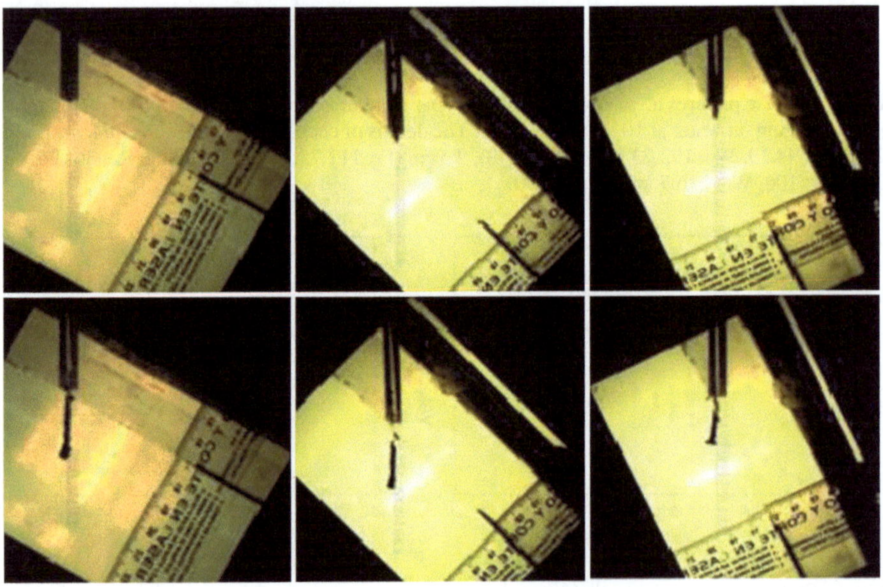

Fig. 8 Effect of the entrance angle, θ, of a 2cm length mercury jet, at room temperature. Details of each jet are the following: from *left* to *right*, 2cm @ 30° (Fr: 178.8, We: 366 and Re: 45,675), 2 cm @ 45° (Fr: 61.5, We: 125.7, and Re: 26,776) y 2 cm @ 60° (Fr: 30.5, We: 62.4 and Re: 18,859)

2. The temperature effect could be observed, and with similar lengths of jets, that variation (although small) of the density of water due to its temperature (10 and 5 °C) the ratio (U/V) is much greater than when mercury jet entries at room temperature (25 °C).

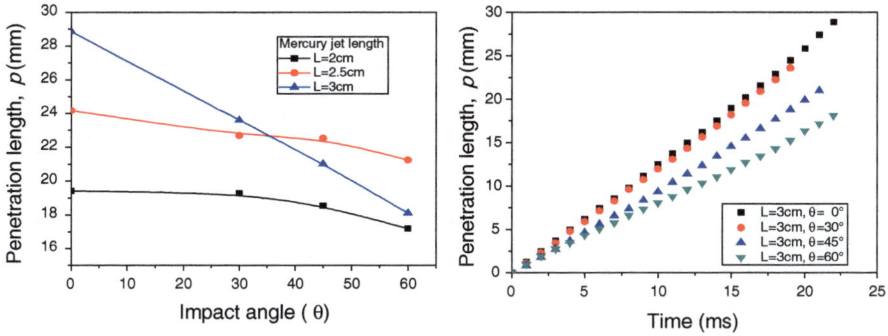

Fig. 9 *LHS*: Plot of the penetration length as a function of θ, *RHS*: plot of the penetration length as a function of time for several angles

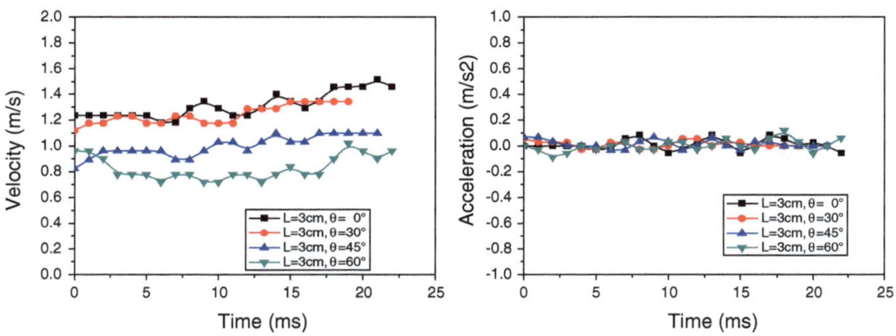

Fig. 10 *LHS*: Velocity variations of a 3cm mercury jet impinging on water at different entrance angles, *RHS*: variations of accelerations in water

Oblique Impinging

It was determined that the mercury jet penetration is strongly affected depending on the angle at which it enters. Plots of these events determine that the greater angle of incidence the lower penetration is obtained by a column of the same length on water at the same temperature.

Taking into account the above, we can justify the following:

1. Compared to the normal impacts to the surface, in the oblique impact, independent of the ratio (U/ V), the penetration length is (according the experiments) always smaller when the angle of incidence increases with respect to the vertical (Figs. 9 and 10).

References

Birkhoff G, MacDougall DP, Pugh EM, Taylor GI (1948) J Appl Phys 19:563

Allen WA, Rogers JW (1961) Penetration of a rod into a semi-infinite target. J Franklin Inst 272:275

Zukas JA, Nicholas T, Swift HF, Greszczuk LB, Curran DR (1982) Impact dynamics. Wiley, New York

Zukas JA (ed) (1990) High velocity impact dynamics. Wiley, New York

Kersten B, Ohl CD, Prosperetti A (2003) Transient impact of a liquid column on a miscible liquid surface. Phys Fluids 15(3):821-824

Velocity Estimation of a Vertically Falling Soap Film

Salvador Galindo-Uribarri, María Dolores Duran-García and Jaime Klapp-Escribano

Abstract The continuously soap film tunnel is a device suitable for experimental studies of the hydrodynamic evaluation of cross sections. The velocity of the undisturbed film is among the important flow parameters of these particular devices. This paper suggests an indirect method for quickly estimating the velocity of vertically falling soap films.

1 Introduction

In recent years Soap film tunnels have gained a more widespread usage in the experimental study of fluids (Bandi et al. 2013). The concept of using flowing soap film in experimental work has been around for some decades (Gharib and Derango 1989). These tunnels are employed primarily for hydrodynamic evaluation of cross sections. They have the advantage of attaining low flow speeds allowing a more detailed observation and associated photographic and video recording than has been achieved in wind tunnels. In addition, experimental information useful for solving hydrodynamic problems may be obtained when the flow characteristics of these particular devices are known. Among them the film speed and its thickness are the most important parameters to be determined of the flowing film. This paper suggests an inexpensive alternative method for quickly estimating the speed of such films.

S. Galindo-Uribarri (✉) · J. Klapp-Escribano
Instituto Nacional de Investigaciones Nucleares, Carretera México-Toluca Km. 36.5, 52750 La Marquesa, MEX, México
e-mail: salvador.galindo@inin.gob.mx

M. D. Duran-García
Facultad de Ingeniería,UAEMex, Ciudad Universitaria, 50100 Toluca, MEX, México
e-mail: mddg_2210@hotmail.com

J. Klapp-Escribano
Departamento de Matemáticas, Cinvestav del I.P.N., 07360 México, D.F., México
e-mail: jaime.klapp@inin.gob.mx

J. Klapp and A. Medina (eds.), *Experimental and Computational Fluid Mechanics*, Environmental Science and Engineering, DOI: 10.1007/978-3-319-00116-6_38,

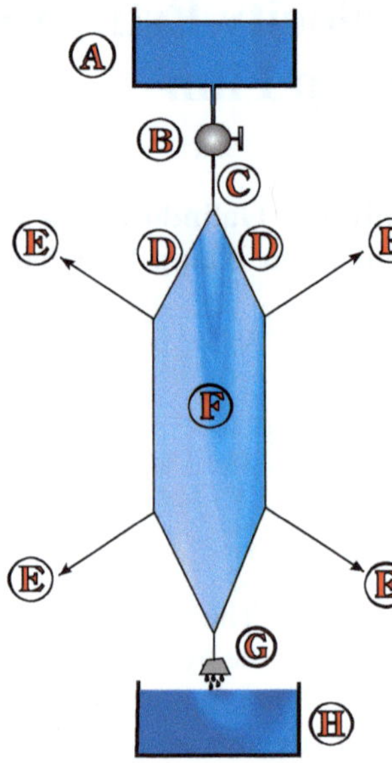

Fig. 1 Diagram of the vertically flowing soap film apparatus. The terminology used is as follows: *A* Upper supply reservoir, *B* metering valve, *C* injection nozzle, *D* guide wires, *E* pull lines, *F* test section, *G* tensioning weight, and *H* collecting deposit

We begin this paper, by describing the apparatus we used to create robust flowing films and then we briefly discuss some of the available techniques for measuring speeds. Next we explain the foundations of our speed estimation method and finally we show our results.

2 The Experimental Set-up

The apparatus used is a fairly conventional but efficient one based on a previous design by Rutgers et al. (Rutgers et al. 2001). The device is schematically shown in Fig. 1. Letters used in the following description refer to those in Fig. 1.

The arrangement consists of a top reservoir (A) of a constant level soap solution. The solution is gravity driven downwards into a plastic metering valve (B) that regulates the flow rate into a nozzle (C). Inside the nozzle there is a pair of nylon fishing lines serving as guide wires once they spread out (D). At this stage a soap film is formed. To branch out the guide wires, two pairs of thinner nylon pull lines, separated at a certain distance, are attached to each of the guide wires (E), as shown in Fig. 1. The pull lines are kept in tension to hold the guide lines parallel to each other. In this way a rectangular "test" section is formed (F). In this section the fluid reaches its speed due to the balance between the gravity pull and air drag resistance. This terminal speed depends on the fluid injection rate which is regulated by the opening of the metering valve (B). Past the test section, the distance between the guide wires reduces until both wires contact each other at the bottom of the instrument. Wires remain tensioned by a hanging weight (G). At the contact point of the guide wires, the soap film ruptures and the fluid is collected at a bottom reservoir (H).

3 Velocity Measurement Techniques

As already stated, it is very important to know the velocity of the soap film at the test section. There are several methods that can be employed to measure the film velocity, namely:

3.1 Laser Doppler Velocimetry (LDV)

The laser Doppler velocimeter sends a monochromatic laser beam toward the target and collects the reflected radiation by seeded particles within the flow. These particles will reflect light with a Doppler shift corresponding to their velocities. A major disadvantage, apart from the high cost of the equipment, is that the fluid needs to be seeded (Durst et al. 1981).

3.2 Hot Wire Anemometry

This instrument uses a very fine wire heated up to some temperature above the ambient. The wire is introduced into the fluid to be measured. The flow past the wire cools the wire changing its electrical resistance. To determine the velocity of the fluid, the calibration between the resistance of the wire and the flow velocity must be known beforehand. This implies that a second method of velocity measurement must be employed to obtain the calibration.

3.3 Fiber Velocimeter

The tip of an optical fiber is plunged through the flowing soap film. Due to the drag force of the surrounding fluid, the tip will experience a small deflection. The resulting deflection can be measured by a position-sensitive detector when a laser light is coupled to the fiber from the other end. The deflection is related to the fluid velocity. A disadvantage is that the inserted tip may have a perturbing effect on the flow (Kellay et al. 1995).

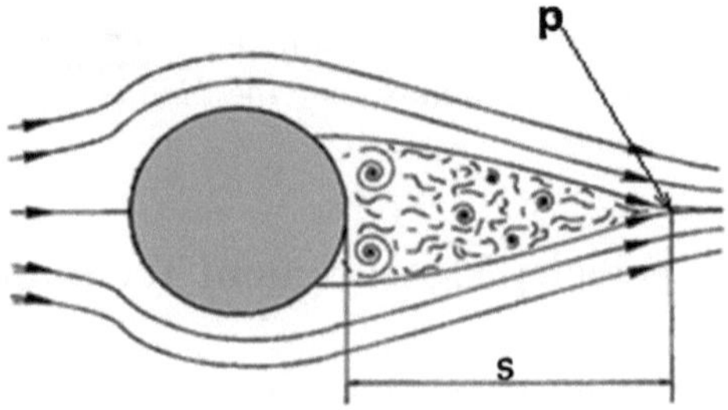

Fig. 2 Diagram showing the distance S and the point p where the stream lines that limit the wake region join together

3.4 Particle Imaging Velocimetry (PIV)

PIV as LDV is another example of a tracer method. The basic idea is very straightforward. In simple words two pictures are taken of a large number of particles with a known time interval between exposures and the distance traveled by each particle during that time interval is measured. One difficulty in applying this technique to soap films is that the films undulate. This can be suppressed by enclosing the soap film into two parallel glass plates, each on each side of the film. This PIV is ideally suited for measuring soap film velocity, however the apparatus is relatively high-priced and also image treatment software is needed (Rivera et al. 1998).

4 Outline of an Alternative Method

We have developed an alternative method to those described above for estimating the soap film velocity. The method indirectly estimates the falling speed of the soap film by comparing a feature that appears in photos of the soap flow pattern with reliable computer calculations. The procedure involves the insertion of a small circular cylinder of known diameter through the test area of the soap film. By photographically recording the flow pattern that sets up in the soap film following the insertion, we can measure the distance "S" from the edge of the circular cross section of the cylinder to the point "p" where the stream lines—that limit the wake region—join together (see Fig. 2). The position of p depends among other factors on the flow speed.

In particular this point p (shown in Fig. 2) corresponds to the place along the horizontal symmetry axis where the flow is no longer turbulent and consequently the angular component of the velocity at that point becomes null. Given that it has been well established that computer simulations adequately predict the simple case of a flow past a circular cylinder, it follows then that we can compute a graph of the distance S as function of the velocity (v_∞) of the undisturbed soap film. Hence, a simple comparison of the experimentally measured distance S to the calculated value of S versus v_∞ graph, leads us to a straightforward estimation of the required value of the undisturbed flow velocity v_∞.

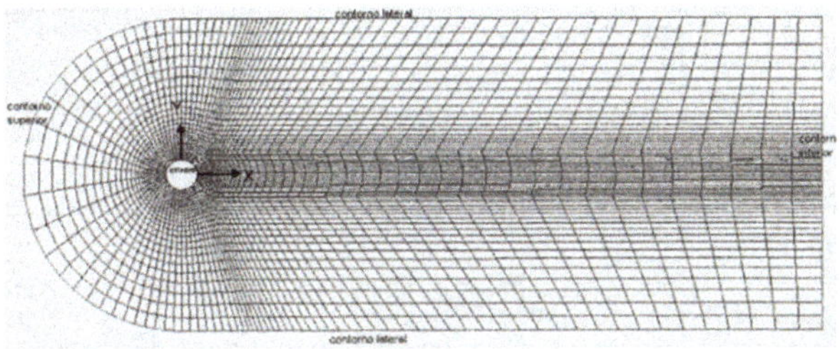

Fig. 3 Grid display showing the origin of coordinates

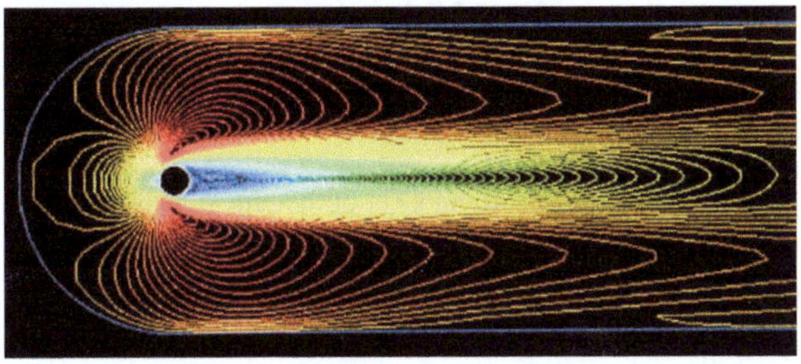

Fig. 4 Example of a fluent display showing the contours of velocity for a small cylinder

5 Details of the Alternative Method

Computer calculations can be performed by using one of the several available commercial computational fluid dynamics codes. We have chosen to simulate the flow past the cylinder the Fluent^{TM} 6.3 code, developed and marketed by Fluent Inc., (Fluent 2011). Our choice is based on reports asserting that Fluent is able to accurately simulate the fluid behavior around geometrically simple and complicated shapes over a large range of Reynolds numbers (Jones and Clarke 2008). The Fluent code solves the mass, momentum and energy conservation equations using a cell-centered finite-volume method. First the fluid domain is divided into a number of discrete control cells using a pre-processor code which creates a computational mesh on which the equations can be solved. The Fluent code includes the preprocessor Gambit^{TM} for mesh generation. The latter software allows the use of several types of computational cells including triangular, quadrilateral, hexahedral, tetrahedral, pyramidal, prismatic and hybrid meshes. Figure 3 shows the mesh we have used for computing the flow past a circular cylinder. For our purposes the origin of coordinates is at the center of the circular section of the cylinder.

Once the fluid domain has been meshed, the code applies the governing equations (in integral form) to each discrete control cell. These are a set of non-linear algebraic equations for the discrete dependent variables. Fluent then offers the user a number of choices for the algorithm used to solve these equations, including coupled explicit, coupled implicit, and segregated solvers. In the calculations reported here only the

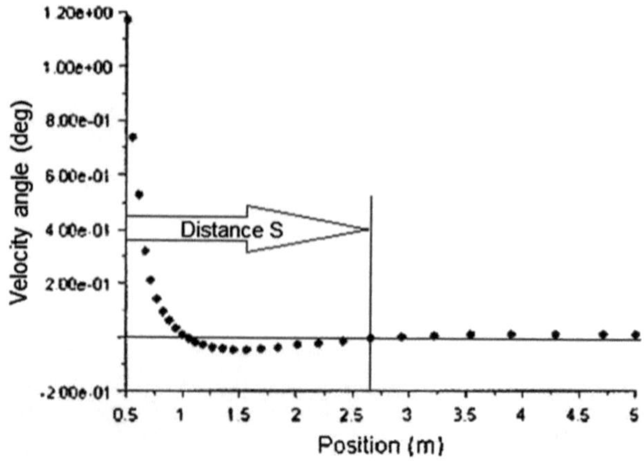

Fig. 5 Example of the angular velocity component of the film along the x-axis

segregated solver has been used. In this approach the governing equations are solved sequentially. Since these equations are non-linear they are first linearized using an implicit method. This produces a scalar system of equations containing only one equation per computational cell per degree of freedom. A point implicit (Gauss-Siedel) linear equation solver is then used in conjunction with an algebraic multi-grid method to solve the resultant scalar system of equations for the dependent variable in each cell. Since the equations are non-linear several iterations of the solution loop must be performed before a converged solution is obtained. Figure 4 shows an example of the velocity contours produced by the code.

6 Procedure

Using the mentioned approach we entered the appropriate input values into the code. These are: the cylinder diameter, fluid density ρ, viscosity μ, a trial value for v_∞, the velocity at infinity and the pressure at infinity p_∞ , which is considered to have the atmospheric value. Several runs were performed changing each time the value of v_∞. Every time that the value for v_∞ is changed, we obtained a resultant value for p, being the latter the point where the wake region ends (see Fig. 2) and thus the distance d is found. To obtain p we refer the reader to Fig. 5. This figure shows the dependence of the angular component the soap film velocity along the line passing through the cylinder center and parallel to the original flow direction (i.e. the x-axis as defined in Fig. 3). As stated before, the point p—where the flow is no longer turbulent—corresponds to a null angular velocity. The intersection of the horizontal and vertical lines shown in Fig. 5, marks point p and it's x-coordinate minus the cylinder radius, indicates the value of distance S.

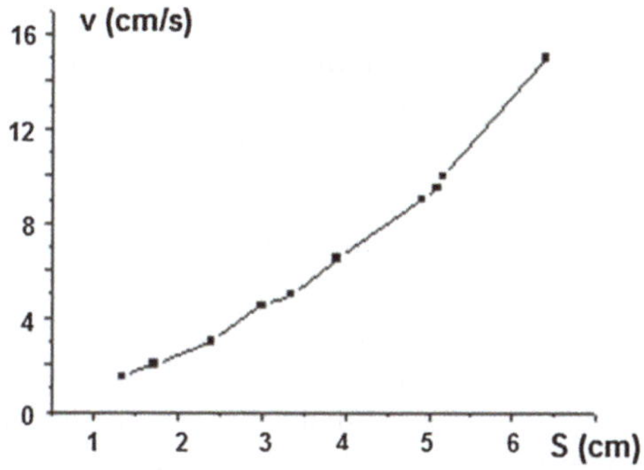

Fig. 6 Example of a graph of the velocity v_∞ versus distance S, for a cylinder of 1 cm diameter

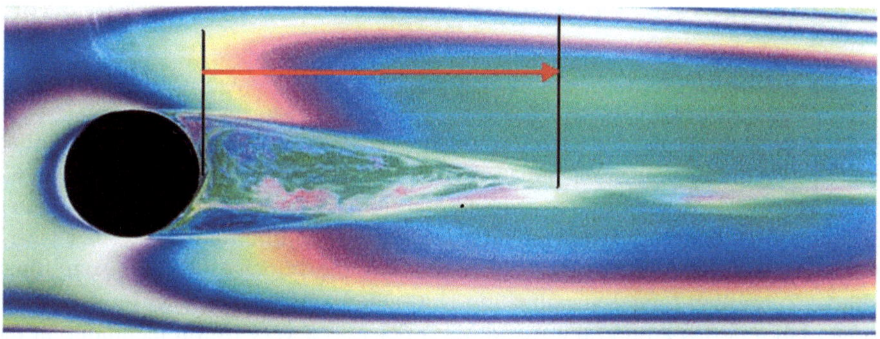

Fig. 7 Snapshot of a small cylinder (1 cm diameter) placed in the soap film. The *arrow* shows the distance S

By repeating the same procedure several times for different values of v_∞, we get a set of pairs of values (S and v_∞) and with them a graph of d versus v_∞ is drawn. This graph serves as a calibration curve (Fig. 6).

7 Experimental Details

To make the liquid for the soap film, we used different mixtures finding the optimum—for stability and durability—to be that composed of: 90 % distilled water and 10 % surfactant. The latter prepared with sodium alkyl sulfate, sodium alkyl ethoxy sulfate and amine oxide. The measured characteristics of this liquid are: for its density $8.75 \times 10^{-4}\,\mathrm{kg/m^3}$ and for its dynamic viscosity $2.56 \times 10^{-3}\,\mathrm{kg/ms}$.

Finally, the last step requires taking a photograph of the flow past the cylinder (see Fig. 7). A comparison of the distance S measured in the photo with the calibration curve yields the desired value of v_∞, i.e. the velocity of the undisturbed soap film.

To test the consistency of our procedure, we made use of three cylinders (diameters 1, 2 and 2.5 cm.) and the soap solution described above. The cylinders were carefully inserted through the film one each time, and a snapshot was taken. Then the distance S was measured for each cylinder and compared to the corresponding velocity in the v_∞ versus S graph.

The values obtained for velocity v_∞ in the tests of cylinders of diameters 1, 2 and 2.5 cm, for our experimental set up, were respectively: 5.3, 6.2 and 4.8 cm/s. This gives an average of $5.4 \pm 0.8\,\text{cm/s}$ were we have taken the error to be the difference between the average value and the farthermost experimental value (6.2 cm/s in this case). From this consideration, we estimate that our method can estimate velocities with an inaccuracy in the order of 15 %. For better results PIV techniques are recommended.

8 Conclusions

We have presented a simple way of estimating the velocity magnitude of free falling soap films. Of course in some situations our procedure does not substitute accurate methods such as PIV measurements. In addition we have presented in this paper, a summary of the basic design and construct of a soap film tunnel, together with a brief appraisal of the techniques used to measure the velocity of falling soap films. Although the soap tunnel visualization is a rather simple technique, its use in low budget laboratories around the world has a place as a very useful technique in the study of fluid dynamics and objects in motion.

Acknowledgments Work partially supported by ABACUS, CONACyT grant EDOMEX-2011-C01-165873.

References

Bandi MM, Concha A, Wood R, Mahadevan L (2013) Phys Fluids 25:1702
Durst F, Melling A, Whitelaw JH (1981) Principles and practice of laser doppler anemometry, 2nd edn. Academic, New York
Fluent (2011) product of Fluent Inc., Centerra Resource Park, 10 Cavendish Court, Lebanon, New Hampshire US 03766
Gharib M, Derango P (1989) Physica D 37:406
Jones DA, Clarke DB (2008) Simulation of the flow past a sphere using the FLUENT code DSTO–TR-2232 Department of Defense, Australia
Kellay H, Wu X, Goldburg W (1995) Phys Rev Lett 74:3875
Rivera M, Vorobeiff P, Ecke R (1998) Phys Rev Lett 81:1417
Rutgers MA, Wu XL, Daniel WB (2001) Rev Sci Instrum 72:3025

Designing Biodegradable Surfactants and Effective Biomolecules with Dissipative Particle Dynamics

Armando Gama Goicochea

Abstract The design of a biodegradable, environmentally friendly surfactant is carried out, taking the structure of a known surfactant that lacks these qualities as the starting point, using mesoscopic computer simulations. The newly designed surfactant is found to perform at least as well as its predecessor, without the latter's inimical characteristics. In the second part of this work, a comparative study of model proteins with different amino acid sequence interacting with surfaces is undertaken. The results show that, all other aspects being equal, this sequence is the key factor determining the optimal activity of the proteins near surfaces. These conclusions are found to be in agreement with recent experiments from the literature.

1 Introduction

Amphiphilic molecules such as surfactants, polymers and proteins are very important in biological processes such as drug delivery, adsorption on living tissue, and molecular association (Jönsson et al. 1998). Although the basic interactions between the various molecules are often well known (van der Waals, electrostatic), the complex interplay that emerges from the many—body manifestations of these forces with the specific structures of the amphiphilic molecules is not. Detailed knowledge of how these interactions play in complex fluids composed of those molecules, biological membranes and water is not easily accessible from a theoretical point of view because it represents a scenario with vastly different length and time scales (Israelachvili 2011). Fortunately, the recent advances in the speed of microprocessors have made it possible to solve computer models of biologically relevant systems

A. Gama Goicochea(✉)
Departamento de Ciencias Naturales, DCNI,
Universidad Autónoma Metropolitana Unidad Cuajimalpa,
Av. Pedro Antonio de los Santos 84, 11850 México, D. F., Mexico
e-mail: agama@alumni.stanford.edu; agama@correo.cua.uam.mx

J. Klapp and A. Medina (eds.), *Experimental and Computational Fluid Mechanics*, Environmental Science and Engineering, DOI: 10.1007/978-3-319-00116-6_39,

in reasonable times, thus providing important information to understand, improve and design new biomolecules tailored to solve specific needs.

The work presented here reports the importance of the structure of biomolecules in their performance in environments of current interest, by means of mesoscopic, dissipative particle dynamics (DPD) computer simulations. The first part is devoted to showing how an environmentally friendly non ionic surfactant was designed starting from one that was not, without detriment to its performance. The prediction of the thermodynamic properties of the new surfactant led to the conclusion that they were at least of equal quality as those of its predecessor, with the structure of the surfactants playing the major role. In the second part, I show how a model protein with different amino acid sequence can lead to entirely different thermodynamic conditions when placed near a lipid membrane, which defines a criterion for choosing the best one before synthesizing one in laboratory. The underlying thesis of this contribution is that, all things being equal, the structure of the molecules defines their function in a complex biological environment.

The rest of this chapter is organized as follows. In Sect. 2 I introduce the basics of the DPD model and simulation details. The Sect. 3 is devoted to the presentation of a newly designed, environmentally friendly and biodegradable surfactant and the comparison of its performance to a commercially available (not environmentally friendly), similar surfactant. Section 4 is dedicated to the study of the influence that the amino acid sequence has on the thermodynamic behaviour of model proteins interacting with biologically relevant surfaces. The conclusions are drawn in Sect. 5.

2 The DPD Model

Atomistically detailed computer simulations (Allen and Tildesley 1987) have proved to be very successful, but in order to achieve scales comparable to those accessible to experiments they require considerable computational resources. When modelling large molecules and long time scales as it is befitting in biological systems, one needs tools that go beyond the atomistic regime, and DPD (Hoogerbrugge and Koelman 1992) is one of them. The reason relies on the fact that DPD involves short—range, linearly decaying forces which can be integrated using a time step that is at least three orders of magnitude larger than that used in microscopic simulations, allowing the study of phenomena at the mesoscopic scale.

The basic structure of DPD consists of three forces, conservative $\boldsymbol{F}_{ij}^{C}$, dissipative $\boldsymbol{F}_{ij}^{D}$ and random $\boldsymbol{F}_{ij}^{R}$. All forces between particles *i* and *j* have simple spatial dependences and vanish beyond a finite cutoff radius R_C, which represents the intrinsic length scale of the DPD model and it is usually chosen as the reduced unit of length, $R_C = 1$. Their functional expressions are:

$$\boldsymbol{F}_{ij}^{C} = a_{ij}\left(1 - r_{ij}/R_C\right)\hat{\boldsymbol{e}}_{ij} \tag{1}$$

$$\boldsymbol{F}_{ij}^{D} = -\gamma \left(1 - r_{ij}/R_C\right)^2 \left[\hat{\boldsymbol{e}}_{ij} \cdot \boldsymbol{v}_{ij}\right] \hat{\boldsymbol{e}}_{ij} \quad (2)$$

$$\boldsymbol{F}_{ij}^{R} = \sigma \left(1 - r_{ij}/R_C\right) \hat{\boldsymbol{e}}_{ij} \xi_{ij}. \quad (3)$$

In the expressions above, $\boldsymbol{r}_{ij} = \boldsymbol{r}_i - \boldsymbol{r}_j$ is the relative position vector, is the unit vector in the direction of $\boldsymbol{r}_{ij}$, and $\boldsymbol{v}_{ij} = \boldsymbol{v}_i - \boldsymbol{v}_j$ is the relative velocity, with $\boldsymbol{r}_i$, $\boldsymbol{v}_i$ the position and velocity of particle *i*, respectively. The variable ξ_{ij} is generated randomly between 0 and 1 with a Gaussian distribution of unit variance; a_{ij}, γ and σ are the strength of the conservative, dissipative and random forces, respectively; all forces are zero for $r_{ij} > R_C$. All beads are the same size, but the difference between beads of different chemical nature is defined by the constant a_{ij}, and all thermodynamic properties are dependent on it. The strengths of the random and dissipative forces are related as follows:

$$\frac{\sigma^2}{2\gamma} = k_B T \quad (4)$$

which represents the fulfilment of the fluctuation—dissipation theorem for DPD and it is one of the defining qualities of the method. Another key feature is that the forces in Eqs. (1–3) are pairwise additive, therefore local and global momentum is conserved, which in turn means that all hydrodynamic modes present in the fluid shall be preserved. The conservative force parameter for particles of the same type, a_{ii}, is given by $a_{ii} = [(\kappa^{-1} N_m - 1)/2\alpha\rho] k_B T$, where N_m is the coarse-graining degree (number of water molecules grouped in a DPD particle), α is a numerical constant equal to 0.101, ρ is the DPD number density, κ^{-1} is the inverse compressibility of the water at room temperature. I choose a coarse-graining degree equal to 3 water molecules in a DPD particle and used the dimensionless water compressibility at standard conditions, $\kappa^{-1} \approx 16$, so that the parameter above is $a_{ij} = 78.3$. The parameter for different types of particles, a_{ij}, is calculated from its Flory-Huggins coefficient χ_{ij} using the relation $a_{ij} = a_{ii} + 3.27\chi_{ij}$. For further details, see (Groot and Warren 1997). The DPD method has enjoyed considerable success over a wide range of applications, including biological systems. Some of the most recent ones have been reviewed extensively by (Murtola et al. 2009).

3 Modelling of a Biodegradable Surfactant

The ecological impact of surfactants has become increasingly important in most contemporary formulations. The rate at which surfactants will biodegrade at some sewer plant, determines to a large extent the preference for one or another surfactant. Some of the aspects that are relevant when monitoring surfactants environmental impact are aquatic toxicity in fish mainly; bioaccumulation resulting from the built up of organic compounds in fish also, and biodegradability. The latter results typically from a series of enzymatic reactions that break down the original composition of the surfactant and turn it into a mix of water, oxides and other products (Jönsson et al. 1998).

$CH_3(CH_2CH_2)_4$-C_6H_4-$O(CH_2CH_2O)_{10}H$

Fig. 1 Chemical structure of the surfactant modelled in this work. It is a nonylphenol ethoxylate with 10 moles of ethylene oxide

Table 1 Conservative DPD interaction parameters (a_{ij}) used in the simulations of adsorption of NP10. The numbers identify each type of particle in the simulations, starting with water (1), and the surfactant DPD particles defined in Fig. 2: C (2), D (3), E (4), F (5), G (6) and H (7). The last column lists the interaction (a_{iw}) of each type of DPD particle with the surfaces, see Eq. 5

a_{ij}	1	2	3	4	5	6	7	a_{iw}
1	78.3	39.15	39.15	39.15	155	171.8	167.5	31.2
2		78.3	79.3	86.3	79	81.3	80.5	31.5
3			78.3	92.9	78.3	78.8	78.6	31.5
4				78.3	91.5	99	97	31.5
5					78.3	79.1	78.8	17.5
6						78.3	78.4	21
7							78.3	21

Among the most frequently used surfactants in modern day applications are the non ionic, nonylphenol ethoxylates which are used in emulsion polymerization processes, as detergents and pesticides to name a few. In this section I shall be primarily interested in a particular example of this class of surfactants, whose structure is shown in Fig. 1.

Surfactants of the type shown in Fig. 1 are useful for many applications but in recent years their use has been considered somewhat deleterious for the environment due to the fact that a linear molecule, such as the one shown in Fig. 1, is easily biodegradable. Under typical circumstances this would be a favourable aspect, except for the fact that the surfactant shown in the figure above in what follows will be called NP10 (meaning nonylphenol with 10 moles of ethylene oxide), contains a benzene ring in its structure. The release of free benzene may lead to the disruption of the balance of hormones in fish and other organisms (Boogaard and van Sittert (1995)), although it appears to be a not too strong effect. Nevertheless, this is an aspect that deserves attention and as such, in this section I shall study the adsorption properties of NP10 in a model biological environment.

To proceed, the first step is to map the structure of NP10 shown in Fig. 1 to DPD beads. Then, a model for the adsorbing surfaces must be introduced. Following the standard procedure (Groot and Warren 1997) for a coarse graining degree equal to 3 water molecules per DPD bead, one obtains the coarse—graining mapping of surfactant NP10 as shown in Fig. 2. The matrix of conservative interaction parameters a_{ij} (see Eq. 1) obtained after applying such procedure is presented in Table 1, where the interaction parameters of each DPD particle of the NP10 surfactant (shown in Fig. 2) with a model biological membrane, a_{iw}, are listed as well.

To model a soft biological membrane on which NP10 will adsorb, the following short range, effective force is added:

DPD Particle	Atoms in DPD particle
C	OCH_2CH_2O
D	$CH_2CH_2OCH_2$
E	$HOCH_2CH_2$
F	C_6H_4
G	$CH_2CH_2CH_2$
H	$CH_3CH_2CH_2$

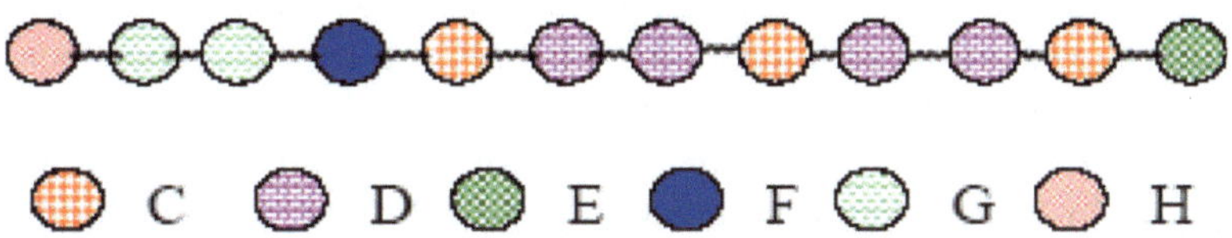

Fig. 2 DPD mapping of the surfactant. Each bead represents a group of atoms. The *top* part of the figure shows the specific atomic grouping associated with each DPD bead shown in the *bottom* part of this figure

$$\boldsymbol{F}_{iw}\left(z_{i}\right)=a_{iw}\left(1-z_{i}/R_{C}\right)\hat{\boldsymbol{e}}_{k}. \tag{5}$$

In Eq. 5, the force between a particle i (either water or the NP10 surfactant) and the wall representing a smooth biomembrane placed parallel to the XY—plane, at the ends of the simulation box in the z—direction, is shown to depend only on the component of such particle's position along the Z-axis. The wall interaction constant a_{iw}, whose values for the various DPD particles in the simulation box have been listed in Table 1, were obtained following the same procedure as that used for the fluid's particle—particle interactions. The symbol $\hat{\boldsymbol{e}}_k$ represents the unit vector in the direction perpendicular to the XY—plane. Equation 5 represent a soft surface, in much the same spirit as the soft interactions given by Eq. 1, as it is appropriate for example, for surfaces formed of lipid bilayers (Israelachvili 2011). More sophisticated, self—consistent DPD surface models are available (Gama Goicochea and Alarcón 2011), but they are better suited to study harder solid walls, which are not of biological relevance, within the context of the present study.

For the modeling of the adsorption of NP10 on the membranes the following procedure is followed. A given concentration of NP10 is chosen and allowed to reach equilibrium while keeping the chemical potential of the solvent constant. Once equilibrium has been reached, one determines the amount of NP10 that was adsorbed and what was left in the supernatant, i.e. not adsorbed, if any. This procedure is followed for as many surfactant concentrations as one wishes, constructing the adsorption isotherm from the data collected. Experimental adsorption isotherms are obtained following precisely the same method (Kronberg 2001). Monte Carlo (MC) simulations in the Grand Canonical (GC) statistical mechanical ensemble (which fixes the chemical potential, μ, as well as the volume and temperature) were carried out to obtain the equilibrium state for each NP10 concentration, using a code hybridized with DPD, see (Gama Goicochea 2007) for full details, including the integration algorithm for the equation of motion, with a time step $\delta t = 0.03$. Only the number of solvent particles was allowed to fluctuate, to keep the chemical potential fixed. The dimensions of the simulation box were fixed at $L_x = 7$, $L_y = 7$ and $L_z = 14$ adimensional DPD units. The temperature was fixed at $T = 1$ by choosing $\sigma = 3$ and $\gamma = 4.5$ (see Eqs. 2 and 3), and the solvent's chemical potential was chosen as $\mu = 37.7$, which leads to an average total density $\rho = 3$; by doing so one assures that the results are invariant under changes of conservative force interaction parameters (Groot and Warren 1997). For the equilibrium phase, 30 blocks of 10^4 MC steps were run, followed by an additional 100×10^4 MC steps for the production phase. The solvent is modeled as monomeric DPD particles, while the surfactant is made up of 12 DPD units, with the sequence as shown in Fig. 2, and with the surfactant beads joined by freely rotating harmonic springs with spring constant $k_0 = 100$ and equilibrium length $r_0 = 0.7$ (Gama Goicochea 2007).

The NP10 concentration was varied in the range of $4 \leq [c] \leq 90$ surfactant molecules in the simulation box, with the number of solvent monomers being adjusted by the GCMC ensemble to keep the chemical potential always fixed. At each concentration the density profile of the monomers that make up NP10, $\rho(z)$, was computed, and the amount of adsorbed NP10, Γ, was obtained from the equation (Gama Goicochea 2007):

$$\Gamma = \int_0^{L_z} [\rho(z) - \rho_B] dz. \tag{6}$$

In Eq. 6, ρ_B is the density of NP10 monomers in the bulk, namely those that were not adsorbed on the confining surfaces. If all NP10 molecules were adsorbed, $\rho_B = 0$. Fig. 3 shows typical equilibrium configurations for two NP10 concentrations, [c] = 14 molecules/volume (left image), and [c] = 30 molecules/volume.

As it is evident from Fig. 3, the surfactant is subjected to two competing factors: on the one hand it has a strong tendency to adsorb at the interfaces (see left image in this figure), especially at relatively low concentrations. But, as the concentration is increased some surfactants find it more favorable to form micelles (right image in Fig. 3) while others are adsorbed. In the particular example shown in Fig. 3, those NP10 molecules forming the micelle (right image) would be the ones that contribute to the bulk surfactant density, ρ_B, in Eq. 6, while those adsorbed would be accounted for in $\rho(z)$. It is also of notice that the surfactant appears to adsorb in monolayers, in much the same way as assumed by the Langmuir adsorption model (Kronberg 2001).

From the series of simulations previously described the adsorption isotherm of the NP10 surfactant in water was obtained, and it is presented in Fig. 4. It is clearly a Langmuir—type isotherm, from which one can easily obtain the saturation concentration, i. e. the amount of surfactant that has saturated the available adsorption sites

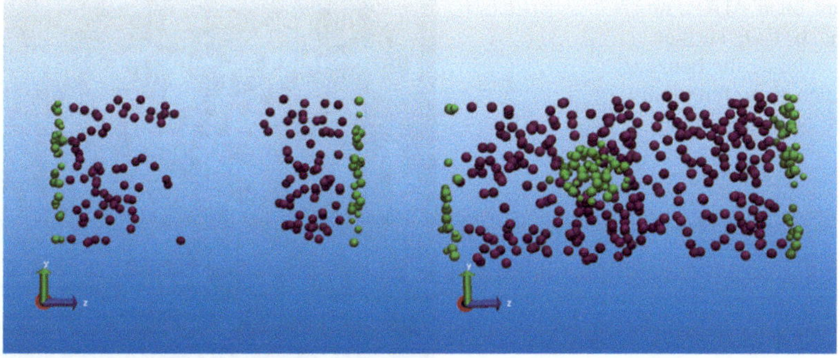

Fig. 3 Adsorption of NP10 on model biomembranes. Green beads represent DPD beads named F, G and H (see mapping in upper panel of Fig. 2). Purple beads are hydrophilic surfactant beads, namely C, D, and E. The solvent beads have been removed for clarity. The left image corresponds to a NP10 concentration [c] = 14 molecules/volume, while the one on the right is for [c] = 30. Notice the incipient formation of a micelle in the latter

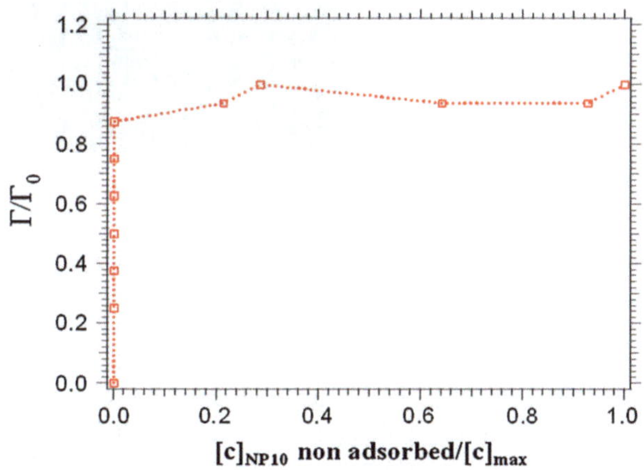

Fig. 4 Adsorption isotherm of NP10 on model biomembranes. The axes have been normalized to allow for comparisons with other surfactants. The *X*-axis is the number of non adsorbed NP10 molecules, normalized by the maximum amount of non adsorbed surfactant, $[c]_{max}$. The *Y*-axis represents the number of adsorbed NP10 molecules for each concentration added to the dispersion, normalized by the maximum adsorbed amount, at saturation, Γ_0

on the surfaces, so that whatever additional amount of surfactant added to the system is not likely to be adsorbed. This is an important quantity because it is obtained from experiments. therefore it represents a basis for comparison.

The curve in Fig. 4 is a typical example of a Langmuir isotherm, whose basic feature is a rapid raise to the saturation value, followed by an essentially constant adsorption after it. From the isotherm in Fig. 4 one obtains a saturation adsorption value equal to $\Gamma_0 = 1.85$ g NP10 per gram of membrane. Now, it is instructive to

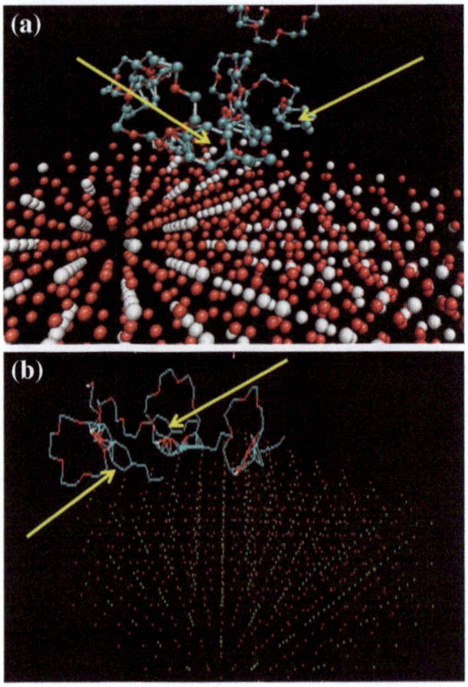

Fig. 5 Adsorption of the surfactant on **a** Al_2O_3 surface, and **b** TiO_2 surface. Arrows indicate the benzene ring, which is preferentially adsorbed on both surfaces

ask ourselves not only how much surfactant is adsorbed, but also *how* it adsorbs. As the following figure will show, the NP10 surfactant adsorbs primarily through the benzene ring (which corresponds to the DPD bead called F, as clearly indicated in Fig. 2) on most hydrophobic surfaces. Recalling the arguments expressed above, about the potential environmental concerns regarding the release of benzene after the break down of surfactant molecules that include it, such as NP10, it is therefore of paramount importance to ascertain to what extent is the presence of benzene indispensable for the performance of the surfactant. This is the key element in the design of a new, environmentally friendly surfactant undertaken in the present work.

Figure 5 shows images obtained from atomistically detailed, microscopic (namely, not DPD) computer simulations of NP10 adsorbed on some typical metal oxide surfaces, carried out using the *Materials Studio* suite of *Accelrys* (see http://www.accelrys.com). It is important to carry out those studies to confirm the hypothesis that, on entropic or free energy grounds, it is through the benzene ring where NP10 preferentially adsorbs on hydrophobic surfaces, thereby discarding the possibility that it is the loss of atomistic detail what is responsible for the predicted behavior of the surfactant.

Based on the microscopic calculations from which Fig. 5 was obtained, as well as those of Fig. 3, it is logically sound to speculate as to what the thermodynamic properties of a surfactant such as NP10 *without* the benzene ring would be. If properties such as adsorption isotherms of the newly modified surfactant turn out to be entirely

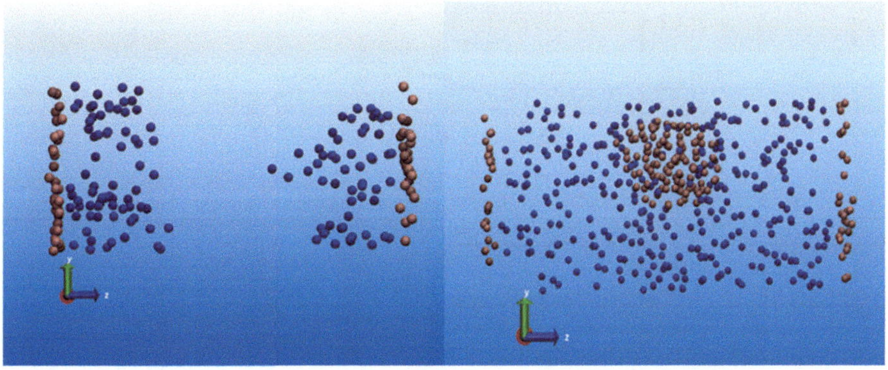

Fig. 6 Adsorption of B10 on model biomembranes. Brown beads represent DPD units named G and H (see mapping in upper panel of Fig. 2). Blue beads are the hydrophilic beads, called C, D, and E. The solvent beads have been removed for clarity. The left image corresponds to a B10 concentration [c] = 16 molecules/volume, while the one on the right is for [c] = 40

different from those of the proven NP10, then that would be a strong indication that a new strategy ought to be sought for a new surfactant.

The structure of this newly designed surfactant, which will be called B10 (for "biodegradable surfactant with 10 moles of ethylene oxide") henceforth, is almost the same as that of NP10, with the F-bead (benzene ring) replaced by a G-bead (see Fig. 2). It is made up of 12 beads also, with the sequence of beads as follows: H-G-G-G-C-D-D-C-D-D-C-E. The interaction parameters, surfactant concentrations, simulation box volume, interaction with the surfaces, and all other simulation details were chosen exactly as those used in the prediction of the adsorption isotherm of NP10. Fig. 6 shows two equilibrium configurations obtained for surfactant B10, one below the micelle formation (left image), and one above it (right image).

By comparing the configurations in Fig. 6 for B10 with those in Fig. 3 for NP10, it is clear that the same qualitative behaviour is obtained for B10, i.e., the surfactant is driven by two factors. At low enough concentrations, it adsorbs readily on the surfaces, forming brushes. As its concentration is increased, it associates with other B10 molecules, forming micelles, like the one shown on the right in Fig. 6, although some of the molecules continue to adsorb on the substrates. With this information I have calculated the adsorption isotherm for B10, which is to be found in Fig. 7.

The adsorption isotherm shown in Fig. 7 is also of the Langmuir type which, as the images in Fig. 6 show, is the result of the fact that B10 adsorbs in monolayers. It should be noted that the competition between adsorption and micelle formation starts before the surfaces have been saturated, and it appears to be driven by the surfactant concentration. The data obtained from the adsorption isotherm allows us to extract the saturation value, which turns out to be $\Gamma_0 = 2.46$ g B10 per gram of membrane. It is slightly larger than that for NP10 (1.85 g), but this has a simple explanation. B10 adsorbs on the surfaces through beads H and G (see Fig. 2 for their atomistic mapping), while NP10 does mainly through bead F, of which there is only one per

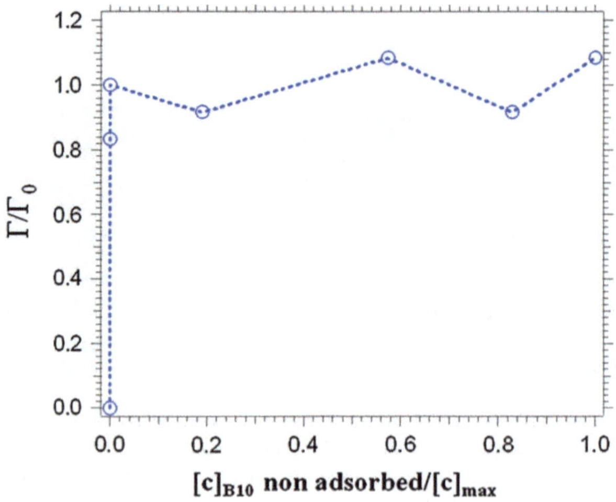

Fig. 7 Adsorption isotherm of surfactant B10, designed in this work. The axes have been normalized to allow for comparisons with other surfactants. The *X*-axis is the number of non adsorbed B10 molecules, normalized by the maximum amount of non adsorbed surfactant, $[c]_{max}$. The *Y*-axis represents the number of adsorbed B10 molecules for each concentration added to the dispersion, normalized by the maximum adsorbed amount, at saturation, Γ_0. The line is only a guide to the eye

molecule. Other than that, both surfactants have essentially the same thermodynamic behavior, but the newly designed B10 has the bonus that, when used in formulations that interact with the environment, it will biodegrade as easily as NP10, but it will not release benzene freely, because its structure does not include it. It is to be concluded that B10 is an excellent candidate to substitute NP10, given their similar molecular structures, which in turn give rise to very similar function and performance.

4 Modelling Protein Activity Through Amino Acid Sequence

The interaction of proteins with biosurfaces is known to be of paramount importance for an ample range of situations (Israelachvili 2011). More often than not, the specific amino acid sequence determines the activity of the protein in a given environment, even for two proteins of the same molecular weight which are otherwise identical, except for their amino acid sequence. This is a problem that is entirely amenable to be approached by the techniques described here, as shall be shown in what follows. The purpose of this section is to determine the effectiveness of model proteins (meaning that the model does not represent an exact mapping of any given real protein or amino acid) interacting on soft, model biomembranes as a function only their amino acid sequence, within the DPD model. The activity of the proteins with the surfaces is monitored through the calculation of the interfacial tension between them, in an aqueous environment.

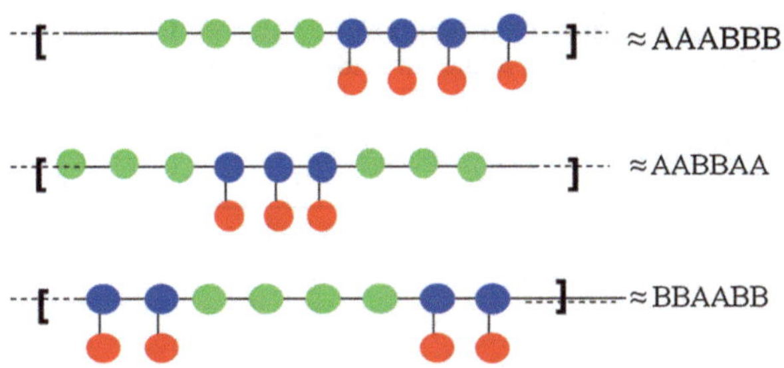

Fig. 8 Three model amino acid sequences for the biomolecules studied in this section. The differently coloured DPD beads are meant to represent different amino acids. The letters on the right symbolize the sequencing, to shorten the notation. The blue—red dimmers are represented in such lettering as B; however, it is the red bead that one that interacts more strongly with the surfaces. Although all molecules are made up of 24 beads, only 12 are shown in each case for simplicity

Let us start by considering model proteins made up of 24 DPD beads each, with each bead representing a model amino acid. Only three amino acid sequences shall be studied, for brevity. These are shown schematically in Fig. 8. All beads have the same size; the molecules are linear, with beads joined by freely rotating harmonic springs of the same type as those used in the previous section, namely with $k_0 = 100$ and $r_0 = 0.7$. Soft DPD-like effective surfaces given by Eq. 5 are placed at the ends of the simulation box in the z-direction. Periodic boundary conditions were implemented in the x- and y-directions, where the fluid is free, but not in the z- direction since the walls are impenetrable. The fluid is made up of water monomers and the proteins only. Simulations are performed using a GCMC algorithm, hybridized with DPD, the same used in the previous section. The length of the simulations, time step, box volume, fixed chemical potential, and temperature are exactly the same as used for the prediction of the adsorption isotherms in the previous section.

Figure 8 shows schematically the three spatial arrangements, or sequences of the DPD beads; these constitute the only difference between the three cases considered here. The different colouring of the DPD beads is meant to represent the different chemical compositions of the amino acids. The interaction with the surfaces is driven primarily by the red beads as was the case with surfactant NP10 of the previous section. The number of proteins in the simulation box was varied from $[c] = 40 - 90$ molecules per volume. As for the conservative force constants (see Eq. 1), they were chosen following the same procedure as in the previous section, and they are given by the matrix shown in Table 2, where the interaction with the walls is listed also.

Once a sequence among those shown in Fig. 8 is chosen, with the appropriate interaction parameters, displayed in Table 2, one proceeds to carry out MC simulations at fixed chemical potential, volume and temperature, at the end of which the components of the pressure tensor, $P_{\alpha\beta}$, are obtained. When equilibrium is reached, these components are calculated using the virial theorem (Allen and Tildesley 1987). Then, the interfacial interaction between the proteins and the membranes, γ, is calculated by means of Eq. 7 (Gama Goicochea et al. 2007):

Table 2 Conservative DPD interaction parameters (a_{ij}) used for the beads shown in Fig. 8. The numbers identify each type of particle in the simulations, starting with water (1), and the protein DPD beads depicted in Fig. 8: green (2), blue (3), and red (4). The last column lists the interaction (a_{iw}) of each type of DPD particle with the surfaces, see Eq. 5

a_{ij}	1	2	3	4	a_{iw}
1	78.3	78.3	171.8	155	30.75
2		78.3	81.3	79.1	31.0
3			78.3	79.1	45.0
4				78.3	25.0

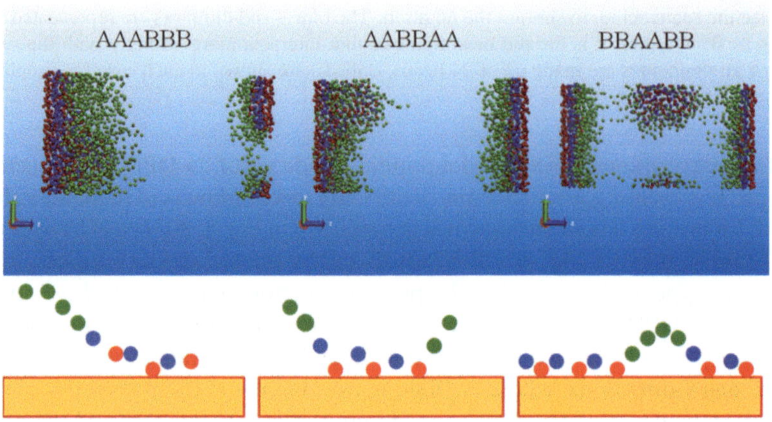

Fig. 9 Interaction of model proteins with membranes. The top part shows the final equilibrium configuration ([c] = 90 mol./vol.) for each sequence. The lower is the corresponding schematic of adsorption. The colour code is the same as in Fig. 8. Water was removed for simplicity

$$\gamma = L_z \left[\langle P_{zz} \rangle - \frac{1}{2} \left(\langle P_{xx} \rangle + \langle P_{yy} \rangle \right) \right] \tag{7}$$

where $\langle\rangle$ represents an average over the ensemble, L_z is the length of the simulation box perpendicular to the surfaces, and only the diagonal elements of the pressure tensor are needed. In Fig. 9 one finds the final configurations obtained for each sequence.

The configurations shown in the top part of Fig. 9 clearly indicate that, although the number of molecules is same for all three sequences (equal to 90 protein mol./vol.), and although the number of amino acids is equal in all three cases, as well as all other input details, the specific structure of the protein determines its thermodynamic interaction with the surface, and its self association. After proper transformation of DPD units, the following values of the interfacial tension between the proteins and the surfaces are obtained: for the sequence called AAABBB, $\gamma = 44.5 \pm 0.8$ dyn/cm; for AABBAA, $\gamma = 37.5 \pm 0.8$ dyn/cm. Finally, for BBAABB, $\gamma = 59.0 \pm 0.3$ dyn/cm.

The conclusion to be drawn from these results is that the most favourable sequence from the thermodynamic point of view between the surfaces and the proteins is given by the one called AABBAA, which corresponds to the image in the middle of Fig. 9, with the minimum value of γ. By contrast, the most unfavourable structure is given by the one labelled BBAABB, because it is the one that requires the most energy investment per unit area, i.e., the largest γ, when interacting with the surface. From inspection of Fig. 9 one sees that the configuration on the far left ("AAABBB") cannot be optimal because it does not cover all the surfaces, which means that the activity of the protein is diminished. On the other hand, the configuration on the far right ("BBAABB") is not the best either, although it appear to adsorb uniformly on the surfaces, because some proteins self—associate in the bulk of the fluid, forming a micelle—like structure, which again means a detriment of the protein function, whose purpose is interacting with the surfaces. The central image in Fig. 9 is the best one because all proteins are fully interacting with the surfaces through physical adsorption, with none of them in the bulk fluid, which is why the interfacial tension was found to be the minimum for this case. It is to be emphasized that these differences arise purely from a structural difference between the proteins, as laid out by their amino acid sequence.

As for the physical reason behind the results shown in Fig. 9 and discussed in the previous paragraph, the caricature adsorption model shown in the lower images in Fig. 9 gives us some insight. The structure with the minimum interfacial tension ("AABBAA") has the particular feature that it groups together as nearest neighbours, all the amino acids that preferentially adsorb on the surfaces (shown in red in Fig. 8). This maximises the area covered on the surface by the molecule. If the red amino acid were surrounded by green or blue ones as nearest neighbours, which do not adsorb as favourable as red on the surfaces (as is the case for structures AAABBB and BBAABB) then the energy cost per unit area of the protein—surface interaction turns out to be more expensive, and it is therefore not preferred by thermodynamics. These results and their interpretation are in complete agreement with those of recent experimental studies (Jhon et al. 2009).

5 Conclusions

The present work reports the design and thermodynamic testing of a biodegradable surfactant and model proteins in aqueous solutions confined by soft surfaces, intended to model biologically relevant membranes, by means of mesoscopic DPD computer simulations. In the first part of this report it was shown that a newly designed linear, non ionic and environmentally friendly surfactant performs at least as well as its benzene—containing (and therefore, *not* environmentally friendly) counterpart. The performance of both surfactants was tested with the calculation of adsorption isotherms and surface saturation. The new surfactant structure was proposed after a careful microscopic analysis of the role played by each part of the original surfactant's structure. The second part of this research was devoted to the design of differently sequenced but otherwise equal model proteins, with the purpose of determining which one had the optimal activity when interacting with a biologically important surface. Predictions of their interfacial (protein—surface) tension led to the conclusion that the proteins that adsorbed uniformly on the surfaces without leaving any of them in the bulk to form energy—consuming micelles were the ones preferred on thermodynamic grounds. These conclusions were found to be fully supported by recent experiments reported in the literature.

The advantages of carrying out coarse—grained, mesoscopic DPD simulations like the ones reported here are numerous. Not only can one reach length scales of the order of μm and times scales of ms, but because of the momentum conserving structure of the DPD forces, one can capture complex mesoscopic hydrodynamic behaviour that is crucial for the study of the association of biomolecules. Other advantages include the fact that simulations can be carried out with many chains at once, and with the solvent included explicitly. The latter is critical to incorporate excluded volume interactions at short distances.

The two cases studied in this work have a common thread, which can perhaps be summarized as follows: when comparing the performance of two molecules whose only difference is their structure, then any difference in their thermodynamic performance must be attributed precisely to their structural dissimilarities. This has been more eloquently stated by Crick, in his famous dictum "*if you want to understand function, study structure*" (Crick 1988). These studies can be considered as a stepping stone toward the construction of more atomistically detailed, albeit more computationally expensive models.

Acknowledgments The author wishes to thank M. Maciel, S. Viale, F. Zaldo, and especially N. López for enlightening discussions during the early stage of this project. Additional insightful comments from G. Pérez Hernandez are gratefully acknowledged. This work was initially sponsored by the Centro de Investigación en Polímeros (CIP, COMEX Group) and subsequently by PROMEP, Project 47310286-912025.

References

Allen MP, Tildesley DJ (1987) Computer simulation of liquids. Oxford University Press, Oxford

Boogaard PJ, van Sittert NJ (1995) Biological monitoring of exposure to benzene: a comparison between S—phenylmercapturic acid, trans, trans—muconic acid and phenol. Occup Environ Med 52:611–620

Crick F (1988) What mad pursuit. Basic Books, New York

Gama Goicochea A (2007) Adsorption and disjoining pressure isotherms of confined polymers using dissipative particle dynamics. Langmuir 23:11656–11663

Gama Goicochea A, Romero-Bastida M, López-Rendón R (2007) Dependence of thermodynamic properties of model systems on some dissipative particle dynamics parameters. Mol Phys 105:2375–2381

Gama Goicochea A, Alarcón F (2011) Solvation force induced by short range, exact dissipative particle dynamics effective surfaces on a simple fluid and on polymer brushes. J Chem Phys 134:014703

Groot RD, Warren PB (1997) Dissipative particle dynamics: bridging the gap between atomistic and mesoscopic simulation. J Chem Phys 107:4423

Hoogerbrugge PJ, Koelman J (1992) Simulating microscopic hydrodynamic phenomena with dissipative particle dynamics. Europhys Lett 19:155–160

Israelachvili JN (2011) Intermolecular and surface forces, 3rd edn. Academic Press, London

Jhon YK, Semler JJ, Genzer J, Beevers M, Gus'kova OA, Khalatur PG, Khokhlov AR (2009) Effect of comonomer sequence distribution on the adsorption of random copolymers onto impenetrable flat surfaces. Macromolecules 42:2843–2853

Jönsson B, Lindman B, Holmberg K, Kronberg B (1998) Surfactants and polymers in aqueous solution. John Wiley & Sons, New York

Kronberg B (2001) Measuring adsorption. In: Holmberg K (ed) Handbook of applied surface and colloid Chemistry 2, Chapter 22. John Wiley & Sons, New York

Murtola T, Bunker A, Vattulainen I, Deserno M, Karttunen M (2009) Multiscale modeling of emergent materials: biological and soft matter. Phys Chem Chem Phys 11:1869–1892

Uranyl Transport in the Unsaturated Zone of the Toluca Valley, State of Mexico

Lázaro Raymundo Reyes-Gutiérrez, Elizabeth Teresita Romero-Guzmán and Jaime Lázaro Klapp-Escribano

Abstract The aim of this work was to simulate the transport of uranium in the unsaturated zone of an experimental parcel located in the community of San Cayetano, State of Mexico. The implanted cultivation is corn, fertilized with granulate triple superphosphate. During the transport of fertilizers through the unsaturated media, the physicochemical processes can produce the exchange of the mass of the fertilizers among the solid and aqueous phases, for example the sorption for the geologic materials, the precipitation or dissolution and the complexation. The evolution of the humidity was evaluated using tensiometers and the concentration and uranyl, [U(VI)] mobility was determined using a cup of porous ceramic. It is important to indicate that phosphate and nitrogen fertilizers are the most employed in that area for corn growing, for this the human health risk associated with the presence of widespread uranium contamination of soils, sediments, and groundwater.

1 Introduction

The unsaturated zone has a great hydrogeological relevance because in it the infiltration, evaporation, erosion processes, recharge of the underground and attenuation-

L. R. Reyes-Gutiérrez (✉) · J. L. Klapp-Escribano
Departamento de Matemáticas, Cinvestav del I.P.N., 07360 México, D.F, México
e-mail: raregu@gmail.com

L. R. Reyes-Gutiérrez
Universidad Autónoma Metropolitana, Hidalgo Poniente 46. Col. La Estación, 52006 Lerma de Villada, Estado de México, México

J. L. Klapp-Escribano · E. T. Romero-Guzm
Instituto Nacional de Investigaciones Nucleares, CarreteraMéxico-Toluca S/N, La Marquesa, 52750 Ocoyoacac, Estado de México, México
e-mail: jaime.klapp@inin.gob.mx

E. T. Romero-Guzm
e-mail: elizabeth.romero@inin.gob.mx

J. Klapp and A. Medina (eds.), *Experimental and Computational Fluid Mechanics*, Environmental Science and Engineering, DOI: 10.1007/978-3-319-00116-6_40,

transformation of liquid and gassy contaminants is presented. This zone plays an important role as source of recharge for the aquifers and for the main processes of contamination to reach the groundwater, these processes can be understood by means of the modeling to determine the mechanisms of solute transport predominant in the soil.

The uranyl ion is an oxycation of uranium in the oxidation state +6, with the chemical formula $[UO_2]^{2+}$. It has a linear structure with short U-O bonds, indicative of the presence of multiple bonds between uranium and oxygen. Four or more ligands are bound to the uranyl ion in an equatorial plane. The uranyl ion forms many complexes, particularly with ligands that have oxygen donor atoms. Complexes of the uranyl ion are important in the extraction of uranium from its ores and in nuclear fuel reprocessing.

One of the main mechanisms is when a contaminant can infiltrate to the groundwater and to reach an aquifer and to spread affecting areas that in principle were considered far from influence and sometimes, very complex. In this group the cases of contaminants transport are included from the surface of the land by the infiltration waters (poured on the land, use of fertilizers, etc.) and when an infiltration exists polluting from superficial waters, rivers, canals, etc. Probably, the cause of the deterioration of the groundwater quality as consequence of the human action is the agricultural practice. The contamination caused by these practices is characterized except in concrete cases, for its diffuse character.

The risk of increase of salinity grows, logically, for: (a) increase of phosphates and nitrates in the surface water, soil water and groundwater in agricultural areas (Vighi et al. 1991; Hall 1992; Eghball et al. 1996), (b) eutrophication of water bodies (He et al. 1995), and (c) increment in the uranium concentration in land and water resulting from the use of phosphated fertilizers (Barisic et al. 1992; Zielinski et al. 1997). Phosphate fertilizers are made of soluble phosphate complexes that contain uranium, ranging between 50 and 200 $mg\,kg^{-1}$ (Romero et al. 1995, 2006). These values are high compared with the average uranium content in the earth's crust, which is of 1.8 $mg\,kg^{-1}$ (Taylor and McClennan 1985). The uranium is deposited in the soil together with the phosphates during the fertilization period. In time, the uranium and phosphorus are dispersed in the soil and can be moved, either by the runoff reaching the water bodies (rivers, lakes, reservoirs) or by soil water until they reach the groundwater; this plays an important role in the migration and redistribution of uranium in nature.

The inadequate use of phosphate fertilizers in watering areas with permeable soil and unsaturated aquifers, could have adverse effects in the readiness of some minor elements and to be a source of contamination of the soil and groundwater, at high phosphorus loading rates and repeated applications of fertilizer, together with manure applications under favorable conditions (Eghball et al. 1996), or when the soil fixation capacity has been completely saturated (Gerritsen 1993).

Uranium is a trace constituent of many phosphate bearing fertilizers. The increase in their concentration found in most of the profiles of fertilized soil has been documented for decades (Rothbaum et al. 1979); this is also the case for superficial water (runoff) or drainage coming from agricultural lands (Barisic et al. 1992;

Zielinski et al. 1997). Uranium is recognized by the World Health Organization World Health Organization and OMS (1993) as a potentially harmful constituent of drinking water; the U.S. Environmental Protection Agency, USEPA (1996) proposed the implementation of a 20 μg L^{-1} uranium drinking water quality standard; likewise, the Mexican government established 1.44 mg L^{-1} as the standard in the official regulations (DOF 1989). The migration of uranium in systems is largely controlled by uranium solution-mineral equilibria and sorption reactions. Uranyl complexes containing hydroxide, carbonate, fluoride, sulfate, nitrate or phosphate can exist and even predominate under conditions depending on water and soil properties (Romero et al. 1995). The mobility of the uranium in oxic groundwater is generally controlled by the adsorption, and it is regulated by the solution chemistry, the surface properties and predominant phases of the minerals.

The aim of the work was to simulate the transport of uranium in the unsaturated zone of an experimental parcel located in the community of San Cayetano, State of Mexico. A brief description of the main physical-chemical processes that operate in the unsaturated zone and how they affect the distribution and transport of solutes is presented. The main focuses are described to approach the quantification of these processes and the solution of the equations.

2 The Experimental Setup

2.1 Site Description

The Toluca Valley is located in the highlands of Central Mexico at an average altitude of 2,600 msnm and at some 60 km of Mexico City (Fig. 1); the Lerma River (upper course of the Lerma River) runs through it. The study site is located to the west, in the outskirts at 5 km of the town of Tlachaloya, 15 km to the north of the city of Toluca (Capital of the State of Mexico), among the latitude Y = 2,142,500 m UTM and X = 422,600 m UTM. It is inside the meridian master 14Q, covering an area of 0.4 has. The sampling site is confined to the portion of the drainage area for the Tejalpa and the San Cayetano rivers, natural influents of the Lerma River. The average rainfall is of about 920 mm year^{-1} with a peak in July, and there is a dry season from October to April. The mean annual temperature is 15–18 °C. The topography of the area is that of a flat land; the main crop is corn, which is grown on 90 % of the arable land. This experimental site was chosen for the study given its accessibility, the long history of agricultural development, the effective drainage toward the river, besides having the typical crop and soil of the Valley. Agriculture in the Valley of Toluca is seasonal, benefiting from the raining period for the cultivation of crops. Sometimes surface water coming from rivers and water reservoirs, sometimes groundwater obtained from wells and/or springs is used for irrigation from March to November.

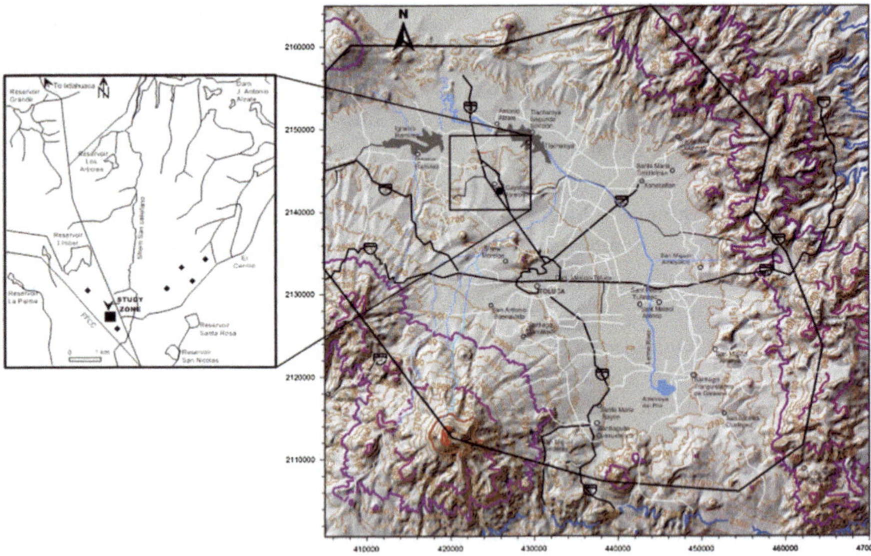

Fig. 1 Localization of the study area

2.2 Field Methodology

Documental information of the study area was obtained including government reports, of scientific articles to have a hydrogeological description of the study area. It was carried out later to program the measurement of the parameters in field including geologic recognition to describe the means, as well as the measure of the hydraulic conductivity of field with permeameter type Guelph, and was made to model conceptual hydrogeologic of the study area that constitutes the bases for the model of numeric simulation of the flow and transport in the unsaturated zone.

Jet-filled tensiometers manufactured by Soilmoisture Equipment Corporation Model 2725 (Soilmoisture 1997) were installed at study area where changes in soil water matric potential indicate water and/or contaminant movement during the annual hydrologic cycle. Recharge events could indicate a high potential for contaminant migration to deep depths. Tensiometers and samplers of porous cups were installed in shallow soils to different depths of 30, 60, 90, 120, 150 and 180 cm (placed within T15, T30, T60, T90, T120, T150 and T180 cm of the soil surface) to determine the layers moisture and collect the interstitial water, at the same levels of the tensiometers and to determines the composition of chemical parameters and the concentration of U(VI).

2.3 Hydrogeological Characterization

The hydraulic conductivity was determined especially by means of a permeameter type Guelph, this is an important physical property of the soil, for modeling of the water flow and solute transport in the soil, for the irrigation design and drainage, modeling of the groundwater and other agricultural processes: engineering and environmental.

2.4 Transport and Flow Simulation of Uranium Through Unsaturated Zone

The characterization of the unsaturated zone includes, to measure key parameters such as the moisture content, the pressure potential, the matric potential, the saturated hydraulic conductivity of field and solute concentrations. The knowledge of these variables is important to understand the processes that are presented in the unsaturated zone and to feed the numeric models of simulation (Gee et al. 1994). These parameters are related with such meteorological variables as the precipitation, evapotranspiration and infiltration; it is possible to have a better understanding of the dynamics of the flow in the unsaturated zone.

2.5 The Mathematical Model

The simulations and subsequent analysis were carried out for a vertical section to field scale and a one-dimensional model (1D). The reactions of superficial complexation of uranium were instantaneous reactions and the supposition was made that alone they are presented inside the interstitial porosity. The simulations of flow and transport were made with the simulator HYDRUS (Simunek et al. 1998), developed by the Department of Environmental Sciences, University of California Riverside, USA, imposing the initial and boundary conditions in the unsaturated zone. The water flow in an unsaturated medium is described traditionally with the equation of Richards (1931) that only keeps in mind the suction gradients and of elevation, expressing the saturated hydraulic conductivity of field, Ks, like a function of the volumetric soil water content, θ, Eq. 1:

$$\frac{\partial}{\partial z}\left(D(\theta)\frac{\partial \theta}{\partial z} - K(\theta)\right) = \frac{\partial \theta}{\partial t} \tag{1}$$

where D(θ) is the function of the diffusivity of the soil water (Guymon 1994).

The fundamental mechanisms of transport are the advection, the molecular diffusion and dispersion. The equation of the transport of mass in the used program is Eq. 2:

$$\theta\frac{\partial C}{\partial t} = \theta D\frac{\partial^2 c}{\partial x^2} - \theta v\frac{\partial c}{\partial x} - \sum\nolimits_{i=1}^{N} \alpha_i \rho_b[f_i K_d C - S_i] \tag{2}$$

$$\frac{\partial S_i}{\partial t} = \alpha_i (f_i K_d C - S_i) \tag{3}$$

where C is the solute concentration [mL^{-3}]; D is the tensor of hydrodynamic dispersion [L^2T^{-1}]; it is the vector of flow velocity [LT^{-1}] and (3) is a source/sink term for transport [$mL^{-3}T - 1$].

3 Results and Analysis

When moisture is added to the soil, the reverse process takes place. Moisture from the soil moves back into the tensiometer through the porous cup until the vacuum level is reduced to a value corresponding with the lower soil suction value. Then, the water flow stops. If enough water is added to the soil until complete saturation, the gauge reading on the tensiometer will drop to zero. As the water can flow in and out of the tube through the pores in the porous ceramic cups, the gauge reading is always in "balance" with the soil suction (Cassel and Klute 1986).

Tensiometers gave readings of 0–10 and 10–26 centibars (cb) for 15 cm of depth, of 0–4 for 30 cm, of 0–4 for 60 cm, of 0–8 to 90 cm, of 4–10 to 120 cm and of 12–20 for 180 cm (Fig. 2). A gauge reading of zero cb means the surrounding soil is completely saturated with water, regardless of the type of soil. A persistent zero reading after irrigation or rain indicates poor drainage conditions which should be investigated and corrected. Gauge readings in the range of 0–10 cb indicated a surplus of water for plant growth. Water held by the soil in this range drains off within a few days. Gauge readings in the range of 10–20 cb indicate that there is ample moisture and also air in the soil for healthy plant growth in all types of soils. This range is often referred to as the "field capacity" range for soils, which means that the soil has reached its "capacity" and cannot hold any more water for future plant growth. When soils are at "field capacity", any additional water that is added drains out of the root zone within a day or two before it can be used by the growing plant. If irrigation has been in process, it should be stopped when gauge drops to this level, since any further additional water will be quickly drained from the root zone and wasted, carrying with it valuable fertilizer. These soils, however, have a very limited water storage capacity and therefore soil suction values increase very rapidly as moisture is removed by the plant after soil suction values reach 15–20 cb. If water-sensitive plants, such as potatoes, are planted in coarse, sandy soils, irrigation may need to be started between 15–20 cb to allow time to apply the irrigation water before damaging stress conditions develop.

As shown in Fig. 2, the measured suctions from each tensiometer significantly vary with time, and the tensiometer embedded at a depth of 60 cm recorded a suction of zero. It is acknowledged that the decrease of suction recorded by the deep instruments could be attributed to a downward redistribution of the infiltrated water. However, the

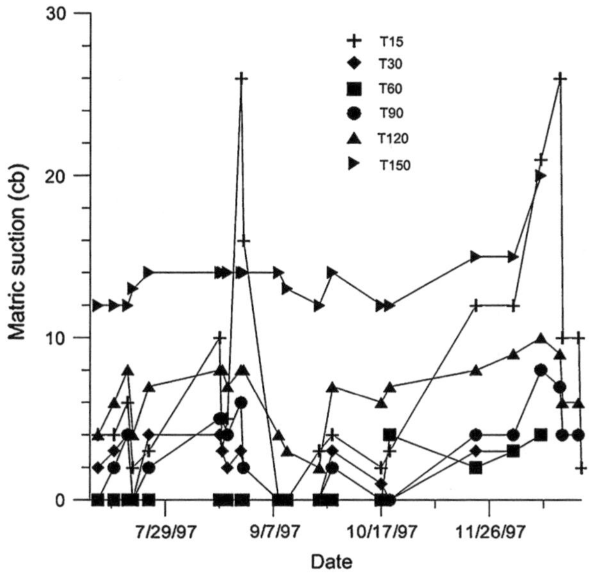

Fig. 2 Tensiometer readings with the elapsed time

significant decrease of suction recorded by the shallow equipments indicated there was a bypass water flow through the gap between the tensiometer tube and the wall of the installation hole.

The concentrations of phosphates and uranium were measures in unsaturated zone where clay, silt and sand samples were collected to different depths, Fig. 3.

Figure 4 shows the profiles of humidity and the concentrations of chlorides, measured along the vertical, which corresponds to the first sampling the day 23 of July of 1997. In the profile, the volumetric content of humidity increases from 0.14 near the surface to 0.28–1.8 m of depth. The depth of the phreatic level is ignored. The profile of the chlorides content present high values measured in interstitial water with an average of 14.8 mg/L, regarding the values measured in superficial water (rivers, preys, borders) and underground water (wells) with values average of 10.13 mg/L, 4.34 mg/L. The rain water values were measured between 1.2 and 3.1 mg/L with an average value of 2.15 mg/L. The pursuit and the balance of chlorides for the study date, of 7 months, the recharge supposes 8 % of the rain registered in this period. For an average precipitation between 550 and 600 mm/year, the balance of ion chloride leads to an annual recharge between 44 and 86 mm. It is worth mentioning that in the study area the data of the rain water is very scarce and variable and due to that it is necessary to have a sufficiently long series to define the average value. The high contents of Cl^- observed in the unsaturated area could be attributed to the contribution of present Cl^- in areas of geologic media, not very accessible to water flow or to the dissolution of accumulated salts, in clay lens, like it can be observed in Fig. 4

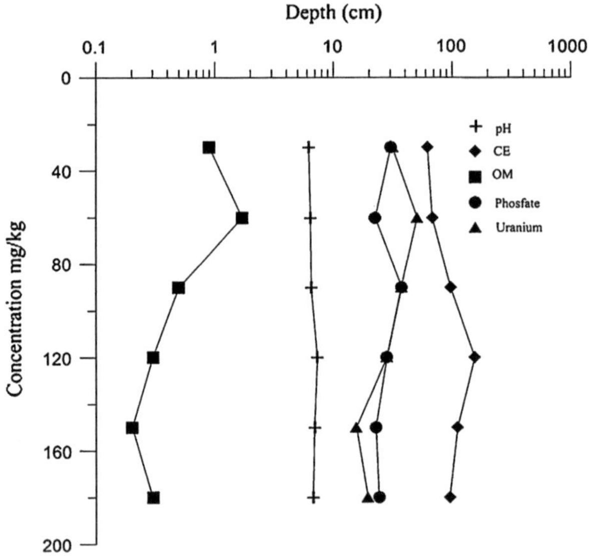

Fig. 3 Distribution of the concentration of phosphates and uranium in mg/kg, organic matter (OM), pH and the CE in the geologic materials to different depths

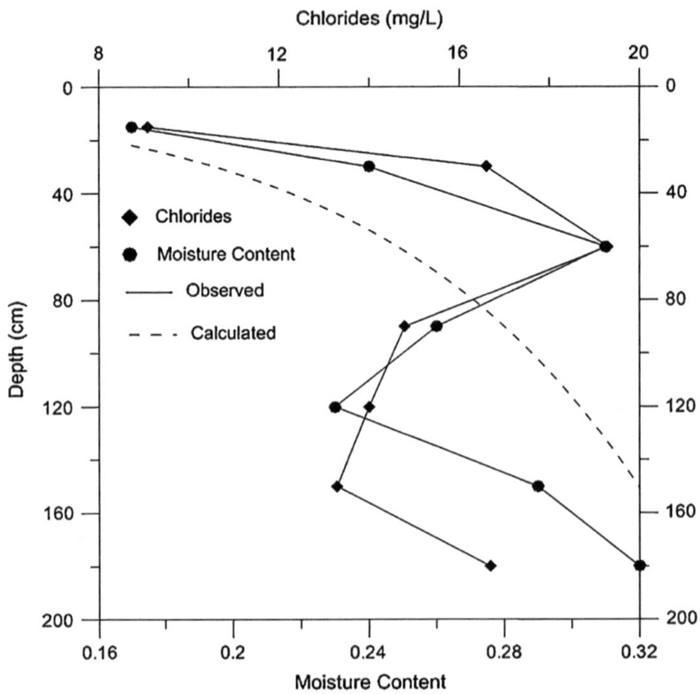

Fig. 4 Profile of moisture content (%) and concentrations of chlorides (mg/L)

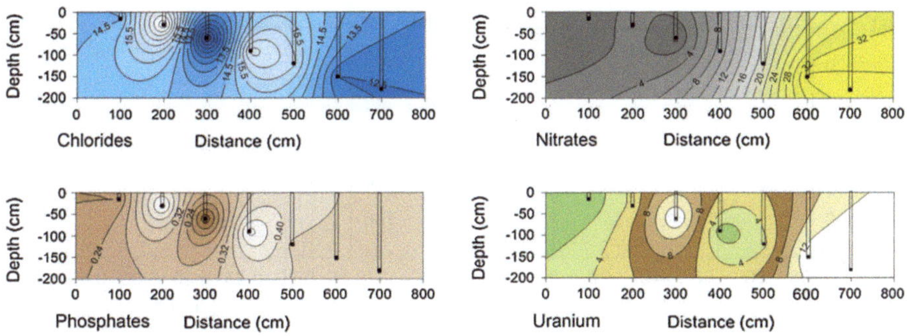

Fig. 5 Cross section of distribution chlorides, nitrates, phosphates and uranyl. Contours of equal concentration (mg/L).

from the 60 to 150 cm of depth. This indicates that in the dilution process present salts can be dissolved in the accustomed to solid phase.

The evolution of phosphates and uranium in the unsaturated zone measured along the vertical in six dates, the first which it corresponds to the first sampling the day 23 of July of 1997 (Romero et al. 1995).

Figure 5 shows cross section of distribution of chemicals measures in the unsaturated zone. If a chemical sorbs to the aquifer media then its residence time is increased and it is retarded with respect to water. Nitrate is negatively charged and generally does not adsorbed to soil and unsaturated zone which usually has a negative charge as well. If allophane or similar minerals, which can have variable charges (sometimes positive) charge, are present in the profile then nitrate can be retarded with respect to water movement. However, these minerals are generally found in association with volcanic soils. In the study area, nitrate can be expected to move with the transporting water, although it can still diffuse with the water into zones of immobile water and exhibit non-equilibrium transport behaviour. If nitrogen is applied to the soil surface as urea, as ammonia fertilizer, then transformation from urea to ammonia to nitrate needs to occur before the leaching of nitrate takes place.

Phosphorous (P) is strongly sorbed to clay minerals and therefore is usually retarded. The P can also sorb to colloidal sized particles and be transported via macropores. Some soils have very low P-retention and significant P loss can occur through soil macropores, predominately co-transported with mobile colloids. If significant P transport occurs through the unsaturated zone, it is to be associated with sorption to colloids and will occur via macropores under saturated flow conditions. Under these conditions P transport could be rapid. Otherwise P transport times will be large.

Cultivation of corn begins in March applying a fertilizer in the proportion 18-40-00 of nitrogen-phosphate-potassium kg/ha. The watering begins when the soil is dry, after the rain time. Soil profile, contains a clay lens at 60 cm which accumulates the uranium, because it is added with the phosphate fertilizer, a proportional concentration of phosphates exists. Therefore, the decrease of uranium is due to the decreasing

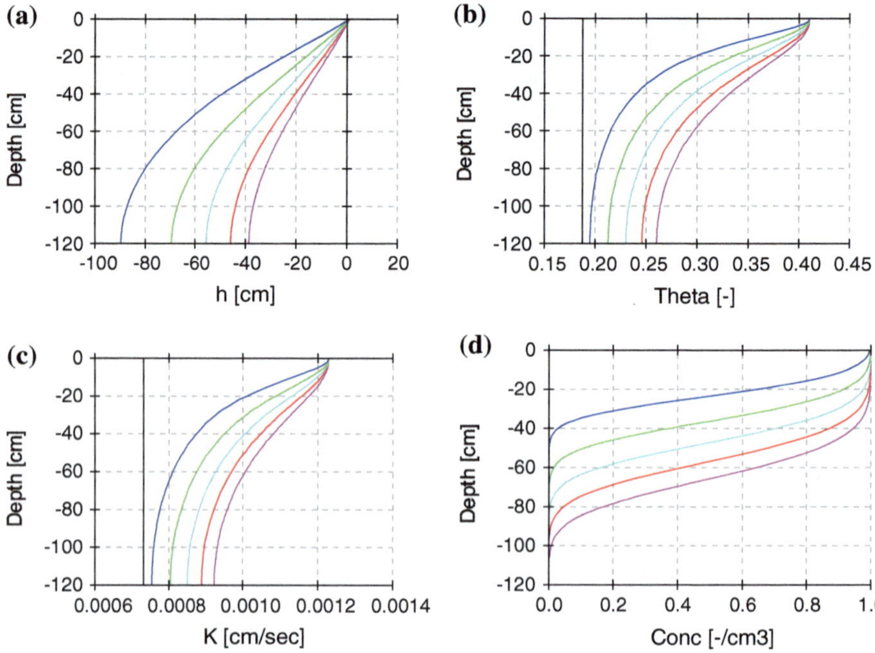

Fig. 6 **a** Pressure head, **b** Water content, **c** Hydraulic conductivity and **d** Relative concentration to t = 0, 30, 60, 90, 120 and 150 days

of phosphates to more depths, for the type of geologic material that prevails. In general, lateral spreading of the U(VI)-plume shown significant vertical spreading. U(VI) is transported downward within the unsaturated zone as it is released from the contaminated sediments during the relatively wet season. Generally, total U concentrations were at low detection ($<10^{-9}$ mg/L) in the size fraction ranging from 2.0 to 75 mm.

3.1 Solute Transport

The model HYDRUS is applied to simulate the processes of transfer of water and solute in the soil in a medium one-dimensional (1D) of 120 cm of depth. The profile of the simulated soil was considered an equivalent single layer of a porous medium. The solute transport is simulated under the consideration that the uranium does not suffer phenomena of adsorption in the soil. It is solved as formulate Simunek et al. (1998) and, for it, takes the value of the coefficient of molecular diffusion (Dw) of the 37.2 cm^2/year and the value of the longitudinal dispersivity (D_L) to 1.55 cm.

The soil hydraulic parameters of the porous domain were taken as: $\theta r = 0.0, \theta s = 0.200, \alpha 1 = 0.041\, cm^{-1}, n = 1.964, Ks = 0.000255\, cm\, s^{-1}$. Initial conditions were set equal to a pressure head of −150 cm. For simplicity we considered only convective solute mass transfer between the two pore regions (i.e. no diffusive transfer), with the dispersivity again fixed at 1.55 cm. Pressure head measurements using tensiometers are, on the other hand, often dominated by the macropore presence that reaches equilibrium relatively quickly. The application of a dual-porosity model can explain nonequilibrium between pressure heads and water contents.

Figure 6 shows computed water and solute (uranium) distributions during infiltration obtained with the model HYDRUS.

The simulations were made for 150 days. Figure 6a shows the distribution of the pressure head, it is observed that it diminishes with depth and with water content. The water content depends, mainly, of the water retained in the porous medium that non percolate for gravity (Fig. 6b). This way, the superior levels of the medium presents a high saturation degree concerning the inferior. The drop saturation of the inferior levels indicates that they are capable of draining with enough speed the water coming from the superior layers. The Fig. 6d shows the front of advance of the concentration of uranium and also this in function of the hydraulic conductivity of the medium.

4 Conclusions

A brief review of uranium contamination in corn field cultivation has been presented in this work. The results of applying the model HYDRUS in a one-dimensional medium, have allowed establishing a first step to know the implications that presents the lixiviation process with superficial irrigation. The present work should be supplemented with a study but detailed considering different strata that constitute the porous means until a depth of 2.0 m. It is considered convenient to simulate the interactions of the soil, plant and atmosphere and their influence on the flow of water and transport of uranyl during a watering campaign.

Acknowledgments Work partially supported by ABACUS, CONACyT grant EDOMEX-2011-C01-165873.

References

Barisic D, Lulic S, Miletic P (1992) Radium and uranium in phosphate fertilizers and their impact on the radioactivity of waters. Water Resour 26:607–611

Cassel DK, Klute A (1986) Water potential: tensiometry. In Klute A (ed) Methods of soil analysis: Part 1. Physical and mineralogical methods, 9. 2nd edn. Madison, Wisconsin, American Society of Agronomy-Soil Science Society of America. pp 563–596. ISBN 0-89118-088-5

Diario Oficial de la Federaci6n (DOF), 13 de Diciembre (1989) Criterios Ecolo' gicos de Calidad del Agua. México

Eghball B, Binford GD, Baltensperger DD (1996) Phosphorus movement and adsorption in a soil receiving long term manure and fertilizer application. J Environ Qual 25:1339–1343

Gee GW, Wierenga PJ, Andraski BJ, Young MH, Fayer MJ, Rockhold ML (1994) Variations in water balance and recharge potential at three western desert sites. Soil Sci Soc Am J 58(1):63–72
Guymon GL (1994) Unsaturated zone hydrology. PTR Prentice Hall, Englewood Cliffs
Gerritsen RG (1993) Prediction of travel times of phosphate in soils at a disposal site for wastewater. Water Resour 27:263–267
Hall DW (1992) Effects of nutrient management on nitrate levels in groundwater near Epharata Pennsylvania. Groundwater 30:720–730
He ZL, Wilson MJ, Campell CO, Edwards AC, Chapman SJ (1995) Distribution of phosphorus in soil aggregate fractions and its significance with regard to phosphorus transport in agricultural runoff. Water Air Soil Pollut 83:69–84
Romero GET, Solache M, Iturbe JL, Ordoñez E (1995) Uranium in phosphate rock and derivates. J Radioanal Nucl Chem 189:301–306
Romero GET, Ordoñez RE, Reyes GLR, Esteller AMV, Rojas HA, Ordoñez RE (2006) Contamination of corn growing areas due to intensive fertilization in the high plane oh Mexico. Water Air Soil Pollut 175:117–98
Richards LA (1931) Capillary conduction of liquids through porous mediums. Physics 1:318–331
Rothbaum HP, McGaveston DA, Wall T, Johston A, Mattingly G (1979) Uranium accumulation in soils from long continued applications of superphosphate. J Soil Sci 30:147–153
Simunek J, Sejna M, van Genuchten M Th (1998) The HYDRUS-1D software package for simulating one-dimensional movement of water, heat, and multiple solutes in variably-saturated media. Version 2.02. International Ground Water Modeling Center (IGWMC)-TPS-70. Colorado School of Mines. Golden, p 202
Soilmoisture(1997) Model 2725A jet fill tensiometer operating instructions. Soilmoisture Equipment Corp, Santa Barbara
Taylor SR, McClennan SM (1985) The continental crust: its composition and evolution. Blackwell Scientific, Oxford
USEPA (1996) National primary drinking water regulations for radionuclides. Proposed rules. Washington D.C. 570 9–71-700. Report
Vighi M, Soprani S, Puzzarini P, Menghi G (1991) Water quality phosphorus loads from selected watersheds in the drainage area of the northern adriatic sea. J Environ Qual 20:439–444
World Health Organization, OMS (1993) Guidelines for drinking water quality. WHO Publications, Ginebra
Zielinski RA, Asher BS, Meier AL, Johnson CA, Szabo BJ (1997) Natural or fertilizer derived uranium in irrigation drainage: a case study in southeasters Colorado, USA. Appl Geochem 12:9–21

Aromatics Extraction in Hydrocarbon Mixtures

Miguel Ángel Balderas Altamirano, Blanca E. García and Arturo Trejo

Abstract This work is devoted to the study of liquid-liquid equilibrium of systems of industrial interest, particularly the purification of oil and gas in the oil refining industry. We present experimental results on the solubility of aromatics in the liquid-liquid equilibrium obtained by gas - liquid chromatography at different temperatures. We used a mixture of solvents to improve the characteristics of selectivity and aromatics extraction. The solvent is a binary mixture of *N*-methyl 2-pyrrolidone (NMP) and diethylene glycol (DEG). The results presented are for two mixtures consisting of 90 % NMP + 10 % DEG, and 70 % NMP + 30 % DEG. Heptane was used as hydrocarbon solvent and toluene as aromatic compound. Furthermore, we applied the equation of state NRTL (Non Random Two Liquids) and UNIQUAC (Universal Quasi Chemical) for coexistence curves and the critical point.

1 Introduction

Oil is a source of energy in the modern world and a non-renewable resource. It is a liquid that is below the surface of the earth embedded in rock cavities, also a complex mixture of hydrocarbons (alkanes, cycloalkanes, aromatics) and is mainly composed of oxygen, sulfur and nitrogen atoms, (Erij et al. 1988). The colour of the oil depends on its content of hydrocarbon structures. Known oils are black, clear or colorless and are lighter and more viscous than water. The origin of petroleum is not

M. Á. Balderas Altamirano (✉)
Universidad Autónoma Metropolitana, Unidad Iztapalapa ,
San Rafael Atlixco No. 186, Col. Vicentina,
09340 Iztapalapa, D.F., Mexico
e-mail: nyfg@xanum.uam.mx

B. E. García · A. Trejo
Instituto Mexicano del Petróleo , Eje Central Lázaro Cárdenas 152,
San Bartolo Atepehuacan, 07730 Gustavo A. Madero, D.F., Mexico

J. Klapp and A. Medina (eds.), *Experimental and Computational Fluid Mechanics*,
Environmental Science and Engineering, DOI: 10.1007/978-3-319-00116-6_41,

well established and is one of the most debated problems of modern science. Once extracted, oil must be refined in order to use its individual components. The process used to separate the constituents of oil is called distillation. Both atmospheric and vacuum distillation are used to refine oil. In the first distillation, the most volatile fractions are obtained: 180 °C gasoline, kerosene from 180 to 240 °C and diesel fraction at 350 °C. The residue from atmospheric distillation is distilled under vacuum and the fractions taken are 350 to 420 °C for lighter distillate oils, 420 to 500 °C for heavy distillate oils. The residue in this process, depending on their viscosity is called tar or semi tar. Tar feedstock has been used to obtain highly viscous lubricating oils and bitumens. The composition of distillate fractions is a way of distinguishing the types of oil.

The fraction in which gasoline distillates is composed mainly of three classes of hydrocarbons: alkanes, cycloalkanes and aromatics. The fractions of kerosene and diesel contain most of the bi- and tri-cyclic hydrocarbons. The distillate fraction, with molecular weights less than 305 g/mol and which are distilled to temperatures close to 350 °C containing alkanes, alkenes, cycloalkanes, bicycloalkanes, bicyclic aromatic, etc. The chemical composition of the distillate fraction above the 350 °C is poorly studied. This part contains macromolecular structures and in practice, is the fraction of oil, lubricants and tar. The molecular mass of these fractions slays between 300 and 1000 g/mol. This hydrocarbon mixture is composed of alkanes, cycloalkanes with mono and polycyclic alkane chains, asphaltenes, resins, etc. The next steps for the use of oil are intended to improve distillate fractions. Thus, there are dearomatization, dewaxing, etc., which are used to produce components or mixtures with fewer impurities.

Dearomatization process (Kirk et al. 1961) involves adding one or more polar solvents to a mixture containing alkanes and aromatics compounds. In the first stage the system is agitated and then allowed to stand until equilibrium is reached. In this state the system is separated into two liquid phases, one rich in solvent the other rich in alkanes. The aromatic compounds are distributed in the two phases. The solvent should extract as much of the aromatics as possible. Aromatics are subsequently purified for use in other applications. Gasoline is mainly composed by benzene, toluene and xylenes. In the case of lubricants the aromatic molecular structure is more complex.

Solvents such as NMP, *N*-formylmorpholine (NFM), glycols and sulfolanes (Saha et al. 1998; Taysser et al. 1990; Baird and Henson 1991; Nagpal and Rawat 1981) are used in multi component hydrocarbon mixtures due to its extraction capabilities, and must have, besides high selectivity towards aromatics, high distribution coefficient, fast separation of phases (high density and low viscosity), good thermal stability, to be non-corrosive and non-reactive. An alternative to using the pure solvent is the use of mixtures of solvents to improve the extraction of aromatics and decrease energy and solvent consumption. On the other hand, analyzing multi component hydrocarbon oil directly from the plant has several disadvantages: it is expensive, its molecular composition and the structure of each component are not well known, the boiling point is high and solubility is difficult to analyze. But some studies have been reported using such mixtures (De Lucas et al. 1993). To make a quantitative analysis

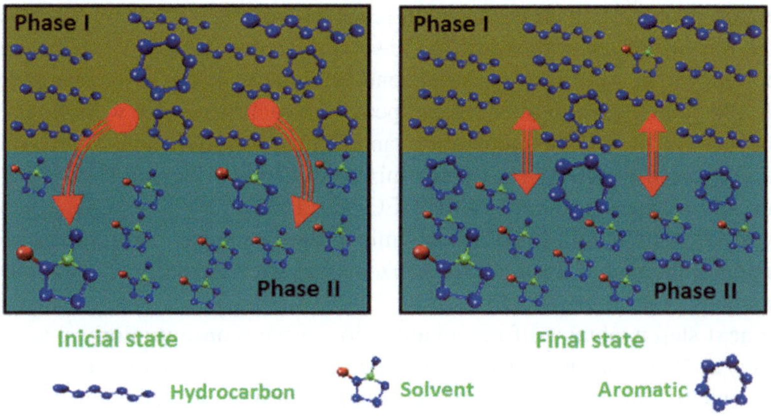

Fig. 1 Liquid-liquid equilibrium between two phases. *Phase I* contains a mixture of hydrocarbons, paraffinic and aromatic compounds. *Phase II* is a polar solvent. The *left* side shows the initial mixture and the *right* when they have reached thermodynamic equilibrium

of the solubility of a component in a mixture, it is convenient to use a mixture model in order to control the variables that determine the liquid-liquid equilibrium, and then extrapolate the results to problems of industrial interest.

In a model system we can perform a systematic study of the solubility and control the effect of temperature, pressure, solvent concentration, hydrocarbon molecular structure, etc. The results obtained at laboratory level can be used to extrapolate the behavior in industrial hydrocarbon mixtures. Examples of results obtained in this way are reported in references (Letcher and Naicker 1998; Bernabe et al. 1988; Romero-Martínez and Trejo 1995). Figure 1 describes the extraction process. Phase I initially contains hydrocarbons and aromatics while Phase II contains only polar solvent. The aromatic is soluble in both the hydrocarbon and the solvent, therefore, when the system reaches equilibrium, the concentration of aromatic in phase I will be less than the initial, i.e., there will be a selective separation of the mixture of aromatics hydrocarbons.

The purpose of this work is to determine the thermodynamic conditions to remove as much of aromatics using polar solvents. The rest of the work is divided into three sections: first we describe the methodology used, second the results section and finally conclusions drawn from the results.

2 Methodology

As an initial step, the purity of the compounds used was examined in the laboratory with the aid of a gas chromatograph, Varian model 3400. The chromatogram reported in all cases areas greater than 99.8 % and the purity reported by the manufacturer was greater than 99 % in all cases. In all reagents we added 4A diameter molecular sieves to remove traces of moisture in the solvent. Before chromatographic analysis it is necessary to find the best operating conditions to obtain the best chromatograms, they should show high and well defined peaks for each substance without any overlapping between them. The temperature range which was more consistent with these conditions was from 40 to 160 °C for mixing n-heptane-toluene-NMP in a DB-5 column, and n-heptane-toluene-NMP-DEG was 100 to 280 °C in a SE-30 column, both with a flow of helium gas of 12 cm^3/min as the carrier gas. To eliminate possible chromatographic signal fluctuations, benzene was used as the internal standard for each analysis.

The next step is obtained from a calibration curve (concentration vs. area under the curve) at 25 °C for each system component, so we prepared mixtures of the three components making sure that the mixture is in the homogeneous region of the phase diagram. The mass of each component and its concentration in the mixture is known. To these mixtures the internal standard (benzene) was added and the concentration is obtained using the chromatograph, with a calibration curve constructed for each component. Once the calibration curve for each component is done, it is necessary to study multi component mixtures. The first is a ternary system, the second and third are mixtures of solvents; the first with a percentage of 90 % NMP + 10 % of DEG and second with 70 % DEG + 30 % NMP. Two equilibria were analyzed at three different temperatures, 15, 25 and 40 °C.

Each mixture was deposited in a cell designed for the IMP (Instituto Mexicano del Petróleo, Mexico) where the separation of the system into two liquid phases can be visually assessed. The cell allows the passage of water, similar to the mechanism of a coolant to control the temperature of the experiment. The system is stirred for at least 4 h with a Teflon-coated magnetic stirrer, the stirring is stopped and allowed to stand until equilibrium; this requires at least 4 h. When the phases have separated and reached equilibrium, each phase is extracted with hypodermic syringes (one for each phase) and transferred in a separate vial (container) containing 0.2 g of benzene as internal standard. Finally the chromatograms for each phase are obtained and the calibration curve of each component is used to determine its concentration. The same procedure is carried out in mixtures of various concentrations to generate the coexistence curve at a single temperature. In this paper the coexistence curves were obtained for three different temperatures. The experimental systems studied in this work can be summarized as follows:

(a) Heptane + Toluene + solvent I at 25 °C
(b) Heptane + Toluene + solvent II at 15, 25 and 40 °C
(c) Heptane + Toluene + solvent III at 15, 25 and 40 °C

Solvent concentration: I = 100 % solvent NMP, solvent II 90 %, NMP + 10 % DEG and solvent III 70 % NMP + 30 % DEG.

The experimental data were correlated using NRTL and UNIQUAC models. (Sorensen et al. 1986) Several versions of these models can be found in the literature with different expressions to describe the temperature dependency of the binary interaction parameters. The ones used here are outlined below. The excess

Gibbs free energy (g^E) for the NRTL model (Renon and Prausnitz 1968) is given by:

$$\frac{g^E}{RT} = \sum_{j=1}^{c} x_i \frac{\left(\sum_{j=1}^{c} \tau_{ji} G_{ji} x_j\right)}{\left(\sum_{k=1}^{c} G_{ki} x_k\right)} \tag{1}$$

where x_i is the mole fraction of component i, τ_{ij} is the interaction parameter, G_{ji} is the interaction energy, R the gas constant and T the temperature. For a binary mixture, the NRTL model contains five parameters, two binary interaction parameters for each component and the non randomness parameter for the binary mixture. The excess Gibbs energy for the UNIQUAC model (Abrams and Prausnitz 1975) is given by

$$\frac{g_E}{RT} = \frac{\left(g^C + g^R\right)}{RT} \tag{2}$$

where g^E and g^R are the combinatorial and the residual contributions.

$$\frac{g_c}{RT} = \sum_{i=1}^{c} x_i \ln\left(\frac{\Phi_i}{x_i}\right) + 5 \sum_{i=1}^{c} q_i x_i \ln\left(\frac{\theta_i}{\Phi_i}\right) \tag{3}$$

$$\frac{g^R}{RT} = \sum_{i=1}^{c} q_i x_i \ln\left(\sum_{i=1}^{c} \theta_j \tau_{ij}\right) \tag{4}$$

where q_i are the pure component surface area parameter, τ_{ij} is the interaction parameter, Φ_i the segment fraction and the area fraction θ_i. The parameters in both models were found by minimizing the objective function, where N is the number of experimental points.

$$F = \sum_{k=i}^{N} \sum_{j=1}^{2} \sum_{i=1}^{3} \left(x'_{ijk} - x_{ijk}\right)^2 \tag{5}$$

where x'_{ijk} y x_{ijk} are the molar fraction calculated by the program and experimentally, respectively. The sums take into account the experimental data k, the phases j and the components i. The computer program used in this work was done at the IMP (García-Sánchez et al. 1996).

3 Results

The data generated for ternary systems are best depicted in a triangular diagram as shown in Fig. 2 for the system heptane-toluene-NMP. Once equilibrium concentrations are obtained from our system, we can draw using these data the triangular

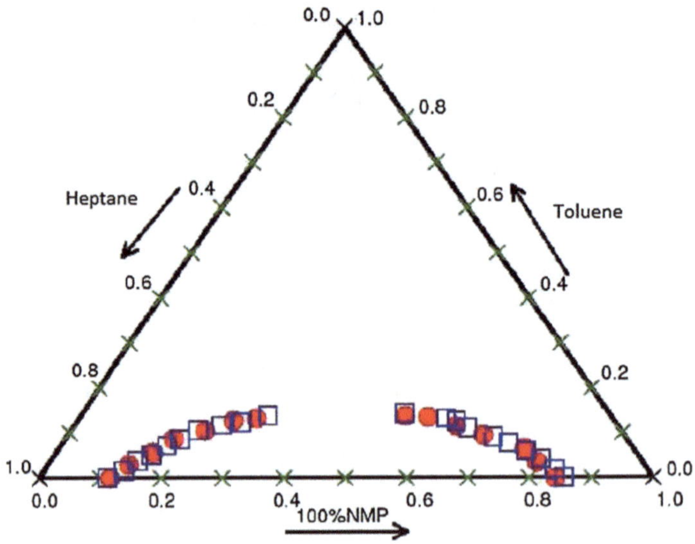

Fig. 2 Ternary diagram for the system heptane toluene NMP. The *squares* represent the values of Ferreira et al. (1984a,b) *filled circles* are the values obtained in this work. These data correspond to measurements made at 25C

diagram or phase diagram. For this case seven experiments were made at constant temperature and varying concentrations of aromatic. Heptane-NMP binary system is in the base of the triangle, the left side is the phase rich in heptane and the right in NMP. Heptane mixtures with concentrations greater than 86 % and lower than 16 % are in a liquid phase. States below the symbols correspond to the two-phase region in the liquid-liquid equilibrium. States outside this region correspond to a liquid phase where the three components are present. Mixtures with toluene concentration greater than 15 % are in a single stage, this is the critical region. Near the critical point it is difficult to measure the concentration due to large density fluctuations in the phases. In this system, with NMP as a solvent, the size of the curve is small. We want to find the way to increase this curve and improve the extraction of aromatics. There is evidence to show that a mixture of solvents could extend this region, for more details see works reported in (Nagpal and Rawat 1981; Naidoo et al. 2001). These results were compared with those reported by (Ferreira et al. 1984a,b) to validate the experimental technique and verify the proper operation of the equipment. To study the effect of a solvent mixture in the aromatic extraction we decide to vary the amount of DEG. For practical purposes, decided these systems were treated as ternary systems, where the extractant mixture NMP - DEG was considered a single component; a system like this is known as a pseudo-ternary system. In what follows in all triangular diagrams the legend 90 % NMP + 10 % DEG or 70 % NMP + 30 % DEG means that the solvent NMP was prepared with *X* mass % DEG, not to be

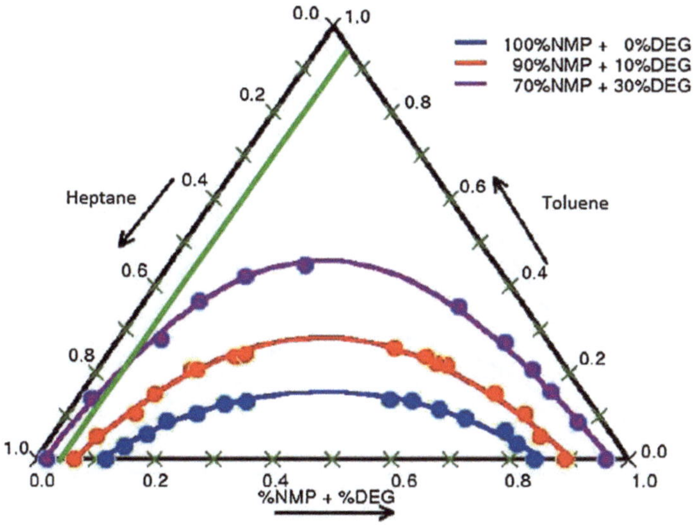

Fig. 3 Effect of DEG concentration in the solvent. The *line* parallel to heptane axis shows the area where we can find homogeneous equilibrium. In the case of the mixture with 30 % DEG, the *line* shows the region where we can find two liquid phases. The experiments were done at 25C. The *continuum line* was obtained with the NRTL model. The *circles* are experimental data

confused with the concentration of the solvent in equilibrium expressed as a molar fraction.

(i) *Effect of concentration of DEG in the solvent*

Figure 3 shows the results at 25 °C and various concentrations of DEG. Immiscibility region increases with the addition of DEG. The maximum amount of the aromatic compound can participate as the liquid-liquid equilibrium narrows the top of the region of insolubility (top of the curve). In this study the limit was 0.15, 0.30 and 0.45 (molar fraction of toluene) for solvent mixtures with 0, 10 and 30 % of DEG, respectively. The points on the curve divide the region in which we can find the liquid - liquid equilibrium, above the curve it will have total miscibility and under it partial miscibility. Application of the NRTL model allows one to obtain the continuous line in Fig. 3 for each of the three sets of data. UNIQUAC model results are similar to those of NRTL and are not reported here.

(ii) *Effect of temperature on the extractant mixture*

The effects of temperature on the mixture with 10 % of DEG are shown in Fig. 4, and at 30 % of DEG in Fig. 5. The increased temperature promotes the solubility of the components and the bimodal curve is smaller. In the system with 10 % DEG, the bimodal curve size is different; it shows a greater displacement to the left side. While the system with 30 % DEG, cornering bimodal separation is larger in the middle part, but decays into points very close to the equilibrium binary line.

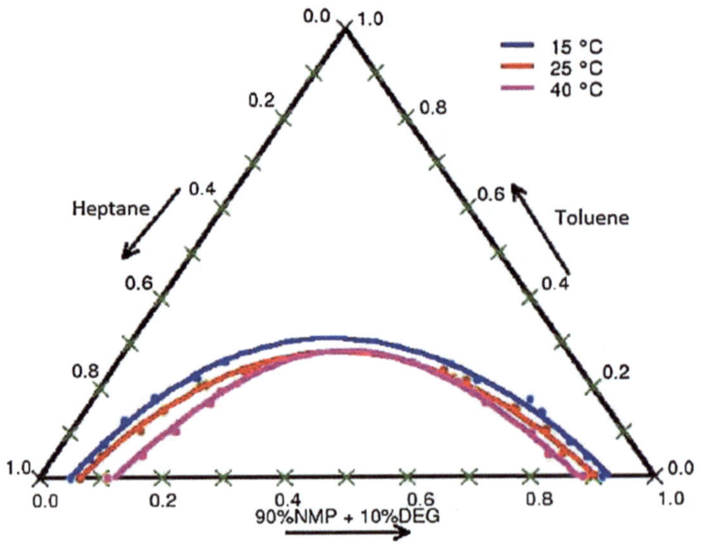

Fig. 4 Temperature effect on the mixture 10 % of DEG in NMP. The *curves* were obtained with the NRTL correlation model. The *circles* are experimental data

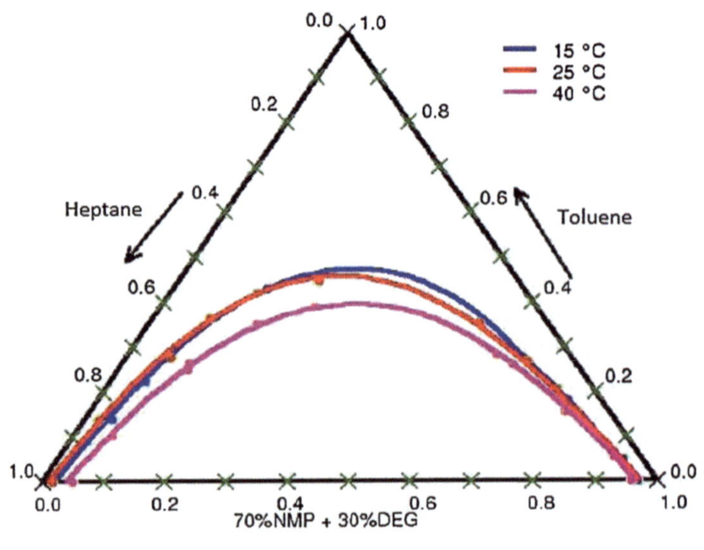

Fig. 5 Temperature effect on the mixture 30 % of DEG in NMP. The *curves* were obtained with the NRTL correlation model. The *circles* are experimental data

(iii) *Distribution coefficient*

A parameter to consider is the distribution coefficient K, (Ferreira et al. 1984a,b) which is defined for the particular case of the aromatic content as follows.

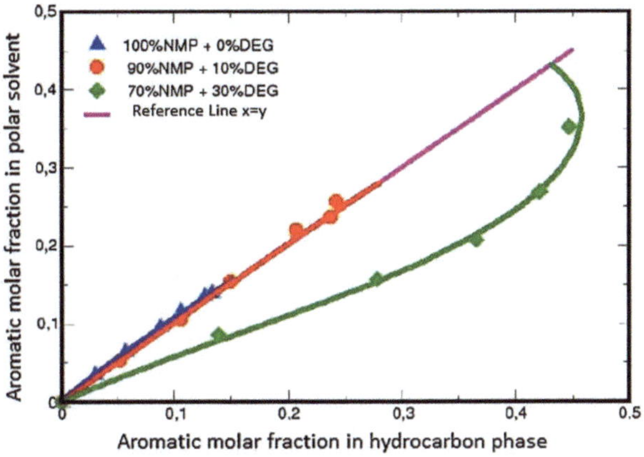

Fig. 6 Distribution *curves* at 25C. The experimental data are indicated by symbols and the continuum *lines* correspond to results obtained by the NRTL model

$$K = \frac{x_A^{II}}{x_A^{I}} \tag{6}$$

x_A^{II} and x_A^{I} are the molar fractions of the phase I aromatic (hydrocarbon) and phase II (solvent), respectively. The distribution coefficients show the distribution of aromatic in both phases, see Fig. 6.

The line of reference indicates that the aromatic is distributed in the same proportion in phases I and II. A value above the reference line points out to higher extraction capacity and a value below it indicates that the greatest amount of aromatic remained in phase I. In the systems studied in this work, the mixture with 0 % DEG extracted at slightly higher concentration of aromatic, mixture with 10 % DEG separates slightly less than that of the pure solvent, while the mixture with 30 % DEG obtained fewer aromatic. The results of the distribution coefficient plotted in Fig. 6 show that the increase of the immiscible region reduces the aromatic distribution in the mixture. The Aromatic concentration is greater in Phase II with 10 % DEG compared with 30 % DEG system. By increasing the bimodal curve a greater amount of aromatic hydrocarbon mixtures may be involved in the extraction process. The K distribution coefficient in 0 and 10 % of DEG mixtures, for all practical purposes are equal and their use as extracting solvent is indistinct, but because the mixture with 10 % DEG has a higher insoluble region, it can be considered that it is better solvent for extraction systems.

(iv) *Selectivity (selectivity factor)*

Another criterion for evaluating a solvent Selectivity is defined as:

$$\beta = \frac{YI}{XI} \tag{7}$$

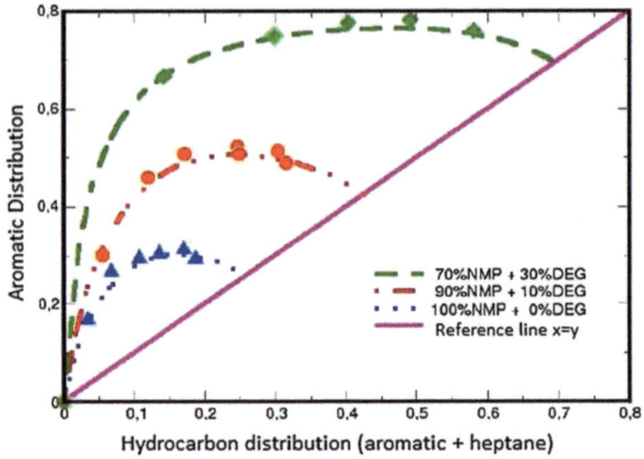

Fig. 7 Graphic selectivity distribution of 25C at different concentrations of DEG. Symbols are experimental data and *trend lines* were obtained with the NRTL model

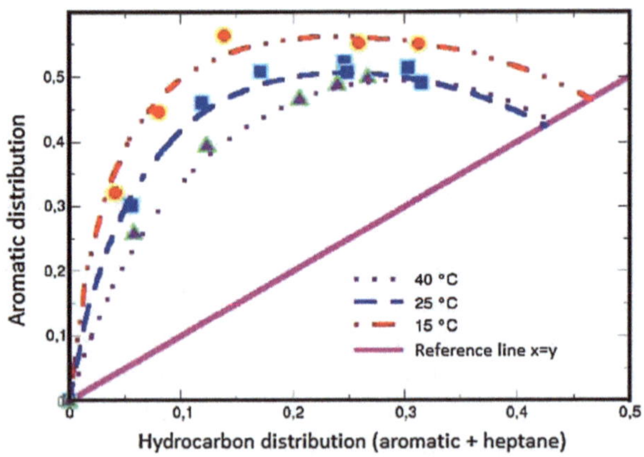

Fig. 8 Graphic distribution of selectivities at different temperatures for the system heptane + toluene + 90 % NMP + 10 % DEG. Symbols are experimental data and *trend lines* were obtained by the NRTL model

where *XI* and *YI* are defined as:

$$XI = \frac{x_A^I}{\left(x_A^I + x_H^I\right)} \quad , \quad YI = \frac{x_A^{II}}{\left(x_A^{II} + x_H^{II}\right)}, \tag{8}$$

where x_A and x_H are the mole fractions of aromatic and hydrocarbon, respectively. Superscripts I and II refer to the hydrocarbon rich phase and solvent respectively. Selectivity is defined as the measure of the ability of the solvent to separate the aromatic compound with a mixture of alkane.

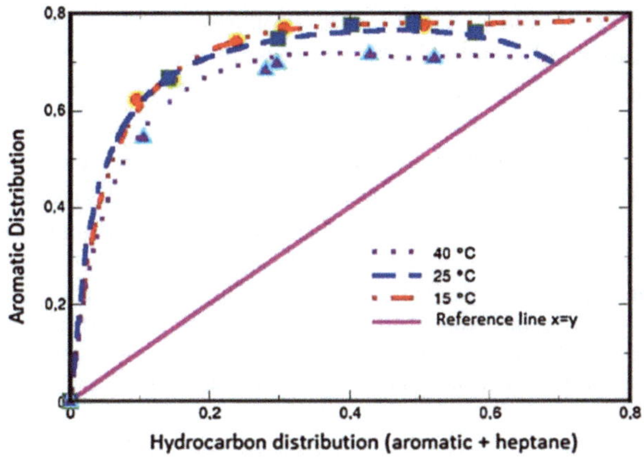

Fig. 9 Graphic distribution of selectivities at different temperatures for the system heptane + toluene + 70 % NMP + 30 % DEG. Symbols are experimental data and *trend lines* were obtained by the NRTL model

For each of the systems obtained in this work, we made a selectivity diagram. Selectivities are shown in Figs. 7, 8 and 9; they indicate that our solvents are highly selective. The solvent is more selective towards the toluene if we added DEG. Selectivity is shown in an ascendant way, the combination of solvent mixture: 100 % NMP+0 % DEG < 90 % NMP + 10 % DEG <70 % NMP + 30 % DEG; that is, the mixture with the highest selectivity curve was the 70 % DEG + 30 % NMP mixture. The selectivity is reduced by increasing temperature for both systems studied, at 10 % DEG and 30 % DEG. However, its effect is rather small. The solid lines in each figure were obtained using the correlation data obtained by NRTL for each experimental system. The curves obtained by the UNIQUAC model are almost the same, and therefore are not reported in this work.

4 Conclusions

The effect of adding DEG to the solvent (NMP) is to increase the region of immiscibility, and decrease the solubility of the aromatic compound. Temperature increase has little effect in the solubility of the aromatic compound. The distribution coefficient increases as the selectivity decreases and vice versa. The solvent with 70 % NMP + 30 % DEG is the most selective and 90 % NMP + 10 % DEG which had a higher extraction capacity (distribution coefficient with a high value). The distribution coefficient, in mixtures with 10 and 0 % DEG, is similar; selectivity and immiscibility regions are greater in the first case, so that the mixture with 10 % DEG is more useful for extraction of aromatics. Analysis of the region of immiscibility, selectivity and distribution coefficient is insufficient to decide, from the mixtures at 0 and 30 % DEG, which may be used in the extraction of aromatics. To make a better choice we should consider other properties such as viscosity, degree of corrosion, thermal stability, toxicity, vapor pressure, etc. Interaction parameters used in the NRTL and UNIQUAC equations fit well the experimental values and the critical point. This is an ongoing research (Balderas Altamirano 2003), so one would have to continue mixing solvents in order to look for the best mixture that can result in the best selectivity (β) with the best distribution constant (K). This research is important for those interested in the separation processes or in purification of compounds.

Acknowledgments MABA is grateful to CONACYT (Consejo Nacional de Ciencia y Tecnología, Mexico) for a grant during this work. Also, thanks are due to the IMP (Instituto Mexicano del Petróleo), to the group of Thermodynamics at IMP for providing the substances, equipment and guidance to develop this research, and finally to Armando Gama and Miguel Gama, for discussions and encouragement.

References

Abrams DS, Prausnitz JM (1975) Statistical thermodynamics of liquid mixtures: a new expression for the excess Gibbs energy of partly or completely miscible systems. AIChe J 21:116

Balderas Altamirano MA (2003) Extracción de aromáticos en mezclas de hidrocarburos. Tesis de Maestría, Universidad Autónoma Metropolitana, Unidad Iztapalapa, Asesores, Arturo Trejo and José Alejandre, México, D.F.

Baird MHI, Henson C (1991) Handbook of solvent extraction. Krieguer, Florida. E.U.A

Bernabe D, Romero-Martínez A, Trejo A (1988) Liquid-liquid coexistence curves for binary systems. Fluid Phase Equilib 40:279

De Lucas A, Rodríguez L, Sánchez P, Carnicer A (1993) Sep Sci Technol 28:2465

Erij V, Rásina M, Rudin M (1988) Química y Tecnología del gas. Mir, U.R.S.S

Ferreira PO, Barbosa D, Medina AG (1984a) Phase equilibria for the separation of aromatic and non-aromatic compounds using mixed solvents. Part I. The system n-heptane - toluene- N-methyl pyrrolidone/monoethyleneglycol. Fluid Phase Equilib 15:309

Ferreira PO, Barbosa D, Medina AG (1984b) Liquid - liquid equilibria for the system N-methyl pyrrolidone + toluene + n-heptane: UNIFAC interaction parameters for N-methyl pyrrolidone. Fluid Phase Equilib 16:369

García-Sánchez F, Schwartzentruber J, Ammar MN, Renon H (1996) Thermodynamics of multiphase liquid equilibria in multicomponent mixtures. Rev Mex Fis 43:59

Kirk ER, Othmer FD, Scott DJ, Standein A (1961) Enciclopedía de Tecnología Química. Unión Tipográfica editorial Hispanoaméricana, D.F., México

Letcher TM, Naicker PK (1998) Ternary liquid-liquid equilibria for mixtures for n-alkane + an Aromatic Hydrocarbon + N-methyl-2-pyrrolidone. J Chem Eng Data 43:1034

Nagpal JM, Rawat BS (1981) Liquid-liquid equilibria for toluene-heptane-N-methyl-pyrrolidone and benzene-heptane solvents. J Chem Technol Biotechnol 31:145

Naidoo RD, Letcher TM, Ramjugernath D (2001) Ternary liquid liquid equilibria for pseudo ternary mixtures containing an n-alkane + an aromatic hydrocarbon + N-methyl-2-pyrrolidone+solvent at 298.2 K and 1 atm. J Chem Eng Data 46:1375

Renon H, Prausnitz JM (1968) Local compositions in thermodynamic excess functions for liquid mixtures. AIChe J 14:135

Romero-Martínez A, Trejo A (1995) Equilibrio Líquido Líquido: Su correlación y medición parte II Sistemas binarios. IMP; D.F, Mexico

Saha M, Rawat BS, Khanna MK, Nautiyal BR (1998) Liquid-liquid equilibrium studies on toluene + heptane + solvent. J Chem Eng Data 43:422

Sorensen JM, Lichtenhaler RN, Gomes-de-Azevedo E (1986) Molecular thermodynamics of fluid phase equilibria. Prentice Hall, New Jersey, E.U.A

Taysser A, Al-Zayled TA, Al-Sahhaf Fahim MA (1990) Measurements of phase equilibrium in multicomponent systems of aromatics with n-methylpyrrolidone and predictions with UNIFAC. Fluid Phase Equilib 61:131

MIX
Papier aus verantwortungsvollen Quellen
Paper from responsible sources
FSC® C105338

If you have any concerns about our products,
you can contact us on
ProductSafety@springernature.com

In case Publisher is established outside the EU,
the EU authorized representative is:
Springer Nature Customer Service Center GmbH
Europaplatz 3, 69115 Heidelberg, Germany

Printed by Libri Plureos GmbH
in Hamburg, Germany